Statistics *for* Environmental Engineers

Second Edition

Paul Mac Berthouex
Linfield C. Brown

LEWIS PUBLISHERS

A CRC Press Company
Boca Raton   London   New York   Washington, D.C.

Library of Congress Cataloging-in-Publication Data

Catalog record is available from the Library of Congress

This book contains information obtained from authentic and highly regarded sources. Reprinted material is quoted with permission, and sources are indicated. A wide variety of references are listed. Reasonable efforts have been made to publish reliable data and information, but the author and the publisher cannot assume responsibility for the validity of all materials or for the consequences of their use.

Neither this book nor any part may be reproduced or transmitted in any form or by any means, electronic or mechanical, including photocopying, microfilming, and recording, or by any information storage or retrieval system, without prior permission in writing from the publisher.

The consent of CRC Press LLC does not extend to copying for general distribution, for promotion, for creating new works, or for resale. Specific permission must be obtained in writing from CRC Press LLC for such copying.

Direct all inquiries to CRC Press LLC, 2000 N.W. Corporate Blvd., Boca Raton, Florida 33431.

Trademark Notice: Product or corporate names may be trademarks or registered trademarks, and are used only for identification and explanation, without intent to infringe.

Visit the CRC Press Web site at www.crcpress.com

© 2002 by CRC Press LLC
Lewis Publishers is an imprint of CRC Press LLC

No claim to original U.S. Government works
International Standard Book Number 1-56670-592-4
Printed in the United States of America 1 2 3 4 5 6 7 8 9 0
Printed on acid-free paper

Preface to 1st Edition

When one is confronted with a new problem that involves the collection and analysis of data, two crucial questions are: How will using statistics help solve this problem? And, Which techniques should be used? This book is intended to help environmental engineers answer these questions in order to better understand and design systems for environmental protection.

The book is not about the environmental systems, except incidentally. It is about how to extract information from data and how informative data are generated in the first place. A selection of practical statistical methods is applied to the kinds of problems that we encountered in our work. We have not tried to discuss every statistical method that is useful for studying environmental data. To do so would mean including virtually all statistical methods, an obvious impossibility. Likewise, it is impossible to mention every environmental problem that can or should be investigated by statistical methods. Each reader, therefore, will find gaps in our coverage; when this happens, we hope that other authors have filled the gap. Indeed, some topics have been omitted precisely because we know they are discussed in other well-known books.

It is important to encourage engineers to see statistics as a professional tool used in familiar examples that are similar to those faced in one's own work. For most of the examples in this book, the environmental engineer will have a good idea how the test specimens were collected and how the measurements were made. The data thus have a special relevance and reality that should make it easier to understand special features of the data and the potential problems associated with the data analysis.

The book is organized into short chapters. The goal was for each chapter to stand alone so one need not study the book from front to back, or in any other particular order. Total independence of one chapter from another is not always possible, but the reader is encouraged to "dip in" where the subject of the case study or the statistical method stimulates interest. For example, an engineer whose current interest is fitting a kinetic model to some data can get some useful ideas from Chapter 25 without first reading the preceding 24 chapters. To most readers, Chapter 25 is not conceptually more difficult than Chapter 12. Chapter 40 can be understood without knowing anything about t-tests, confidence intervals, regression, or analysis of variance.

There are so many excellent books on statistics that one reasonably might ask, Why write another book that targets environmental engineers? A statistician may look at this book and correctly say, "Nothing new here." We have seen book reviews that were highly critical because "this book is much like book X with the examples changed from biology to chemistry." Does "changing the examples" have some benefit? We feel it does (although we hope the book does something more than just change the examples).

A number of people helped with this book. Our good friend, the late William G. Hunter, suggested the format for the book. He and George Box were our teachers and the book reflects their influence on our approach to engineering and statistics. Lars Pallesen, engineer and statistician, worked on an early version of the book and is in spirit a co-author. A. (Sam) James provided early encouragement and advice during some delightful and productive weeks in northern England. J. Stuart Hunter reviewed the manuscript at an early stage and helped to "clear up some muddy waters." We thank them all.

P. Mac Berthouex
Madison, Wisconsin

Linfield C. Brown
Medford, Massachusetts

Preface to 2nd Edition

This second edition, like the first, is about how to generate informative data and how to extract information from data. The short-chapter format of the first edition has been retained. The goal is for the reader to be able to "dip in" where the case study or the statistical method stimulates interest without having to study the book from front to back, or in any particular order.

Thirteen new chapters deal with experimental design, selecting the sample size for an experiment, time series modeling and forecasting, transfer function models, weighted least squares, laboratory quality assurance, standard and specialty control charts, and tolerance and prediction intervals. The chapters on regression, parameter estimation, and model building have been revised. The chapters on transformations, simulation, and error propagation have been expanded.

It is important to encourage engineers to see statistics as a professional tool. One way to do this is to show them examples similar to those faced in one's own work. For most of the examples in this book, the environmental engineer will have a good idea how the test specimens were collected and how the measurements were made. This creates a relevance and reality that makes it easier to understand special features of the data and the potential problems associated with the data analysis.

Exercises for self-study and classroom use have been added to all chapters. A solutions manual is available to course instructors. It will not be possible to cover all 54 chapters in a one-semester course, but the instructor can select chapters that match the knowledge level and interest of a particular class. Statistics and environmental engineering share the burden of having a special vocabulary, and students have some early frustration in both subjects until they become familiar with the special language. Learning both languages at the same time is perhaps expecting too much. Readers who have prerequisite knowledge of both environmental engineering and statistics will find the book easily understandable. Those who have had an introductory environmental engineering course but who are new to statistics, or vice versa, can use the book effectively if they are patient about vocabulary.

We have not tried to discuss every statistical method that is used to interpret environmental data. To do so would be impossible. Likewise, we cannot mention every environmental problem that involves statistics. The statistical methods selected for discussion are those that have been useful in our work, which is environmental engineering in the areas of water and wastewater treatment, industrial pollution control, and environmental modeling. If your special interest is air pollution control, hydrology, or geostatistics, your work may require statistical methods that we have not discussed. Some topics have been omitted precisely because you can find an excellent discussion in other books. We hope that whatever kind of environmental engineering work you do, this book will provide clear and useful guidance on data collection and analysis.

P. Mac Berthouex
Madison, Wisconsin

Linfield C. Brown
Medford, Massachusetts

The Authors

Paul Mac Berthouex is Emeritus Professor of civil and environmental engineering at the University of Wisconsin-Madison, where he has been on the faculty since 1971. He received his M.S. in sanitary engineering from the University of Iowa in 1964 and his Ph.D. in civil engineering from the University of Wisconsin-Madison in 1970. Professor Berthouex has taught a wide range of environmental engineering courses, and in 1975 and 1992 was the recipient of the Rudolph Hering Medal, American Society of Civil Engineers, for most valuable contribution to the environmental branch of the engineering profession. Most recently, he served on the Government of India's Central Pollution Control Board.

In addition to *Statistics for Environmental Engineers, 1st Edition* (1994, Lewis Publishers), Professor Berthouex has written books on air pollution and pollution control. He has been the author or co-author of approximately 85 articles in refereed journals.

Linfield C. Brown is Professor of civil and environmental engineering at Tufts University, where he has been on the faculty since 1970. He received his M.S. in environmental health engineering from Tufts University in 1966 and his Ph.D. in sanitary engineering from the University of Wisconsin-Madison in 1970. Professor Brown teaches courses on water quality monitoring, water and wastewater chemistry, industrial waste treatment, and pollution prevention, and serves on the U.S. Environmental Protection Agency's Environmental Models Subcommittee of the Science Advisory Board. He is a Task Group Member of the American Society of Civil Engineers' National Subcommittee on Oxygen Transfer Standards, and has served on the Editorial Board of the *Journal of Hazardous Wastes and Hazardous Materials*.

In addition to *Statistics for Environmental Engineers, 1st Edition* (1994, Lewis Publishers), Professor Brown has been the author or co-author of numerous publications on environmental engineering, water quality monitoring, and hazardous materials.

Table of Contents

1 Environmental Problems and Statistics 1
2 A Brief Review of Statistics 7
3 Plotting Data 25
4 Smoothing Data 41
5 Seeing the Shape of a Distribution 47
6 External Reference Distributions 55
7 Using Transformations 61
8 Estimating Percentiles 71
9 Accuracy, Bias, and Precision of Measurements 77
10 Precision of Calculated Values 87
11 Laboratory Quality Assurance 97
12 Fundamentals of Process Control Charts 103
13 Specialized Control Charts 113
14 Limit of Detection 119
15 Analyzing Censored Data 129
16 Comparing a Mean with a Standard 141
17 Paired *t*-Test for Assessing the Average of Differences 147
18 Independent *t*-Test for Assessing the Difference of Two Averages 157
19 Assessing the Difference of Proportions 161
20 Multiple Paired Comparisons of *k* Averages 169

21	Tolerance Intervals and Prediction Intervals	175
22	Experimental Design	185
23	Sizing the Experiment	197
24	Analysis of Variance to Compare k Averages	215
25	Components of Variance	223
26	Multiple Factor Analysis of Variance	233
27	Factorial Experimental Designs	239
28	Fractional Factorial Experimental Designs	249
29	Screening of Important Variables	261
30	Analyzing Factorial Experiments by Regression	271
31	Correlation	281
32	Serial Correlation	289
33	The Method of Least Squares	295
34	Precision of Parameter Estimates in Linear Models	303
35	Precision of Parameter Estimates in Nonlinear Models	311
36	Calibration	319
37	Weighted Least Squares	327
38	Empirical Model Building by Linear Regression	337
39	The Coefficient of Determination, R^2	345
40	Regression Analysis with Categorical Variables	355
41	The Effect of Autocorrelation on Regression	365
42	The Iterative Approach to Experimentation	373
43	Seeking Optimum Conditions by Response Surface Methodology	379

44	Designing Experiments for Nonlinear Parameter Estimation	389
45	Why Linearization Can Bias Parameter Estimates	397
46	Fitting Models to Multiresponse Data	403
47	A Problem in Model Discrimination	411
48	Data Adjustment for Process Rationalization	419
49	How Measurement Errors Are Transmitted into Calculated Values	425
50	Using Simulation to Study Statistical Problems	433
51	Introduction to Time Series Modeling	441
52	Transfer Function Models	453
53	Forecasting Time Series	459
54	Intervention Analysis	467
Appendix — Statistical Tables		477
Index		481

1

Environmental Problems and Statistics

There are many aspects of environmental problems: economic, political, psychological, medical, scientific, and technological. Understanding and solving such problems often involves certain quantitative aspects, in particular the acquisition and analysis of data. Treating these quantitative problems effectively involves the use of statistics. Statistics can be viewed as the prescription for making the quantitative learning process effective.

When one is confronted with a new problem, a two-part question of crucial importance is, "How will using statistics help solve this problem and which techniques should be used?" Many different substantive problems arise and many different statistical techniques exist, ranging from making simple plots of data to iterative model building and parameter estimation.

Some problems can be solved by subjecting the available data to a particular analytical method. More often the analysis must be stepwise. As Sir Ronald Fisher said, "...a statistician ought to strive above all to acquire versatility and resourcefulness, based on a repertoire of tried procedures, always aware that the next case he wants to deal with may not fit any particular recipe."

Doing statistics on environmental problems can be like coaxing a stubborn animal. Sometimes small steps, often separated by intervals of frustration, are the only way to progress at all. Even when the data contains bountiful information, it may be discovered in bits and at intervals.

The goal of statistics is to make that discovery process efficient. Analyzing data is part science, part craft, and part art. Skills and talent help, experience counts, and tools are necessary. This book illustrates some of the statistical tools that we have found useful; they will vary from problem to problem. We hope this book provides some useful tools and encourages environmental engineers to develop the necessary craft and art.

Statistics and Environmental Law

Environmental laws and regulations are about toxic chemicals, water quality criteria, air quality criteria, and so on, but they are also about statistics because they are laced with statistical terminology and concepts. For example, *the limit of detection* is a statistical concept used by chemists. In environmental biology, *acute and chronic toxicity criteria* are developed from complex data collection and statistical estimation procedures, safe and adverse conditions are differentiated through statistical comparison of control and exposed populations, and *cancer potency factors* are estimated by extrapolating models that have been fitted to dose-response data.

As an example, the Wisconsin laws on toxic chemicals in the aquatic environment specifically mention the following statistical terms: *geometric mean, ranks, cumulative probability, sums of squares, least squares regression, data transformations, normalization of geometric means, coefficient of determination, standard F-test at a 0.05 level, representative background concentration, representative data, arithmetic average, upper 99th percentile, probability distribution, log-normal distribution, serial correlation, mean, variance, standard deviation, standard normal distribution,* and *Z value*. The U.S. EPA guidance documents on statistical analysis of bioassay test data mentions *arc-sine transformation, probit analysis, non-normal distribution, Shapiro-Wilks test, Bartlett's test, homogeneous variance, heterogeneous variance, replicates, t-test with Bonferroni adjustment, Dunnett's test, Steel's rank test,* and *Wilcoxon rank sum test*. Terms mentioned in EPA guidance documents on groundwater monitoring at RCRA sites include *ANOVA, tolerance units, prediction intervals, control charts, confidence intervals, Cohen's adjustment, nonparametric ANOVA, test of proportions, alpha error, power curves*, and *serial correlation*. Air pollution standards and regulations also rely heavily on statistical concepts and methods.

One burden of these environmental laws is a huge investment in collecting environmental data. No nation can afford to invest huge amounts of money in programs and designs that are generated from badly designed sampling plans or by laboratories that have insufficient quality control. The cost of poor data is not only the price of collecting the sample and making the laboratory analyses, but is also investments wasted on remedies for non-problems and in damage to the environment when real problems are not detected. One way to eliminate these inefficiencies in the environmental measurement system is to learn more about statistics.

Truth and Statistics

Intelligent decisions about the quality of our environment, how it should be used, and how it should be protected can be made only when information in suitable form is put before the decision makers. They, of course, want facts. They want truth. They may grow impatient when we explain that at best we can only make inferences about the truth. "Each piece, or part, of the whole of nature is always merely an approximation to the complete truth, or the complete truth so far as we know it....Therefore, things must be learned only to be unlearned again or, more likely, to be corrected" (Feynman, 1995).

By making carefully planned measurements and using them properly, our level of knowledge is gradually elevated. Unfortunately, regardless of how carefully experiments are planned and conducted, the data produced will be imperfect and incomplete. The imperfections are due to unavoidable random variation in the measurements. The data are incomplete because we seldom know, let alone measure, all the influential variables. These difficulties, and others, prevent us from ever observing the truth exactly.

The relation between truth and inference in science is similar to that between guilty and not guilty in criminal law. A verdict of not guilty does not mean that innocence has been proven; it means only that guilt has not been proven. Likewise the truth of a hypothesis cannot be firmly established. We can only test to see whether the data dispute its likelihood of being true. If the hypothesis seems plausible, in light of the available data, we must make decisions based on the likelihood of the hypothesis being true. Also, we assess the consequences of judging a true, but unproven, hypothesis to be false. If the consequences are serious, action may be taken even when the scientific facts have not been established. Decisions to act without scientific agreement fall into the realm of mega-tradeoffs, otherwise known as politics.

Statistics are numerical values that are calculated from imperfect observations. *A statistic estimates a quantity that we need to know about but cannot observe directly.* Using statistics should help us move toward the truth, but it cannot guarantee that we will reach it, nor will it tell us whether we have done so. It can help us make scientifically honest statements about the likelihood of certain hypotheses being true.

The Learning Process

Richard Feynman said (1995), "The principle of science, the definition almost, is the following. The test of all knowledge is experiment. Experiment is the sole judge of scientific truth. But what is the course of knowledge? Where do the laws that are to be tested come from? Experiment itself helps to produce these laws, in the sense that it gives us hints. But also needed is imagination to create from these hints the great generalizations — to guess at the wonderful, simple, but very strange patterns beneath them all, and then to experiment again to check whether we have made the right guess."

An experiment is like a window through which we view nature (Box, 1974). Our view is never perfect. The observations that we make are distorted. The imperfections that are included in observations are "noise." A statistically efficient design reveals the magnitude and characteristics of the noise. It increases the size and improves the clarity of the experimental window. Using a poor design is like seeing blurred shadows behind the window curtains or, even worse, like looking out the wrong window.

Environmental Problems and Statistics

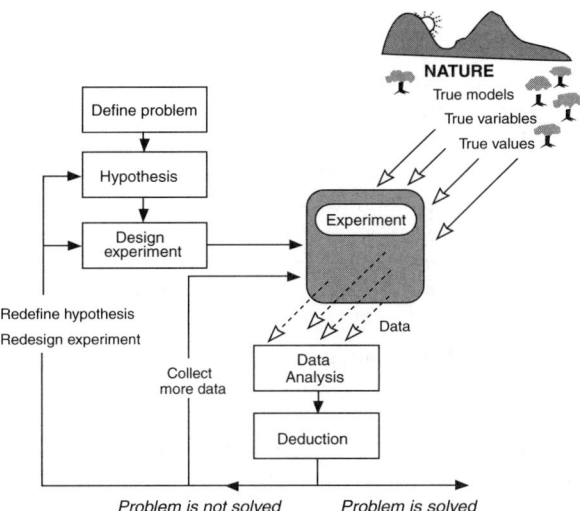

FIGURE 1.1 Nature is viewed through the experimental window. Knowledge increases by iterating between experimental design, data collection, and data analysis. In each cycle the engineer may formulate a new hypothesis, add or drop variables, change experimental settings, and try new methods of data analysis.

Learning is an iterative process, the key elements of which are shown in Figure 1.1. The cycle begins with expression of a working hypothesis, which is typically based on *a priori* knowledge about the system. The hypothesis is usually stated in the form of a mathematical model that will be tuned to the present application while at the same time being placed in jeopardy by experimental verification. Whatever form the hypothesis takes, it must be probed and given every chance to fail as data become available. Hypotheses that are not "put to the test" are like good intentions that are never implemented. They remain hypothetical.

Learning progresses most rapidly when the experimental design is statistically sound. If it is poor, so little will be learned that intelligent revision of the hypothesis and the data collection process may be impossible. A statistically efficient design may literally let us learn more from eight well-planned experimental trials than from 80 that are badly placed. Good designs usually involve studying several variables simultaneously in a group of experimental runs (instead of changing one variable at a time). Iterating between data collection and data analysis provides the opportunity for improving precision by shifting emphasis to different variables, making repeated observations, and adjusting experimental conditions.

We strongly prefer working with experimental conditions that are statistically designed. It is comparatively easy to arrange designed experiments in the laboratory. Unfortunately, in studies of natural systems and treatment facilities it may be impossible to manipulate the independent variables to create conditions of special interest. A range of conditions can be observed only by spacing observations or field studies over a long period of time, perhaps several years. We may need to use historical data to assess changes that have occurred over time and often the available data were not collected with a view toward assessing these changes. A related problem is not being able to replicate experimental conditions. These are huge stumbling blocks and it is important for us to recognize how they block our path toward discovery of the truth. Hopes for successfully extracting information from such historical data are not often fulfilled.

Special Problems

Introductory statistics courses commonly deal with linear models and assume that available data are normally distributed and independent. There are some problems in environmental engineering where these fundamental assumptions are satisfied. Often the data are not normally distributed, they are serially or spatially correlated, or nonlinear models are needed (Berthouex et al., 1981; Hunter, 1977, 1980, 1982). Some specific problems encountered in data acquisition and analysis are:

Aberrant values. Values that stand out from the general trend are fairly common. They may occur because of gross errors in sampling or measurement. They may be mistakes in data recording. If we think only in these terms, it becomes too tempting to discount or throw out such values. However, rejecting any value out of hand may lead to serious errors. Some early observers of stratospheric ozone concentrations failed to detect the hole in the ozone layer because their computer had been programmed to screen incoming data for "outliers." The values that defined the hole in the ozone layer were disregarded. This is a reminder that rogue values may be real. Indeed, they may contain the most important information.

Censored data. Great effort and expense are invested in measurements of toxic and hazardous substances that should be absent or else be present in only trace amounts. The analyst handles many specimens for which the concentration is reported as "not detected" or "below the analytical method detection limit." This method of reporting censors the data at the limit of detection and condemns all lower values to be qualitative. This manipulation of the data creates severe problems for the data analyst and the person who needs to use the data to make decisions.

Large amounts of data (which are often observational data rather than data from designed experiments). Every treatment plant, river basin authority, and environmental control agency has accumulated a mass of multivariate data in filing cabinets or computer databases. Most of this is *happenstance data*. It was collected for one purpose; later it is considered for another purpose. Happenstance data are often ill suited for model building. They may be ill suited for detecting trends over time or for testing any hypothesis about system behavior because (1) the record is not consistent and comparable from period to period, (2) all variables that affect the system have not been observed, and (3) the range of variables has been restricted by the system's operation. In short, happenstance data often contain surprisingly little information. No amount of analysis can extract information that does not exist.

Large measurement errors. Many biological and chemical measurements have large measurement errors, despite the usual care that is taken with instrument calibration, reagent preparation, and personnel training. There are efficient statistical methods to deal with *random errors*. Replicate measurements can be used to estimate the random variation, averaging can reduce its effect, and other methods can compare the random variation with possible real changes in a system. *Systematic errors* (bias) cannot be removed or reduced by averaging.

Lurking variables. Sometimes important variables are not measured, for a variety of reasons. Such variables are called lurking variables. The problems they can cause are discussed by Box (1966) and Joiner (1981). A related problem occurs when a truly influential variable is carefully kept within a narrow range with the result that the variable appears to be insignificant if it is used in a regression model.

Nonconstant variance. The error associated with measurements is often nearly proportional to the magnitude of their measured values rather than approximately constant over the range of the measured values. Many measurement procedures and instruments introduce this property.

Nonnormal distributions. We are strongly conditioned to think of data being symmetrically distributed about their average value in the bell shape of the normal distribution. Environmental data seldom have this distribution. A common asymmetric distribution has a long tail toward high values.

Serial correlation. Many environmental data occur as a sequence of measurements taken over time or space. The order of the data is critical. In such data, it is common that the adjacent values are not statistically independent of each other because the natural continuity over time (or space) tends to make neighboring values more alike than randomly selected values. This property, called serial correlation, violates the assumptions on which many statistical procedures are based. Even low levels of serial correlation can distort estimation and hypothesis testing procedures.

Complex cause-and-effect relationships. The systems of interest — the real systems in the field — are affected by dozens of variables, including many that cannot be controlled, some that cannot be measured accurately, and probably some that are unidentified. Even if the known variables were all controlled, as we try to do in the laboratory, the physics, chemistry, and biochemistry of the system are complicated and difficult to decipher. Even a system that is driven almost entirely by inorganic chemical reactions can be difficult to model (for example, because of chemical complexation and amorphous solids formation). The situation has been described by Box and Luceno (1997): "All models are wrong but some are useful." Our ambition is usually short of trying to discover all causes and effects. We are happy if we can find a *useful* model.

Environmental Problems and Statistics

The Aim of this Book

Learning statistics is not difficult, but engineers often dislike their introductory statistics course. One reason may be that the introductory course is largely a sterile examination of textbook data, usually from a situation of which they have no intimate knowledge or deep interest. We hope this book, by presenting statistics in a familiar context, will make the subject more interesting and palatable.

The book is organized into short chapters, each dealing with one essential idea that is usually developed in the context of a case study. We hope that using statistics in relevant and realistic examples will make it easier to understand peculiarities of the data and the potential problems associated with its analysis. The goal was for each chapter to stand alone so the book does not need to be studied from front to back, or in any other particular order. This is not always possible, but the reader is encouraged to "dip in" where the subject of the case study or the statistical method stimulates interest.

Most chapters have the following format:

- **Introduction** to the general kind of engineering problem and the statistical method to be discussed.
- **Case Study** introduces a specific environmental example, including actual data.
- **Method** gives a brief explanation of the statistical method that is used to prepare the solution to the case study problem. Statistical theory has been kept to a minimum. Sometimes it is condensed to an extent that reference to another book is mandatory for a full understanding. Even when the statistical theory is abbreviated, the objective is to explain the broad concept sufficiently for the reader to recognize situations when the method is likely to be useful, although all details required for their correct application are not understood.
- **Analysis** shows how the data suggest and influence the method of analysis and gives the solution. Many solutions are developed in detail, but we do not always show all calculations. Most problems were solved using commercially available computer programs (e.g., MINITAB, SYSTAT, Statview, and EXCEL).
- **Comments** provide guidance to other chapters and statistical methods that could be useful in analyzing a problem of the kind presented in the chapter. We also attempt to expose the sensitivity of the statistical method to assumptions and to recommend alternate techniques that might be used when the assumptions are violated.
- **References** to selected articles and books are given at the end of each chapter. Some cover the statistical methodology in greater detail while others provide additional case studies.
- **Exercises** provides additional data sets, models, or conceptual questions for self-study or classroom use.

Summary

To gain from what statistics offer, we must proceed with an attitude of letting the data reveal the critical properties and of selecting statistical methods that are appropriate to deal with these properties. Environmental data often have troublesome characteristics. If this were not so, this book would be unnecessary. All useful methods would be published in introductory statistics books. This book has the objective of bringing together, primarily by means of examples and exercises, useful methods with real data and real problems. Not all useful statistical methods are included and not all widely encountered problems are discussed. Some problems are omitted because they are given excellent coverage in other books (e.g., Gilbert, 1987). Still, we hope the range of material covered will contribute to improving the state-of-the-practice of statistics in environmental engineering and will provide guidance to relevant publications in statistics and engineering.

References

Berthouex, P. M., W. G. Hunter, and L. Pallesen (1981). "Wastewater Treatment: A Review of Statistical Applications," *ENVIRONMETRICS 81—Selected Papers,* pp. 77–99, Philadelphia, SIAM.
Box, G. E. P. (1966). "The Use and Abuse of Regression," *Technometrics,* 8, 625–629.
Box, G. E. P. (1974). "Statistics and the Environment," *J. Wash. Academy Sci.,* 64, 52–59.
Box, G. E. P., W. G. Hunter, and J. S. Hunter (1978). *Statistics for Experimenters: An Introduction to Design, Data Analysis, and Model Building,* New York, Wiley Interscience.
Box, G. E. and A. Luceno (1997). *Stastical Control by Monitoring and Feedback Adjustment,* New York, Wiley Interscience.
Feynman, R. P. (1995). *Six Easy Pieces,* Reading, Addison-Wesley.
Gibbons, R. D. (1994). *Statistical Methods for Groundwater Monitoring,* New York, John Wiley.
Gilbert, R. O. (1987). *Statistical Methods for Environmental Pollution Monitoring,* New York, Van Nostrand Reinhold.
Green, R. (1979). *Sampling Design and Statistical Methods for Environmentalists,* New York, John Wiley.
Hunter, J. S. (1977). "Incorporating Uncertainty into Environmental Regulations," in *Environmental Monitoring,* Washington, D.C., National Academy of Sciences.
Hunter, J. S. (1980). "The National Measurement System," *Science,* 210, 869–874.
Hunter, W. G. (1982). "Environmental Statistics," in *Encyclopedia of Statistical Sciences,* Vol. 2, Kotz and Johnson, Eds., New York, John Wiley.
Joiner, B. L. (1981). "Lurking Variables: Some Examples," *Am. Statistician,* 35, 227–233.
Millard, S. P. (1987). "Environmental Monitoring, Statistics, and the Law: Room for Improvement," *Am. Statistician,* 41, 249–259.

Exercises

1.1 Statistical Terms. Review a federal or state law on environmental protection and list the statistical terms that are used.

1.2 Community Environmental Problem. Identify an environmental problem in your community and list the variables (factors) for which data should be collected to better understand this problem. What special properties (nonnormal distribution, nonconstant variance, etc.) do you think data on these variables might have?

1.3 Incomplete Scientific Information. List and briefly discuss three environmental or public health problems where science (including statistics) has not provided all the information that legislators and judges needed (wanted) before having to make a decision.

2

A Brief Review of Statistics

KEY WORDS *accuracy, average, bias, central limit effect, confidence interval, degrees of freedom, dot diagram, error, histogram, hypothesis test, independence, mean, noise, normal distribution, parameter, population, precision, probability density function, random variable, sample, significance, standard deviation, statistic,* t *distribution,* t *statistic, variance.*

It is assumed that the reader has some understanding of the basic statistical concepts and computations. Even so, it may be helpful to briefly review some notations, definitions, and basic concepts.

Population and Sample

The person who collects a specimen of river water speaks of that specimen as a sample. The chemist, when given this specimen, says that he has a sample to analyze. When people ask, "How many samples shall I collect?" they usually mean, "On how many specimens collected from the population shall we make measurements?" They correctly use "sample" in the context of their discipline. The statistician uses it in another context with a different meaning. The *sample* is a group of n observations actually available. A *population* is a very large set of N observations (or data values) from which the sample of n observations can be imagined to have come.

Random Variable

The term random variable is widely used in statistics but, interestingly, many statistics books do not give a formal definition for it. A practical definition by Watts (1991) is "the value of the next observation in an experiment." He also said, in a plea for terminology that is more descriptive and evocative, that "A random variable is the soul of an observation" and the converse, "An observation is the birth of a random variable."

Experimental Errors

A guiding principle of statistics is that any quantitative result should be reported with an accompanying estimate of its error. Replicated observations of some physical, chemical, or biological characteristic that has the true value η will not be identical although the analyst has tried to make the experimental conditions as identical as possible. This relation between the value η and the observed (measured) value y_i is $y_i = \eta + e_i$, where e_i is an error or disturbance.

Error, experimental error, and *noise* refer to the fluctuation or discrepancy in replicate observations from one experiment to another. In the statistical context, error does not imply fault, mistake, or blunder. It refers to variation that is often unavoidable resulting from such factors as measurement fluctuations due to instrument condition, sampling imperfections, variations in ambient conditions, skill of personnel, and many other factors. Such variation always exists and, although in certain cases it may have been minimized, it should not be ignored entirely.

Example 2.1

A laboratory's measurement process was assessed by randomly inserting 27 specimens having a known concentration of $\eta = 8.0$ mg/L into the normal flow of work over a period of 2 weeks. A large number of measurements were being done routinely and any of several chemists might be assigned any sample specimen. The chemists were 'blind' to the fact that performance was being assessed. The 'blind specimens' were outwardly identical to all other specimens passing through the laboratory. This arrangement means that observed values are random and independent. The results in order of observation were 6.9, 7.8, 8.9, 5.2, 7.7, 9.6, 8.7, 6.7, 4.8, 8.0, 10.1, 8.5, 6.5, 9.2, 7.4, 6.3, 5.6, 7.3, 8.3, 7.2, 7.5, 6.1, 9.4, 5.4, 7.6, 8.1, and 7.9 mg/L.

The *population* is all specimens having a known concentration of 8.0 mg/L. The *sample* is the 27 observations (measurements). The *sample size* is $n = 27$. The *random variable* is the measured concentration in each specimen having a known concentration of 8.0 mg/L. *Experimental error* has caused the observed values to vary about the true value of 8.0 mg/L. The errors are $6.9 - 8.0 = -1.1$, $7.8 - 8.0 = -0.2$, $+0.9$, -2.8, -0.3, $+1.6$, $+0.7$, and so on.

Plotting the Data

A useful first step is to plot the data. Figure 2.1 shows the data from Example 2.1 plotted in time order of observation, with a dot diagram plotted on the right-hand side. Dots are stacked to indicate frequency.

A dot diagram starts to get crowded when there are more than about 20 observations. For a large number of points (a large sample size), it is convenient to group the dots into intervals and represent a group with a bar, as shown in Figure 2.2. This plot shows the empirical (realized) distribution of the data. Plots of this kind are usually called *histograms*, but the more suggestive name of *data density plot* has been suggested (Watts, 1991).

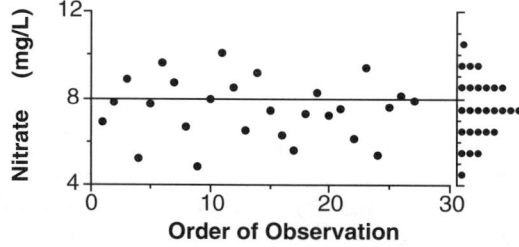

FIGURE 2.1 Time plot and dot diagram (right-hand side) of the nitrate data in Example 2.1.

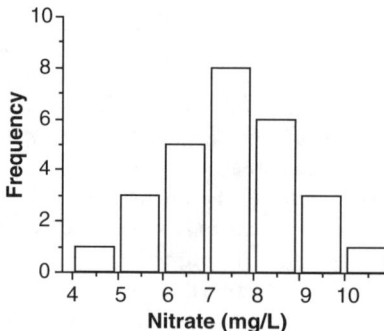

FIGURE 2.2 Frequency diagram (histogram).

A Brief Review of Statistics

The ordinate of the histogram can be the actual count (n_i) of occurrences in an interval or it can be the relative frequency, $f_i = n_i/n$, where n is the total number of values used to construct the histogram. Relative frequency provides an estimate of the probability that an observation will fall within a particular interval.

Another useful plot of the raw data is the cumulative frequency distribution. Here, the data are rank ordered, usually from the smallest (rank = 1) to the largest (rank = n), and plotted versus their rank. Figure 2.3 shows this plot of the nitrate data from Example 2.1. This plot serves as the basis of the *probability plots* that are discussed in Chapter 5.

Probability Distributions

As the sample size, n, becomes very large, the frequency distribution becomes smoother and approaches the shape of the underlying *population frequency distribution*. This distribution function may represent discrete random variables or continuous random variables. A *discrete random variable* is one that has only point values (often integer values). A *continuous random variable* is one that can assume any value over a range. A continuous random variable may appear to be discrete as a manifestation of the sensitivity of the measuring device, or because an analyst has rounded off the values that actually were measured.

The mathematical function used to represent the population frequency distribution of a continuous random variable is called the *probability density function*. The ordinate $p(y)$ of the distribution is not a probability itself; it is the probability density. It becomes a probability when it is multiplied by an interval on the horizontal axis (i.e., $P = p(y)\Delta$ where Δ is the size of the interval). Probability is always given by the area under the probability density function. The laws of probability require that the area under the curve equal one (1.00). This concept is illustrated by Figure 2.4, which shows the probability density function known as the *normal distribution*.

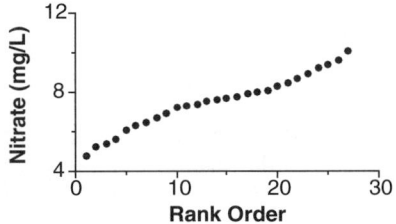

FIGURE 2.3 Cumulative distribution plot of the nitrate data from Example 2.1.

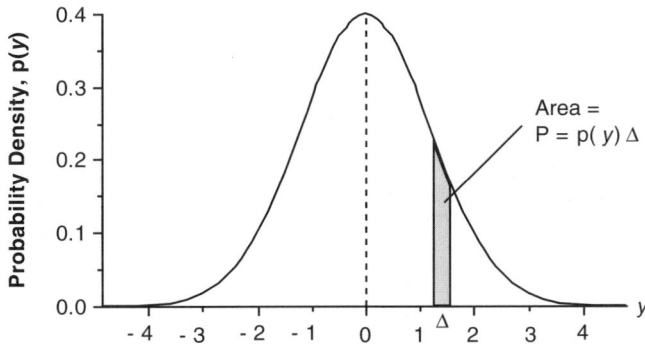

FIGURE 2.4 The normal probability density function.

The Average, Variance, and Standard Deviation

We distinguish between a quantity that represents a population and a quantity that represents a sample. A *statistic* is a realized quantity calculated from data that are taken to represent a population. A *parameter* is an idealized quantity associated with the population. Parameters cannot be measured directly unless the entire population can be observed. Therefore, *parameters are estimated by statistics*. Parameters are usually designated by Greek letters (α, β, γ, etc.) and statistics by Roman letters (a, b, c, etc.). Parameters are constants (often unknown in value) and statistics are random variables computed from data.

Given a population of a very large set of N observations from which the sample is to come, the *population mean* is η:

$$\eta = \frac{\sum y_i}{N}$$

where y_i is an observation. The summation, indicated by Σ, is over the population of N observations. We can also say that the mean of the population is the expected value of y, which is written as $E(y) = \eta$, when N is very large.

The *sample of n observations* actually available from the population is used to calculate the *sample average*:

$$\bar{y} = \frac{1}{n}\sum y_i$$

which estimates the mean η.

The *variance* of the population is denoted by σ^2. The measure of how far any particular observation is from the mean η is $y_i - \eta$. The variance is the mean value of the square of such deviations taken over the whole population:

$$\sigma^2 = \frac{\sum(y_i - \eta)^2}{N}$$

The *standard deviation* of the population is a measure of spread that has the same units as the original measurements and as the mean. The standard deviation is the square root of the variance:

$$\sigma = \sqrt{\frac{\sum(y_i - \eta)^2}{N}}$$

The true values of the population parameters σ and σ^2 are often unknown to the experimenter. They can be estimated by the *sample variance*:

$$s^2 = \frac{\sum(y_i - \bar{y})^2}{n-1}$$

where n is the size of the sample and $\bar{y}$ is the sample average. The *sample standard deviation* is the square root of the sample variance:

$$s = \sqrt{\frac{\sum(y_i - \bar{y})^2}{n-1}}$$

Here the denominator is $n-1$ rather than n. The $n-1$ represents the *degrees of freedom* of the sample. One degree of freedom (the -1) is consumed because the average must be calculated to estimate s. The deviations of n observations from their sample average must sum exactly to zero. This implies that any

A Brief Review of Statistics

$n - 1$ of the deviations or *residuals* completely determines the one remaining residual. The n residuals, and hence their sum of squares and sample variance, are said therefore to have $n - 1$ degrees of freedom. Degrees of freedom will be denoted by the Greek letter v. For the sample variance and sample standard deviation, $v = n - 1$.

Most of the time, "sample" will be dropped from sample standard deviation, sample variance, and sample average. It should be clear from the context that the calculated statistics are being discussed. The Roman letters, for example s^2, s, and $\bar{y}$, will indicate quantities that are statistics. Greek letters (σ^2, σ, and η) indicate parameters.

Example 2.2

For the 27 nitrate observations, the sample average is

$$\bar{y} = \frac{6.9 + 7.8 + \cdots + 8.1 + 7.9}{27} = 7.51 \text{ mg/L}$$

The sample variance is

$$s^2 = \frac{(6.9 - 7.51)^2 + \cdots + (7.9 - 7.51)^2}{27 - 1} = 1.9138 \text{ (mg/L)}^2$$

The sample standard deviation is

$$s = \sqrt{1.9138} = 1.38 \text{ mg/L}$$

The sample variance and sample standard deviation have $v = 27 - 1 = 26$ degrees of freedom.

The data were reported with two significant figures. The average of several values should be calculated with at least one more figure than that of the data. The standard deviation should be computed to at least three significant figures (Taylor, 1987).

Accuracy, Bias, and Precision

Accuracy is a function of both *bias* and *precision*. As illustrated by Example 2.3 and Figure 2.5, bias measures *systematic errors* and precision measures the degree of scatter in the data. Accurate measurements have good precision and near zero bias. Inaccurate methods can have poor precision, unacceptable bias, or both.

Bias (systematic error) can be removed, once identified, by careful checks on experimental technique and equipment. It cannot be averaged out by making more measurements. Sometimes, bias cannot be identified because the underlying true value is unknown.

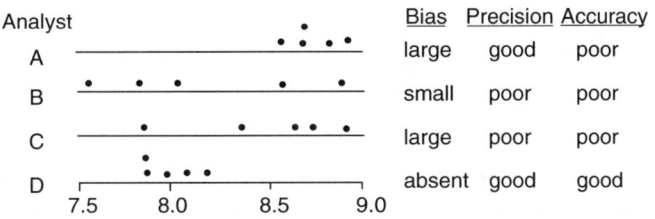

FIGURE 2.5 Accuracy is a function of bias and good precision.

Precision has to do with the scatter between repeated measurements. This scatter is caused by random errors in the measurements. Precise results have small random errors. The standard deviation, s, is often used as an index of precision (or imprecision). When s is large, the measurements are imprecise. Random errors can never be eliminated, although by careful technique they can be minimized. Their effect can be reduced by making repeated measurements and averaging them. Making replicate measures also provides the means to quantify the measurement errors and evaluate their importance.

Example 2.3

Four analysts each were given five samples that were prepared to have a known concentration of 8.00 mg/L. The results are shown in Figure 2.5. Two separate kinds of errors have occurred in A's work: (1) random errors cause the individual results to be 'scattered' about the average of his five results and (2) a fixed component in the measurement error, a systematic error or bias, makes the observations too high. Analyst B has poor precision, but little observed bias. Analyst C has poor accuracy and poor precision. Only Analyst D has little bias and good precision.

Example 2.4

The estimated bias of the 27 nitrate measurements in Example 2.1 is the difference between the sample average and the known value:

$$\text{Bias} = \bar{y} - \eta = 7.51 - 8.00 = -0.49 \text{ mg/L}$$

The precision of the measurements is given by the sample standard deviation:

$$\text{Precision} = s = 1.38 \text{ mg/L}$$

Later examples will show how to assess whether this amount of apparent bias is likely to result just from random error in the measurements.

Reproducibility and Repeatability

Reproducibility and *repeatability* are sometimes used as synonyms for precision. However, a distinction should be made between these words. Suppose an analyst made the five replicate measurements in rapid succession, say within an hour or so, using the same set of reagent solutions and glassware throughout. Temperature, humidity, and other laboratory conditions would be nearly constant. Such measurements would estimate repeatability, which might also be called *within-run precision*. If the same analyst did the five measurements on five different occasions, differences in glassware, lab conditions, reagents, etc., would be reflected in the results. This set of data would give an indication of reproducibility, which might also be called *between-run precision*. We expect that the between-run precision will have greater spread than the within-run precision. Therefore, repeatability and reproducibility are not the same and it would be a misrepresentation if they were not clearly distinguished and honestly defined. We do not want to underestimate the total variability in a measurement process. Error estimates based on sequentially repeated observations are likely to give a false sense of security about the precision of the data. The quantity of practical importance is reproducibility, which refers to differences in observations recorded from *replicate experiments* performed in random sequence.

Example 2.5

Measured values frequently contain multiple sources of variation. Two sets of data from a process are plotted in Figure 2.6. The data represent (a) five repeat tests performed on a single specimen from a batch of product and (b) one test made on each of five different specimens from the same batch. The variation associated with each data set is different.

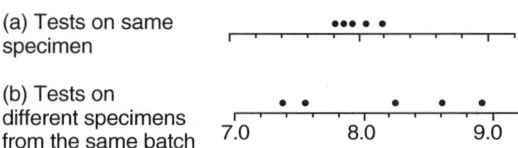

FIGURE 2.6 Repeated tests from (a) a single specimen that reflect variation in the analytical measurement method and (b) five specimens from a single batch that reflect variation due to collecting the test specimens *and* the measurement method.

If we wish to compare two testing methods A and B, the correct basis is to compare five determinations made using test method A with five determinations using test method B with all tests made on portions of the *same test specimen*. These two sets of measurements are not influenced by variation between test specimens or by the method of collection.

If we wish to compare two sampling methods, the correct basis is to compare five determinations made on five different specimens collected using sampling method A with those made on five specimens using sampling method B, with all specimens coming from the *same batch*. These two sets of data will contain variation due to the collection of the specimens and the testing method. They do not contain variation due to differences between batches.

If the goal is to compare two different processes for making a product, the observations used as a basis for comparison should reflect variation due to differences between batches taken from the two processes.

Normality, Randomness, and Independence

The three important properties on which many statistical procedures rest are *normality, randomness*, and *independence*. Of these, normality is the one that seems to worry people the most. It is not always the most important.

Normality means that the error term in a measurement, e_i, is assumed to come from a normal probability distribution. This is the familiar symmetrical bell-shaped distribution. There is a tendency for error distributions that result from many *additive component errors* to be "normal-like." This is the *central limit effect*. It rests on the assumption that there are several sources of error, that no single source dominates, and that the overall error is a linear combination of independently distributed errors. These conditions seem very restrictive, but they often (but not always) exist. Even when they do not exist, lack of normality is not necessarily a serious problem. Transformations are available to make nonnormal errors "normal-like."

Many commonly used statistical procedures, including those that rely directly on comparing averages (such as *t* tests to compare two averages and analysis of variance tests to compare several averages) are *robust* to deviations from normality. Robust means that the test tends to yield correct conclusions even when applied to data that are not normally distributed.

Random means that the observations are drawn from a population in a way that gives every element of the population an equal chance of being drawn. Randomization of sampling is the best form of insurance that observations will be independent.

Example 2.6

> Errors in the nitrate laboratory data were checked for randomness by plotting the errors, $e_i = y_i - \eta$. If the errors are random, the plot will show no pattern. Figure 2.7 is such a plot, showing e_i in order of observation. The plot does not suggest any reason to believe the errors are not random.

Imagine ways in which the errors of the nitrate measurements might be nonrandom. Suppose, for example, that the measurement process drifted such that early measurements tended to be high and later measurements low. A plot of the errors against time of analysis would show a trend (positive errors followed by negative

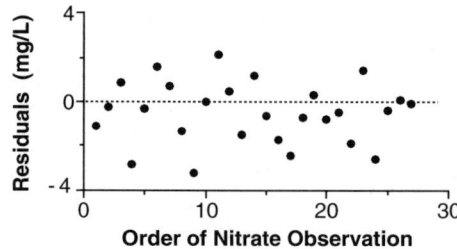

FIGURE 2.7 Plot of nitrate measurement errors indicates randomness.

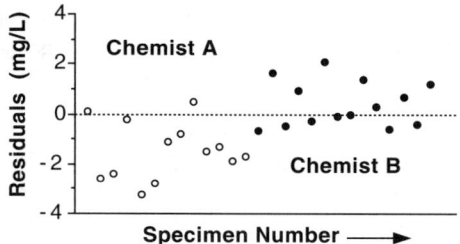

FIGURE 2.8 Plot of nitrate residuals in order of sample number (not order of observation) and differentiated by chemist.

errors), indicating that an element of nonrandomness had entered the measurement process. Or, suppose that two different chemists had worked on the specimens and that one analyst always measured values that tended too high, and the other always too low. A plot like Figure 2.8 reveals this kind of error, which might be disguised if there is no differentiation by chemist. It is a good idea to check randomness with respect to each identifiable factor (day of the week, chemist, instrument, time of sample collection, etc.) that could influence the measurement process.

Independence means that the simple multiplicative laws of probability work (that is, the probability of the joint occurrence of two events is given by the product of the probabilities of the individual occurrence). In the context of a series of observations, suppose that unknown causes produced experimental errors that tended to persist over time so that whenever the first observation y_1 was high, the second observation y_2 was also high. In such a case, y_1 and y_2 are not statistically independent. They are dependent in time, or *serially correlated*. The same effect can result from cyclic patterns or slow drift in a system. Lack of independence can seriously distort the variance estimate and thereby make probability statements based on the normal or t distributions very much in error.

Independence is often lacking in environmental data because (1) it is inconvenient or impossible to randomize the sampling, or (2) it is undesirable to randomize the sampling because it is the cyclic or otherwise dynamic behavior of the system that needs to be studied. We therefore cannot automatically assume that observations are independent. When they are not, special methods are needed to account for the correlation in the data.

Example 2.7

The nitrate measurement errors were checked for independence by plotting y_i against the previous observation, y_{i-1}. This plot, Figure 2.9, shows no pattern (the correlation coefficient is -0.077) and indicates that the measurements are independent of each other, at least with respect to the order in which the measurements were performed. There could be correlation with respect to some other characteristic of the specimens, for example, spatial correlation if the specimens come from different depths in a lake or from different sampling stations along a river.

A Brief Review of Statistics

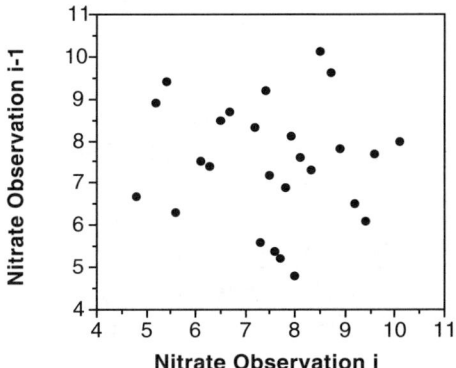

FIGURE 2.9 Plot of measurement y_i vs. measurement y_{i-1} shows a lack of serial correlation between adjacent measurements.

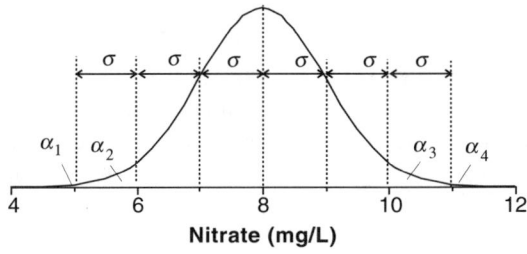

FIGURE 2.10 A normal distribution centered at mean $\eta = 8$. Because of symmetry, the areas $\alpha_1 = \alpha_4$ and $\alpha_1 + \alpha_2 = \alpha_3 + \alpha_4$.

The Normal Distribution

Repeated observations that differ because of experimental error often vary about some central value with a bell-shaped probability distribution that is symmetric and in which small deviations occur much more frequently than large ones. A continuous population frequency distribution that represents this condition is the normal distribution (also sometimes called the *Gaussian distribution*). Figure 2.10 shows a normal distribution for a random variable with $\eta = 8$ and $\sigma^2 = 1$. The normal distribution is characterized completely by its mean and variance and is often described by the notation $N(\eta, \sigma^2)$, which is read "a normal distribution with mean η and variance σ^2."

The geometry of the normal curve is as follows:

1. The vertical axis (probability density) is scaled such that area under the curve is unity (1.0).
2. The standard deviation σ measures the distance from the mean to the point of inflection.
3. The probability that a positive deviation from the mean will exceed one standard deviation is 0.1587, or roughly 1/6. This is the area to the right of 9 mg/L in Figure 2.8. The probability that a positive deviation will exceed 2σ is 0.0228 (roughly 1/40), which is area $\alpha_3 + \alpha_4$ in Figure 2.8. The chance of a positive deviation exceeding 3σ is 0.0013 (roughly 1/750), which is the area α_4.
4. Because of symmetry, the probabilities are the same for negative deviations and $\alpha_1 = \alpha_4$ and $\alpha_1 + \alpha_2 = \alpha_3 + \alpha_4$.
5. The chance that a deviation in either direction will exceed 2σ is $2(0.0228) = 0.0456$ (roughly 1/20). This is the sum of the two small areas under the extremes of the tails, $\alpha_1 + \alpha_2 = \alpha_3 + \alpha_4$.

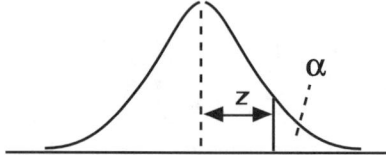

TABLE 2.1

Tail Area Probability (α) of the Unit Normal Distribution

z	0.00	0.01	0.02	0.03	0.04	0.05	0.06	0.07	0.08	0.09
0.0	0.5000	0.4960	0.4920	0.4880	0.4840	0.4801	0.4761	0.4721	0.4681	0.4641
...	...	...	...	...	...	...	...	...	...	...
1.5	0.0668	0.0655	0.0643	0.0630	0.0618	0.0606	0.0594	0.0582	0.0571	0.0559
1.6	0.0548	0.0537	0.0526	0.0516	0.0505	0.0495	0.0485	0.0475	0.0465	0.0455
1.7	0.0446	0.0436	0.0427	0.0418	0.0409	0.0401	0.0392	0.0384	0.0375	0.0367
1.8	0.0359	0.0351	0.0344	0.0366	0.0329	0.0322	0.0314	0.0307	0.0301	0.0294
1.9	0.0287	0.0281	0.0274	0.0268	0.0262	0.0256	0.0250	0.0244	0.0239	0.0233
2.0	0.0228	0.0222	0.0217	0.0212	0.0207	0.0202	0.0197	0.0192	0.0188	0.0183

It is convenient to work with standardized normal deviates, $z = (y - \eta)/\sigma$, where z has the distribution $N(0, 1)$, because the areas under the standardized normal curve are tabulated. This merely scales the data in terms of the standard deviation instead of the original units of measurement (e.g., concentration). A portion of this table is reproduced in Table 2.1. For example, the probability of a standardized normal deviate exceeding 1.57 is 0.0582, or 5.82%.

The t Distribution

Standardizing a normal random variable requires that both η and σ are known. In practice, however, we cannot calculate $z = (y - \eta)/\sigma$ because σ is unknown. Instead, we substitute s for σ and calculate the t *statistic*:

$$t = \frac{y - \eta}{s}$$

The value of η may be known (e.g., because it is a primary standard) or it may be assumed when constructing a hypothesis that will be tested (e.g., the difference between two treatments is assumed to be zero). Under certain conditions, which are given below, t has a known distribution, called the *Student's t distribution*, or simply the t *distribution*.

The t distribution is bell-shaped and symmetric (like the normal distribution), but the tails of the t distribution are wider than tails of the normal distribution. The width of the t distribution depends on the degree of uncertainty in s^2, which is measured by the degrees of freedom ν on which this estimate of s^2 is based. When the sample size is infinite ($\nu = \infty$), there is no uncertainty in s^2 (because $s^2 = \sigma^2$) and the t distribution becomes the standard normal distribution. When the sample size is small ($\nu \leq 30$), the variation in s^2 increases. This is reflected by the spread of the t distribution increasing as the number of degrees of freedom of s^2 decreases. The tail area under the bell-shaped curve of the t distribution is the probability of t exceeding a given value. A portion of the t table is reproduced in Table 2.2.

The conditions under which the quantity $t = (y - \eta)/s$ has a t distribution with ν degrees of freedom are:

1. y is normally distributed about η with variance σ^2.
2. s is distributed independently of the mean; that is, the variance of the sample does not increase or decrease as the mean increases or decreases.
3. The quantity s^2, which has ν degrees of freedom, is calculated from normally and independently distributed observations having variance σ^2.

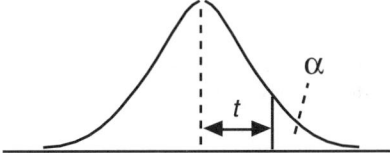

TABLE 2.2
Values of t for Several Tail Probabilities and Degrees of Freedom

	Tail Area Probability				
n	$\alpha = 0.1$	0.05	0.025	0.01	0.005
2	1.886	2.920	4.303	6.965	9.925
4	1.533	2.132	2.776	3.747	4.604
6	1.440	1.943	2.447	3.143	3.707
10	1.372	1.812	2.228	2.764	3.169
20	1.325	1.725	2.086	2.528	2.845
25	1.316	1.708	2.060	2.485	2.787
26	1.315	1.706	2.056	2.479	2.779
27	1.314	1.703	2.052	2.473	2.771
40	1.303	1.684	2.021	2.423	2.704
∞	1.282	1.645	1.960	2.326	2.576

Sampling Distribution of the Average and the Variance

All calculated statistics are random variables and, as such, are characterized by a probability distribution having an expected value (mean) and a variance. First we consider the *sampling distribution* of the average $\bar{y}$. Suppose that many random samples of size n were collected from a population and that the average was calculated for each sample. Many different average values would result and these averages could be plotted in the form of a probability distribution. This would be the sampling distribution of the average (that is, the distribution of $\bar{y}$ values computed from different samples). If discrepancies in the observations y_i about the mean are random and independent, then the sampling distribution of $\bar{y}$ has mean η and variance, σ^2/n. The quantity σ^2/n is the *variance of the average*. Its square root is called the *standard error of the mean*:

$$\sigma_{\bar{y}} = \frac{\sigma}{\sqrt{n}}$$

A *standard error* is an estimate of variation of a statistic. In this case the statistic is the mean and the subscript $\bar{y}$ is a reminder of that. The standard error of the mean describes the spread of sample averages about η, while the standard deviation, σ, describes the spread of the sample observations y about η. That is, $\sigma_{\bar{y}}$ indicates the spread we would expect to observe in calculated average values if we could repeatedly draw samples of size n at random from a population that has mean η and variance σ^2. We note that the sample average has smaller variability about η than does the sample data.

The *sample standard deviation* is:

$$s = \sqrt{\frac{\sum(y_i - \bar{y})^2}{n-1}}$$

The estimate of the standard error of the mean is:

$$s_{\bar{y}} = \frac{s}{\sqrt{n}}$$

Example 2.8

The average for the $n = 27$ nitrate measurements is $\bar{y} = 7.51$ and the sample standard deviation is $s = 1.38$. The estimated standard error of the mean is:

$$s_{\bar{y}} = \frac{1.38}{\sqrt{27}} = 0.27 \text{ mg/L}$$

If the parent distribution is normal, the sampling distribution of $\bar{y}$ will be normal. If the parent distribution is nonnormal, the distribution of $\bar{y}$ will be more nearly normal than the parent distribution. As the number of observations n used in the average increases, the distribution of $\bar{y}$ becomes increasingly more normal. This fortunate property is the *central limit effect*. This means that we can use the normal distribution with mean η and variance σ^2/n as the reference distribution to make probability statements about $\bar{y}$ (e.g., that the probability that $\bar{y}$ is less than or greater than a particular value, or that it lies in the interval between two particular values).

Usually the population variance, σ^2, is not known and we cannot use the normal distribution as the reference distribution for the sample average. Instead, we substitute $s_{\bar{y}}$ for $\sigma_{\bar{y}}$ and use the t distribution. If the parent distribution is normal and the population variance is estimated by s^2, the quantity:

$$t = \frac{\bar{y} - \eta}{s/\sqrt{n}}$$

which is known as the *standardized mean* or as the *t statistic*, will have a t distribution with $\nu = n - 1$ degrees of freedom. If the parent population is not normal but the sampling is random, the t statistic will tend toward the t distribution (just as the distribution of $\bar{y}$ tends toward being normal).

If the parent population is $N(\eta, \sigma^2)$, and assuming once again that the observations are random and independent, the sample variance s^2 has especially attractive properties. For these conditions, s^2 is distributed independently of y in a scaled χ^2 (*Chi-square*) distribution. The scaled quantity is:

$$\chi^2 = \nu \frac{s^2}{\sigma^2}.$$

This distribution is skewed to the right. The exact form of the χ^2 distribution depends on the number of degrees of freedom, ν, on which s^2 is based. The spread of the distribution increases as ν increases. The tail area under the Chi-square distribution is the probability of a value of $\chi^2 = \nu s^2/\sigma^2$ exceeding a given value.

Figure 2.11 illustrates these properties of the sampling distributions of $\bar{y}$, s^2, and t for a random sample of size $n = 4$.

Example 2.9

For the nitrate data, the sample mean concentration of $\bar{y} = 7.51$ mg/L lies a considerable distance below the true value of 8.00 mg/L (Figure 2.12). If the true mean of the sample is 8.0 mg/L and the laboratory is measuring accurately, an estimated mean as low as 7.51 would occur by chance only about four times in 100. This is established as follows. The value of the t statistic is:

$$t = \frac{\bar{y} - \eta}{s/\sqrt{n}} = \frac{7.51 - 8.00}{1.38/\sqrt{27}} = -1.842$$

with $\nu = 26$ degrees of freedom. Find the probability of such a value of t occurring by referring to the tabulated tail areas of the t distribution in Appendix A. Because of symmetry, this table

A Brief Review of Statistics

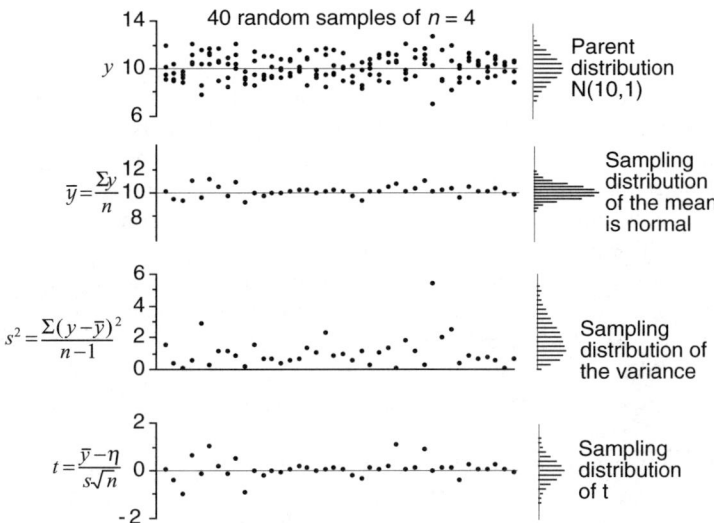

FIGURE 2.11 Forty random samples of $n = 4$ from a N(10,1) normal distribution to produce the sampling distributions of $\bar{y}$, s^2, and t. (Adapted from Box, G. E. P., W. G. Hunter, and J. S. Hunter (1978). *Statistics for Experimenters: An Introduction to Design, Data Analysis, and Model Building*, New York, Wiley Interscience.)

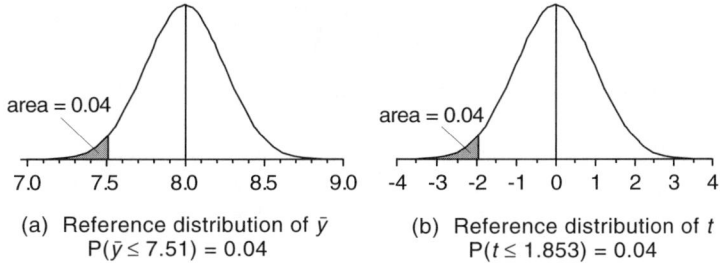

FIGURE 2.12 The $\bar{y}$ and t reference distributions for the sample average of the nitrate data of Example 2.1.

serves for negative as well as positive values. For $\nu = 26$, the tail areas are 0.05 for $t = -1.706$, 0.025 for $t = -2.056$, and 0.01 for $t = -2.479$. Plotting these and drawing a smooth curve as an aid to interpolation gives Prob($t < -1.853$) ≈ 0.04, or only about 4%. This low probability suggests that there may be a problem with the measurement method in this laboratory.

The assessment given in Example 2.9 can also be made by examining the reference distributions of $\bar{y}$. The distribution of $\bar{y}$ is centered about $\eta = 8.0$ mg/L with standard deviation $s = 0.266$ mg/L. The value of $\bar{y}$ observed for this particular experiment is 7.51 mg/L. The shaded area to the left of $\bar{y} = 7.51$ in Figure 2.12(a) is the same as the area to the left of $t = -1.853$ in Figure 2.12(b). Thus, P($t \leq -1.853$) = P($\bar{y} \leq 7.51$) ≈ 0.04.

In the context of Example 2.9, the investigator is considering the particular result that $\bar{y} = 7.51$ mg/L in a laboratory assessment based on 27 blind measurements on specimens known to have concentration $\eta = 8.00$ mg/L. A relevant reference distribution is needed in order to decide whether the result is easily explained by mere chance variation or whether it is exceptional. This reference distribution represents the set of outcomes that could occur by chance. The t distribution is a relevant reference distribution under certain conditions which have already been identified. An outcome that falls on the tail of the distribution can be considered exceptional. If it is found to be exceptional, it is declared statistically significant. Significant in this context does not refer to scientific importance, but only to its statistical plausibility in light of the data.

Significance Tests

In Example 2.9 we knew that the nitrate population mean was truly 8.0 mg/L, and asked, "How likely are we to get a sample mean as small as $\bar{y} = 7.51$ mg/L from the analysis of 27 specimens?" If this result is highly unlikely, we might decide that the sample did not represent the population, probably because the measurement process was biased to yield concentrations below the true value. Or, we might decide that the result, although unlikely, should be accepted as occurring due to chance rather than due to an assignable cause (like bias in the measurements).

Statistical inference involves making an assessment from experimental data about an unknown population parameter (e.g., a mean or variance). Consider that the true mean is unknown (instead of being known as in Example 2.9) and we ask, "If a sample mean of 7.51 mg/L is estimated from measurements on 27 specimens, what is the likelihood that the true population mean is 8.00 mg/L?" Two methods for making such statistical inferences are to make a *significance test* and to examine the *confidence interval* of the population parameter.

The significance test typically takes the form of a *hypothesis test*. The hypothesis to be tested is often designated H_o. In this case, H_o is that the true value of the population mean is $\eta = 8.0$ mg/L. This is sometimes more formally written as H_o: $\eta = 8.0$. This is the *null hypothesis*. The alternate hypothesis is H_a: or $\eta \neq 8.0$, which could be either $\eta < 8.0$ or $\eta > 8.0$. A significance level, α, is selected at which the null hypothesis will be rejected. The significance level, α, represents the risk of falsely rejecting the null hypothesis.

The relevant t statistic is:

$$t = \frac{\text{statistic} - E(\text{statistic})}{\sqrt{V(\text{statistic})}}$$

where $E(\text{statistic})$ denotes the expected value of the statistic being estimated and $V(\text{statistic})$ denotes the variance of this statistic.

A t statistic with v degrees of freedom and significance level α is written as $t_{v,\alpha}$.

Example 2.10

Use the nitrate data to test the hypothesis that $\eta = 8.0$ at $\alpha = 0.05$. The appropriate hypotheses are H_o: $\eta = 8.0$ and H_a: $\eta < 8.0$. This is a one-sided test because the alternate hypothesis involves η on the lower side of 8.0.

The hypothesis test is made using:

$$t = \frac{\text{statistic} - E(\text{statistic})}{\sqrt{V(\text{statistic})}} = \frac{\bar{y} - \eta}{s_{\bar{y}}} = \frac{7.51 - 8.0}{0.266} = -1.842$$

The null hypothesis will be rejected if the computed t is less than the value of the lower tail t statistic having probability of $\alpha = 0.05$. The value of t with $\alpha = 0.05$ and $v = 26$ degrees of freedom obtained from tables is $t_{v=26, \alpha=0.05} = -1.706$. The computed value of $t = -1.853$ is smaller than the table value of -1.706. The decision is to reject the null hypothesis in favor of the alternate hypothesis.

Examples 2.9 and 2.10 are outwardly different, but mathematically and statistically equivalent. In Example 2.9, the experimenter assumes the population parameter to be known and asks whether the sample data can be construed to represent the population. In Example 2.10, the experimenter assumes the sample data are representative and asks whether the assumed population value is reasonably supported by the data. In practice, the experimental context will usually suggest one approach as the more comfortable interpretation.

Example 2.10 illustrated a *one-sided hypothesis test*. It evaluated the hypothesis that the sample mean was truly to one side of 8.0. This particular example was interested in the mean being below the true value. A *two-sided hypothesis test* would consider the statistical plausibility of both the positive and negative deviations from the mean.

Example 2.11

Use the nitrate data to test the null hypothesis that H_o: $\eta = 8.0$ and H_a: $\eta \neq 8.0$. Here the alternate hypothesis considers deviations on both the positive and negative sides of the population mean, which makes this a two-sided test. Both the lower and upper tail areas of the t reference distribution must be used. Because of symmetry, these tail areas are equal. For a test at the $\alpha = 0.05$ significance level, the sum of the upper and lower tail areas equals 0.05. The area of each tail is $\alpha/2 = 0.05/2 = 0.025$. For $\alpha/2 = 0.025$ and $v = 26$, $t_{v=26, \alpha/2=0.025} = \pm 2.056$. The computed t value is the same as in Example 2.9; that is, $t = -1.852$. The computed t value is not outside the range of the critical t values. There is insufficient evidence to reject the null hypothesis at the stated level of significance.

Notice that the hypothesis tests in Examples 2.10 and 2.11 reached different conclusions although they used the same data, the same significance level, and the same null hypothesis. The only difference was the alternate hypothesis. The two-sided alternative hypothesis stated an interest in detecting both negative and positive deviations from the assumed mean by dividing the rejection probability α between the two tails. Thus, a decision to reject the null hypothesis takes into account differences between the sample mean and the assumed population mean that are both significantly smaller and significantly larger than zero. The consequence of this is that in order to be declared *statistically significant*, the deviation must be larger in a two-sided test than in a one-sided test.

Is the correct test one-sided or the two-sided? The question cannot be answered in general, but often the decision-making context will indicate which test is appropriate. In a case where a positive deviation is undesirable but a negative deviation is not, a one-sided test would be indicated. Typical situations would be (1) judging compliance with an environmental protection limit where high values indicate a violation, and (2) an experiment intended to investigate whether adding chemical A to the process *increases* the efficiency. If the experimental question is whether adding chemical A *changes* the efficiency (either for better or worse), a two-sided test would be indicated.

Confidence Intervals

Hypothesis testing can be overdone. It is often more informative to state an interval within which the value of a parameter would be expected to lie. A $1 - \alpha$ confidence interval for the population mean can be constructed using the appropriate value of t as:

$$\bar{y} - s_{\bar{y}} t_{\alpha/2} < \eta < \bar{y} + s_{\bar{y}} t_{\alpha/2}$$

where $t_{\alpha/2}$ and $s_{\bar{y}}$ have $v = n - 1$ degrees of freedom. This confidence interval is bounded by a lower and an upper limit. The meaning of the $1 - \alpha$ confidence level is "If a series of random sets of n observations is sampled from a normal distribution with mean η and fixed σ, and a $1 - \alpha$ confidence interval $\bar{y} \pm s_{\bar{y}} t_{\alpha/2}$ is constructed from each set, a proportion, $1 - \alpha$, of these intervals will include the value η and a proportion, α, will not" (Box et al., 1978). (Another interpretation, a Bayesian interpretation, is that there is a $1 - \alpha$ probability that the true value falls within this confidence interval.)

Example 2.12

The confidence limits for the true mean of the test specimens are constructed for $\alpha/2 = 0.05/2 = 0.025$, which gives a 95% confidence interval. For $t_{v=26, \alpha/2=0.025} = 2.056$, $\bar{y} = 7.51$ and $s_{\bar{y}} = 0.266$, the upper and lower 95% confidence limits are:

$$7.51 - 2.056(0.266) < \eta < 7.51 + 2.056(0.266)$$
$$6.96 < \eta < 8.05$$

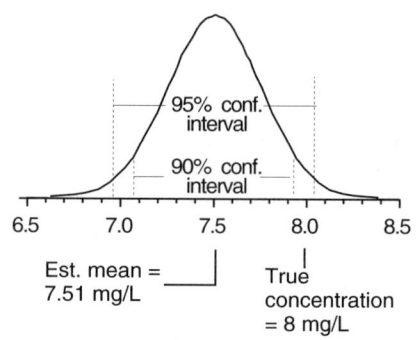

FIGURE 2.13 The t distribution for the estimated mean of the nitrate data with the 90% and 95% confidence intervals.

This interval contains $\eta = 8.0$, so we conclude that the difference between $\bar{y}$ and η is not so large that random measurement error should be rejected as a plausible explanation.

This use of a confidence interval is equivalent to making a two-sided test of the null hypothesis, as was done in Example 2.11. Figure 2.13 shows the two-sided 90% and 95% confidence intervals for η.

Summary

This chapter has reviewed basic definitions, assumptions, and principles. The key points are listed below.

A sample is a sub-set of a population and consists of a group of n observations taken for analysis. Populations are characterized by parameters, which are usually unknown and unmeasurable because we cannot measure every item in the population. Parameters are estimated by statistics that are calculated from the sample. Statistics are random variables and are characterized by a probability distribution that has a mean and a variance.

All measurements are subject to experimental (measurement) error. Accuracy is a function of both bias and precision. The role of statistics in scientific investigations is to quantify and characterize the error and take it into account when the data are used to make decisions.

Given a *normal parent distribution* with mean η and variance σ^2 and for random and independent observations, the sample average $\bar{y}$ has a normal distribution with mean η and variance σ^2/n. The sample variance s^2 has expected value σ^2. The statistic $t = (y - \eta)/(s/\sqrt{n})$ with $\nu = n - 1$ degrees of freedom has a t distribution.

Statistical procedures that rely directly on comparing means, such as t tests to compare two means and analysis of variance tests to compare several means, are robust to nonnormality but may be adversely affected by a lack of independence.

Hypothesis tests are useful methods of statistical inference but they are often unnecessarily complicated in making simple comparisons. Confidence intervals are statistically equivalent alternatives to hypothesis testing, and they are simple and straightforward. They give the interval (range) within which the population parameter value is expected to fall.

These basic concepts are discussed in any introductory statistics book (Devore, 2000; Johnson, 2000). A careful discussion of the material in this chapter, with special attention to the importance regarding normality and independence, is found in Chapters 2, 3, and 4 of Box et al. (1978).

References

Box, G. E. P., W. G. Hunter, and J. S. Hunter (1978). *Statistics for Experimenters: An Introduction to Design, Data Analysis, and Model Building,* New York, Wiley Interscience.

Devore, J. (2000). *Probability and Statistics for Engineers,* 5th ed., Duxbury.

Johnson, R. A. (2000). *Probability and Statistics for Engineers,* 6th ed., Englewood Cliffs, NJ, Prentice-Hall.

Taylor, J. K. (1987). *Quality Assurance of Chemical Measurements,* Chelsea, MI: Lewis Publishers, Inc.
Watts, D. G. (1991). "Why Is Introductory Statistics Difficult to Learn? And What Can We Do to Make It Easier?" *Am. Statistician,* 45, 4, 290–291.

Exercises

2.1 Concepts I. Define (a) population, (b) sample, and (c) random variable.

2.2 Concepts II. Define (a) random error, (b) noise, and (c) experimental error.

2.3 Randomization. A laboratory receives 200 water specimens from a city water supply each day. This exceeds their capacity so they randomly select 20 per day for analysis. Explain how you would select the sample of $n = 20$ water specimens.

2.4 Experimental Errors. The measured concentration of phosphorus (P) for $n = 20$ identical specimens of wastewater with known concentration of 2 mg/L are:

$$\begin{array}{cccccccccc}
1.8 & 2.2 & 2.1 & 2.3 & 2.1 & 2.2 & 2.1 & 2.1 & 1.8 & 1.9 \\
2.4 & 2.0 & 1.9 & 1.9 & 2.2 & 2.3 & 2.2 & 2.3 & 2.1 & 2.2
\end{array}$$

Calculate the experimental errors. Are the errors random? Plot the errors to show their distribution.

2.5 Summary Statistics. For the phosphorus data in Exercise 2.4, calculate the average, variance, and standard deviation. The average and standard deviation are estimated with how many degrees of freedom?

2.6 Bias and Precision. What are the precision and bias of the phosphorus data in Exercise 2.4?

2.7 Concepts III. Define reproducibility and repeatability. Give an example to explain each. Which of these properties is more important to the user of data from a laboratory?

2.8 Concepts IV. Define normality, randomness, and independence in sampled data. Sketch plots of "data" to illustrate the presence and lack of each characteristic.

2.9 Normal Distribution. Sketch the normal distribution for a population that has a mean of 20 and standard deviation of 2.

2.10 Normal Probabilities. What is the probability that the standard normal deviate z is less than 3; that is, $P(z \leq 3.0)$? What is the probability that the absolute value of z is less than 2; that is, $P(|z| \leq 2)$? What is the probability that $z \geq 2.2$?

2.11 t Probabilities. What is the probability that $t \leq 3$ for $v = 4$ degrees of freedom; that is, $P(t \leq 3.0)$? What is the probability that the absolute value t is less than 2 for $v = 30$; that is, $P(|t| \leq 2)$? What is the probability that $t > 6.2$ for $v = 2$?

2.12 t Statistic I. Calculate the value of t for sample size $n = 12$ that has a mean of $\bar{y} = 10$ and a standard deviation of 2.2, for (a) $\eta = 12.4$ and (b) $\eta = 8.7$.

2.13 Sampling Distributions I. Below are eight groups of five random samples drawn from a normal distribution which has mean $\eta = 10$ and standard deviation $\sigma = 1$. For each sample of five (i.e., each column), calculate the average, variance, and t statistic and plot them in the form of Figure 2.11.

1	2	3	4	5	6	7	8
9.1	9.1	8.9	12.1	11.7	11.7	8.4	10.4
9.5	9.0	9.2	7.8	11.1	9.0	10.9	9.7
10.1	10.4	11.2	10.4	10.4	10.6	12.1	9.3
11.9	9.7	10.3	8.6	11.3	9.2	11.2	8.7
9.6	9.4	10.6	11.6	10.6	10.4	10.0	9.1

2.14 Sampling Distributions II. Below are ten groups of five random samples drawn from a lognormal distribution. For each sample of five (i.e., each column), calculate the average and variance and plot them in the form of Figure 2.11. Does the distribution of the averages seem to be approximately normal? If so, explain why.

1	2	3	4	5	6	7	8	9	10
2.3	24.4	62.1	37.1	10.4	34.6	14.7	7.1	15.4	26.9
24.4	12.0	6.4	25.2	111.8	3.0	56.4	3.3	9.7	8.4
12.6	12.3	4.1	38.5	9.3	2.7	2.5	28.0	11.3	5.9
28.3	3.0	4.2	10.8	0.4	1.3	3.4	17.8	2.4	20.7
15.1	4.8	17.5	16.2	32.4	14.9	8.8	13.9	3.3	11.4

2.15 Standard Error I. Calculate the standard error of the mean for a sample of size $n = 16$ that has a variance of 9.

2.16 Standard Error II. For the following sample of $n = 6$ data values, calculate the standard error of the mean, $s_{\bar{y}}$.

$$3.9, 4.4, 4.2, 3.9, 4.2, 4.0$$

2.17 t Statistic II. For the phosphorus data in Exercise 2.4, calculate the value of t. Compare the calculated value with the tabulated value for $\alpha = 0.025$. What does this comparison imply?

2.18 Hypothesis Test I. For the phosphorus data of Exercise 2.4, test the null hypothesis that the true average concentration is not more than 2 mg/L. Do this for the risk level of $\alpha = 0.05$.

2.19 Hypothesis Test II. Repeat Exercise 2.18 using a two-sided test, again using a risk level of $\alpha = 0.05$.

2.20 Confidence Interval I. For the phosphorus data of Exercise 2.4, calculate the 95% confidence interval for the true mean concentration. Does the confidence interval contain the value 2 mg/L? What does this result imply?

2.21 Confidence Interval II. Ten analyses of chemical in soil gave a mean of 20.92 mg/kg and a standard deviation of 0.45 mg/kg. Calculate the 95% confidence interval for the true mean concentration.

2.22 Confidence Interval III. For the data in Exercise 2.16, calculate the mean $\bar{y}$, the standard deviation s, and the standard error of the mean $s_{\bar{y}}$, and the two-sided 95% confidence interval for population mean.

2.23 Soil Contamination. The background concentration of a chemical in soil was measured on ten random specimens of soil from an uncontaminated area. The measured concentrations, in mg/kg, are 1.4, 0.6, 1.2, 1.6, 0.5, 0.7, 0.3, 0.8, 0.2, and 0.9. Soil from neighboring area will be declared "contaminated" if test specimens contain a chemical concentration higher than the upper 99% confidence limit of "background" level. What is the cleanup target concentration?

3

Plotting Data

KEY WORDS *box plot, box-and-whisker plot, chartjunk, digidot plot, error bars, matrix scatterplot, percentile plot, residual plots, scatterplot, seasonal subseries plot, time series plot.*

"The most effective statistical techniques for analyzing environmental data are graphical methods. They are useful in the initial stage for checking the quality of the data, highlighting interesting features of the data, and generally suggesting what statistical analyses should be done. Interesting enough, graphical methods are useful again after intermediate quantitative analyses have been completed, and again in the final stage for providing complete and readily understood summaries of the main findings of investigations (Hunter, 1988)."

The first step in data analysis should be to plot the data. Graphing data should be an interactive experimental process (Chatfield, 1988, 1991; Tukey, 1977). Do not expect your first graph to reveal all interesting aspects of the data. Make a variety of graphs to view the data in different ways. Doing this may:

1. reveal the answer so clearly that little more analysis is needed
2. point out properties of the data that would invalidate a particular statistical analysis
3. reveal that the sample contains unusual observations
4. save time in subsequent analyses
5. suggest an answer that you had not expected
6. keep you from doing something foolish

The time spent making some different plots almost always rewards the effort. Many top-notch statisticians like to plot data by hand, believing that the physical work of the hand stimulates the mind's eye. Whether you adopt this work method or use one of the many available computer programs, the goal is to free your imagination by trying a variety of graphical forms. Keep in mind that some computer programs offer a restricted set of plots and thus could limit rather than expand the imagination.

Make the Original Data Record a Plot

Because the best way to display data is in a plot, it makes little sense to make the primary data record a table of values. Instead, plot the data directly on a *digidot plot*, which is Hunter's (1988) innovative combination of a *time-sequence plot* with a *stem-and-leaf plot* (Tukey, 1977) and is extremely useful for a modest-sized collection of data.

The graph is illustrated in Figure 3.1 for a time series of 36 hourly observations (time, in hours, is measured from left to right).

30	27	41	38	44	29	43	21	15
33	33	28	49	16	22	17	17	23
27	32	47	71	46	42	34	34	34
44	27	32	28	25	36	22	29	24

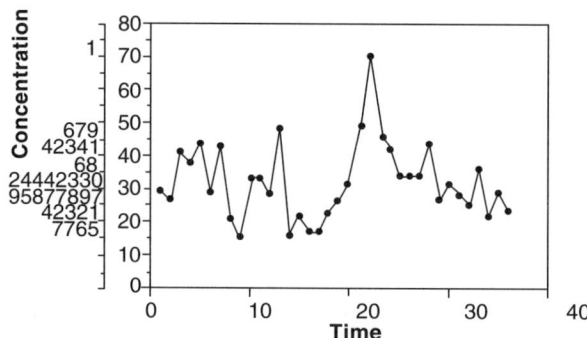

FIGURE 3.1 Digidot plot shows the sequence and distribution of the data.

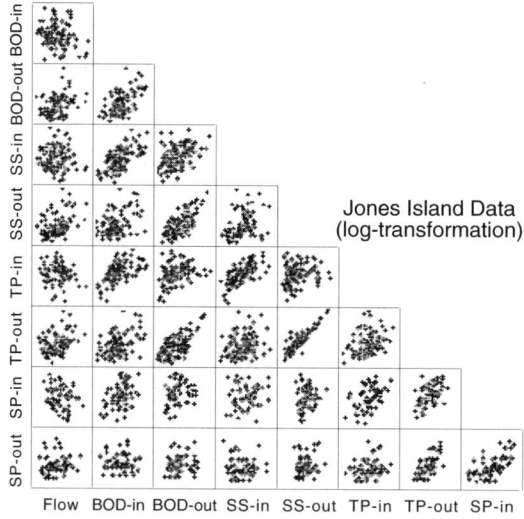

FIGURE 3.2 Multiple two-variable scatterplots of wastewater treatment plant data.

As each observation arrives, it is placed as a dot on the time-sequence plot and simultaneously recorded with its final digit on a stem-and-leaf plot. For example, the first observation was 30. The last digit, a zero, is written in the "bin" between the tick marks for 30 and 35. As time goes on, this bin also accumulates the last digits of observations having the values of 30, 33, 33, 32, 34, 34, 34, and 32. The analyst thus generates a complete visual record of the data: a display of the data distribution, a display of the data time history, and a complete numerical record for later detailed arithmetic analysis.

Scatterplots

It has been estimated that 75% of the graphs used in science are scatterplots (Tufte, 1983). Simple scatterplots are often made before any other data analysis is considered. The insights gained may lead to more elegant and informative graphs, or suggest a promising model. Linear or nonlinear relations are easily seen, and so are outliers or other aberrations in the data.

The use of scatterplots is illustrated with data from a study of how phosphorus removal by a wastewater treatment plant was related to influent levels of phosphorus, flow, and other characteristics of wastewater. The *matrix scatterplots* (sometimes called *draftsman's plots*), shown in Figure 3.2, were made as a guide to constructing the first tentative models. There are no scales shown on these plots because we are

Plotting Data

looking for patterns; the numerical levels are unimportant at this stage of work. The computer automatically scales each two-variable scatterplot to best fill the available area of the graph. Each paired combination of the variables is plotted to reveal possible correlations. For example, it is discovered that effluent total phosphorus (TP-out) is correlated rather strongly with effluent suspended solids (SS-out) and effluent BOD (BOD-out), moderately correlated with flow, BOD-in, and not correlated with SS-in and TP-in. Effluent soluble phosphorus (SP-out) is correlated only with SP-in and TP-out. These observations provide a starting point for model building.

The values plotted in Figure 3.2 are logarithms of the original variables. Making this transformation was advantageous in showing extreme values, and it simplified interpretation by giving linear relations between variables. It is often helpful to use transformations in analyzing environmental data. The logarithmic and other transformations are discussed in Chapter 7.

In Search of Trends

Figure 3.3 is a *time series plot* of 558 pH observations on a small stream in the Smokey Mountains. The data cover the period from mid-1971 to mid-1981, as shown across the top of the plot. Time is measured in weeks on the bottom abcissa.

The data were submitted (on computer tape) to an agency that intended to do a trend analysis to assess possible changes in water quality related to acid precipitation. The data were plotted before any regression analysis or time series modeling was begun. This plot was not expected to be useful in showing a trend because any trend would be small (subsequent analysis indicated that there was no trend). The purpose of plotting the data was to reveal any peculiarities in it.

Two features stand out: (1) the lowest pH values were observed in 1971–1974 and (2) the variation, which was large early in the series, decreased at about 150 weeks and seemed to decrease again at about 300 weeks. The second observation prompted the data analyst to ask two questions. Was there any natural phenomenon to explain this pattern of variability? Is there anything about the measurement process that could explain it? From this questioning, it was discovered that different instruments had been used to measure pH. The original pH meter was replaced at the beginning of 1974 with a more precise instrument, which was itself replaced by an improved model in 1976.

The change in variance over time influenced the subsequent data analysis. For example, if ordinary linear regression were used to assess the existence of a trend, the large variance in 1971–1973 would have given the early data more "weight" or "strength" in determining the position and slope of the trend line. This is not desirable because the latter data are the most precise.

Failure to plot the data initially might not have been fatal. The nonconstant variance might have been discovered later in the analysis, perhaps by plotting the residual errors (with respect to the average or to a fitted model), but by then considerable work would have been invested. However, this feature of the data might be overlooked because an analyst who does not start by plotting the data is not likely to make residual plots either. If the problem is overlooked, an improper conclusion is reported.

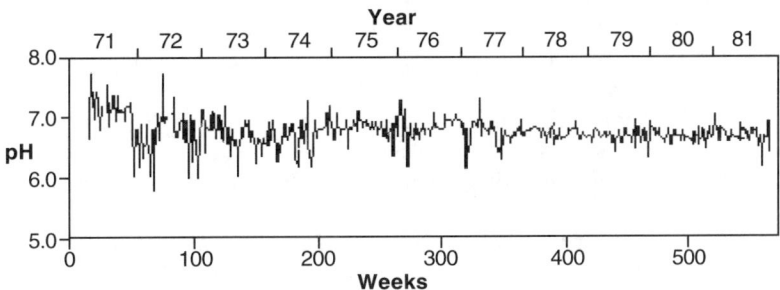

FIGURE 3.3 Time series plot of pH data measured on a small mountain stream.

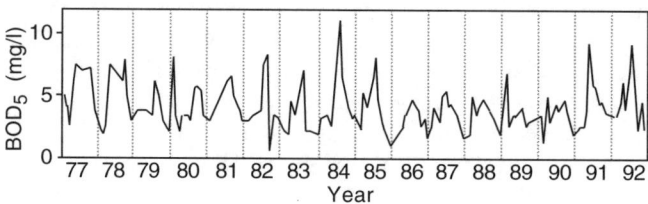

FIGURE 3.4 Time series plot of BOD_5 concentration in the Fox River, Wisconsin.

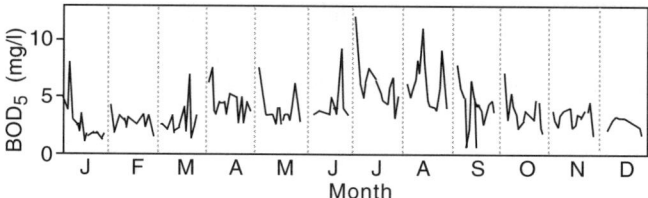

FIGURE 3.5 Seasonal subseries plot of BOD_5 concentration in the Fox River, Wisconsin.

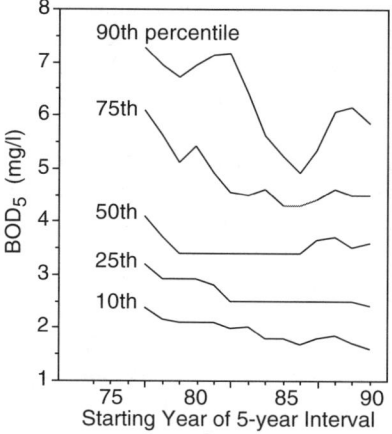

FIGURE 3.6 Percentile plot of the Fox River BOD_5 data.

Figure 3.4 is a time series plot of a 16-year record of monthly average BOD_5 concentrations measured at one of many monitoring stations in the Fox River, Wisconsin. This is part of the data record that was analyzed to assess improvements in the river due to a massive investment in pollution control facilities along this heavily industrialized river. The fishermen in the area knew that water quality had improved, but improvement was not apparent in these BOD data or in time series plots of other water quality data.

Figure 3.5 shows another way of looking at the same data. This is a *seasonal subseries plot* (Cleveland, 1994). The original times series is divided into a time series for each month. (These have unequal numbers of data values because the monitoring was not complete in all years.) The annual time sequence is preserved within each subseries. It does appear that BOD_5 in the summer months may be decreasing after about the mid-1980s.

Figure 3.6 is a *percentile plot* of Fox River BOD_5 data. The values plotted at 1977 are percentiles of monthly averages of BOD_5 concentrations for the 5-year period of 1975–1979. The reason for aggregating data over 5-year periods is that a reliable estimate of the 90th percentile cannot be made from just the 12 monthly averages from 1975. This plot shows that the median (50th percentile) BOD_5 concentration has not changed over the period of record, but there has been improvement at the extremes. The highest

BODs in the 1980s are not as high as in the past. This reduction is what has improved the fishery, because the highest BODs were occurring in the summer when stream flow was minimal and water temperature was high. Several kinds of plots were needed to extract useful information from these data. This is often the case with environmental data.

Showing Statistical Variation and Precision

Measurements vary and one important function of graphs is to show the variation. There are three very different ways of showing variation: a *histogram*, a *box plot* (or *box-and-whisker plot*), and with *error bars* that represent statistics such as standard deviations, standard errors, or confidence intervals.

A histogram shows the shape of the frequency distribution and the range of values; it also gives an impression of central tendency and shows symmetry or lack of it. A box plot is a designed to convey a few primary features of a set of data. One form of box plot, the so-called box-and-whisker plot, is used in Figure 3.7 to compare the effluent quality of 12 identical trickling filter pilot plants that received the same influent and were operated in parallel for 35 weeks (Gameson, 1961). It shows the median (50th percentile) as a center bar, and the quartiles (25th and 75th percentiles) as a box. The box covers the middle 50% of the data; this 50% is called the *interquartile range*. Plotting the median instead of the average has this advantage: the median is not affected by the extreme values. The "whiskers" cover all but the most extreme values in the data set (the whiskers are explained in Cleveland, 1990, 1994). Extreme values beyond the whiskers are plotted as individual points. If the data come from a normal distribution, the fraction of observations expected to lie beyond the whiskers is slightly more than 1%. The simplicity of the plot makes a convenient comparison of the performance of the 12 replicate filters.

Figure 3.8 summarizes and compares the trickling filter data of Figure 3.7 by showing the average with error bars that are plus and minus two standard errors (the standard error is an estimate of the standard deviation of the average). This has some weaknesses. The standard error bars are symmetrical about the average, which may lead the viewer to assume that the data are also distributed symmetrically about the mean. Figure 3.7 showed that this is not the case. Also, Figure 3.8 makes the 12 trickling filters appear more different than Figure 3.7 does. This happens because in a few cases the averages are

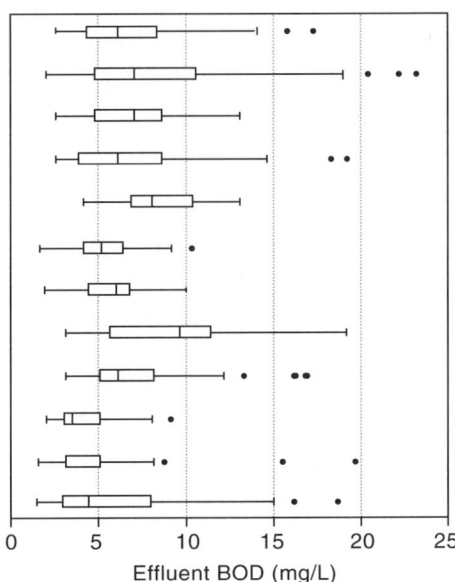

FIGURE 3.7 Box-and-whisker plots to compare the performance of 12 identical trickling filters operating in parallel. Each panel summarizes 35 measurements.

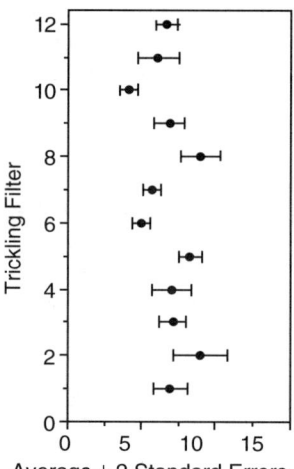

FIGURE 3.8 The trickling filter data of Figure 3.7 plotted to show the average, and plus and minus two standard errors.

strongly influenced by the few extreme values. If the purpose of using error bars is to show the empirical distributions of the data, consider using box plots. That is, Figure 3.8 is better for showing the precision with which the mean is estimated, but Figure 3.7 reveals more about the data.

Often, repeated observations of the dependent variable are made at the settings of the independent variable. In this case it is desirable that the plot show the average value of the replicate measured values and some indication of their precision or variation. This is done by plotting a symbol to locate the sample average and adding to it error bars to show statistical variation.

Authors often fail to tell the reader what the error bars represent. Error bars can convey several possibilities: (1) sample standard deviation, (2) an estimate of the standard deviation (standard error) of the statistical quantity, or (3) a confidence interval. Whichever is used, the meaning of the error bars must be clear or the author will introduce confusion when the intent is to clarify. The text and the label of the graph should state clearly what the error bars mean; for example,

- The error bars show plus and minus one sample standard deviation.
- The error bars show plus and minus an estimate of the standard deviation (or one standard error) of the statistic that is graphed.
- The error bars show a confidence interval for the parameter that is graphed.

If the error bars are intended to show the precision of the average of replicate values, one can plot the standard error or a confidence interval. This has weaknesses as well. Bars marking the sample standard deviation are symmetrical above and below the average, which tends to imply that the data are also distributed symmetrically about the mean. This is somewhat less a problem if the errors bars represent standard errors because averages of replicates do tend to be normally distributed (and symmetrical). Nevertheless, it is better to show confidence intervals. If all plotted averages were based on the same number of observations, one-standard-error bars would convey an approximate 68% confidence interval. This is not a particularly interesting interval. If the averages are calculated from different numbers of values, the confidence intervals would be different multiples of the standard error bars (according to the appropriate degrees of freedom of the *t*-distribution). Cleveland (1994) suggests two-tiered error bars. The inner error bars would show the 50% confidence interval, a middle range analogous to the box of a box plot. The outer of the two-tiered error bars would reflect the 95% confidence interval.

Plotting data on a log scale or transforming data by taking logarithms is often a useful procedure (see Chapters 4 and 7), but this is usually done when the process creates symmetry. Figure 3.9 shows how error bars that are constant and symmetrical on an arithmetic scale become variable and asymmetric when transformed to a logarithmic scale.

Plotting Data

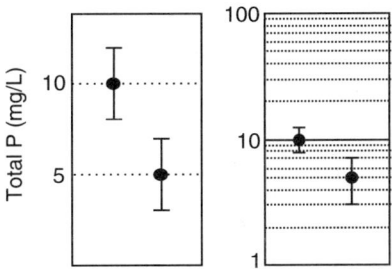

FIGURE 3.9 Illustration of how error bars that are symmetrical on the arithmetic scale become unsymmetrical on the log scale.

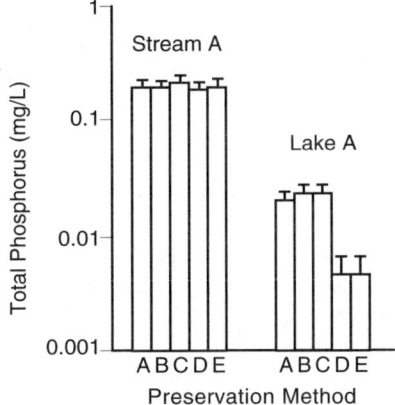

FIGURE 3.10 This often-used chart format hides and obscures information. The T-bar on top of the column shows the upper standard error of the mean. The plot hides the lower standard error bar. Plotting the data on a log scale is convenient for comparing the stream data with the lake data, but it obscures the important comparison, which is between sample preservation methods. Also, error bars on a log scale are not symmetrical.

Figure 3.10 shows a graph with error bars. The graph in the left-hand panel copies a style of graph that appears often in print.[1] The plot conveys little information and distorts part of what it does display. The T on top of the column shows the upper standard error of the mean. The lower standard-error-bar is hidden by the column. Because the data are plotted on a log scale, the lower bar (hidden) is not symmetrical. A small table would convey the essential information more clearly and in less space.

Plots of Residuals

Graphing residuals is an important method that has applications in all areas of data analysis and model building. Residuals are the difference between the observed values and the smooth curve constructed from a model of the data. If the model fits the data, the residuals represent the measurement error. Measurement error is usually assumed to be random. A lack of randomness in the residuals therefore indicates some weakness in the fitted model.

The visual impression in the top panel in Figure 3.11 is that the curve fits the data fairly well but the vertical deviations of points from the fitted curve are smaller for low values of time than for longer times. The graph of residuals in the bottom plot shows the opposite is true. The curve does not fit well

---

[1] A recent issue of *Water Research* contained 12 graphs with error bars. Only three of the twelve graphs had error bars that were fully informative. Six did not say what the error bars represented, six were column graphs with error bars half hidden by the columns, and four of these were on a log scale. One article had five pages of graphs in this style.

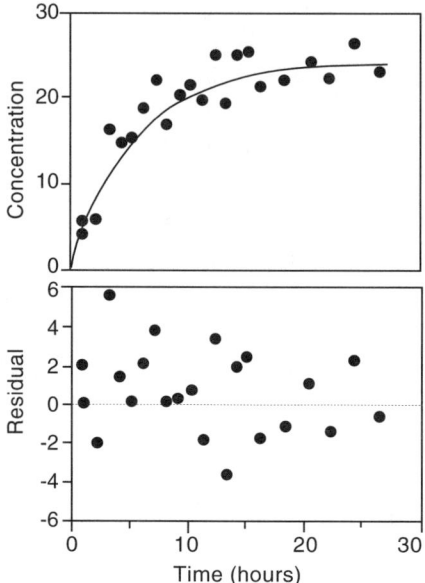

FIGURE 3.11 Graphing residuals. The visual impression from the top plot is that the vertical deviations are greater for large values of time, but the residual plot (bottom) shows that the curve does not fit the points at low times.

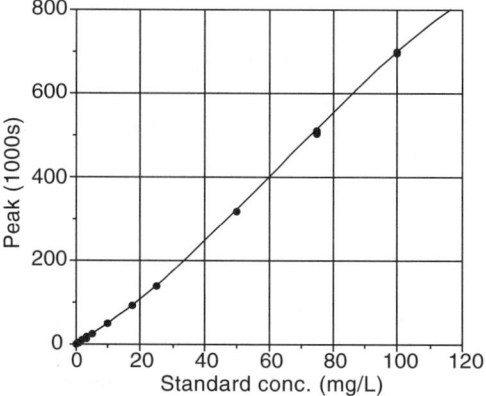

FIGURE 3.12 Calibration curve for measuring chloride with an ion chromatograph. There are three replicate measurements at each of the 13 levels of chloride.

at the shorter times and in this region the residuals are large and predominantly positive. Tukey (1977) calls this process of plotting residuals *flattening the data*. He emphasizes its power to shift our attention from the fitted line to the discrepancies between prediction and observation. It is these discrepancies that contain the information needed to improve the model.

Make it a habit to examine the residuals of a fitted model, including deviations from a simple mean. Check for normality by making a dot diagram or histogram. Plot the residuals against the predicted values, against the predictor variables, and as a function of the time order in which the measurements were made. Residuals that appear to be random and to have uniform variance are persuasive evidence that the model has no serious deficiencies. If the residuals show a trend, it is evidence that the model is inadequate. If the residuals spread out, it suggests that a data transformation is probably needed.

Figure 3.12 is a calibration curve for measuring chloride using an ion chromatograph. There are three replicate measures at each concentration level. The hidden variation of the replicates is revealed in Figure 3.13,

Plotting Data 33

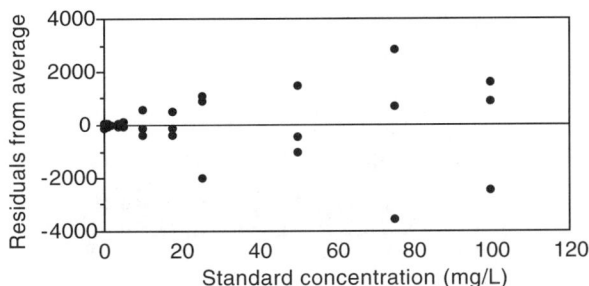

FIGURE 3.13 Residuals of the chloride data with respect to the average peak value at each concentration level.

which has flattened the data by looking at deviations from the average of the three values at each level. An important fact is revealed: the measurement error (variation) tends to increase as the concentration increases. This must be taken into account when fitting the calibration curve to the data.

A Note on Clarity and Style

Here are the words of some people who have devoted their talent and energy to improving the quality of graphical presentations of statistical data.

> *"Excellence in statistical graphics consists of complex ideas communicated with clarity, precision, and efficiency."* Edward Tufte (1983)
> *"The greatest possibilities of visual display lie in vividness and inescapability of the intended message."* John Tukey (1990)
> *"Graphing data should be an iterative experiment process."* Cleveland (1994)

Tufte (1983) emphasizes clarity and simplicity in graphics. Wainer (1997) uses elegance, grace, and impact to describe good graphics. Cleveland (1994) emphasizes clarity, precision, and efficiency. William Playfair (1786), a pioneer and innovator in the use of statistical graphics, desires to tell a story graphically as well as dramatically.

Vividness, drama, elegance, grace, clarity, and impact are not technical terms and the ideas they convey are not easy to capture in technical rules, but Cleveland (1994) and Tufte (1983) have suggested basic principles that will produce better graphics. Tufte (1983) says that graphical excellence:

- is the well-designed presentation of interesting data: a matter of *substance*, of *statistics*, and of *design*
- consists of complex ideas communicated with clarity, precision, and efficiency
- is that which gives the viewer the greatest number of ideas in the shortest time with the least ink in the smallest space
- is almost always multivariate
- requires telling the truth about the data

These guidelines discourage fancified graphs with multiple fonts, cross hatching, and 3-D effects. They do not say that color is necessary or helpful. A poor graph does not become better because color is added.

Style is to a large extent personal. Let us look at five graphical versions of the same data in Figure 3.14. The graphs show how the downward trend in the average number of bald eagle hatchlings in northwestern Ontario reversed after DDT was banned in 1973. The top graphic (so easily produced by computer graphics) does not facilitate understanding the data. It is loaded with what Tufte (1983) calls *chartjunk*—three-dimensional boxes and shading. "Every bit of ink on a graphic requires a reason. And nearly always that reason should be that the ink presents new information (Tufte, 1983)." The two bar charts in the

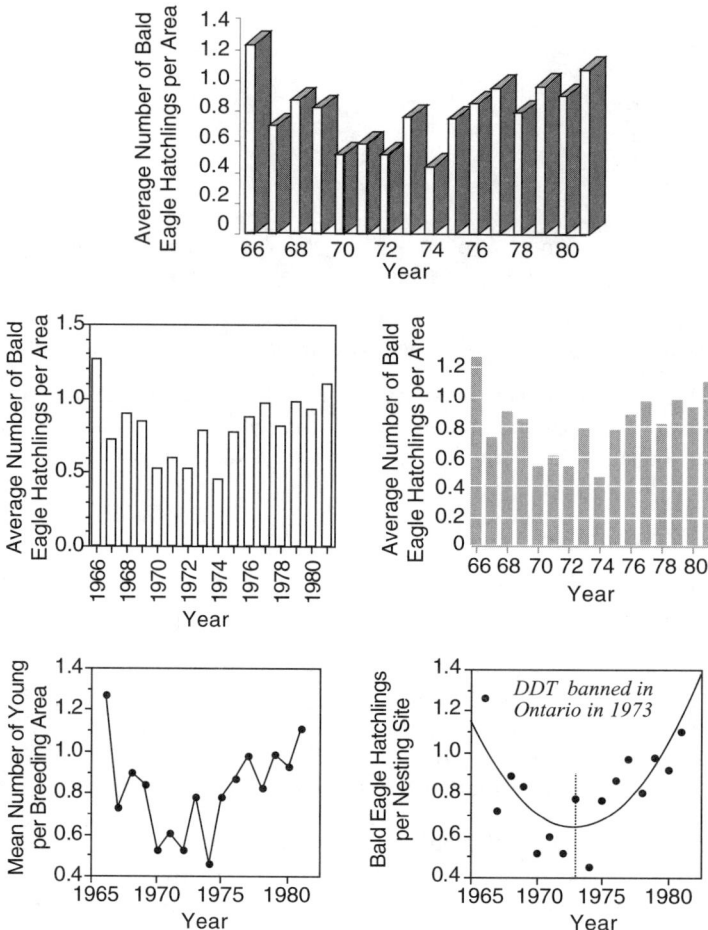

FIGURE 3.14 Several versions of plots that show how banning DDT helped the recovery of the bald eagle population in northwestern Ontario.

middle row are clear. The version on the right is cleaner and clearer (the box frame is not needed). The white lines through the bars serve as the vertical scale. The two graphs in the bottom row are better yet. The bars become dots with a line added to emphasize the trend. The version on the right smoothes the trend with a curve and adds a note to show when DDT was banned.

Most data sets, like this simple one, can be plotted in a variety of ways. The viewer will appreciate the effort required to explore variations and present one that is clear, precise, and efficient in the presentation of the essential information.

Should We Always Plot the Data?

According to Farquhar and Farquhar (1891), two 19th century economists, "Getting information from a table is like extracting sunlight from a cucumber." A virtually perfect rule of statistics is "Plot the data." There are times, however, when a plot is unnecessary. Figure 3.15 is an example. This is a simplified reproduction (shading removed) of a published graph that showed five values.

pH = 5   COD = 2300 mg/L   BOD = 1500 mg/L   TSS = 875 mg/L   TDS = 5700 mg/L

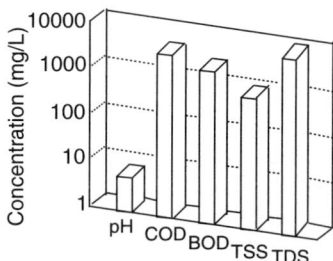

FIGURE 3.15 This unnecessary graph, which shows just five values, should be replaced by a table.

These five values say it all, and better than the graph. Do not use an axe to hack your way through an open door. Aside from being unnecessary, this chart has three major faults. It confuses units—pH is not measured in mg/L. Three-dimensional effects make it more difficult to read the numerical values. Using a log scale makes the values seem nearly the same when they are much different. The 875 mg/L TSS and the 1500 mg/L COD have bars that are nearly the same height.

Summary

Graphical methods are obviously useful for both initial and exploratory data analyses, but they also serve us well in the final analysis. "A picture is worth a thousand words" is a cliché, but still powerfully true. The right graph may reveal all that is important. If it only tells part of the story, that is the part that is most likely to be remembered.

Tables of numbers camouflage the interesting features of data. The human mind, which is remarkably well adapted to so many and varied tasks, is simply not capable of extracting useful information from tabulated figures. Putting these same numbers in appropriate graphical form completely changes the situation. The informed human mind can then operate efficiently with these graphs as inputs. In short, suitable graphs of data and the human mind are an effective combination; endless tables of data and the mind are not.

It is extremely important that plots be kept current because the first purpose of keeping these plots is to help monitor and, if necessary, to troubleshoot difficulties as they arise. The plots do not have to be beautiful, or computer drafted, to be useful. Make simple plots by hand as the data become available. If the plots are made at some future date to provide a record of what happened in the distant past, it will be too late to take appropriate action to improve performance. The second purpose is to have an accurate record of what has happened in the past, especially if the salient information is in such a form that it is easily communicated and readily understood. If they are kept up-to-date and used for the first purpose, they can also be used for the second. On the other hand, if they are not kept up-to-date, they may be useful for the second purpose only. In the interest of efficiency, they ought to serve double duty.

Intelligent data analysis begins with plotting the data. Be imaginative. Use a collection of different graphs to see different aspects of the data. Plotting graphs in a notebook is not as useful as making plots large and visible. Plots should be displayed in a prominent place so that those concerned with the environmental system can review them readily.

We close with Tukey's (1977) declaration: "**The greatest value of a picture** is when it *forces* us to *notice what we never expected to see*." (Emphasis and italics in the original.)

References

Anscombe, F. J. (1973). "Graphs in Statistical Analysis," *American Statistician,* 27, 17–21.
Chatfield, C. (1988). *Problem Solving: A Statistician's Guide,* London, Chapman & Hall.
Chatfield, C. (1991). "Avoiding Statistical Pitfalls," *Stat. Sci.,* 6(3), 240–268.
Cleveland, W. S. (1990). *The Elements of Graphing Data,* 2nd ed., Summit, NJ, Hobart Press.
Cleveland, W. S. (1994), *Visualizing Data,* Summit, NJ, Hobart Press.

Farquhar, A. B. and H. Farquhar (1891). "*Economic and Industrial Delusions: A Discourse of the Case for Protection*," New York, Putnam.

Gameson, A. L. H., G. A. Truesdale, and M. J. Van Overdijk (1961). "Variation in Performance of Twelve Replicate Small-Scale Percolating Filters," *Water and Waste Treatment J.,* 9, 342–350.

Hunter, J. S. (1988). "The Digidot Plot," *Am. Statistician,* 42, 54.

Tufte, E. R. (1983). *The Visual Display of Quantitative Information,* Cheshire, CN, Graphics Press.

Tufte, E. R. (1990). *Envisioning Information,* Cheshire, CN, Graphics Press.

Tufte, E. R. (1997). *Visual Explanations,* Cheshire, CN, Graphics Press.

Tukey, J. W. (1977). *Exploratory Data Analysis,* Reading, MA, Addison-Wesley.

Tukey, J. W. (1990). "Data Based Graphics: Visual Display in the Decades to Come," *Stat. Sci.,* 5, 327–329.

Wainer, H. (1997). *Visual Revelations: Graphical Tales of Fate and Deception from Napoleon Boneparte to Ross Perot,* New York, Copernicus, Springer-Verlag.

Exercises

3.1 Box-Whisker Plot. For the 11 ordered observations below, make the box-whisker plot to show the median, the upper and lower quartiles, and the upper and lower cut-off.

36 37 45 52 56 58 66 68 75 90 100

3.2 Phosphorus in Sludge. The values below are annual average concentrations of total phosphorus in municipal sewage sludge, measured as percent of dry weight solids. Time runs from right to left. The first value is for 1979. Make several plots of the data to discover any trends or patterns. Try to explain any patterns you discover.

2.7 2.5 2.3 2.4 2.6 2.7 2.6 2.7 2.3 2.9 2.8
2.5 2.6 2.7 2.8 2.6 2.4 2.7 3.0 4.5 4.5 4.3

3.3 Waste Load Survey Data Analysis. The table gives 52 weekly average flow and BOD_5 data for wastewater. Plot the data in variety of ways that might interest an engineer who needs to base a treatment plant design on these data. As a minimum, (a) make the time series plots for BOD concentration, flow, and BOD mass load (lb/day); and (b) determine whether flow and BOD are correlated.

Week	Flow (MGD)	BOD (mg/L)	Week	Flow (MGD)	BOD (mg/L)	Week	Flow (MGD)	BOD (mg/L)
1	3.115	1190	18	3.42	1143	35	2.434	1167
2	3.08	1211	19	3.276	1213	36	2.484	1042
3	4.496	1005	20	3.595	1300	37	2.466	1116
4	3.207	1208	21	4.377	1245	38	2.69	1228
5	3.881	1349	22	3.28	1211	39	2.026	1156
6	4.769	1221	23	3.986	1148	40	1.004	1073
7	3.5	1288	24	3.838	1258	41	1.769	1259
8	5.373	1193	25	3.424	1289	42	1.63	1337
9	3.779	1380	26	3.794	1147	43	2.67	1228
10	3.113	1168	27	2.903	1169	44	1.416	1107
11	4.008	1250	28	1.055	1102	45	2.164	1298
12	3.455	1437	29	2.931	1000	46	2.559	1284
13	3.106	1105	30	2.68	1372	47	1.735	1064
14	3.583	1155	31	2.048	1077	48	2.073	1245
15	3.889	1278	32	2.548	1324	49	1.641	1199
16	4.721	1046	33	1.457	1063	50	2.991	1279
17	4.241	1068	34	1.68	1242	51	3.031	1203
						52	2.972	1197

3.4 Effluent Suspended Solids. The data below are effluent suspended solids data for one year of a wastewater treatment plant operation. Plot the data and discuss any patterns or characteristics of the data that might interest plant management or operators.

Day	Jan	Feb	Mar	Apr	May	Jun	Jul	Aug	Sept	Oct	Nov	Dec
1	33	32	16	25	21	20	29	26	6	23	18	11
2	37	22	14	26	30	11	36	33	18	13	16	9
3	23	22	12	26	31	22	68	21	13	13	20	12
4	28	38	14	30	28	21	32	21	8	9	15	15
5	41	10	13	27	31	21	24	19	9	9	12	18
6	37	34	23	24	35	27	22	19	22	5	12	17
7	23	32	15	20	26	58	22	15	38	6	12	19
8	24	33	22	25	39	40	31	20	11	10	39	6
9	22	42	17	22	13	25	32	25	10	5	8	10
10	28	19	22	15	21	24	21	11	8	10	16	9
11	27	29	28	12	20	26	24	13	9	10	14	10
12	16	28	11	18	41	24	20	14	10	5	12	12
13	18	31	41	30	25	32	23	26	12	4	13	9
14	24	29	30	23	25	9	24	19	14	16	15	11
15	21	20	34	23	21	13	26	21	15	7	16	14
16	26	21	24	33	20	12	23	16	9	8	14	8
17	21	12	19	38	19	15	22	7	18	11	18	9
18	31	37	35	31	21	20	28	13	20	8	15	4
19	28	25	27	31	18	24	14	10	29	11	12	17
20	13	19	20	32	24	20	28	10	33	5	10	21
21	14	16	60	36	42	9	36	11	14	10	16	28
22	30	23	68	24	28	8	50	11	11	15	18	42
23	41	9	46	20	35	15	24	14	26	47	10	22
24	30	17	27	20	25	20	22	11	34	20	11	16
25	47	20	23	22	11	19	24	16	38	13	3	9
26	33	24	26	25	22	28	26	13	40	28	9	12
27	41	28	26	25	25	29	11	12	41	41	8	9
28	44	13	42	24	23	25	19	20	18	29	9	7
29	30		20	33	27	28	21	10	28	29	29	10
30	15		21	13	33	25	21	13	38	18	11	7
31	32		22		26		24	14		25		12

3.5 Solid Waste Fuel Value. The table gives fuel values (Btu/lb) for typical solid waste from 35 countries. (The United States is number 35). Make a matrix scatterplot of the five waste characteristics and any other plots that might help to identify a plausible model to relate fuel value to composition.

Country	MSW Components					Fuel Value (Btu/lb)
	Paper	Metal	Glass	Food	Plastic	
1	38	11	18	13	0.1	3260
2	35	10	9	24	6	3951
3	2	1	9	40	1	1239
4	20	5.3	8	40	5	3669
5	10	1.7	1.6	54	1.7	2208
6	1	3	6	80	4	2590
7	22	1	2	56	5	3342
8	13.4	6.2	6.6	41.8	4.2	2637
9	32.9	4.1	6.1	44	6.8	4087
10	37	8	8	28	2	3614
11	55	5	6	20	6	4845
12	30	4	4	30	1	2913
13	6	5	9	77	3	2824
14	20	5	10	21	2	2237

15	32	2	10	9	11	3596
16	3	1	8	36	1	1219
17	10	2	1	72	6	3136
18	17.2	1.8	2.1	69.8	3.8	3708
19	31	7	3	36	7	3945
20	21	5.7	3.9	50	6.2	3500
21	12.2	2.7	1.3	42.6	1	2038
22	22.2	3.2	11.9	50	6.2	3558
23	28	6	7	48	0.1	3219
24	15.5	4.5	2.5	51.5	2	2651
25	38.2	2	7.5	30.4	6.5	3953
26	2.2	2.2	1.75	52.5	1.2	1574
27	17	2	5	43	4	2649
28	24	9	8	53	2	3462
29	43	3	1	5	6	3640
30	18	4	3	50	4	2555
31	8	1	6	80	1	2595
32	4	3		30	2.6	1418
33	50	7	8	15	8	4792
34	8	1	3	25	2	1378
35	28.9	9.3	10.4	17.8	3.4	3088

Source: Khan, et al., *J. Envir. Eng.,* ASCE, 117, 376, 1991.

3.6 Highway TPH Contamination. Total petroleum hydrocarbons (TPH) in soil specimens collected at 30 locations alongside a 44.8-mile stretch of major highway are given in the table below. The length was divided into 29 segments of 1.5 miles and one segment of 1.3 miles. The sampling order for these segments was randomized, as was the location within each segment. Also, the sample collection was randomized with respect to the eastbound or westbound lane of the roadway. There are duplicate measurements on three specimens. Plot the data in a variety of ways to check for randomness, independence, trend, and other interesting patterns.

Distance (mile)	Location	Sample Order	TPH (mg/kg)	Distance (mile)	Location	Sample Order	TPH (mg/kg)
0.7	West	14	19	23.4	West	7	101
1.5	East	15	40	25.3	West	6	119
4.1	West	13	48	25.6	West	5	129
4.8	East	17	23	28.2	East	23a	114
7.2	East	30	79	28.2	East	23b	92
8.7	West	12	19	28.7	East	24a	43
9	West	11	118	28.7	East	24b	62
11.7	East	18	8	30.9	East	25	230
12.6	West	10	91	32.8	West	4	14
14.5	East	19	21	33.2	West	3	272
15.6	East	20	36	34.6	West	2	242
16.6	West	9	16	36.7	West	1	30
19.4	West	8	44	38.9	East	26	76
20.5	East	21	44	39.8	East	27	208
21.7	East	22a	160	40.6	East	28	196
21.7	East	22b	153	42.7	West	16	125
				44.3	East	29	167

Source: Phillips, I. (1999). Unpublished paper, Tufts University.

Plotting Data 39

3.7 Heavy Metals. Below are 100 daily observations of wastewater influent and effluent lead (Pb) concentration, measured as μg/L, in wastewater. State your expectation for the relation between influent and effluent and then plot the data to see whether your ideas need modification.

Obs	Inf	Eff	Obs	Inf	Eff	Obs	Inf	Eff	Obs	Inf	Eff
1	47	2	26	16	7	51	29	1	76	13	1
2	30	4	27	32	9	52	21	1	77	14	1
3	23	4	28	19	6	53	18	1	78	18	1
4	29	1	29	22	4	54	19	1	79	10	1
5	30	6	30	32	4	55	27	1	80	4	1
6	28	1	31	29	7	56	36	2	81	5	1
7	13	6	32	48	2	57	27	1	82	60	2
8	15	3	33	34	1	58	28	1	83	28	1
9	30	6	34	22	1	59	31	1	84	18	1
10	52	6	35	37	2	60	6	1	85	8	11
11	39	5	36	64	19	61	18	1	86	11	1
12	29	2	37	24	15	62	97	1	87	16	1
13	33	4	38	33	36	63	20	1	88	15	1
14	29	5	39	41	2	64	17	2	89	25	3
15	33	4	40	28	2	65	9	3	90	11	1
16	42	7	41	21	3	66	12	6	91	8	1
17	36	10	42	27	1	67	10	5	92	7	1
18	26	4	43	30	1	68	23	5	93	4	1
19	105	82	44	34	1	69	41	4	94	3	1
20	128	93	45	36	3	70	28	4	95	4	1
21	122	2	46	38	2	71	18	4	96	6	1
22	170	156	47	40	2	72	5	1	97	5	2
23	128	103	48	10	2	73	2	1	98	5	1
24	139	128	49	10	1	74	19	10	99	5	1
25	31	7	50	42	1	75	24	10	100	16	1

4

Smoothing Data

KEY WORDS   *moving average, exponentially weighted moving average, weighting factors, smoothing, and median smoothing.*

Smoothing is drawing a smooth curve through data in order to eliminate the roughness (scatter) that blurs the fundamental underlying pattern. It sharpens our focus by unhooking our eye from the irregularities.

Smoothing can be thought of as a decomposition of the data. In curve fitting, this decomposition has the general relation: *data = fit + residuals*. In smoothing, the analogous expression is: *data = smooth + rough*. Because the *smooth* is intended to be smooth (as the "fit" is smooth in curve fitting), we usually show its points connected. Similarly, we show the *rough* (or residuals) as separated points, if we show them at all. We may choose to show only those rough (residual) points that stand out markedly from the smooth (Tukey, 1977).

We will discuss several methods of smoothing to produce graphs that are especially useful with time series data from treatment plants and complicated environmental systems. The methods are well established and have a long history of successful use in industry and econometrics. The methods are effective and economical in terms of time and money. They are simple; they are useful to everyone, regardless of statistical expertise. Only elementary arithmetic is needed. A computer may be helpful, but is not needed, especially if one keeps the plot up-to-date by adding points daily or weekly as they become available.

In statistics and quality control literature, one finds mathematics and theory that can embellish these graphs. A formal statistical analysis, such as adding control limits, can become quite complex because often the assumptions on which such tests are usually based are violated rather badly by environmental data. These embellishments are discussed in another chapter.

Smoothing Methods

One method of smoothing would be to fit a straight line or polynomial curve to the data. Aside from the computational bother, this is not a useful general procedure because the very fact that smoothing is needed means that we cannot see the underlying pattern clearly enough to know what particular polynomial would be useful.

The simplest smoothing method is to plot the data on a logarithmic scale (or plot the logarithm of *y* instead of *y* itself). Smoothing by plotting the moving averages (MA) or exponentially weighted moving averages (EWMA) requires only arithmetic.

A moving average (MA) gives equal weight to a sequence of past values; the weight depends on how many past values are to be remembered. The EWMA gives more weight to recent events and progressively forgets the past. How quickly the past is forgotten is determined by one parameter. The EWMA will follow the current observations more closely than the MA. Often this is desirable but this responsiveness is purchased by a loss in smoothing.

The choice of a smoothing method might be influenced by the application. Because the EWMA forgets the past, it may give a more realistic representation of the actual threat of the pollutant to the environment.

For example, the BOD discharged into a freely flowing stream is important the day it is discharged. A 2- or 3-day average might also be important because a few days of dissolved oxygen depression could be disastrous while one day might be tolerable to aquatic organisms. A 30-day average of BOD could be a less informative statistic about the threat to fish than a short-term average, but it may be needed to assess the long-term trend in treatment plant performance.

For suspended solids that settle on a stream bed and form sludge banks, a long-term average might be related to depth of the sludge bed and therefore be an informative statistic. If the solids do not settle, the daily values may be more descriptive of potential damage. For a pollutant that could be ingested by an organism and later excreted or metabolized, the exponentially weighted moving average might be a good statistic.

Conversely, some pollutants may not exhibit their effect for years. Carcinogens are an example where the long-term average could be important. Long-term in this context is years, so the 30-day average would not be a particularly useful statistic. The first ingested (or inhaled) irritants may have more importance than recently ingested material. If so, perhaps past events should be weighted more heavily than recent events if a statistic is to relate source of pollution to present effect. Choosing a statistic with the appropriate weighting could increase the value of the data to biologists, epidemiologists, and others who seek to relate pollutant discharges to effects on organisms.

Plotting on a Logarithmic Scale

The top panel of Figure 4.1 is a plot of influent copper concentration at a wastewater treatment plant. This plot emphasizes the few high values, especially those at days 225, 250, and 340. The bottom panel shows the same data on a logarithmic scale. Now the process behavior appears more consistent. The low values are more evident, and the high values do not seem so extreme. The episode around day 250 still looks unusual, but the day 225 and 340 values are above the average (on the log scale) by about the same amount that the lowest values are below average.

Are the high values so extraordinary as to deserve special attention? Or are they rogue values (outliers) that can be disregarded? This question cannot be answered without knowing the underlying distribution of the data. If the underlying process naturally generates data with a lognormal distribution, the high values fit the general pattern of the data record.

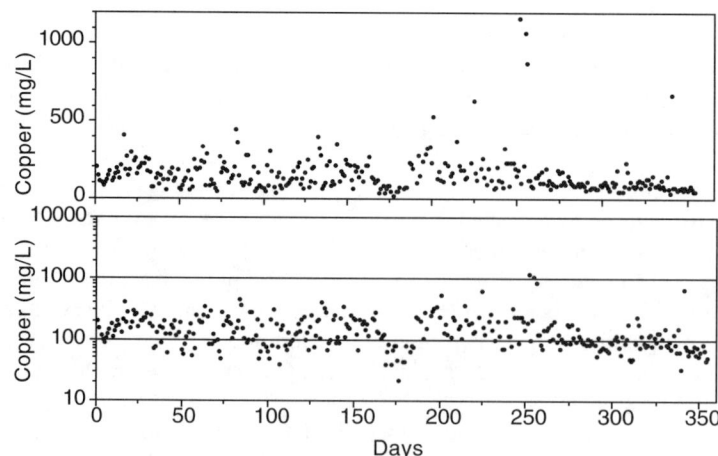

FIGURE 4.1 Copper data plotted on arithmetic and logarithmic scales give a different impression about the high values.

Smoothing Data

The Moving Average

Many standards for environmental quality have been written for an average of 30 consecutive days. The language is something like the following: "Average daily values for 30 consecutive days shall not exceed...." This is commonly interpreted to mean a monthly average, probably because dischargers submit monthly reports to the regulatory agencies, but one should note the great difference between the moving 30-day average and the monthly average as an effluent standard. There are only 12 monthly averages in a year of the kind that start on the first day of a month, but there are a total of 365 moving 30-day averages that can be computed. One very bad day could make a monthly average exceed the limit. This same single value is used to calculate 30 other moving averages and several of these might exceed the limit. These two statistics — the strict monthly average and the 30-day moving average — have different properties and imply different effects on the environment, although the effluent and the environment are the same.

The length of time over which a moving average is calculated can be adjusted to represent the memory of the environmental system as it responds to pollutants. This is done in ambient air pollution monitoring, for example, where a short averaging time (one hour) is used for ozone.

The moving average is the simple average of the most recent k data points, that is, the sum of the most recent k data divided by k:

$$\bar{y}_i(k) = \frac{1}{k} \sum_{j=i-k+1}^{i} y_j \quad i = k, k+1, \ldots, n$$

Thus, a seven-day moving average (MA7) uses the latest seven daily values, a ten-day average (MA10) uses 10 points, and so on. Each data point is given equal weight in computing the average.

As each new observation is made, the summation will drop one term and add another term, giving the simple updating formula:

$$\bar{y}_i(k) = \bar{y}_{i-1}(k) + \frac{1}{k}y_i - \frac{1}{k}y_{i-k} = \bar{y}_{i-1}(k) + \frac{1}{k}(y_i - y_{i-k})$$

By smoothing random fluctuations, the moving average sharpens the focus on recent performance levels. Figure 4.2 shows the MA7 and MA30 moving averages for some PCB data. Both moving averages help general trends in performance show up more clearly because random variations are averaged and smoothed.

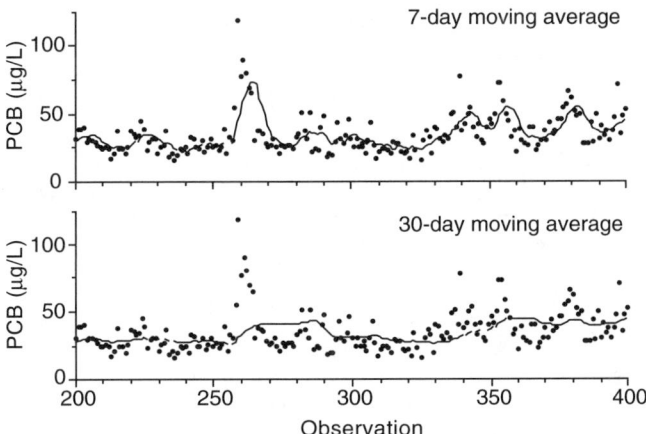

FIGURE 4.2 Seven-day and thirty-day moving averages of PCB data.

The MA7, which is more reflective of short-term variations, has special appeal in being a weekly average. Notice how the moving average lags behind the daily variation. The peak day is at 260, but the MA7 peaks three to four days later (about $k/2$ days later). This does not diminish its value as a smoother, but it does limit its value as a predictor. The longer the smoothing period (the larger k), the more the average will lag behind the daily values.

The MA30 highlights long-term changes in performance. Notice the lack of response in the MA30 at day 255 when several high PCB concentrations occurred. The MA30 did not increase by very much — only from 25 μg/L to about 40 μg/L — but it stayed at the 40 μg/L level for almost 30 days after the elevated levels had disappeared. High concentrations of PCBs are not immediately harmful, but the chemical does bioaccumulate in fish and other organisms and the long-term average is probably more reflective of the environmental danger than the more responsive MA7.

Exponentially Weighted Moving Average

In the simple moving average, recent values and long-past values are weighted equally. For example, the performance four weeks ago is reflected in an MA30 to the same degree as yesterday's, although the receiving environment may have "forgotten" the event of 4 weeks ago. The exponentially weighted moving average (EWMA) weights the most recent event heavily, and each event going into the past proportionately less.

The EWMA is calculated as:

$$\bar{Z}_i = (1-\phi)\sum_{j=0}^{\infty} \phi^j y_{i-j} \quad i = 1, 2, \ldots$$

where ϕ is a suitably chosen constant between 0 and 1 that determines the length of the EWMA's memory and how much smoothing is done.

Why do we call the EWMA an average? Because it has the property that if all the observations are increased by some fixed amount, then the EWMA is also increased by that same amount. The weights must add up to one (unity) for this to happen. Obviously this is true for the weights of the equally weighted average, as well as the EWMA.

Figure 4.3 shows how the weight given to past times depends on the selected value of ϕ. The parameter ϕ indicates how much smoothing is done. As ϕ increases from 0 to 1, the smoothing increases and long-term cycles and trends stand out more clearly. When ϕ is small, the "memory" of the EWMA is short

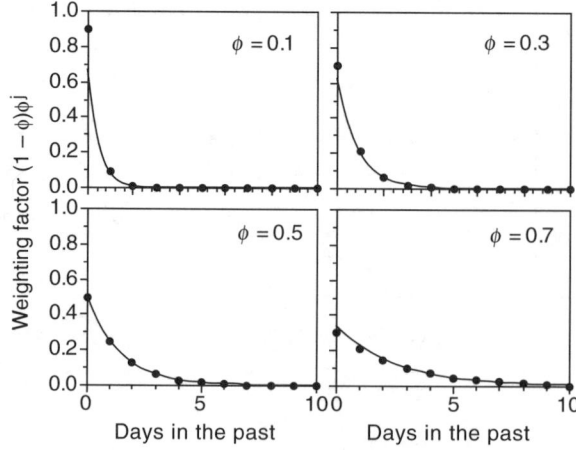

FIGURE 4.3 Weights for exponentially weighted moving average (EWMA).

Smoothing Data

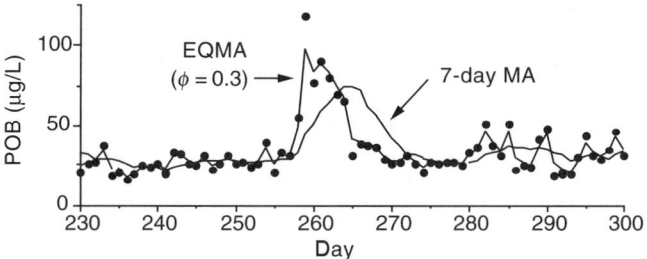

FIGURE 4.4 Comparison of 7-day moving average and an exponentially weighted moving average with $\phi = 0.3$.

and the weights a few days past rapidly shrink toward zero. A value of $\phi = 0.5$ to 0.3 often gives a useful balance between smoothing and responsiveness. Values in this range will roughly approximate a simple seven-day moving average, as shown in Figure 4.4, which shows a portion of the PCB data from Figure 4.2. Note that the EWMA ($\phi = 0.3$) increases faster and recovers to normal levels faster than the MA7. This is characteristic of EWMAs.

Mathematically, the EMWA has an infinite number of terms, but in practice only five to ten are needed because the weight $(1 - \phi)\phi^j$ rapidly approaches 0 as j increases. For example, if $\phi = 0.3$:

$$\bar{Z}_i = (1 - 0.3)y_i + (1 - 0.3)(0.3)y_{i-1} + (1 - 0.3)(0.3)^2 y_{i-2} + \cdots$$
$$\bar{Z}_i = 0.7y_i + 0.21y_{i-1} + 0.063y_{i-2} + 0.019y_{i-3} + \cdots$$

The small coefficient of y_{i-3} shows that values more than three days into the past are essentially forgotten because the weighting factor is small.

The EWMA can be easily updated using:

$$\bar{Z}_i = \phi \bar{Z}_{i-1} + (1 - \phi)y_i$$

where $\bar{Z}_{i-1}$ is the EWMA at the previous sampling time and $\bar{Z}_i$ is the updated average that is computed when the new observation of y_i becomes available.

Comments

Suitable graphs of data and the human mind are an effective combination. A suitable graph will often show the smooth along with the rough. This prevents the eye from being distracted by unimportant details. The smoothing methods illustrated here are ideal for initial data analysis (Chatfield, 1988, 1991) and exploratory data analysis (Tukey, 1977). Their application is straightforward, fast, and easy.

The simple moving averages (7-day, 30-day, etc.) effectively smooth out random and other high-frequency variation. The longer the averaging period, the smoother the moving average becomes and the more slowly it reacts to changes in the underlying pattern. That is, to gain smoothness, response to short-term change is sacrificed.

Exponentially weighted moving averages can smooth effectively while also being responsive. This is because they give more relative weight (influence) to recent events and dilute or forget the past. The rate of forgetting is determined by the value of the smoothing factor, ϕ. We have not tried to identify the best value of ϕ in the EWMA. It is possible to do this by fitting time series models (Box et al., 1994; Cryer, 1986). This becomes important if the smoothing function is used to predict future values, but it is not necessary if we just want to clarify the general underlying pattern of variation.

An alternate to the moving average smoothers is the nonparametric median smooth (Tukey, 1977). A median-of-3 smooth is constructed by plotting the middle value of three consecutive observations. It can be constructed without computations and it is entirely resistant to occasional extreme values. The computational simplicity is an insignificant advantage, however, because the moving averages are so easy to compute.

Missing values in the data series might seem to be a barrier to smoothing, but for practical purposes they usually can be filled in using some simple ad hoc method. For purposes of smoothing to clarify the general trend, several methods of filling in missing values can be used. The simplest is linear interpolation between adjacent points. Other alternatives are to fill in the most recent moving average value, or to replicate the most recent observation. The general trend will be nearly the same regardless of the choice of method, and the user should not be unduly worried about this so long as missing values occur only occasionally.

References

Box, G. E. P., G. M. Jenkins, and G. C. Reinsel (1994). *Time Series Analysis, Forecasting and Control*, 3rd ed., Englewood Cliffs, NJ, Prentice-Hall.

Chatfield, C. (1988). *Problem Solving: A Statistician's Guide*, London, Chapman & Hall.

Chatfield, C. (1991). "Avoiding Statistical Pitfalls," *Stat. Sci.*, 6(3), 240–268.

Cryer, J. D. (1986). *Time Series Analysis*, Duxbury Press, Boston.

Tukey, J. W. (1977). *Exploratory Data Analysis*, Reading, MA, Addison-Wesley.

Exercises

4.1 Cadmium. The data below are influent and effluent cadmium at a wastewater treatment plant. Use graphical and smoothing methods to interpret the data. Time runs from left to right.

Inf. Cd (μg/L)	2.5	2.3	2.5	2.8	2.8	2.5	2.0	1.8	1.8	2.5	3.0	2.5
Eff. Cd (μg/L)	0.8	1.0	0.0	1.0	1.0	0.3	0.0	1.3	0.0	0.5	0.0	0.0
Inf. Cd (μg/L)	2.0	2.0	2.0	2.5	4.5	2.0	10.0	9.0	10.0	12.5	8.5	8.0
Eff. Cd (μg/L)	0.3	0.5	0.3	0.3	1.3	1.5	8.8	8.8	0.8	10.5	6.8	7.8

4.2 PCBs. Use smoothing methods to interpret the series of 26 PCB concentrations below. Time runs from left to right.

29	62	33	189	289	135	54	120	209	176	100	137	112
120	66	90	65	139	28	201	49	22	27	104	56	35

4.3 EWMA. Show that the exponentially weighted moving average really is an average in the sense that if a constant, say $\alpha = 2.5$, is added to each value, the EWMA increases by 2.5.

5

Seeing the Shape of a Distribution

KEY WORDS *dot diagram, histogram, probability distribution, cumulative probability distribution, frequency diagram.*

The data in a sample have some frequency distribution, perhaps symmetrical or perhaps skewed. The statistics (mean, variance, etc.) computed from these data also have some distribution. For example, if the problem is to establish a 95% confidence interval on the mean, it is not important that the sample is normally distributed because the distribution of the mean tends to be normal regardless of the sample's distribution. In contrast, if the problem is to estimate how frequently a certain value will be exceeded, it is essential to base the estimate on the correct distribution of the sample. This chapter is about the shape of the distribution of the data in the sample and not the distribution of statistics computed from the sample.

Many times the first analysis done on a set of data is to compute the mean and standard deviation. These two statistics fully characterize a normal distribution. They do not fully describe other distributions. We should not assume that environmental data will be normally distributed. Experience shows that stream quality data, wastewater treatment plant influent and effluent data, soil properties, and air quality data typically do not have normal distributions. They are more likely to have a long tail skewed toward high values (positive skewness). Fortunately, one need not assume the distribution. It can be discovered from the data.

Simple plots help reveal the sample's distribution. Some of these plots have already been discussed in Chapters 2 and 3. *Dot diagrams* are particularly useful. These simple plots have been overlooked and underused. Environmental engineering references are likely to advise, by example if not by explicit advice, the construction of a *probability plot* (also known as the *cumulative frequency plot*). Probability plots can be useful. Their construction and interpretation and the ways in which such plots can be misused will be discussed.

Case Study: Industrial Waste Survey Data Analysis

The BOD (5-day) data given in Table 5.1 were obtained from an industrial wastewater survey (U.S. EPA, 1973). There are 99 observations, each measured on a 4-hr composite sample, giving six observations daily for 16 days, plus three observations on the 17th day. The survey was undertaken to estimate the average BOD and to estimate the concentration that is exceeded some small fraction of the time (for example, 10%). This information is needed to design a treatment process. The pattern of variation also needs to be seen because it will influence the feasibility of using an equalization process to reduce the variation in BOD loading. The data may have other interesting properties, so the data presentation should be complete, clear, and not open to misinterpretation.

Dot Diagrams

Figure 5.1 is a time series plot of the data. The concentration fluctuates rapidly with more or less equal variation above and below the average, which is 687 mg/L. The range is from 207 to 1185 mg/L. The BOD may change by 1000 mg/L from one sampling interval to the next. It is not clear whether the ups and downs are random or are part of some cyclic pattern. There is little else to be seen from this plot.

TABLE 5.1

BOD Data from an Industrial Survey

Date	4 am	8 am	12 N	4 pm	8 pm	12 MN
2/10	717	946	623	490	666	828
2/11	1135	241	396	1070	440	534
2/12	1035	265	419	413	961	308
2/13	1174	1105	659	801	720	454
2/14	316	758	769	574	1135	1142
2/15	505	221	957	654	510	1067
2/16	329	371	1081	621	235	993
2/17	1019	1023	1167	1056	560	708
2/18	340	949	940	233	1158	407
2/19	853	754	207	852	318	358
2/20	356	847	711	1185	825	618
2/21	454	1080	440	872	294	763
2/22	776	502	1146	1054	888	266
2/23	619	691	416	1111	973	807
2/24	722	368	686	915	361	346
2/25	1110	374	494	268	1078	481
2/26	472	671	556	—	—	—

Source: U.S. EPA (1973). Monitoring Industrial Wastewater, Washington, D.C.

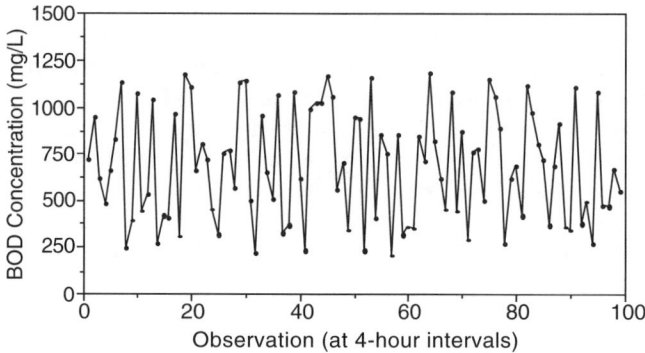

FIGURE 5.1 Time series plot of the BOD data.

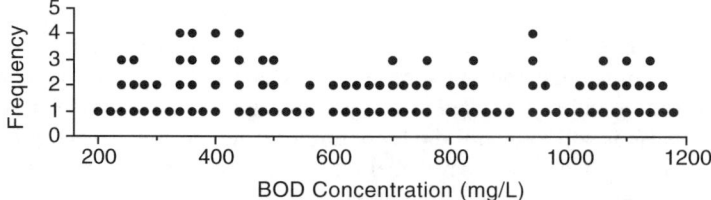

FIGURE 5.2 Dot diagram of the 99 BOD observations.

A *dot diagram* shown in Figure 5.2 gives a better picture of the variability. The data have a *uniform distribution* between 200 and 1200 mg/L. Any value within this range seems equally likely. The dot diagrams in Figure 5.3 subdivide the data by time of day. The observed values cover the full range regardless of time of day. There is no regular cyclic variation and no time of day has consistently high or consistently low values.

Given the uniform pattern of variation, the extreme values take on a different meaning than if the data were clustered around the average, as they would be in a normal distribution. If the distribution were

Seeing the Shape of a Distribution

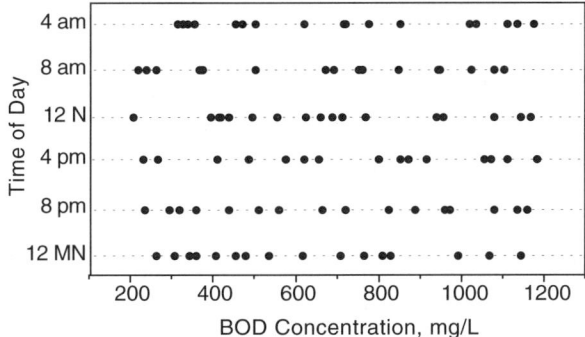

FIGURE 5.3 Dot diagrams of the data for each sampling time.

normal, the extreme values would be relatively rare in comparison to other values. Here, they are no more rare than values near the average. The designer may feel that the rapid fluctuation with no tendency to cluster toward one average or central value is the most important feature of the data.

The elegantly simple dot diagram and the time series plot have beautifully described the data. No numerical summary could transmit the same information as efficiently and clearly. Assuming a "normal-like" distribution and reporting the average and standard deviation would be very misleading.

Probability Plots

A probability plot is not needed to interpret the data in Table 5.1 because the time series plot and dot diagrams expose the important characteristics of the data. It is instructive, nevertheless, to use these data to illustrate how a probability plot is constructed, how its shape is related to the shape of the frequency distribution, and how it could be misused to estimate population characteristics.

The *probability plot*, or *cumulative frequency distribution*, shown in Figure 5.4 was constructed by ranking the observed values from small to large, assigning each value a rank, which will be denoted by i, and calculating the plotting position of the probability scale as $p = i/(n + 1)$, where n is the total number of observations. A portion of the ranked data and their calculated plotting positions are shown in Table 5.2. The relation $p = i/(n + 1)$ has traditionally been used by engineers. Statisticians seem to prefer $p = (i - 0.5)/n$, especially when n is small.[1] The major differences in plotting position values computed from these formulas occur in the tails of the distribution (high and low ranks). These differences diminish in importance as the sample size increases.

Figure 5.4(top) is a normal probability plot of the data, so named because the probability scale (the ordinate) is arranged in a special way to give a straight line plot when the data are normally distributed. Any frequency distribution that is not normal will plot as a curve on the normal probability scale used in Figure 5.4(top). The abcissa is an arithmetic scale showing the BOD concentration. The ordinate is a cumulative probability scale on which the calculated p values are plotted to show the probability that the BOD is less than the value shown on the abcissa.

Figure 5.4 shows that the BOD data are distributed symmetrically, but not in the form of a normal distribution. The S-shaped curve is characteristic of distributions that have more observations on the tails than predicted by the normal distribution. This kind of distribution is called "heavy tailed." A data set that is light-tailed (peaked) or skewed will also have an S-shape, but with different curvature (Hahn and Shapiro, 1967).

There is often no reason to make the probability plot take the form of a straight line. If a straight line appears to describe the data, draw such a line on the graph "by eye." If a straight line does not appear to describe the points, and you feel that a line needs to be drawn to emphasize the pattern, draw a

---

[1] There are still other possibilities for the probability plotting positions (see Hirsch and Stedinger, 1987). Most have the general form of $p = (i - a)/(n + 1 - 2a)$, where a is a constant between 0.0 and 0.5. Some values are: a = 0 (Weibull), a = 0.5 (Hazen), and a = 0.375 (Blom).

TABLE 5.2

Probability Plotting Positions for the $n = 99$ Values in Table 5.1

BOD Value (mg/L)	Rank i	Plotting Position $p = 1/(n+1)$
207	1	$1/100 = 0.01$
221	2	0.02
223	3	0.03
235	4	0.04
...	...	...
1158	96	0.96
1167	97	0.97
1174	98	0.98
1185	99	0.99

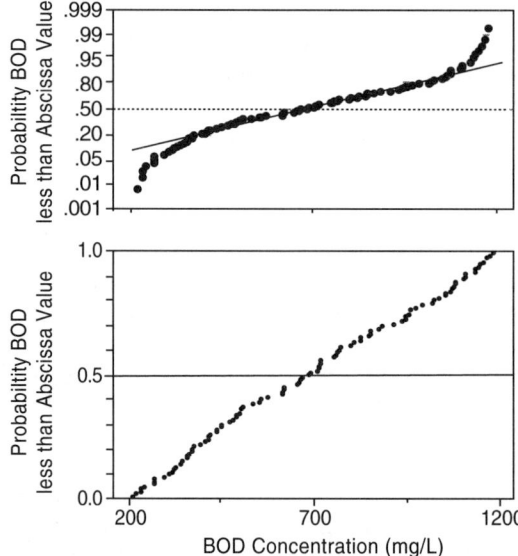

FIGURE 5.4 Probability plots of the uniformly distributed BOD data. The top panel is *a normal probability plot*. The ordinate is scaled so that normally distributed data would plot as a straight line. The bottom panel is scaled so the BOD data plot as a straight line. These BOD data are not normally distributed. They are uniformly distributed.

smooth curve. If the plot is used to estimate the median and the 90th percentile value, a curve like Figure 5.4(top) is satisfactory.

If a straight-line probability plot were wanted for this data, a simple arithmetic plot of p vs. BOD will do, as shown by Figure 5.4(bottom). The linearity of this plot indicates that the data are uniformly distributed over the range of observed values, which agrees with the impression drawn from the dot plots.

A probability plot can be made with a logarithmic scale on one axis and the normal probability scale on the other. This plot will produce a straight line if the data are lognormally distributed. Figure 5.5 shows the dot diagram and normal probability plot for some data that has a lognormal distribution. The left-hand panel shows that the logarithms are normally distributed and do plot as a straight line.

Figure 5.6 shows normal probability plots for four samples of $n = 26$ observations, each drawn at random from a pool of observations having a mean $\eta = 10$ and standard deviation $\sigma = 1$. The sample data in the two top panels plot neat straight lines, but the bottom panels do not. This illustrates the difficulty in using probability plots to prove normality (or to disprove it).

Figure 5.7 is a probability plot of some industrial wastewater COD data. The ordinate is constructed in terms of *normal scores*, also known as *rankits*. The shape of this plot is the same as if it were made

Seeing the Shape of a Distribution

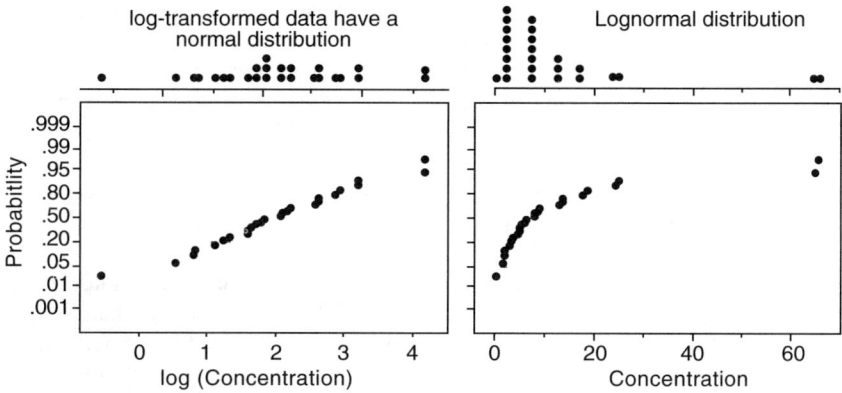

FIGURE 5.5 The logarithms of the lognormal data on the right will plot as a straight line on a normal probability plot (left-hand panel).

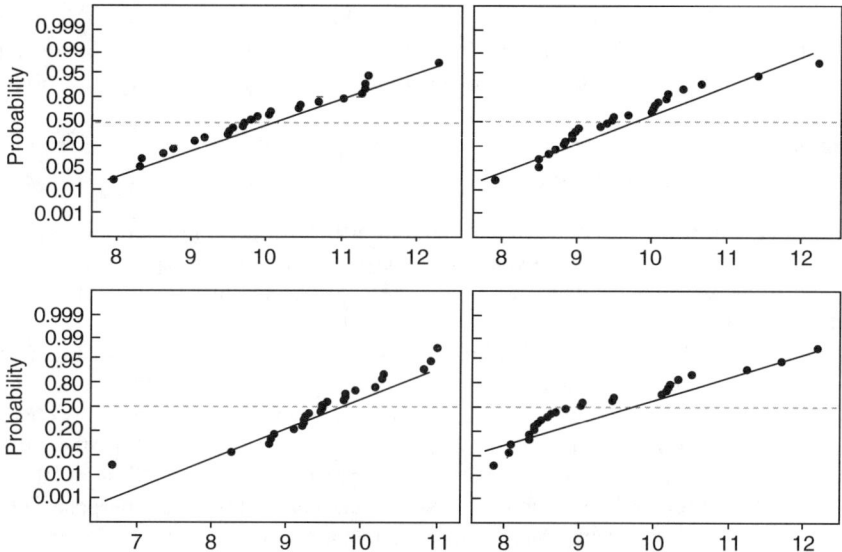

FIGURE 5.6 Normal probability plots, each constructed with $n = 26$ observations drawn at random from a normal distribution with $\eta = 10$ and $\sigma = 1$. Notice the difference in the range of values in the four samples.

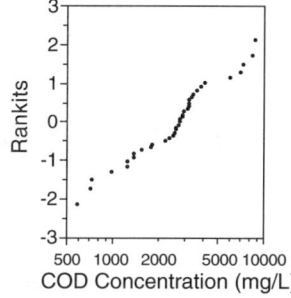

FIGURE 5.7 Probability plot constructed in terms of normal order scores, or *rankits*. The ordinate is the normal distribution measured in standard deviations; 1 rankit = 1 standard deviation. Rankit = 0 is the median (50th percentile).

on normal probability paper. Normal scores or rankits can be generated in many computer software packages (such as Microsoft Excel) and can be looked up in standard statistical tables (Sokal and Rohlf, 1969). This is handy because some graphics programs do not draw probability plots. Another advantage of using rankits is that linear regression can be done on the rankit scores (see the example of censored data analysis in Chapter 15).

The Use and Misuse Probability Plots

Engineering texts often suggest estimating the mean and sample standard deviations of a sample from a probability plot, saying that the mean is located at $p = 50\%$ on a normal probability graph and the standard deviation is the distance from $p = 50\%$ to $p = 84.1\%$ (or, because of symmetry, from $p = 15.9\%$ to $p = 50\%$). These graphical estimates are valid only when the data are normally distributed. Because few environmental data sets are normally distributed, this graphical estimation of the mean and standard deviation is not recommended. A probability plot is useful, however, to estimate the median ($p = 50\%$) and to read directly any percentile of special interest.

One way that probability plots are misused is to make the graphical estimates of sample statistics when the distribution is not normal. For example, if the data are lognormally distributed, $p = 50\%$ is the median and not the arithmetic mean, and the distance from $p = 50\%$ to $p = 84.1\%$ is not the sample standard deviation. If the data have a uniform distribution, or any other symmetrical distribution, $p = 50\%$ is the median *and* the average, but the standard deviation cannot be read from the probability plot.

Randomness and Independence

Data can be normally distributed without being random or independent. Furthermore, randomness and independence cannot be perceived or proven using a probability plot. This plot does not provide any information regarding serial dependence or randomness, both of which may be more critical than normality in the statistical analysis.

The histogram of the 52 weekly BOD loading values plotted on the right side of Figure 5.8 is symmetrical. It looks like a normal distribution and the normal probability plot will be a straight line. It could be said therefore that the sample of 52 observations is normally distributed. This characterization is uninteresting and misleading because the data are not randomly distributed about the mean and there is a strong trend with time (i.e., serial dependence). The time series plot, Figure 5.8, shows these important features. In contrast, the probability plot and dot plot, while excellent for certain purposes, obscure these features. To be sure that all important features of the data are revealed, a variety of plots must be used, as recommended in Chapter 3.

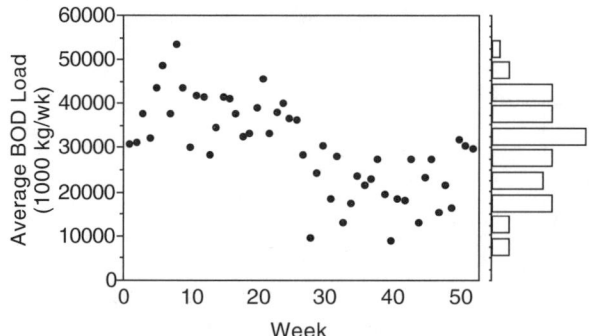

FIGURE 5.8 This sample of 52 observations will give a linear normal probability plot, but such a plot would hide the important time trend and the serial correlation.

Comments

We are almost always interested in knowing the shape of a sample's distribution. Often it is important to know whether a set of data is distributed symmetrically about a central value, or whether there is a tail of data toward a high or a low value. It may be important to know what fraction of time a critical value is exceeded.

Dot plots and probability plots are useful graphical tools for seeing the shape of a distribution. To avoid misinterpreting probability plots, use them only in conjunction with other plots. Make dot diagrams and, if the data are sequential in time, a time series plot. Sometimes these graphs provide all the important information and the probability plot is unnecessary.

Probability plots are convenient for estimating percentile values, especially the median (50th percentile) and extreme values. It is not necessary for the probability plot to be a straight line to do this. If it is straight, draw a straight line. But if it is not straight, draw a smooth curve through the plotted points and go ahead with the estimation.

Do not use probability plots to estimate the mean and standard deviation except in the very special case when the data give a linear plot on normal probability paper. This special case is common in textbooks, but rare with real environmental data. If the data plot as a straight line on log-probability paper, the 50th percentile value is not the mean (it is the geometric mean) and there is no distance that can be measured on the plot to estimate the standard deviation.

Probability plots may be useful in discovering the distribution of the data in a sample. Sometimes the analysis is not clear-cut. Because of random sampling variation, the curve can have a substantial amount of "wiggle" when the data actually are normally distributed. When the number of observations approaches 50, the shape of the probability distribution becomes much more clear than when the sample is small (for example, 20 observations). Hahn and Shapiro (1967) point out that:

1. The variance of points in the tails (extreme low or high plotted values) will be larger than that of points at the center of the distribution. Thus, the relative linearity of the plot near the tails of the distribution will often seem poorer than at the center even if the correct model for the probability density distribution has been chosen.

2. The plotted points are ordered and hence are not independent. Thus, we should not expect them to be randomly scattered about a line. For example, the points immediately following a point above the line are also likely to be above the line. Even if the chosen model is correct, the plot may consist of a series of successive points (known as runs) above and below the line.

3. A model can never be proven to be adequate on the basis of sample data. Thus, the probability of a small sample taken from a near-normal distribution will frequently not differ appreciably from that of a sample from a normal distribution.

If the data have positive skew, it is often convenient to use graph paper that has a log scale on one axis and a normal probability scale on the other axis. If the logarithms of the data are normally distributed, this kind of graph paper will produce a straight-line probability plot. The log scale may provide a convenient scaling for the graph even if it does not produce a straight-line plot; for example, when the data are bacterial counts that range from 10 to 100,000.

References

Hahn, G. J. and S. S. Shapiro (1967). *Statistical Methods for Engineers,* New York, John Wiley.

Hirsch, R. M. and J. D. Stedinger (1987). "Plotting Positions for Historical Floods and Their Precision," *Water Resources Research,* 23(4), 715–727.

Mage, D. T. (1982). "An Objective Graphical Method for Testing Normal Distributional Assumptions Using Probability Plots," *Am. Statistician,* 36, 116–120.

Sokal, R. R. and F. J. Rohlf (1969). *Biometry: The Principles and Practice of Statistics in Biological Research,* New York, W.H. Freeman & Co.

U.S. EPA (1973). *Monitoring Industrial Wastewater,* Washington, D.C.

Exercises

5.1 Normal Distribution. Graphically determine whether the following data could have come from a normal distribution.

Data Set A	13	21	13	18	27	16	17	18	22	19
	15	21	18	20	23	25	5	20	20	21
Data Set B	22	24	19	28	22	23	20	21	25	22
	18	21	35	21	36	24	24	23	23	24

5.2 Flow and BOD. What is the distribution of the weekly flow and BOD data in Exercise 3.3?

5.3 Histogram. Plot a histogram for these data and describe the distribution.

0.02	0.18	0.34	0.50	0.65	0.81
0.04	0.20	0.36	0.51	0.67	0.83
0.06	0.22	0.38	0.53	0.69	0.85
0.08	0.24	0.40	0.55	0.71	0.87
0.10	0.26	0.42	0.57	0.73	0.89
0.12	0.28	0.44	0.59	0.75	0.91
0.14	0.30	0.46	0.61	0.77	0.93
0.16	0.32	0.48	0.63	0.79	0.95

5.4 Wastewater Lead. What is the distribution of the influent lead and the effluent lead data in Exercise 3.7?

6

External Reference Distributions

KEY WORDS *histogram, reference distribution, moving average, normal distribution, serial correlation,* t *distribution.*

When data are analyzed to decide whether conditions are as they should be, or whether the level of some variable has changed, the fundamental strategy is to compare the current condition or level with an appropriate reference distribution. The reference distribution shows how things should be, or how they used to be. Sometimes an external reference distribution should be created, instead of simply using one of the well-known and nicely tabulated statistical reference distributions, such as the normal or t distribution. Most statistical methods that rely upon these distributions assume that the data are random, normally distributed, and independent. Many sets of environmental data violate these requirements.

A specially constructed reference distribution will not be based on assumptions about properties of the data that may not be true. It will be based on the data themselves, whatever their properties. If serial correlation or nonnormality affects the data, it will be incorporated into the external reference distribution.

Making the reference distribution is conceptually and mathematically simple. No particular knowledge of statistics is needed, and the only mathematics used are counting and simple arithmetic. Despite this simplicity, the concept is statistically elegant, and valid judgments about statistical significance can be made.

Constructing an External Reference Distribution

The first 130 observations in Figure 6.1 show the natural background pH in a stream. Table 6.1 lists the data. Suppose that a new effluent has been discharged to the stream and someone suggests it is depressing the stream pH. A survey to check this has provided ten additional consecutive measurements: 6.66, 6.63, 6.82, 6.84, 6.70, 6.74, 6.76, 6.81, 6.77, and 6.67. Their average is 6.74. We wish to judge whether this group of observations differs from past observations. These ten values are plotted as open circles on the right-hand side of Figure 6.1. They do not appear to be unusual, but a careful comparison should be made with the historical data.

The obvious comparison is the 6.74 average of the ten new values with the 6.80 average of the previous 130 pH values. One reason not to do this is that the standard procedure for comparing two averages, the *t*-test, is based on the data being independent of each other in time. Data that are a time series, like these pH data, usually are not independent. Adjacent values are related to each other. The data are serially correlated (autocorrelated) and the *t*-test is not valid unless something is done to account for this correlation. To avoid making any assumption about the structure of the data, the average of 6.74 should be compared with a reference distribution for averages of sets of ten consecutive observations.

Table 6.1 gives the 121 averages of ten consecutive observations that can be calculated from the historical data. The ten-day moving averages are plotted in Figure 6.2. Figure 6.3 is a reference distribution for these averages. Six of the 121 ten-day averages are as low as 6.74. About 95% of the ten-day averages are larger than 6.74. Having only 5% of past ten-day averages at this level or lower indicates that the river pH may have changed.

TABLE 6.1

Data Used to Plot Figure 6.1 and the Associated External Reference Distributions

6.79	6.84	6.85	6.47	6.67	6.76	6.75	6.72	6.88	6.83	6.65	6.77	6.92	6.73	6.94	6.84	6.71
6.88	6.66	6.97	6.63	7.06	6.55	6.77	6.99	6.70	6.65	6.87	6.89	6.92	6.74	6.58	6.40	7.04
6.95	7.01	6.97	6.78	6.88	6.80	6.77	6.64	6.89	6.79	6.77	6.86	6.76	6.80	6.80	6.81	6.81
6.80	6.90	6.67	6.82	6.68	6.76	6.77	6.70	6.62	6.67	6.84	6.76	6.98	6.62	6.66	6.72	6.96
6.89	6.42	6.68	6.90	6.72	6.98	6.74	6.76	6.77	7.13	7.14	6.78	6.77	6.87	6.83	6.84	6.77
6.76	6.73	6.80	7.01	6.67	6.85	6.90	6.95	6.88	6.73	6.92	6.76	6.68	6.79	6.93	6.86	6.87
6.95	6.73	6.59	6.84	6.62	6.77	6.53	6.94	6.91	6.90	6.75	6.74	6.74	6.76	6.65	6.72	6.87
6.92	6.98	6.70	6.97	6.95	6.94	6.93	6.80	6.84	6.78	6.67						

Note: Time runs from left to right.

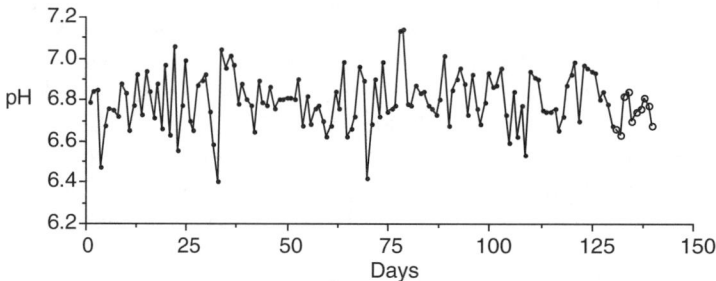

FIGURE 6.1 Time series plot of the pH data with the moving average of ten consecutive values.

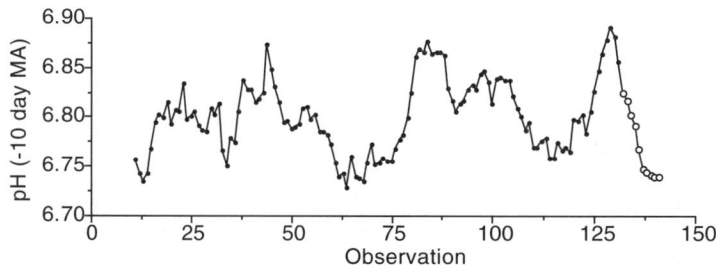

FIGURE 6.2 Ten-day moving averages of pH.

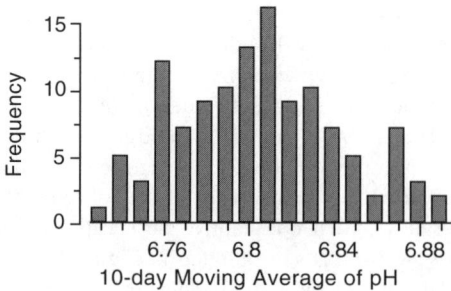

FIGURE 6.3 External reference distribution for ten-day moving averages of pH.

External Reference Distributions 57

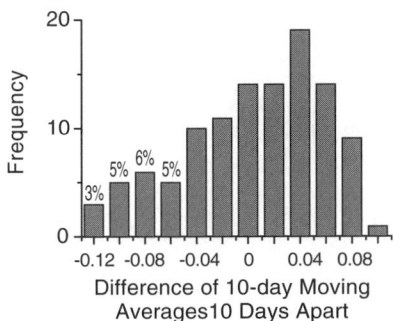

FIGURE 6.4 External reference distribution for differences of 10-day moving averages of pH.

Using a Reference Distribution to Compare Two Mean Values

Let the situation in the previous example change to the following. An experiment to evaluate the effect of an industrial discharge into a treatment process consists of making 10 observations consecutively before any addition and 10 observations afterward. We assume that the experiment is not affected by any transients between the two operating conditions. The average of 10 consecutive pre-discharge samples was 6.80, and the average of the 10 consecutive post-discharge samples was 6.86. Does the difference of 6.80 − 6.86 = −0.06 represent a significant shift in performance?

A reference distribution for the difference between batches of 10 consecutive samples is needed. There are 111 differences of MA10 values that are 10 days apart that can be calculated from the data in Table 6.1. For example, the difference between the averages of the 10th and 20th batches is 6.81 − 6.76 = 0.05. The second value is the difference between the 11th and 21st is 6.74 − 6.80 = −0.06. Figure 6.4 is the reference distribution of the 111 differences of batches of 10 consecutive samples. A downward difference as large as −0.06 has occurred frequently. We conclude that the new condition is not different than the recent past.

Looking at the 10-day moving averages suggests that the stream pH may have changed. Looking at the differences in averages indicates that a noteworthy change has not occurred. Looking at the differences uses more information in the data record and gives a better indication of change.

Using a Reference Distribution for Monitoring

Treatment plant effluent standards and water quality criteria are usually defined in terms of 30-day averages and 7-day averages. The effluent data themselves typically have a lognormal distribution and are serially correlated. This makes it difficult to derive the statistical properties of the 30- and 7-day averages. Fortunately, if historical data are readily available at all treatment plants and we can construct external reference distributions, not only for 30- and 7-day averages, but also for any other statistics of interest.

The data in this example are effluent 5-day BOD measurements that have been made daily on 24-hour flow-weighted composite samples from an activated sludge treatment plant. We realize that BOD data are not timely for process control decisions, but they can be used to evaluate whether the plant has been performing at its normal level or whether effluent quality has changed. A more complete characterization of plant performance would include reference distributions for other variables, such as suspended solids, ammonia, and phosphorus.

A long operating record was used to generate the top histogram in Figure 6.5. From the operator's log it was learned that many of the days with high BOD had some kind of assignable problem. These days were defined as unstable performance, the kind of performance that good operation could eliminate. Eliminating these poor days from the histogram produces the target stable performance shown by the reference distribution in the bottom panel of Figure 6.5. "Stable" is the kind of performance of which

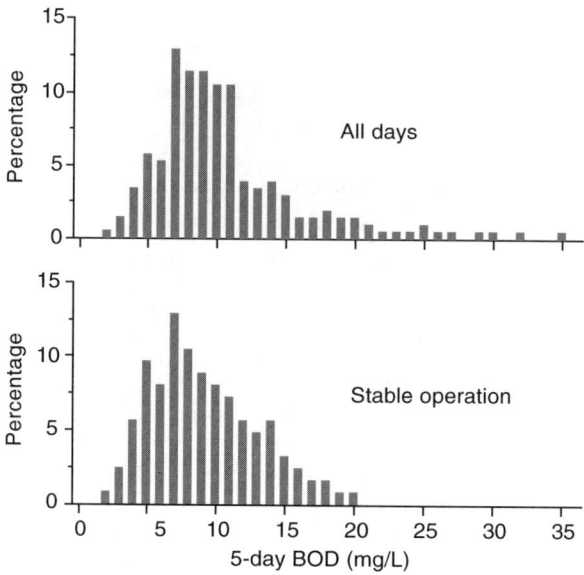

FIGURE 6.5 External reference distributions for effluent 5-day BOD (mg/L) for the complete record and for the stable operating conditions.

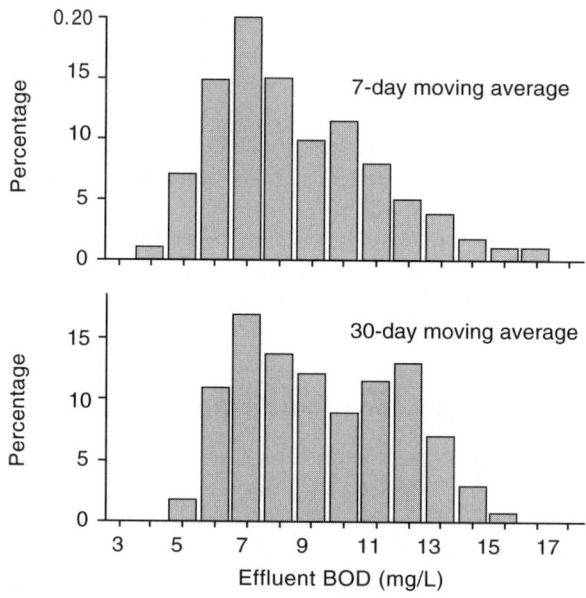

FIGURE 6.6 External reference distributions for 7- and 30-day moving averages of effluent 5-day BOD during periods of stable treatment plant operation.

the plant is capable over long stretches of time (Berthouex and Fan, 1986). This is the reference distribution against which new daily effluent measurements should be compared when they become available, which is five or six days after the event in the case of BOD data.

If the 7-day moving average is used to judge effluent quality, a reference distribution is required for this statistic. The periods of stable operation were used to calculate the 7-day moving averages that produce the reference distribution shown in Figure 6.6 (top). Figure 6.6 (bottom) is the reference distribution of 30-day moving averages for periods of stable operation. Plant performance can now be monitored by comparing, as they become available, new 7- or 30-day averages against these reference distributions.

Setting Critical Levels

The reference distribution shows at a glance which values are exceptionally high or low. What is meant by "exceptional" can be specified by setting critical decision levels that have a specified probability value. For example, one might specify exceptional as the level that is exceeded p percent of the time. The reference distribution for daily observations during stable operation (bottom panel in Figure 6.5) is based on 1150 daily values representing stable performance. The critical upper 5% level cut is a BOD concentration of 33 mg/L. This is found by summing the frequencies, starting from the highest BOD observed during stable operation, until the accumulated percentage equals or exceeds 5%. In this case, the probability that the BOD is 20 is P(BOD = 20) = 0.8%. Also, P(BOD = 19) = 0.8%, P(BOD = 18) = 1.6%, and P(BOD = 17) = 1.6%. The sum of these percentages is 4.8%. So, as a practical matter, we can say that the BOD exceeds 16 mg/L only about 5% of the time when operation is stable.

Upper critical levels can be set for the MA(7) reference distribution as well. The probability that a 7-day MA(7) of 14 mg/L or higher will occur when the treatment plant is stable is 4%. An MA(7) greater than 13 mg/L serves warning that the process is performing poorly and may be upset. By definition, 5% of such warnings will be false alarms. A two-level warning system could be devised, for example, by using the upper 1% and the upper 5% levels. The upper 1% level, which is about 16 mg/L, is a signal that something is almost certainly wrong; it will be a false in only 1 out of 100 alerts.

There is a balance to be found between having occasional false alarms and no false alarms. Setting a warning at the 5% level, or perhaps even at the 10% level, means that an operator is occasionally sent to look for a problem when none exists. But it also means that many times a warning is given before a problem becomes too serious and on some of these occasions action will prevent a minor upset from becoming more serious. An occasional wild goose chase is the price paid for the early warnings.

Comments

Consider why the warning levels were determined empirically instead of by calculating the mean and standard deviation and then using the normal distribution. People who know some statistics tend to think of the bell-shaped, symmetrical normal distribution when they hear that "the mean is X and the standard deviation is Y." The words "mean" and "standard deviation" create an image of approximately 95% of the values falling within two standard deviations of the mean.

A glance at Figure 6.6 reveals why this is an inappropriate image for the reference distribution of moving averages. The distributions are not symmetrical and, furthermore, they are truncated. These characteristics are especially evident in the MA(30) distribution. By definition, the effluent BOD values are never very high when operation is stable, so MA cannot take on certain high values. Low values of the MA do not occur because the effluent BOD cannot be less than zero and values less than 2 mg/L were not observed. The normal distribution, with its finite probability of values occurring far out on the tails of the distribution (and even into negative values), would be a terrible approximation of the reference distribution derived from the operating record.

The reference distribution for the daily values will always give a warning before the MA does. The MA is conservative. It flattens one-day upsets, even fairly large ones, and rolls smoothly through short intervals of minor disturbances without giving much notice. The moving average is like a shock absorber on a car in that it smooths out the small bumps. Also, just as a shock absorber needs to have the right stiffness, a moving average needs to have the right length of memory to do its job well. A 30-day MA is an interesting statistic to plot only because effluent standards use a 30-day average, but it is too sluggish to usefully warn of trouble. At best, it can confirm that trouble has existed. The seven-day average is more responsive to change and serves as a better warning signal. Exponentially weighted moving averages (see Chapter 4) are also responsive and reference distributions can be constructed for them as well.

Just as there is no reason to judge process performance on the basis of only one variable, there is no reason to select and use only one reference distribution for any particular single variable. One statistic and its reference distribution might be most useful for process control while another is best for judging

References

Berthouex, P. M. and W. G. Hunter, (1983). "How to Construct a Reference Distribution to Evaluate Treatment Plant Performance," *J. Water Poll. Cont. Fed.,* 55, 1417–1424.

Berthouex, P. M. and R. Fan (1986). "Treatment Plant Upsets: Causes, Frequency, and Duration," *J. Water Poll. Cont. Fed.,* 58, 368–375.

Exercises

6.1 BOD Tests. The table gives 72 duplicate measurements of wastewater effluent 5-day BOD measured at 2-hour intervals. (a) Develop reference distrbutions that would be useful to the plant operator. (b) Develop a reference distribution for the difference between duplicates that would be useful to the plant chemist.

Time	BOD	(mg/L)	Time	BOD	(mg/L)	Time	BOD	(mg/L)	Time	BOD	(mg/L)
2	185	193	38	212	203	74	124	118	110	154	139
4	116	119	40	167	158	76	166	157	112	142	129
6	158	156	42	116	118	78	232	225	114	142	137
8	185	181	44	122	129	80	220	207	116	157	174
10	140	135	46	119	116	82	220	214	118	196	197
12	179	174	48	119	124	84	223	210	120	136	124
14	173	169	50	172	166	86	133	123	122	143	138
16	119	119	52	106	105	88	175	156	124	116	108
18	119	116	54	121	124	90	145	132	126	128	123
20	113	112	56	163	162	92	139	132	128	158	161
22	116	115	58	148	140	94	148	130	130	158	150
24	122	110	60	184	184	96	133	125	132	194	190
26	161	171	62	175	172	98	190	185	134	158	148
28	110	116	64	172	166	100	187	174	136	155	145
30	176	166	66	118	117	102	190	171	138	137	129
32	197	191	68	91	98	104	115	102	140	152	148
34	167	165	70	115	108	106	136	127	142	140	127
36	179	178	72	124	119	108	154	141	144	125	113

6.2 Wastewater Effluent TSS. The histogram shows one year's total effluent suspended solids data ($n = 365$) for a wastewater treatment plant (data from Exercise 3.5). The average TSS concentration is 21.5 mg/L. (a) Assuming the plant performance will continue to follow this pattern, indicate on the histogram the upper 5% and upper 10% levels for out-of-control performance. (b) Calculate (approximately) the annual average effluent TSS concentration if the plant could eliminate all days with TSS > upper 10% level specified in (b).

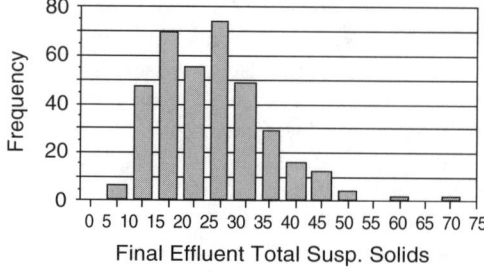

7

Using Transformations

KEY WORDS antilog, arcsin, bacterial counts, Box-Cox transformation, cadmium, confidence interval, geometric mean, transformations, linearization, logarithm, nonconstant variance, plankton counts, power function, reciprocal, square root, variance stabilization.

There is usually no scientific reason why we should insist on analyzing data in their original scale of measurement. Instead of doing our analysis on y it may be more appropriate to look at $\log(y)$, $\sqrt{y}$, $1/y$, or some other function of y. These re-expressions of y are called transformations. Properly used transformations eliminate distortions and give each observation equal power to inform.

Making a transformation is not cheating. It is a common scientific practice for presenting and interpreting data. A pH meter reads in logarithmic units, $pH = -\log_{10}[H^+]$ and not in hydrogen ion concentration units. The instrument makes a data transformation that we accept as natural. Light absorbency is measured on a logarithmic scale by a spectrophotometer and converted to a concentration with the aid of a calibration curve. The calibration curve makes a transformation that is accepted without hesitation. If we are dealing with bacterial counts, N, we think just as well in terms of $\log(N)$ as N itself.

There are three technical reasons for sometimes doing the calculations on a transformed scale: (1) to make the spread equal in different data sets (to make the variances uniform); (2) to make the distribution of the residuals normal; and (3) to make the effects of treatments additive (Box et al., 1978).[1] Equal variance means having equal spread at the different settings of the independent variables or in the different data sets that are compared. The requirement for a normal distribution applies to the measurement errors and not to the entire sample of data. Transforming the data makes it possible to satisfy these requirements when they are not satisfied by the original measurements.

Transformations for Linearization

Transformations are sometimes used to obtain a straight-line relationship between two variables. This may involve, for example, using reciprocals, ratios, or logarithms. The left-hand panel of Figure 7.1 shows the exponential growth of bacteria. Notice that the variance (spread) of the counts increases as the population density increases. The right-hand panel shows that the data can be described by a straight line when plotted on a log scale. Plotting on a log scale is equivalent to making a log transformation of the data.

The important characteristic of the original data is the nonconstant variance, not nonlinearity. This is a problem when the curve or line is fitted to the data using regression. Regression tries to minimize the distance between the data points and the line described by the model. Points that are far from the line exert a strong effect because the regression mathematics wants to reduce the square of this distance. The result is that the precisely measured points at time $t = 1$ will have less influence on the position of the regression line than the poorly measured data at $t = 3$. This gives too much influence to the least reliable data. We would prefer for each data point to have about the same amount of influence on the location of the line. In this example, the log-transformed data have constant variance at the different population levels. Each data

---

[1] For example, if $y = x^a z^b$, a log transformation gives $\log y = a \log x + b \log z$. Now the effects of factors x and z are additive. See Box et al. (1978) for an example of how this can be useful.

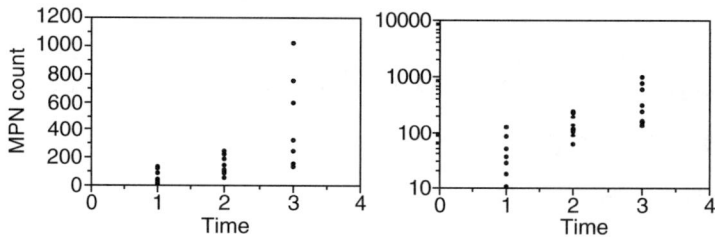

FIGURE 7.1 An example of how a transformation can create constant variance. Constant variance at all levels is important so each data point will carry equal weight in locating the position of the fitted curve.

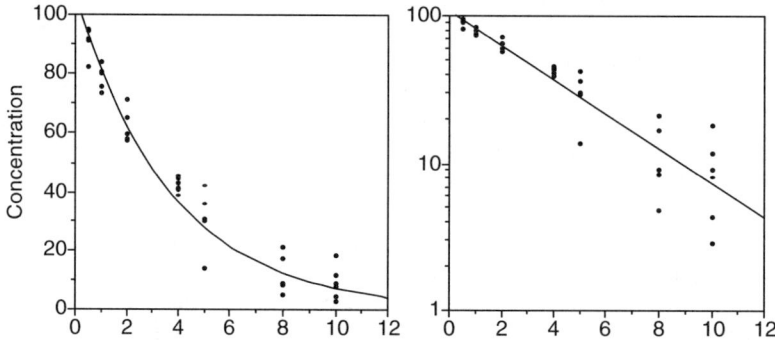

FIGURE 7.2 An example of how a transformation could create nonconstant variance.

value has roughly equal weight in determining the position of the line. The log transformation is used to achieve this equal weighting and not because it gives a straight line.

A word of warning is in order about using transformations to obtain linearity. A transformation can turn a good situation into a bad one by distorting the variances and making them unequal (see Chapter 45). Figure 7.2 shows a case where the constant variance of the original data is destroyed by an inappropriate logarithmic transformation.

In the examples above it was easy to check the variances at the different levels of the independent variables because the measurements had been replicated. If there is no replication, this check cannot be made. This is only one reason why replication is always helpful and why it is recommended in experimental and monitoring work.

Lacking replication, should one assume that the variances are originally equal or unequal? Sometimes the nature of the measurement process gives a hint as to what might be the case. If dilutions or concentrations are part of the measurement process, or if the final result is computed from the raw measurements, or if the concentration levels are widely different, it is not unusual for the variances to be unequal and to be larger at high levels of the independent variable. Biological counts frequently have nonconstant variance. These are not justifications to make transformations indiscriminately. Do not avoid making transformations, but use them wisely and with care.

Transformations to Obtain Constant Variance

When the variance changes over the range of experimental observations, the variance is said to be nonconstant, or unstable. Common situations that tend to create this pattern are (1) measurements that involve making dilutions or other steps that introduce multiplicative errors, (2) using instruments that read out on a log scale which results in low values being recorded more precisely than high values, and (3) biological counts. One of the transformations given in Table 7.1 should be suitable to obtain constant variance.

TABLE 7.1

Transformations that are Useful to Obtain Uniform Variance

Condition		Replace y by
σ uniform over range of y		no transformation needed
$\sigma \propto \bar{y}^2$		$x = 1/y$
$\sigma \propto \bar{y}^{3/2}$		$x = 1/\sqrt{y}$
$\sigma \propto \bar{y} \ (\sigma^2 > \bar{y})$	all $y > 0$	$x = \log(y)$
	some $y = 0$	$x = \log(y + c)$
$\sigma \propto \bar{y}^{1/2}$	all $y > 0$	$x = \sqrt{y}$
	some $y < 0$	$x = \sqrt{y + c}$
$\sigma^2 > \bar{y}$	p = ratio or percentage	$x = \arcsin \sqrt{p}$

Source: Box, G. E. P., W. G. Hunter, and J. S. Hunter (1978). *Statistics for Experimenters: An Introduction to Design, Data Analysis, and Model Building,* New York, Wiley Interscience.

TABLE 7.2

Plankton Counts on 20 Replicate Water Samples from Five Stations in a Reservoir

Station 1	0	2	1	0	0	1	1	0	1	1	0	2	1	0	0	2	3	0	1	1
Station 2	3	1	1	1	4	0	1	4	3	3	5	3	2	2	1	1	2	2	2	0
Station 3	6	1	5	7	4	1	6	5	3	3	5	3	4	3	8	4	2	2	4	2
Station 4	7	2	6	9	5	2	7	6	4	3	5	3	6	4	8	5	2	3	4	1
Station 5	12	7	10	15	9	6	13	11	8	7	10	8	11	8	14	9	6	7	9	5

Source: Elliot, J. (1977). *Some Methods for the Statistical Analysis of Samples of Benthic Invertebrates,* 2nd ed., Ambleside, England, Freshwater Biological Association.

TABLE 7.3

Statistics Computed from the Data in Table 7.2

Station		1	2	4	5	6
Untransformed data		$\bar{y} = 0.85$	2.05	3.90	4.60	9.25
		$s_y^2 = 0.77$	1.84	3.67	4.78	7.57
Transformed $x = \sqrt{y+c}$		$\bar{x} = 1.10$	1.54	2.05	2.20	3.09
		$s_x^2 = 0.14$	0.20	0.22	0.22	0.19

The effect of square root and logarithmic transformations is to make the larger values less important relative to the small ones. For example, the square root converts the values (0, 1, 4) to (0, 1, 2). The 4, which tends to dominate on the original scale, is made relatively less important by the transformation.

The log transformation is a stronger transformation than the square root transformation. "Stronger" means that the range of the transformed variables is relatively smaller for a log transformation that for the square root. When the sample contains some zero values, the log transformation is $x = \log(y + c)$, where c is a constant. Usually the value of c is arbitrarily chosen to be 1 or 0.5. The larger the value of c, the less severe the transformation. Similarly, for square root transformations, $\sqrt{y + c}$ is less severe than $\sqrt{y}$.

The arcsin transformation is used for decimal fractions and is most useful when the sample includes values near 0.00 and 1.00. One application is in bioassys where the data are fractions of organisms showing an effect.

Example 7.1

Twenty replicate samples from five stations were counted for plankton, with the results given in Table 7.2. The computed averages and variances are in Table 7.3. The computed means and variance on the original data show that variance is not uniform; it is ten times larger at station 5 than at station 1. Also, the variance increases as the average increases and s_y^2 seems to be proportional to $\bar{y}$. This indicates that a square root transformation may be suitable. Because most of the counts

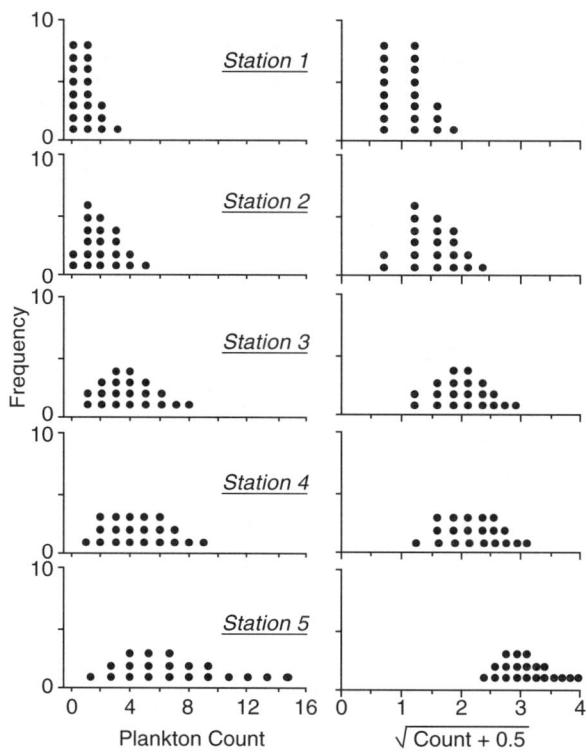

FIGURE 7.3 Original and transformed plankton counts.

TABLE 7.4

Eight Replicate Measurements on Bacteria at Three Sampling Stations

y = Bacteria/100 mL			x = log₁₀(Bacteria/100 mL)		
1	2	3	1	2	3
27	225	1020	1.431	2.352	3.009
11	99	136	1.041	1.996	2.134
48	41	317	1.681	1.613	2.501
36	60	161	1.556	1.778	2.207
120	190	130	2.079	2.279	2.114
85	240	601	1.929	2.380	2.779
18	90	760	1.255	1.954	2.889
130	112	240	2.144	2.049	2.380
$\bar{y}$ = 59.4	132	420.6	$\bar{x}$ = 1.636	2.050	2.502
s_y^2 = 2156	5771	111,886	s_x^2 = 0.151	0.076	0.124

are small, the transform used was $\sqrt{y + 0.5}$. Figure 7.3 shows the distribution of the original and the transformed data. The transformed distributions are more symmetrical and normal-like than the originals. The variances computed from the transformed data are uniform.

Example 7.2

Table 7.4 shows eight replicate measurements of bacterial density that were made at three locations to study the spatial pattern of contamination in an estuary. The data show that $s^2 > \bar{y}$ and that s increases in proportion to $\bar{y}$. Table 7.1 suggests a logarithmic transformation. The improvement due to the log transformation is shown in Table 7.4. Note that the transformation could be done using either $\log_e$ or $\log_{10}$ because they differ by only a constant ($\log_e$ = 2.303 $\log_{10}$).

Confidence Intervals and Transformations

After summary statistics (means, standard deviations, etc.) have been calculated on the transformed scale, it is often desirable to translate the results back to the original scale of measurement. This can create some confusion. For example, if the average $\bar{x}$ has been estimated using $x = \log(y)$, the simple back-transformation of antilog($\bar{x}$) does not give an unbiased estimate of $\bar{y}$. The antilogarithm of $\bar{x}$ is the *geometric mean* of the original data (y); that is, antilog($\bar{x}$) = $\bar{y}_g$. The correct estimate of the arithmetic mean on the original y scale is $\bar{y} = $ antilog($\bar{x} + 0.5\ s^2$) (Gilbert, 1987).

If the transformation produced a near-normal distribution, the standard deviations and standard errors computed from the transformed data will be symmetric about the mean on the transformed scale. But they will be asymmetric on the original scale. The options are to:

1. Quote symmetric confidence limits on the transformed scale.
2. Quote asymmetric confidence limits on the original scale (recognizing that in the case of a log transformation they apply to the geometric mean and not to the arithmetic average).
3. Give two sets of results, one with standard errors and symmetric confidence limits on the transformed scale and a corresponding set of means (arithmetic and geometric) on the original scale. The reader can judge the statistical significance on the transformed scale and the practical importance on the original scale.

Two examples illustrate the use of log-transformed data to construct confidence limits for the geometric mean.

Example 7.3

A sample of $n = 5$ observations [95, 20, 74, 195, 71] gives $\bar{y} = 91$ and $s_y^2 = 4{,}140$. Clearly, $s_y^2 > \bar{y}$ and a log transformation should be tried. The $x = \log_{10}(y)$ values are 1.97772, 1.30103, 1.86923, 2.29003, and 1.85126. This gives $\bar{x} = 1.85786$ and $s_x^2 = 0.12784$. The value of t for $v = n - 1 = 4$ degrees of freedom and $\alpha/2 = 0.025$ is 2.776. Therefore, the 95% confidence interval for the mean on the log-transformed scale, η_x, is:

$$\bar{x} \pm t\sqrt{\frac{s_x^2}{n}} = 1.85786 \pm 2.776\sqrt{\frac{0.1278}{5}} = 1.85786 \pm 0.44381$$

and

$$1.4140 < \eta_x < 2.3017$$

Transforming η_x back to the original scale gives an estimate of the geometric mean of the y's:

$$\bar{y}_g = \text{antilog}(\bar{x}) = \text{antilog}(1.85786) = 72.09$$

The asymmetric 95% confidence limits for the true value of the geometric mean, η_x, are obtained by taking antilogarithms of the upper and lower confidence limits of η_x, which gives:

$$25.94 \le \eta_g \le 200.29$$

Note that the upper and lower confidence limits in the log metric are $\bar{x} + \beta$ and $\bar{x} - \beta$, where $\beta = t_{\alpha/2}\sqrt{s^2/n}$. The upper confidence limit on the original scale is antilog($\bar{x} + \beta$) = antilog($\bar{x}$) · antilog(β), which becomes $\bar{y}_g \cdot \beta'$ where $\beta' = $ antilog(β). Likewise, the lower confidence limit is $\bar{y}_g/\beta'$. For this example, $\beta = 0.44381$, antilog (0.44381) = 2.778, and the 95% confidence limits for the geometric mean on the original scale are 72.09(2.7785) = 200.29 and 72.09/2.7785 = 25.94.

Example 7.4

A log transformation is needed on a sample of $n = 6$ that includes some zeros: $y = [0, 2, 1, 0, 3, 9]$. Because of the zeros, use the transformation $x = \log(y+1)$ to find the transformed values: $x = [0, 0.47712, 0.30103, 0.60206, 1.0]$. This gives $\bar{x} = 0.39670$ with standard deviation $s_x^2 = 0.14730$. For $v = n - 1 = 5$ and $\alpha/2 = 0.025$, $t_{5, 0.025} = 2.571$ and the 95% confidence limits on the transformed scale are:

$$\text{LCL}(x) = 0.39670 - 2.571\sqrt{0.14730/6} = -0.006 \text{ (say zero)}$$

and

$$\text{UCL}(x) = 0.39670 + 2.571\sqrt{0.14730/6} = 0.79954$$

Transforming back to the original metric gives the geometric mean:

$$\bar{y}_g = \text{antilog}_{10}(0.39670) - 1 = 2.5 - 1 = 1.5$$

The -1 is due to using $x = \log(y+1)$. The similar inverse of the confidence limits gives:

$$\text{LCL}(y) = 0 \quad \text{and} \quad \text{UCL}(y) = 5.3$$

Confidence Intervals on the Original Scale

Notice that the above examples are for the geometric mean η_g and not the arithmetic mean η. The work becomes more difficult if we want the asymmetric confidence intervals of the arithmetic mean on the original scale. This will be shown for the lognormal distribution.

A simple method of estimating the mean η and variance σ^2 of the two-parameter lognormal distribution is to use:

$$\hat{\eta} = \exp\left(\bar{x} + \frac{s_x^2}{2}\right) \quad \text{and} \quad \hat{\sigma}^2 = \hat{\eta}^2[\exp(s_x^2) - 1]$$

where $\hat{\eta}$ and $\hat{\sigma}^2$ are the estimated mean and variance. $\bar{x}$ and s_x^2 are calculated in the usual way shown in Chapter 2 (Gilbert, 1987).

These estimates are slightly biased upward by about 5% for $n = 20$ and 1% for $n = 100$. The importance of this when using $\hat{\eta}$ to judge compliance with environmental standards, or when comparing estimates based on unequal n, is discussed by Gilbert (1987) and Landwehr (1978).

Computing confidence limits involves more than simply adding a multiple of the standard deviation (as we do for the normal distribution). Confidence limits for the mean, η, are estimated using the following equations (Land, 1971):

$$\text{UCL}_{1-\alpha} = \exp\left(\bar{x} + 0.5 s_x^2 + \frac{s_y H_{1-\alpha}}{\sqrt{n-1}}\right)$$

$$\text{LCL}_{\alpha} = \exp\left(\bar{x} + 0.5 s_x^2 + \frac{s_y H_{\alpha}}{\sqrt{n-1}}\right)$$

The quantities $H_{1-\alpha}$ and H_{α} depend on s_x, n, and the confidence level α. Land (1975) provides the necessary tables; a subset of these may be found in Gilbert (1987).

The Box-Cox Power Transformations

A power transformation model developed by Box and Cox (1964) can, so far as possible, satisfy the conditions of normality and constant variance simultaneously. The method is applicable for almost any kind of statistical model and any kind of transformation. The transformed value $Y_i^{(\lambda)}$ of the original variable y_i is:

$$Y_i^{(\lambda)} = \frac{y_i^\lambda - 1}{\lambda \bar{y}_g^{\lambda-1}}$$

where $\bar{y}_g$ is the geometric mean of the original data series, and λ expresses the power of the transformation. The geometric mean is obtained by averaging $\ln(y)$ and taking the exponential (antilog) of the result. The special case when $\lambda = 0$ is the log transformation: $Y_i^{(0)} = \bar{y}_g \ln(y_i)$. $\lambda = -1$ is a reciprocal transformation, $\lambda = 1/2$ is a square root transformation, and $\lambda = 1$ is no transformation. Example applications of this transformation are given in Box et al. (1978).

Example 7.5

Table 7.5 lists 36 measurements on cadmium (Cd) in soil, and their logarithms. The Cd concentrations range from 0.005 to 0.094 mg/kg. The limit of detection was 0.01. Values below this were arbitrarily replaced with 0.005. Comparisons must be made with other sets of similar data and some transformation is needed before this can be done. Experience with environmental data suggests that a log transformation may be useful, but something better might be discovered if we make the Box-Cox transformation for several values of λ and compare the variances of the transformed data.

The variance of the log-transformed values in Table 7.5 is $\sigma^2_{\ln(y)} = 0.549$. This cannot be compared directly with the variance from, for instance, a square root transformation unless the calculations are normalized to keep them on the same relative scale. The denominator of the Box-Cox transformation, $\lambda \bar{y}_g^{\lambda-1}$, is a normalizing factor to make the variances comparable across different

TABLE 7.5

Cadmium Concentrations in Soil

Cadmium (mg/kg)			ln [Cadmium]		
0.023	0.005	0.005	−3.7723	−5.2983	−5.2983
0.020	0.005	0.032	−3.9120	−5.2983	−3.4420
0.010	0.005	0.031	−4.6052	−5.2983	−3.4738
0.020	0.013	0.005	−3.9120	−4.3428	−4.2687
0.020	0.005	0.014	−3.9120	−5.2983	−3.9120
0.020	0.094	0.020	−3.9120	−2.3645	−5.2983
0.010	0.011	0.005	−4.6052	−4.5099	−3.6119
0.010	0.005	0.027	−4.6052	−5.2983	−4.1997
0.010	0.005	0.015	−4.6052	−5.2983	−3.3814
0.010	0.028	0.034	−4.6052	−3.5756	−5.2983
0.010	0.010	0.005	−4.6052	−4.6052	−5.2983
0.005	0.018	0.013	−5.2983	−4.0174	−4.3428

Average = 0.0161
Variance = 0.000255

Average of ln[Cd] = −4.42723
Variance = 0.549
Geo. mean = 0.01195

Note: Concentrations in mg/kg.

TABLE 7.6

Transformed Values, Means, and Variances as a Function of λ

	$\lambda = -1$	$\lambda = -0.5$	$\lambda = 0$	$\lambda = 0.5$	$\lambda = 1$
	−0.0061	−0.0146	−0.0451	−0.1855	−0.9770
	−0.0070	−0.0159	−0.0467	−0.1877	−0.9800
	−0.0141	−0.0235	−0.0550	−0.1967	−0.9900
	...	...	...	...	...
	...	...	...	...	...
	−0.0284	−0.0343	−0.0633	−0.2032	−0.9950
	−0.0108	−0.0203	−0.0519	−0.1937	−0.9870
$\bar{y}^\lambda =$	−0.01496	−0.0228	−0.0529	−0.1930	−0.9839
$\text{Var}(\bar{y}^\lambda) =$	0.000093	0.000076	0.000078	0.000113	0.000255

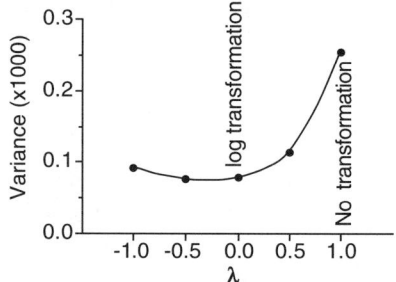

FIGURE 7.4 Variance of the transformed cadmium data as a function of λ.

values of λ. The geometric mean of the untransformed data is $\bar{y}_g = \exp(-4.42723) = 0.01195$. The denominator for $\lambda = 0.5$, for example, is $0.5(0.01195)^{0.5-1} = 4.5744$; the denominator for $\lambda = -0.5$ is -765.747.

Table 7.6 gives some of the power-transformed data for $\lambda = -1, -0.5, 0, 0.5,$ and 1. $\lambda = 1$ is no transformation ($Y_i^{(1)} = y_i - 1$) except for scaling to be comparable to the other transformations. $\lambda = 0$ is the log transformation, calculated from $Y_i^{(0)} = \bar{y}_g \ln(y_i)$, which is again scaled so the variance can be compared directly with variances of other power-transformed values.

The two bottom rows of Table 7.6 give the mean and the variance of the power-transformed values. The suitable transformations give small variances. Rather than pick the smallest value from the table, make a plot (Figure 7.4) that shows how the variance changes with λ. The smooth curve is drawn as a reminder that these variances are estimates and that small differences between them should not be taken seriously. Do not seek an optimal value of λ that minimizes the variance. Such a value is likely to be awkward, like $\lambda = 0.23$. The data do not justify such detail, especially because the censored values ($y < 0.01$) were arbitrarily replaced with 0.005. (This inflates the variance from whatever it would be if the censored values were known.) Values of $\lambda = -0.5$, $\lambda = 0$, or $\lambda = 0.5$ are almost equally effective transformations. Any of these will be better than no transformation ($\lambda = 1$). The log transformation ($\lambda = 0$) is very satisfactory and is our choice as a matter of convenience.

Figure 7.5 shows dot diagrams for the original data, the square root ($\lambda = 0.5$), the logarithms ($\lambda = 0$), and reciprocal square root ($\lambda = -0.5$). The log transformation is most symmetric, but it is not normal because of the 11 non-detect data that were replaced with 0.005 (i.e., 1/2 the MDL).

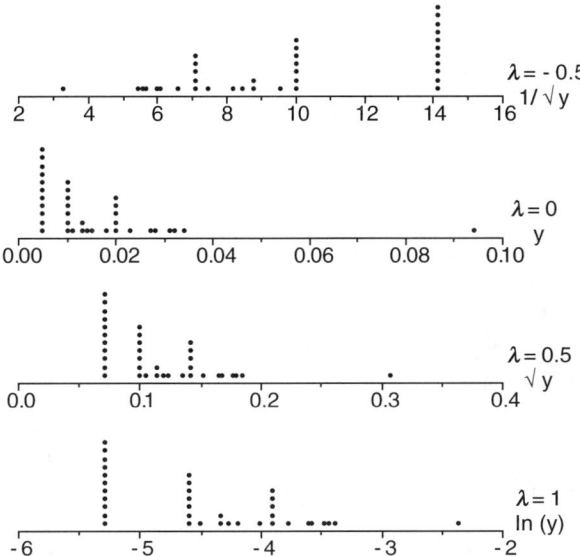

FIGURE 7.5 Dot diagrams of the data and the square root, log, and reciprocal square root transformed values. The eye-catching spike of 11 points are "non-detects" that were arbitrarily assigned values of 0.005 mg/kg.

Comments

Transformations are not tricks to reduce variation or to convert a complicated nonlinear model into a simple linear form. There are often statistical justifications for making transformations and then analyzing the transformed data. They may be needed to stabilize the variance or to make the distribution of the errors normal. The most common and important use is to stabilize (make uniform) the variance.

It can be tempting to use a transformation to make a nonlinear function linear so that it can be fitted using simple linear regression methods. Sometimes the transformation that gives a linear model will coincidentally produce uniform variance. Beware, however, because linearization can also produce the opposite effect of making constant variance become nonconstant (see Chapter 45).

When the analysis has been done on transformed data, the analyst must consider carefully whether to report the final results on the transformed or the original scale of measurement. Confidence intervals that are symmetrical on the transformed scale will not be symmetric when transformed back to the original scale. Care must also be taken when converting the mean on the transformed scale back to the original scale. A simple back-transformation typically does not give an unbiased estimate, as was demonstrated in the case of the logarithmic transformation.

References

Box, G. E. P. and D. R. Cox (1964). "An Analysis of Transformations," *J. Royal Stat. Soc.,* Series B, 26, 211.

Box, G. E. P., W. G. Hunter, and J. S. Hunter (1978). *Statistics for Experimenters: An Introduction to Design, Data Analysis, and Model Building,* New York, Wiley Interscience.

Elliot, J. (1977). *Some Methods for the Statistical Analysis of Samples of Benthic Invertebrates,* 2nd ed., Ambleside, England, Freshwater Biological Association.

Gilbert, R. O. (1987). *Statistical Methods for Environmental Pollution Monitoring,* New York, Van Nostrand Reinhold.

Land, C. E. (1975). "Tables of Confidence Limits for Linear Functions of the Normal Mean and Variance," in Selected Tables in *Mathematical Statistics,* Vol. III, Am. Math. Soc., Providence, RI, 358–419.

Landwehr, J. M. (1978). "Some Properties of the Geometric Mean and its Use in Water Quality Standards," *Water Resources Res.,* 14, 467–473.

Exercises

7.1 Plankton Counts. Transform the plankton data in Table 7.2 using a square root transformation $x = \mathrm{sqrt}(y)$ and a logarithmic transformation $x = \log(y)$ and compare the results with those shown in Figure 7.3.

7.2 Lead in Soil. Examine the distribution of the 36 measurements of lead (mg/kg) in soil and recommend a transformation that will make the data nearly symmetrical and normal.

7.6	32	5	4.2	14	18	2.3	52	10	3.3	38	3.4	4.3	0.10	5.7	0.10	0.10	4.4
0.42	0.10	16	2.0	12	0.10	3.2	0.43	1.4	5.9	0.23	0.10	0.10	0.23	0.29	5.3	2.0	1.0

7.3 Box-Cox Transformation. Use the Box-Cox power function to find a suitable value of λ to transform the 48 lead measurements given below. Note: All < MDL values were replaced by 0.05.

7.6	32	5.0	4.2	14	18	2.3	52	10	3.3	38	3.4	4.3	0.05	0.05	0.10
0.10	0.05	0.05	0.05	0.0	0.05	1.2	0.10	0.10	0.10	0.10	0.10	0.23	4.4	0.42	0.10
16.	2.0	2.0	1.0	3.2	0.43	1.4	0.10	5.9	0.10	0.10	0.23	0.29	5.3	5.7	0.10

7.4 Are Transformations Necessary? Which of the following are correct reasons for transforming data? (a) Facilitate interpretation in a natural way. (b) Promote symmetry in a data sample. (c) Promote constant variance in several sets of data. (d) Promote a straight-line relationship between two variables. (e) Simplify the structure so that a simple additive model can help us understand the data.

7.5 Power Transformations. Which of the following statements about power transformations are correct? (a) The order of the data in the sample is preserved. (b) Medians are transformed to medians, and quartiles are transformed to quartiles. (c) They are continuous functions. (d) Points very close together in the raw data will be close together in the transformed data, at least relative to the scale being used. (e) They are smooth functions. (f) They are elementary functions so the calculations of re-expression are quick and easy.

8

Estimating Percentiles

KEY WORDS *confidence intervals, distribution free estimation, geometric mean, lognormal distribution, normal distribution, nonparametric estimation, parametric estimation, percentile, quantile, rank order statistics.*

The use of percentiles in environmental standards and regulations has grown during the past few years. England has water quality consent limits that are based on the 90th and 95th percentiles of monitoring data not exceeding specified levels. The U.S. EPA has specifications for air quality monitoring that are, in effect, percentile limitations. These may, for example, specify that the ambient concentration of a compound cannot be exceeded more often than once a year (the 364/365th percentile). The U.S. EPA has provided guidance for setting aquatic standards on toxic chemicals that require estimating 99th percentiles and using this statistic to make important decisions about monitoring and compliance. They have also used the 99th percentile to establish maximum daily limits for industrial effluents (e.g., pulp and paper). Specifying a 99th percentile in a decision-making rule gives an impression of great conservatism, or of having great confidence in making the "safe" and therefore correct environmental decision. Unfortunately, the 99th percentile is a statistic that cannot be estimated precisely.

Definition of Quantile and Percentile

The population distribution is the true underlying pattern. Figure 8.1 shows a lognormal *population distribution* of y and the normal distribution that is obtained by the transformation $x = \ln(y)$. The population 50th percentile (the median), and 90th, 95th, and 99th percentiles are shown. The population pth percentile, y_p, is a parameter that, in practice, is unknown and must be estimated from data. The estimate of the percentile is denoted by $\hat{y}_p$. In this chapter, the parametric estimation method and one nonparametric estimation method are shown.

The pth *quantile* is a population parameter and is denoted by y_p. (Chapter 2 stated that parameters would be indicated with Greek letters but this convention is violated in this chapter.) By definition, a proportion p of the population is smaller or equal to y_p and a proportion $1 - p$ is larger than y_p. Quantiles are expressed as decimal fractions.

Quantiles expressed as percentages are called *percentiles*. For example, the 0.5 quantile is equivalent to the 50th percentile; the 0.99 quantile is the 99th percentile. The 95th percentile will be denoted as y_{95}.

A *quartile* of the distribution contains one-fourth of the area under the frequency distribution (and one-fourth of the data points). Thus, the distribution is divided into four equal areas by $y_{0.250}$ (the lower quantile), the median, $y_{0.5}$ (the 0.5 quantile, or median), and $y_{0.75}$ (known as the upper quartile).

Parametric Estimates of Quantiles

If we know or are willing to assume the population distribution, we can use a *parametric method*. Parametric quantile (percentile) estimation will be discussed initially in terms of the normal distribution. The same methods can be used on nonnormally distributed data after transformation to make them approximately normal. This is convenient because the properties of the normal distribution are known and accessible in tables.

71

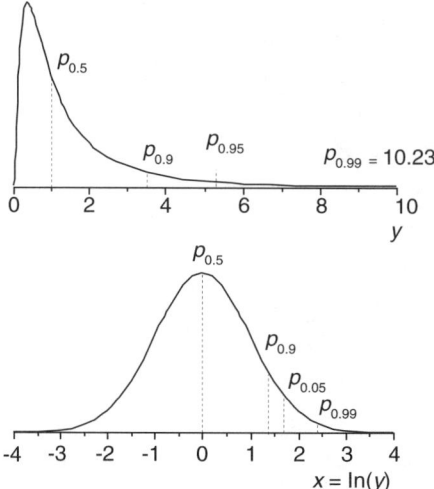

FIGURE 8.1 Correspondence of percentiles on the lognormal and normal distributions. The transformation $x = \ln(y)$ converts the lognormal distribution to a normal distribution. The percentiles also transform.

The normal distribution is completely specified by the mean η and standard deviation σ of the distribution, respectively. The true pth quantile of the normal distribution is $y_p = \eta + z_p \sigma$, where z_p is the pth quantile of the standard normal distribution. Generally, the parameters η and σ are unknown and we must estimate them by the sample average, $\bar{y}$, and the sample standard deviation, s. The quantile, y_p, of a normal distribution is estimated using:

$$\hat{y}_p = \bar{y} + z_p s$$

The appropriate value of z_p can be found in a table of the normal distribution.

Example 8.1

Suppose that a set of data is normally distributed with estimated mean and standard deviation of 10.0 and 1.2. To estimate the 99th quantile, look up $z_{0.99} = 2.326$ and compute:

$$\hat{y}_p = 10 + 2.326(1.2) = 12.8$$

This method can be used even when a set of data indicates that the population distribution is not normally distributed if a transformation will make the distribution normal. For example, if a set of observations y appears to be from a lognormal distribution, the transformed values $x = \log(y)$ will be normally distributed. The pth quantile of y on the original measurement scale corresponds to the pth quantile of x on the log scale. Thus, $x_p = \log(y_p)$ and $y_p = \text{antilog}(x_p)$.

Example 8.2

A sample of observations, y, appears to be from a lognormal distribution. A logarithmic transformation, $x = \ln(y)$, produces values that are normally distributed. The log-transformed values have an average value of 1.5 and a standard deviation of 1.0. The 99th quantile on the log scale is located at $z_{0.99} = 2.326$, which corresponds to:

$$\hat{x}_{0.99} = 1.5 + 2.326(1.0) = 3.826.$$

Estimating Percentiles

The 99th quantile of the lognormal distribution is found by making the transformation in reverse:

$$\hat{y}_p = \text{antilog}(\hat{x}_p) = \exp(\hat{x}_p) = 45.9.$$

An upper $100(1-\alpha)\%$ confidence limit for the true pth quantile, y_p, can be easily obtained if the underlying distribution is normal (or has been transformed to become normal). This upper confidence limit is:

$$\text{UCL}_{1-\alpha}(y_p) = \bar{y} + K_{1-\alpha, p} s$$

where $K_{1-\alpha, p}$ is obtained from a table by Owen (1972), which is reprinted in Gilbert (1987).

Example 8.3

From $n = 300$ normally distributed observations we have calculated $\bar{y} = 10.0$ and $s = 1.2$. The estimated 99th quantile is $\hat{y}_{0.99} = 10 + 2.326(1.2) = 12.79$. For $n = 300$, $1 - \alpha = 0.95$, and $p = 0.99$, $K_{0.95, 0.99} = 2.522$ (from Gilbert, 1987) and the 95% upper confidence limit for the true 99th percentile value is:

$$\text{UCL}_{0.95}(y_{0.99}) = 10 + (1.2)(2.522) = 13.0.$$

In summary, the best estimate of the 99th quantile is 12.79 and we can state with 95% confidence that its true value is less than 13.0.

Sometimes one is asked to estimate a 99th percentile value and its upper confidence limit from samples sizes that are much smaller than the $n = 300$ used in this example. Suppose that we have $\bar{y} = 10$, $s = 1.2$, and $n = 30$, which again gives $\hat{y}_{0.99} = 12.8$. Now, $K_{0.95, 0.99} = 3.064$ (Gilbert, 1987) and:

$$\text{UCL}_{0.95}(y_{0.99}) = 10 + (1.2)(3.064) = 13.7$$

compared with UCL of 13.0 in Example 8.3. This 5% increase in the UCL has no practical importance. A potentially greater error resides in the assumption that the data are normally distributed, which is difficult to verify with a sample of $n = 30$. If the assumed distribution is wrong, the estimated $p_{0.99}$ is badly wrong, although the confidence limit is quite small.

Nonparametric Estimates of Quantiles

Nonparametric estimation methods do not require a distribution to be known or assumed. They apply to all distributions and can be used with any data set. There is a price for being unable (or unwilling) to make a constraining assumption regarding the population distribution. The estimates obtained by these methods are not as precise as we could obtain with a parametric method. Therefore, use the nonparametric method only when the underlying distribution is unknown or cannot be transformed to make it become normal.

The data are ordered from smallest to largest just as was done to construct a probability plot (Chapter 5). Percentile estimates could be read from a probability plot. The method to be illustrated here skips the plotting (but with a reminder that plotting data is always a good idea). The estimated pth quantile, $\hat{y}_p$, is simply the kth largest datum in the set, where $k = p(n + 1)$, n is the number of data points, and p is the quantile level of interest. If k is not an integer, y_p is obtained by linear interpolation between the two closest ordered values.

Example 8.4

A sample of $n = 575$ daily BOD observations is available to estimate the 99th percentile by the nonparametric method for the purpose of setting a maximum limit in a paper mill's discharge permit. The 11 largest ranked observations are:

Rank	575	574	573	572	571	570	569	568	567	566	565
BOD	10565	10385	7820	7580	7322	7123	6627	6289	6261	6079	5977

The 99th percentile is located at observation number $p(n+1) = 0.99(575+1) = 570.24$. Because this is not an integer, interpolate between the 570th and 571st largest observations to estimate $\hat{y}_{0.99} = 7171$.

The disadvantage of this method is that only the few largest observed values are used to estimate the percentile. The lower values are not used, except as they contribute to ranking the large values. Discarding these lower values throws away information that could be used to get more precise parameter estimates if the shape of the population distribution could be identified and used to make a parametric estimate.

Another disadvantage is that the data set must be large enough that extrapolation is unnecessary. A 95th percentile can be estimated from 20 observations, but a 99th percentile cannot be estimated with less than 100 observations. The data set should be much larger than the minimum if the estimates are to be much good. The advisability of this is obvious from a probability plot, which clearly shows that greatest uncertainty is in the location of the extreme quantiles (the tails of the distribution). This uncertainty can be expressed as confidence limits.

The confidence limits for quantiles that have been estimated using the nonparametric method can be determined with the following formulas if $n > 20$ observations (Gilbert, 1987). Compute the rank order of two-sided confidence limits (LCL and UCL):

$$\text{Rank(LCL)} = p(n+1) - z_{\alpha/2}\sqrt{np(1-p)}$$
$$\text{Rank(UCL)} = p(n+1) + z_{\alpha/2}\sqrt{np(1-p)}$$

The rank of the one-sided $1 - \alpha$ upper confidence limit is obtained by computing:

$$\text{Rank(UCL)} = p(n+1) + z_{\alpha}\sqrt{np(1-p)}$$

Because Rank(UCL) and Rank(LCL) are usually not integers, the limits are obtained by linear interpolation between the closest ordered values.

Example 8.5

The 95% two-sided confidence limits for the Example 8.4 estimate of $\hat{y}_{0.99} = 7171$, for $n = 575$ observations and $\alpha = 0.05$, are calculated using $z_{\alpha/2} = z_{0.025} = 1.96$ and

$$\text{Rank(LCL)} = 0.99(576) - 1.96\sqrt{575(0.99)(0.01)} = 565.6$$
$$\text{Rank(UCL)} = 0.99(576) + 1.96\sqrt{575(0.99)(0.01)} = 574.9$$

Interpolating between observations 565 and 566, and between observations 574 and 575, gives LCL = 6038 and UCL = 10,547.

Comments

Quantiles and percentiles can be estimated using parametric or nonparametric methods. The nonparametric method is simple, but the sample must contain more than p observations to estimate the pth quantile (and still more observations if the upper confidence limits are needed). Use the nonparametric method whenever you are unwilling or unable to specify a plausible distribution for the sample. Parametric estimates should be made whenever the distribution can be identified because the estimates will more precise than those

Estimating Percentiles

obtained from the nonparametric method. They also allow estimates of extreme quantiles (e.g., $\hat{y}_{0.99}$) from small data sets ($n < 20$). This estimation involves extrapolation beyond the range of the observed values. The danger in this extrapolation is in assuming the wrong population distribution.

The 50th percentile can be estimated with greater precision than any other can, and precision decreases rapidly as the estimates move toward the extreme tails of the distribution. Neither estimation method produces very precise estimates of extreme percentiles, even with large data sets.

References

Berthouex, P. M. and I. Hau (1991). "Difficulties in Using Water Quality Standards Based on Extreme Percentiles," *Res. Jour. Water Pollution Control Fed.,* 63(5), 873–879.

Bisgaard, S. and W. G. Hunter (1986). "Studies in Quality Improvement: Designing Environmental Regulations," Tech. Report No. 7, Center for Quality and Productivity Improvement, University of Wisconsin–Madison.

Crabtree, R. W., I. D. Cluckie, and C. F. Forster (1987). "Percentile Estimation for Water Quality Data," *Water Res.,* 23, 583–590.

Gilbert, R. O. (1987). *Statistical Methods for Environmental Pollution Monitoring,* New York, Van Nostrand Reinhold.

Hahn, G. J. and S. S. Shapiro (1967). *Statistical Methods for Engineers,* New York, John Wiley.

Exercises

8.1 Log Transformations. The log-transformed values of $n = 90$ concentration measurements have an average value of 0.9 and a standard deviation of 0.8. Estimate the 99th percentile and its upper 95% confidence limit.

8.2 Percentile Estimation. The ten largest-ranked observations from a sample of $n = 365$ daily observations are 61, 62, 63, 66, 71, 73, 76, 78, 385, and 565. Estimate the 99th percentile and its two-sided 95% confidence interval by the nonparametric method.

8.3 Highway TPH Data. Estimate the 95th percentile and its upper 95% confidence limit for the highway TPH data in Exercise 3.6. Use the averages of the duplicated measurements for a total of $n = 30$ observations.

9

Accuracy, Bias, and Precision of Measurements

KEY WORDS *accuracy, bias, collaborative trial, experimental error, interlaboratory comparison, precision, repeatability, reproducibility, ruggedness test, Youden pairs, Youden plots, Youden's rank test.*

> In your otherwise beautiful poem, there is a verse which read,
> *Every moment dies a man, Every moment one is born.*
> It must be manifest that, were this true, the population of the world would be at a standstill.... I suggest that in the next edition of your poem you have it read.
> *Every moment dies a man, Every moment* $1\frac{1}{16}$ *is born.*
> ... The actual figure is a decimal so long that I cannot get it into the line, but I believe 1/16 is sufficiently accurate for poetry.
>
> —Charles Babbage in a letter to Tennyson

The next measurement you make or the next measurement reported to you will be corrupted by experimental error. That is a fact of life. Statistics helps to discover and quantify the magnitude of experimental errors.

Experimental error is the deviation of observed values from the true value. It is the fluctuation or discrepancy between repeated measurements on identical test specimens. Measurements on specimens with true value η will not be identical although the people who collect, handle, and analyze the specimens make conditions as nearly identical as possible. The observed values y_i will differ from the true values by an error ε_i:

$$y_i = \eta + \varepsilon_i$$

The error can have systematic or random components, or both. If e_i is purely random error and τ_i is systematic error, then $\varepsilon_i = e_i + \tau_i$ and:

$$y_i = \eta + (e_i + \tau_i)$$

Systematic errors cause a consistent offset or *bias* from the true value. Measurements are consistently high or low because of poor technique (instrument calibration), carelessness, or outright mistakes. Once discovered, bias can be removed by calibration and careful checks on experimental technique and equipment. Bias cannot be reduced by making more measurements or by averaging replicated measurements. The magnitude of the bias cannot be estimated unless the true value is known.

Once bias has been eliminated, the observations are affected only by random errors and $y_i = \eta + e_i$. The observed e_i is the sum of all discrepancies that slip into the measurement process for the many steps required to proceed from collecting the specimen to getting the lab work done. The collective e_i may be large or small. It may be dominated by one or two steps in the measurement process (drying, weighing, or extraction, for example). Our salvation from these errors is their randomness.

The sign or the magnitude of the random error is not predictable from the error in another observation. If the total random error, e_i, is the sum of a variety of small errors, which is the usual case, then e_i will tend to be normally distributed. The average value of e_i will be zero and the distribution of errors will be equally positive and negative in sign.

Suppose that the final result of an experiment, y, is given by y = a + b, where a and b are measured values. If a and b each have a systematic error of +1, it is clear that the systematic error in y is +2. If, however, a and b each have a random error between zero and ±1, the random error in y is not ±2. This is because there will be occasions when the random error in a is positive and the error in b is negative (or vice versa). The average random error in the true value will be zero if the measurements and calculations are done many times. This means that the expected random error in y is zero. The variance and standard deviation of the error in y will not be zero, but simple mathematical rules can be used to estimate the precision of the final result if the precision of each observation (measurement) is known. The rules for *propagation of errors* are explained in Chapters 10 and 49.

Repeated (replicate) measurements provide the means to quantify the measurement errors and evaluate their importance. The effect of random errors can be reduced by averaging repeated measurements. The error that remains can be quantified and statistical statements can be made about the precision of the final results. *Precision* has to do with the scatter between replicate measurements. Precise results have small random errors. The scatter caused by random errors cannot be eliminated but it can minimized by careful technique. More importantly, it can be averaged and quantified.

The purpose of this chapter is to understand the statistical nature of experimental errors in the laboratory. This was discussed briefly in Chapter 2 and will be discussed more in Chapters 10, 11, and 12. The APHA (1998) and ASTM (1990, 1992, 1993) provide detailed guidance on this subject.

Quantifying Precision

Precision can refer to a measurement or to a calculated value (a statistic). Precision is quantified by the standard deviation (or the variance) of the measurement or statistic.

The average of five measurements [38.2, 39.7, 37.1, 39.0, 38.6] is $\bar{y} = 38.5$. This estimates the true mean of the process that generates the data. The standard deviation of the sample of the five observations is $s = 0.97$. This is a measure of the scatter of the observed values about the sample average. Thus, s quantifies the precision of this collection of measurements. The precision of the estimated mean measured by the *standard deviation of the average*, $s_{\bar{y}} = s/\sqrt{n}$, also known as the *standard error of the mean*. For this example data, $s_{\bar{y}} = 0.97/\sqrt{5} = 0.43$. This is a measure of how a collection of averages, each calculated from a set of five similar observations, would scatter about the true mean of the process generating the data.

A further statement of the precision of the estimated mean is the confidence interval. Confidence intervals of the mean, for large sample size, are calculated using the normal distribution:

95% confidence interval $\qquad \bar{y} - 1.96\dfrac{\sigma}{\sqrt{n}} < \eta < \bar{y} + 1.96\dfrac{\sigma}{\sqrt{n}}$

99.7% confidence interval $\qquad \bar{y} - 2.33\dfrac{\sigma}{\sqrt{n}} < \eta < \bar{y} + 2.33\dfrac{\sigma}{\sqrt{n}}$

99% confidence interval $\qquad \bar{y} - 2.58\dfrac{\sigma}{\sqrt{n}} < \eta < \bar{y} + 2.58\dfrac{\sigma}{\sqrt{n}}$

For small sample size, the confidence interval is calculated using Student's *t-statistic*:

$$\bar{y} - t\dfrac{s}{\sqrt{n}} < \eta < \bar{y} + t\dfrac{s}{\sqrt{n}} \qquad \text{or} \qquad \bar{y} - ts_{\bar{y}} < \eta < \bar{y} + ts_{\bar{y}}$$

The value of t is chosen for $n - 1$ degrees of freedom.

The precision of the mean estimated from the five observations, stated as a 95% confidence interval, is:

$$38.5 - 2.78(0.43) < \eta < 38.5 + 2.78(0.43)$$

$$37.3 < \eta < 39.7$$

Accuracy, Bias, and Precision of Measurements

When confidence limits are calculated, there is no point in giving the value of $ts/\sqrt{n}$ to more than two significant figures. The value of $\bar{y}$ should be given the corresponding number of decimal places.

When several measured quantities are be used to calculate a final result, these quantities should not be rounded off too much or a needless loss of precision will result. A good rule is to keep one digit beyond the last significant figure and leave further rounding until the final result is reached. This same advice applies when the mean and standard deviation are to be used in a statistical test such as the F- and t-tests; the unrounded values of $\bar{y}$ and s should be used.

Relative Errors

The *coefficient of variation* (CV), also known as the *relative standard deviation* (RSD), is defined by $s/\bar{y}$. The CV or RSD, expressed as a decimal fraction or as a percent, is a *relative error*. A relative error implies a proportional error; that is, random errors that are proportional to the magnitude of the measured values. Errors of this kind are common in environmental data. Coliform bacterial counts are one example.

Example 9.1

Total coliform bacterial counts at two locations on the Mystic River were measured on triplicate water samples, with the results shown below. The variation in the bacterial density is large when the coliform count is large. This happens because the high density samples must be diluted before the laboratory bacterial count is done. The counts in the laboratory cultures from locations A and B are about the same, but the error is distorted when these counts are multiplied by the dilution factor. Whatever variation there may be in the counts of the diluted water samples is multiplied when these counts are multiplied by the dilution factor. The result is proportional errors: the higher the count, the larger the dilution factor, and the greater the magnification of error in the final result.

Location	A	B
Total coliform (cfu/100 mL)	13, 22, 14	1250, 1583, 1749
Averages	$\bar{y}_A = 16.3$	$\bar{y}_B = 1527$
Standard deviation (s)	$s_A = 4.9$	$s_B = 254$
Coefficient of variation (CV)	0.30	0.17

We leave this example with a note that the standard deviations will be nearly equal if the calculations are done with the logarithms of the counts. Doing the calculations on logarithms is equivalent to taking the geometric mean. Most water quality standards on coliforms recommend reporting the geometric mean. The geometric mean of a sample $y_1, y_2, \ldots, y_n$ is $\bar{y}_g = \sqrt{y_1 \times y_2 \times \cdots \times y_n}$, or $\bar{y}_g = \text{antilog}[\frac{1}{n}\Sigma \log(y_i)]$.

Assessing Bias

Bias is the difference between the measured value and the true value. Unlike random error, the effect of systematic error (bias) cannot be reduced by making replicate measurements. Furthermore, it cannot be assessed unless the true value is known.

Example 9.2

Two laboratories each were given 14 identical specimens of standard solution that contained $C_S = 2.50 \ \mu g/L$ of an analyte. To get a fair measure of typical measurement error, the analyst was kept blind to the fact that these specimens were not run-of-the-laboratory work. The measured values are:

Laboratory A	2.8	3.5	2.3	2.7	2.3	3.1	2.5
	2.5	2.5	2.7	2.5	2.5	2.6	2.7
	$\bar{y}_A = 2.66$ μg/L						
	Bias = $\bar{y}_A - C_S = 2.66 - 2.50 = 0.16$ μg/L						
	Standard deviation = 0.32 μg/L						
Laboratory B	5.3	4.7	3.6	5.0	3.6	4.5	4.6
	4.3	3.9	4.1	4.2	4.2	4.3	4.9
	$\bar{y}_B = 4.38$ μg/L						
	Bias = $\bar{y}_B - C_S = 4.38 - 2.5 = 1.88$ μg/L						
	Standard deviation = 0.50 μg/L						

The best estimate of the bias is the average minus the concentration of the standard solution. The $100(1-\alpha)\%$ confidence limits for the true bias are:

$$(\bar{y} - C_S) \pm t_{\nu=n-1,\alpha/2} \frac{s}{\sqrt{n}}$$

For Laboratory A, the confidence interval is:

$$0.16 \pm 2.160(0.32/\sqrt{14}) = 0.16 \pm 0.18 = -0.03 \text{ to } 0.34 \text{ μg/L}$$

This interval includes zero, so we conclude with 95% confidence that the true bias is not greater than zero. Laboratory B has a confidence interval of 1.88 ± 0.29, or 1.58 to 2.16 μg/L. Clearly, the bias is greater than zero.

The precision of the two laboratories is the same; there is no significant difference between standard deviations of 0.32 and 0.50 μg/L. Roughly speaking, the ratios of the variances would have to exceed a value of 3 before we would reject the hypothesis that they are the same. The ratio in this example is $0.5^2/0.32^2 = 2.5$. The test statistic (the "roughly three") is the F-statistic and this test is called the F-test on the variances. It will be explained more in Chapter 24 when we discuss analysis of variance.

Having a "blind" analyst make measurements on specimens with known concentrations is the only way to identify bias. Any certified laboratory must invest a portion of its effort in doing such checks on measurement accuracy. Preparing test specimens with precisely known concentrations is not easy. Such standard solutions can be obtained from certified laboratories (U.S. EPA labs, for example).

Another quality check is to split a well-mixed sample and add a known quantity of analyte to one or more of the resulting portions. Example 9.3 suggests how splitting and spiking would work.

Example 9.3

Consider that the measured values in Example 9.2 were obtained in the following way. A large portion of a test solution with unknown concentration was divided into 28 portions. To 14 of the portions a quantity of analyte was added to increase the concentration by exactly 1.8 μg/L. The true concentration is not known for the spiked or the unspiked specimens, but the measured values should differ by 1.8 μg/L. The observed difference between labs A and B is $4.38 - 2.66 = 1.72$ μg/L. This agrees with the true difference of 1.8 μg/L. This is presumptive evidence that the two laboratories are doing good work.

There is a possible weakness in this kind of a comparison. It could happen that both labs are biased. Suppose the true concentration of the master solution was 1 μg/L. Then, although the difference is as expected, both labs are measuring about 1.5 μg/L too high, perhaps because there is some fault in the measurement procedure they were given. Thus, "splitting and spiking" checks work only when one laboratory is known to have excellent precision and low bias. This is the reason for having certified reference laboratories.

Multiple Sources of Variation (or Reproducibility ≠ Repeatability)

The measure of whether errors are becoming smaller as analysts are trained and as techniques are refined is the standard deviation of replicate measurements on identical specimens. This standard deviation must include all sources of variation that affect the measurement process.

Reproducibility and repeatability are often used as synonyms for precision. They are not the same. Suppose that identical specimens were analyzed on five different occasions using different reagents and under different laboratory conditions, and perhaps by different analysts. Measurement variation will reflect differences in analyst, laboratory, reagent, glassware, and other uncontrolled factors. This variation measures the reproducibility of the measurement process.

Compare this with the results of a single analyst who made five replicate measurements in rapid succession using the same set of reagents and glassware throughout, while temperature, humidity, and other laboratory conditions remained nearly constant. This variation measures repeatability.

Expect that reproducibility variation will be greater than repeatability variation. Repeatability gives a false sense of security about the precision of the data. The quantity of practical importance is reproducibility.

Example 9.4

Two analysts (or two laboratories) are each given five identical specimens to test. One analyst made five replicate measurements in rapid succession and obtained these results: 38.2, 39.7, 37.1, 39.0, and 38.6. The average is 38.5 and the variance is 0.97. This measures *repeatability*.

The other analyst made five measurements on five different days and got these results: 37.0, 38.5, 37.9, 41.3, and 39.9. The average is 38.9 and the variance is 2.88. This measures *reproducibility*.

The *reproducibility variance* is what should be expected as the laboratory generates data over a period of time. It consists of the *repeatability variance* plus additional variance due to other factors. Beware of making the standard deviation artificially small by replicating only part of the process.

Interlaboratory Comparisons

A consulting firm or industry that sends test specimens to several different laboratories needs to know that the performance of the laboratories is consistent. This can be checked by doing do an *interlaboratory comparison* or *collaborative trial*. A number of test materials, covering the range of typical values, are sent to a number of laboratories, each of which submits the values it measures on these materials. Sometimes, several properties are studied simultaneously. Sometimes, two or more alternate measuring techniques are covered. Mandel (1964) gives an extended example. One method of comparison is Youden's Rank Test (Youden and Steiner, 1975). The U.S. EPA used these methods to conduct interlaboratory studies for the 600 series analytical methods. See Woodside and Kocurek (1997) for an application of the Rank Test to compare 20 laboratories.

The simplest method is the Youden pairs test (Youden, 1972; Kateman and Buydens, 1993). Youden proposed having different laboratories each analyze two similar test specimens, one having a low concentration and one having a high concentration. The two measurements from each laboratory are a Youden pair.

The data in Table 9.1 are eight Youden pairs from eight laboratories and the differences between the pairs for each lab. Figure 9.1 is the Youden plot of these eight pairs. Each point represents one laboratory. Vertical and horizontal lines are drawn through sample averages of the laboratories, which are 2.11 μg/L for the low concentration and 6.66 μg/L for the high concentration. If the measurement errors are unbiased and random, the plotted pairs will scatter randomly about the intersection of the two lines, with about the same number of points in each quadrant.

TABLE 9.1

Youden Pairs from Eight Laboratories

Lab	Low (2.0 µg/L)	High (6.2 µg/L)	d_i
1	2.0	6.3	4.3
2	3.6	6.6	3.0
3	1.8	6.8	5.0
4	1.1	6.8	5.7
5	2.5	7.4	4.9
6	2.4	6.9	4.5
7	1.8	6.1	4.4
8	1.7	6.3	4.6

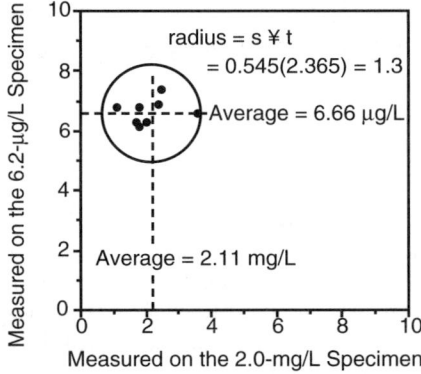

FIGURE 9.1 Plot of Youden pairs to evaluate the performance of eight laboratories.

The center of the circle is located by the average sample concentrations of the laboratories. The radius, which is proportional to the interlaboratory precision, is quantified by the standard deviation:

$$s = \sqrt{\frac{\Sigma(d_i - \bar{d})^2}{2(n-1)}}$$

where d_i is the difference between the two samples for each laboratory, $\bar{d}$ is the average of all laboratory sample pair differences, and n is number of participating laboratories (i.e., number of sample pairs).

It captures the pairs that would be expected from random error alone. Laboratories falling inside the circle are considered to be doing good work; those falling outside have poor precision and perhaps also a problem with bias. For these data:

$$s = \sqrt{\frac{(4.3-4.54)^2 + (3.0-4.54)^2 + \cdots + (4.6-4.54)^2}{2(8-1)}} = \sqrt{\frac{4.18}{14}} = 0.55 \, \mu g/L$$

The radius of the circle is s times Student's t for a two-tailed 95% confidence interval with $\nu = 8 - 1 = 7$ degrees of freedom. Thus, $t_{7,0.025} = 2.365$ and the radius is $0.55(2.365) = 1.3$ µg/L.

A 45° diagonal can be added to the plots to help assess bias. A lab that measures high will produce results that fall in the upper-right quadrant of the plot. Figure 9.2 shows four possible Youden plots. The upper panels show laboratories that have no bias. The lower-left panel shows two labs that are biased high and one lab that is biased low. The lower-right panel shows all labs with high bias, presumably

Accuracy, Bias, and Precision of Measurements

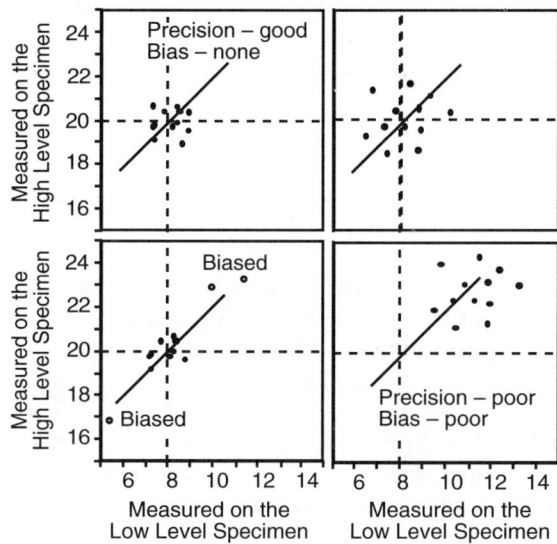

FIGURE 9.2 Four possible Youden plots. Bias is judged with respect to the 45° diagonal. The lower-left panel shows two labs that are biased high specimens and one lab that is biased low. The lower-right panel shows all labs with high bias, presumably because of some weakness in the measurement protocol.

because of some weakness in the measurement protocol. Additional interpretation of the Youden plot is possible. Consult Youden and Steiner (1975), Woodside and Kocurek (1997), or Miller and Miller (1984) for details.

Ruggedness Testing

Before a test method is recommended as a standard for general use, it should be submitted to a ruggedness test. This test evaluates the measurement method when small changes are made in selected factors of the method protocol. For example, a method might involve the pH of an absorbing solution, contact time, temperature, age of solution, holding time-stored test specimens, concentration of suspected interferences, and so on. The number of factors (k) that might influence the variability and stability of a method can be quite large, so an efficient strategy is needed.

One widely used design for ruggedness tests allows a subset of $k = 7$ factors to be studied in $N = 8$ trials, with each factor being set at two levels (or versions). Table 9.2 shows that this so-called 2^{7-4} fractional factorial experimental design can assess seven factors in eight runs (ASTM, 1990). The pluses

TABLE 9.2

A 2^{7-4} Fractional Factorial Design for Ruggedness Testing

Run	\multicolumn{7}{c}{Factor}	Observed Response
	1	2	3	4	5	6	7	
1	−	−	−	+	+	+	−	y_1
2	+	−	−	−	−	+	+	y_2
3	−	+	−	−	+	−	+	y_3
4	+	+	−	+	−	−	−	y_4
5	−	−	+	+	−	−	+	y_5
6	+	−	+	−	+	−	−	y_6
7	−	+	+	−	−	+	−	y_7
8	+	+	+	+	+	+	+	y_8

and minuses indicate the two levels of each factor to be investigated. Notice that each factor is studied four times at a high (+) level and four times at a low (−) level. There is a desirable and unusual balance across all $k = 7$ factors. These designs exist for $N = k + 1$, as long as N is a multiple of four. The analysis of these factorial designs is explained in Chapters 27 to 30.

Comments

An accurate measurement has no bias and high precision. Bias is systematic error that can only be removed by improving the measurement method. It cannot be averaged away by statistical manipulations. It can be assessed only when the true value of the measured quantity is known.

Precision refers to the magnitude of unavoidable random errors. Careful measurement work will minimize, but not eliminate, random error. Small random errors from different sources combine to make larger random errors in the final result. The standard deviation (s) is an index of precision (or imprecision). Large s indicates imprecise measurements. The effect of random errors can be reduced by averaging replicated measurements. Replicate measures provide the means to quantify the measurement errors and evaluate their importance.

Collaborative trials are used to check for and enforce consistent quality across laboratories. The Youden pairs plot is an excellent graphical way to report a laboratory's performance. This provides more information than reports of averages, standard deviations, and other statistics.

A ruggedness test is used to consider the effect of environmental factors on a test method. Systematic changes are made in variables associated with the test method and the associated changes in the test response are observed. The ruggedness test is done in a single laboratory so the effects are easier to see, and should precede the interlaboratory round-robin study.

References

APHA, AWWA, WEF (1998). *Standard Methods for the Examination of Water and Wastewater,* 20th ed., Clesceri, L. S., A. E. Greenberg, and A. D. Eaton, Eds.

ASTM (1990). *Standard Guide for Conducting Ruggedness Tests,* E 1169-89, Washington, D.C., U.S. Government Printing Office.

ASTM (1992). *Standard Practice for Conducting an Interlaboratory Study to Determine the Precision of a Test Method,* E691-92, Washington, D.C., U.S. Government Printing Office.

ASTM (1993). *Standard Practice for Generation of Environmental Data Related to Waste Management Activities: Quality Assurance and Quality Control Planning and Implementation,* D 5283, Washington, D.C., U.S. Government Printing Office.

Kateman, G. and L. Buydens (1993). *Quality Control in Analytical Chemistry,* 2nd ed., New York, John Wiley.

Maddelone, R. F., J. W. Scott, and J. Frank (1988). *Round-Robin Study of Methods for Trace Metal Analysis: Vols. 1 and 2: Atomic Absorption Spectroscopy — Parts 1 and 2,* EPRI CS-5910.

Maddelone, R. F., J. W. Scott, and N. T. Whiddon (1991). *Round-Robin Study of Methods for Trace Metal Analysis, Vol. 3: Inductively Coupled Plasma-Atomic Emission Spectroscopy,* EPRI CS-5910.

Maddelone, R. F., J. K. Rice, B. C. Edmondson, B. R. Nott, and J. W. Scott (1993). "Defining Detection and Quantitation Levels," *Water Env. & Tech.,* Jan., 41–44.

Mandel, J. (1964). *The Statistical Analysis of Experimental Data,* New York, Interscience Publishers.

Miller, J. C. and J. N. Miller (1984). *Statistics for Analytical Chemistry,* Chichester, England, Ellis Horwood Ltd.

Woodside, G. and D. Kocurek (1997). *Environmental, Safety, and Health Engineering,* New York, John Wiley.

Youden, W. J. (1972). *Statistical Techniques for Collaborative Tests,* Washington, D.C., Association of Official Analytical Chemists.

Youden, W. J. and E. H. Steiner (1975). *Statistical Manual of the Association of Official Analytical Chemists,* Arlington, VA, Association of Official Analytical Chemists.

Exercises

9.1 Student Accuracy. Four students (A–D) each performs an analysis in which *exactly* 10.00 mL of *exactly* 0.1M sodium hydroxide is titrated with *exactly* 0.1M hydrochloric acid. Each student performs five replicate titrations, with the results shown in the table below. Comment on the accuracy, bias, and precision of each student.

Student A	Student B	Student C	Student D
10.08	9.88	10.19	10.04
10.11	10.14	9.79	9.98
10.09	10.02	9.69	10.02
10.10	9.80	10.05	9.97
10.12	10.21	9.78	10.04

9.2 Platinum Auto Catalyst. The data below are measurements of platinum auto catalyst in standard reference materials. The known reference concentrations were low = 690 and high = 1130. Make a Youden plot and assess the work of the participating laboratories.

Laboratory	1	2	3	4	5	6	7	8	9
Low Level	700	770	705	718	680	665	685	655	615
High Level	1070	1210	1155	1130	1130	1130	1125	1090	1060

9.3 Lead Measurement. Laboratories A and B made multiple measurements on prepared wastewater effluent specimens to which lead (Pb) had been added in the amount of 1.25 µg/L or 2.5 µg/L. The background lead concentration was low, but not zero. Compare the bias and precision of the two laboratories.

Laboratory A		Laboratory B	
Spike = 1.25	Spike = 2.5	Spike = 1.25	Spike = 2.5
1.1	2.8	2.35	5.30
2.0	3.5	2.86	4.72
1.3	2.3	2.70	3.64
1.0	2.7	2.56	5.04
1.1	2.3	2.88	3.62
0.8	3.1	2.04	4.53
0.8	2.5	2.78	4.57
0.9	2.5	2.16	4.27
0.8	2.5	2.43	3.88

9.4 Split Samples. An industry and a municipal wastewater treatment plant disagreed about wastewater concentrations of total Kjeldahl nitrogen (TKN) and total suspended solids (TSS). Specimens were split and analyzed by both labs, and also by a commercial lab. The results are below. Give your interpretation of the situation.

	TKN			TSS		
Specimen	Muni.	Ind.	Comm.	Muni.	Ind.	Comm.
1	1109	940	1500	1850	1600	1600
2	1160	800	1215	2570	2100	1400
3	1200	800	1215	2080	2100	1400
4	1180	960	1155	2380	1600	1750
5	1160	1200	1120	2730	2100	2800
6	1180	1200	1120	3000	2700	2700
7	1130	900	1140	2070	1800	2000

9.5. Ruggedness Testing. Select an instrument or analytical method used in your laboratory and identify seven factors to evaluate in a ruggedness test.

10

Precision of Calculated Values

KEY WORDS *additive error, alkalinity, calcium, corrosion, dilution, error transmission, glassware, hardness, Langlier stability index, LSI, measurement error, multiplicative error, precision, propagation of variance, propagation of error, random error, relative standard deviation, RSI, Ryzner stability index, saturation index, standard deviation, systematic error, titration, variance.*

Engineers use equations to calculate the behavior of natural and constructed systems. An equation's solid appearance misleads. Some of the variables put into the equation are measurements or estimates, perhaps estimated from an experiment or from experience reported in a handbook. Some of the constants in equations, like π, are known, but most are estimated values. Most of the time we ignore the fact that the calculated output of an equation is imprecise to some extent because the inputs are not known with certainty.

In doing this we are speculating that uncertainty or variability in the inputs will not translate into unacceptable uncertainty in the output. There is no need to speculate. If the precision of each measured or estimated quantity is known, then simple mathematical rules can be used to estimate the precision of the final result. This is called *propagation of errors*. This chapter presents a few simple cases without derivation or proof. They can be derived by the general method given in Chapter 49.

Linear Combinations of Variables

The variance of a sum or difference of *independent* quantities is equal to the sum of the variances. The measured quantities, which are subject to random measurement errors, are $a, b, c, \ldots$:

$$y = a + b + c + \cdots$$

$$\sigma_y^2 = \sigma_a^2 + \sigma_b^2 + \sigma_c^2 + \cdots$$

$$\sigma_y = \sqrt{\sigma_a^2 + \sigma_b^2 + \sigma_c^2 + \cdots}$$

The signs do not matter. Thus, $y = a - b - c - \cdots$ also has $\sigma_y^2 = \sigma_a^2 + \sigma_b^2 + \sigma_c^2 + \cdots$

We used this result in Chapter 2. The estimate of the mean is the average:

$$\bar{y} = \frac{1}{n}(y_1 + y_2 + y_3 + \cdots + y_n)$$

The variance of the mean is the sum of the variances of the individual values used to calculate the average:

$$\sigma_{\bar{y}}^2 = \frac{1}{n^2}(\sigma_1^2 + \sigma_2^2 + \sigma_3^2 + \cdots + \sigma_n^2)$$

Assuming that $\sigma_1 = \sigma_2 = \cdots = \sigma_n = \sigma_y$, $\sigma_{\bar{y}} = \frac{\sigma_y}{\sqrt{n}}$, this is the standard error of the mean.

TABLE 10.1

Standard Deviations from Algebraically Combined Data

x	σ_x	x^2	$2x\sigma_x$
y	σ_y	$\sqrt{x}$	$\dfrac{\sigma_x}{2\sqrt{x}}$
$1/x$	$\dfrac{\sigma_x}{x^2}$	xy	$\sqrt{\sigma_x^2 y^2 + \sigma_y^2 x^2}$
$x+y$	$\sqrt{\sigma_x^2 + \sigma_y^2}$	x/y	$\dfrac{x}{y}\left(\dfrac{\sigma_x^2}{y^2} + \dfrac{\sigma_y^2}{x^2}\right)$
$x-y$	$\sqrt{\sigma_x^2 + \sigma_y^2}$	$\exp(ax)$	$\sqrt{a^2}\exp(ax)$
$\ln x$	$\dfrac{\sigma_x}{x}$	$\log_{10} x$	$\log_{10} e \dfrac{\sigma_x}{x}$

Another common case is the difference of two averages, as in comparative t-tests. The variance of the difference $\delta = \bar{y}_1 - \bar{y}_2$ is:

$$\sigma_\delta^2 = \sigma_{\bar{y}_1} + \sigma_{\bar{y}_2}$$

If the measured quantities are multiplied by a fixed constant:

$$y = k + k_a a + k_b b + k_c c + \cdots$$

the variance and standard deviation of y are:

$$\sigma_y^2 = k_a^2 \sigma_a^2 + k_b^2 \sigma_b^2 + k_c^2 \sigma_c^2 + \cdots$$

$$\sigma_y = \sqrt{k_a^2 \sigma_a^2 + k_b^2 \sigma_b^2 + k_c^2 \sigma_c^2 + \cdots}$$

Table 10.1 gives the standard deviation for a few examples of algebraically combined data.

Example 10.1

In a titration, the initial reading on the burette is 3.51 mL and the final reading is 15.67 mL both with standard deviation of 0.02 mL. The volume of titrant used is $V = 15.67 - 3.51 = 12.16$ mL. The variance of the difference between the two burette readings is the sum of the variances of each reading. The standard deviation of titrant volume is:

$$\sigma_V = \sqrt{(0.02)^2 + (0.02)^2} = 0.03$$

The standard deviation for the final result is larger than the standard deviations of the individual burette readings, although the volume is calculated as the difference, but it is less than the sum of the standard deviations.

Sometimes, calculations produce nonconstant variance from measurements that have constant variance. Another look at titration errors will show how this happens.

Example 10.2

The concentration of a water specimen is measured by titration as $C = 20(y_2 - y_1)$ where y_1 and y_2 are initial and final burette readings. The coefficient 20 converts milliliters of titrant used ($y_2 - y_1$) into a concentration (mg/L). Assuming the variance of a burette reading σ_y^2 is constant for all y,

Precision of Calculated Values

the variance of the computed concentration is:

$$\text{Var}(C) = \sigma_C^2 = (20)^2(\sigma_{y_2}^2 + \sigma_{y_1}^2) = 400(\sigma_y^2 + \sigma_y^2) = 800\sigma_y^2$$

Suppose that the standard deviation of a burette reading is $\sigma_y = 0.02$ mL, giving $\sigma_y^2 = 0.0004$. For $y_1 = 38.2$ and $y_2 = 25.7$, the concentration is:

$$C = 20(38.2 - 25.7) = 250 \text{ mg/L}$$

and the variance and standard deviation of concentration are:

$$\text{Var}(C) = \sigma_C^2 = (20)^2(0.0004 + 0.0004) = 0.32$$

$$\sigma_C = 0.6 \text{ mg/L}$$

Notice that the variance and standard deviation are not functions of the actual burette readings. Therefore, this value of the standard deviation holds for any difference $(y_2 - y_1)$. The approximate 95% confidence interval would be:

$$250 \pm 2(0.6) \text{ mg/L} = 250 \pm 1.2 \text{ mg/L}$$

Example 10.3

Suppose that a water specimen is diluted by a factor D before titration. $D = 2$ means that the specimen was diluted to double its original volume, or half its original concentration. This might be done, for example, so that no more than 15 mL of titrant is needed to reach the end point (so that $y_2 - y_1 \leq 15$). The estimated concentration is $C = 20D(y_2 - y_1)$ with variance:

$$\sigma_C^2 = (20D)^2(\sigma_y^2 + \sigma_y^2) = 800D^2\sigma_y^2$$

$D = 1$ (no dilution) gives the results just shown in Example 10.2. For $D > 1$, any variation in error in reading the burette is magnified by D^2. Var(C) will be uniform over a narrow range of concentration where D is constant, but it will become roughly proportional to concentration over a wider range if D varies with concentration.

It is not unusual for environmental data to have a variance that is proportional to concentration. Dilution or concentration during the laboratory processing will produce this characteristic.

Multiplicative Expressions

The propagation of error is different when variables are multiplied or divided. Variability may be magnified or suppressed. Suppose that $y = ab$. The variance of y is:

$$\sigma_y^2 = \sigma_a^2 a^2 + \sigma_b^2 b^2$$

and

$$\frac{\sigma_y^2}{y^2} = \frac{\sigma_a^2}{a^2} + \frac{\sigma_b^2}{b^2}$$

Likewise, if $y = a/b$, the variance is:

$$\sigma_y^2 = \sigma_a^2 a^2 + \sigma_b^2 \frac{a^2}{b^2}$$

and

$$\frac{\sigma_y^2}{y^2} = \frac{\sigma_a^2}{a^2} + \frac{\sigma_b^2}{b^2}$$

Notice that each term is the square of the relative standard deviation (RSD) of the variables. The RSDs are σ_y/y, σ_a/a, and σ_b/b.

These results can be generalized to any combination of multiplication and division. For:

$$y = kab/cd$$

where a, b, c and d are measured and k is a constant, there is again a relationship between the squares of the relative standard deviations:

$$\frac{\sigma_y}{y} = \sqrt{\left(\frac{\sigma_a}{a}\right)^2 + \left(\frac{\sigma_b}{b}\right)^2 + \left(\frac{\sigma_c}{c}\right)^2 + \left(\frac{\sigma_d}{d}\right)^2}$$

Example 10.4

The sludge age of an activated sludge process is calculated from $\theta = \frac{X_a V}{Q_w X_w}$, where X_a is mixed-liquor suspended solids (mg/L), V is aeration basin volume, Q_w is waste sludge flow (mgd), and X_w is waste activated sludge suspended solids concentration (mg/L). Assume $V = 10$ million gallons is known, and the relative standard deviations for the other variables are 4% for X_a, 5% for X_w, and 2% for Q_w. The relative standard deviation of sludge age is:

$$\frac{\sigma_\theta}{\theta} = \sqrt{4^2 + 5^2 + 2^2} = \sqrt{45} = 6.7\%$$

The RSD of the final result is not so much different than the largest RSD used to calculate it. This is mainly a consequence of squaring the RSDs.

Any efforts to improve the precision of the experiment need to be directed toward improving the precision of the least precise values. There is no point wasting time trying to increase the precision of the most precise values. That is not to say that small errors are unimportant. Small errors at many stages of an experiment can produce appreciable error in the final result.

Error Suppression and Magnification

A nonlinear function can either suppress or magnify error in measured quantities. This is especially true of the quadratic, cubic, and exponential functions that are used to calculate areas, volumes, and reaction rates in environmental engineering work. Figure 10.1 shows that the variance in the final result depends on the variance *and* the level of the inputs, according to the slope of the curve in the range of interest.

Precision of Calculated Values

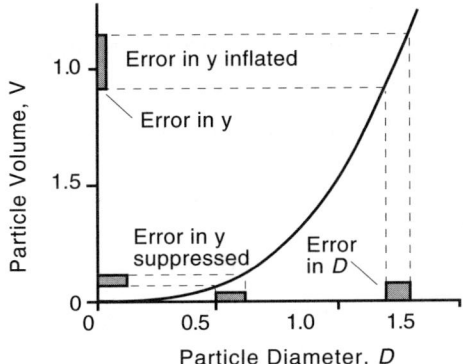

FIGURE 10.1 Errors in the computed volume are suppressed for small diameter (*D*) and inflated for large *D*.

Example 10.5

Particle diameters are to be measured and used to calculate particle volumes. Assuming that the particles are spheres, $V = \pi D^3/6$, the variance of the volume is:

$$\text{Var}(V) = \left(\frac{3\pi}{6}D^2\right)^2 \sigma_D^2 = 2.467 D^4 \sigma_D^2$$

and

$$\sigma_V = 1.571 D^2 \sigma_D$$

The precision of the estimated volumes will depend upon the measured diameter of the particles. Suppose that $\sigma_D = 0.02$ for all diameters of interest in a particular application. Table 10.2 shows the relation between the diameter and variance of the computed volumes.

At $D = 0.798$, the variance and standard deviation of volume equal those of the diameter. For small D (<0.798), errors are suppressed. For larger diameters, errors in D are magnified. The distribution of V will be stretched or compressed according to the slope of the curve that covers the range of values of D.

TABLE 10.2

Propagation of Error in Measured Particle Diameter into Error in the Computed Particle Diameter

D	0.5	0.75	0.798	1	1.25	1.5
V	0.065	0.221	0.266	0.524	1.023	1.767
σ_V^2	0.00006	0.00031	0.00040	0.00099	0.00241	0.00500
σ_V	0.008	0.018	0.020	0.031	0.049	0.071
σ_V/σ_D	0.393	0.884	1.000	1.571	2.454	3.534
σ_V^2/σ_D^2	0.154	0.781	1.000	2.467	6.024	12.491

Preliminary investigations of error transmission can be a valuable part of experimental planning. If, as was assumed here, the magnitude of the measurement error is the same for all diameters, a greater number of particles should be measured and used to estimate *V* if the particles are large.

Case Study: Calcium Carbonate Scaling in Water Mains

A small layer of calcium carbonate scale on water mains protects them from corrosion, but heavy scale reduces the hydraulic capacity. Finding the middle ground (protection without damage to pipes) is a matter of controlling the pH of the water. Two measures of the tendency to scale or corrode are the Langlier saturation index (LSI) and the Ryznar stability index (RSI). These are:

$$\text{LSI} = \text{pH} - \text{pH}_s$$

$$\text{RSI} = 2\text{pH}_s - \text{pH}$$

where pH is the measured value and pH_s the saturation value. pH is a calculated value that is a function of temperature (T), total dissolved solids concentration (TDS), alkalinity [Alk], and calcium concentration [Ca]. [Alk] and [Ca] are expressed as mg/L equivalent $CaCO_3$. The saturation pH is $\text{pH}_s = A - \log_{10}[\text{Ca}] - \log_{10}[\text{Alk}]$, where $A = 9.3 + \log_{10}(K_s/K_2) + \frac{2.5\sqrt{\mu}}{\mu + 5.5}$, in which μ is the ionic strength. K_s, a solubility product, and K_2, an ionization constant, depend on temperature and TDS.

As a rule of thumb, it is desirable to have LSI = 0.25 ± 0.25 and RSI = 6.5 ± 0.3. If LSI > 0, $CaCO_3$ scale tends to deposit on pipes, if LSI < 0, pipes may corrode (Spencer, 1983). RSI < 6 indicates a tendency to form scale; at RSI > 7.0, there is a possibility of corrosion.

This is a fairly narrow range of ideal conditions and one might like to know how errors in the measured pH, alkalinity, calcium, TDS, and temperature affect the calculated values of the LSI and RSI. The variances of the index numbers are:

$$\text{Var(LSI)} = \text{Var}(\text{pH}_s) + \text{Var}(\text{pH})$$

$$\text{Var(RSI)} = 2^2 \text{Var}(\text{pH}_s) + \text{Var}(\text{pH})$$

Given equal errors in pH and pH_s, the RSI value is more uncertain than the LSI value. Also, errors in estimating pH_s are four times more critical in estimating RSI than in estimating LSI.

Suppose that pH can be measured with a standard deviation $\sigma = 0.1$ units and pH_s can be estimated with a standard deviation of 0.15 unit. This gives:

$$\text{Var(LSI)} = (0.15)^2 + (0.1)^2 = 0.0325 \quad \sigma_{\text{LSI}} = 0.18 \text{ pH units}$$

$$\text{Var(RSI)} = 4(0.15)^2 + (0.1)^2 = 0.1000 \quad \sigma_{\text{RSI}} = 0.32 \text{ pH units}$$

Suppose further that the true index values for the water are RSI = 6.5 and LSI = 0.25. Repeated measurements of pH, [Ca], [Alk], and repeated calculation of RSI and LSI will generate values that we can expect, with 95% confidence, to fall in the ranges of:

$$\text{LSI} = 0.25 \pm 2(0.18) \quad -0.11 < \text{LSI} < 0.61$$

$$\text{RSI} = 6.5 \pm 2(0.32) \quad 5.86 < \text{RSI} < 7.14$$

These ranges may seem surprisingly large given the reasonably accurate pH measurements and pH_s estimates. Both indices will falsely indicate scaling or corrosive tendencies in roughly one out of ten calculations even when the water quality is exactly on target. A water utility that had this much variation in calculated values would find it difficult to tell whether water is scaling, stable, or corrosive until after many measurements have been made. Of course, in practice, real variations in water chemistry add to the "analytical uncertainty" we have just estimated.

Precision of Calculated Values

In the example, we used a standard deviation of 0.15 pH units for pH_s. Let us apply the same error propagation technique to see whether this was reasonable. To keep the calculations simple, assume that A, K_s, K_2, and μ are known exactly (in reality, they are not). Then:

$$\mathrm{Var}(pH_s) = (\log_{10}e)^2\{[Ca]^{-2}\mathrm{Var}[Ca] + [Alk]^{-2}\mathrm{Var}[Alk]\}$$

The variance of pH_s depends on the level of the calcium and alkalinity as well as on their variances. Assuming $[Ca] = 36$ mg/L, $\sigma_{[Ca]} = 3$ mg/L, $[Alk] = 50$ mg/L, and $\sigma_{[Alk]} = 3$ mg/L gives:

$$\mathrm{Var}(pH_s) = 0.1886\{[36]^{-2}(3)^2 + [50]^{-2}(3)^2\} = 0.002$$

which converts to a standard deviation of 0.045, much smaller than the value used in the earlier example. Using this estimate of $\mathrm{Var}(pH_s)$ gives approximate 95% confidence intervals of:

$$0.03 < LSI < 0.47$$

$$6.23 < RSI < 6.77$$

This example shows how errors that seem large do not always propagate into large errors in calculated values. But the reverse is also true. Our intuition is not very reliable for nonlinear functions, and it is useless when several equations are used. Whether the error is magnified or suppressed in the calculation depends on the function and on the level of the variables. That is, the final error is not solely a function of the measurement error.

Random and Systematic Errors

The titration example oversimplifies the accumulation of random errors in titrations. It is worth a more complete examination in order to clarify what is meant by multiple sources of variation and additive errors. Making a volumetric titration, as one does to measure alkalinity, involves a number of steps:

1. *Making up a standard solution of one of the reactants.* This involves (a) weighing some solid material, (b) transferring the solid material to a standard volumetric flask, (c) weighing the bottle again to obtain by subtraction the weight of solid transferred, and (d) filling the flask up to the mark with reagent-grade water.
2. *Transferring an aliquot of the standard material to a titration flask with the aid of a pipette.* This involves (a) filling the pipette to the appropriate mark, and (b) draining it in a specified manner into the flask.
3. *Titrating the liquid in the flask with a solution of the other reactant, added from a burette.* This involves filling the burette and allowing the liquid in it to drain until the meniscus is at a constant level, adding a few drops of indicator solution to the titration flask, reading the burette volume, adding liquid to the titration flask from the burette a little at a time until the end point is adjudged to have been reached, and measuring the final level of liquid in the burette.

The ASTM tolerances for grade A glassware are ±0.12 mL for a 250-mL flask, ±0.03 mL for a 25-mL pipette, and ±0.05 mL for a 50-mL burette. If a piece of glassware is within the tolerance, but not exactly the correct weight or volume, there will be a systematic error. Thus, if the flask has a volume of 248.9 mL, this error will be reflected in the results of all the experiments done using this flask. Repetition will not reveal the error. If different glassware is used in making measurements on different specimens, random fluctuations in volume become a random error in the titration results.

The random errors in filling a 250-mL flask might be ±0.05 mL, or only 0.02% of the total volume of the flask. The random error in filling a transfer pipette should not exceed 0.006 mL, giving an error of about 0.024% of the total volume (Miller and Miller, 1984). The error in reading a burette (of the conventional variety graduated in 0.1-mL divisions) is perhaps ±0.02 mL. Each titration involves two such readings (the errors of which are not simply additive). If the titration volume is about 25 mL, the percentage error is again very small. (The titration should be arranged so that the volume of titrant is not too small.)

In skilled hands, with all precautions taken, volumetric analysis should have a relative standard deviation of not more than about 0.1%. (Until recently, such precision was not available in instrumental analysis.)

Systematic errors can be due to calibration, temperature effects, errors in the glassware, drainage errors in using volumetric glassware, failure to allow a meniscus in a burette to stabilize, blowing out a pipette that is designed to drain, improper glassware cleaning methods, and "indicator errors." These are not subject to prediction by the propagation of error formulas.

Comments

The general propagation of error model that applies exactly to all linear models $z = f(x_1, x_2, \ldots, x_n)$ and approximately to nonlinear models (provided the relative standard deviations of the measured variables are less than about 15%) is:

$$\sigma_z^2 \approx (\partial z / \partial x_1)^2 \sigma_1^2 + (\partial z / \partial x_2)^2 \sigma_2^2 + \cdots + (\partial z / \partial x_n)^2 \sigma_n^2$$

where the partial derivatives are evaluated at the expected value (or average) of the x_i. This assumes that there is no correlation between the x's. We shall look at this and some related ideas in Chapter 49.

References

Betz Laboratories (1980). *Betz Handbook of Water Conditioning,* 8th ed., Trevose, PA, Betz Laboratories.
Langlier, W. F. (1936). "The Analytical Control of Anticorrosion in Water Treatment," *J. Am. Water Works Assoc.,* 28, 1500.
Miller, J. C. and J. N. Miller (1984). *Statistics for Analytical Chemistry,* Chichester, England, Ellis Horwood Ltd.
Ryznar, J. A. (1944). "A New Index for Determining the Amount of Calcium Carbonate Scale Formed by Water," *J. Am. Water Works Assoc.,* 36, 472.
Spencer, G. R. (1983). "Program for Cooling-Water Corrosion and Scaling," *Chem. Eng.,* Sept. 19, pp. 61–65.

Exercises

10.1 Titration. A titration analysis has routinely been done with a titrant strength such that concentration is calculated from $C = 20(y_2 - y_1)$, where $(y_2 - y_1)$ is the difference between the final and initial burette readings. It is now proposed to change the titrant strength so that $C = 40(y_2 - y_1)$. What effect will this have on the standard deviation of measured concentrations?

10.2 Flow Measurement. Two flows ($Q_1 = 7.5$ and $Q_2 = 12.3$) merge to form a larger flow. The standard deviation of measurement on flows 1 and 2 are 0.2 and 0.3, respectively. What is the standard deviation of the larger downstream flow? Does this standard deviation change when the upstream flows change?

Precision of Calculated Values

10.3 Sludge Age. In Example 10.4, reduce each relative standard deviation by 50% and recalculate the RSD of the sludge age.

10.4 Friction Factor. The Fanning equation for friction loss in turbulent flow is $\Delta p = \frac{2fV^2 L \rho}{gD}$, where Δp is pressure drop, f is the friction factor, V is fluid velocity, L is pipe length, D is inner pipe diameter, ρ is liquid density, and g is a known conversion factor. f will be estimated from experiments. How does the precision of f depend on the precision of the other variables?

10.5 F/M Loading Ratio. Wastewater treatment plant operators often calculate the food to microorganism ratio for an activated sludge process:

$$\frac{F}{M} = \frac{QS_0}{XV}$$

where Q = influent flow rate, S_0 = influent substrate concentration, X = mixed liquor suspended solids concentration, and V = aeration tank volume. Use the values in the table below to calculate the F/M ratio and a statement of its precision.

Variable	Average	Std. Error
Q = Flow (m³/d)	35000	1500
S_0 = BOD$_5$ (mg/L)	152	15
X = MLSS (mg/L)	1725	150
V = Volume (m³)	13000	600

10.6 TOC Measurements. A total organic carbon (TOC) analyzer is run by a computer that takes multiple readings of total carbon (TC) and inorganic carbon (IC) on a sample specimen and computes the average and standard deviation of those readings. The instrument also computes TOC = TC − IC using the average values, but it does not compute the standard deviation of the TOC value. Use the data in the table below to calculate the standard deviation for a sample of settled wastewater from the anaerobic reactor of a milk processing plant.

Measurement	Mean (mg/L)	Number of Replicates	Standard Deviation (mg/L)
TC	390.6	3	5.09
IC	301.4	4	4.76

10.7 Flow Dilution. The wastewater flow in a drain is estimated by adding to the upstream flow a 40,000 mg/L solution of compound A at a constant rate of 1 L/min and measuring the diluted A concentration downstream. The upstream (background) concentration of A is 25 mg/L. Five downstream measurements of A, taken within a short time period, are 200, 230, 192, 224, and 207. What is the best estimate of the wastewater flow, and what is the variance of this estimate?

10.8 Surface Area. The surface area of spherical particles is estimated from measurements on particle diameter. The formula is $A = \pi D^2$. Derive a formula for the variance of the estimated surface areas. Prepare a diagram that shows how measurement error expands or contracts as a function of diameter.

10.9 Lab Procedure. For some experiment you have done, identify the possible sources of random and systematic error and explain how they would propagate into calculated values.

11

Laboratory Quality Assurance

KEY WORDS *bias, control limit, corrective action, precision, quality assurance, quality control, range, Range chart, Shewhart chart, $\bar{X}$ (X-bar) chart, warning limit.*

Engineering rests on making measurements as much as it rests on making calculations. Soil, concrete, steel, and bituminous materials are tested. River flows are measured and water quality is monitored. Data are collected for quality control during construction and throughout the operational life of the system. These measurements need to be accurate. The measured value should be close to the true (but unknown) value of the density, compressive strength, velocity, concentration, or other quantity being measured. Measurements should be consistent from one laboratory to another, and from one time period to another.

Engineering professional societies have invested millions of dollars to develop, validate, and standardize measurement methods. Government agencies have made similar investments. Universities, technical institutes, and industries train engineers, chemists, and technicians in correct measurement techniques. Even so, it is unrealistic to assume that all measurements produced are accurate and precise. Testing machines wear out, technicians come and go, and sometimes they modify the test procedure in small ways. Chemical reagents age and laboratory conditions change; some people who handle the test specimens are careful and others are not. These are just some of the reasons why systematic checks on data quality are needed.

It is the laboratory's burden to show that measurement accuracy and precision fall consistently within acceptable limits. It is the data user's obligation to evaluate the quality of the data produced and to insist that the proper quality control checks are done. This chapter reviews how $\bar{X}$ and *Range charts* are used to check the accuracy and precision of laboratory measurements. This process is called *quality control* or *quality assurance*.

$\bar{X}$ and Range charts are graphs that show the consistency of the measurement process. Part of their value and appeal is that they are graphical. Their value is enhanced if they can be seen by all lab workers. New data are plotted on the control chart and compared against recent past performance and against the expected (or desired) performance.

Constructing X-Bar and Range Charts

The scheme to be demonstrated is based on multiple copies of prepared control specimens being inserted into the routine work. As a minimum, duplicates (two replicates) are needed. Many labs will work with this minimum number.

The first step in constructing a control chart is to get some typical data from the measurement process *when it is in a state of good statistical control*. Good statistical control means that the process is producing data that have negligible bias and high precision (small standard deviation). Table 11.1 shows measurements on 15 pairs of specimens that were collected when the system had a level and range of variation that were typical of good operation.

Simple plots of data are always useful. In this case, one might plot each measured value, the average of paired values, and the absolute value of the range of the paired values, as in Figure 11.1. These plots

TABLE 11.1

Fifteen Pairs of Measurements on Duplicate Test Specimens

Specimen	1	2	3	4	5	6	7	8	9	10	11	12	13	14	15
X_1	5.2	3.1	2.5	3.8	4.3	3.1	4.5	3.8	4.3	5.3	3.6	5.0	3.0	4.7	3.7
X_2	4.4	4.6	5.3	3.7	4.4	3.3	3.8	3.2	4.5	3.7	4.4	4.8	3.6	3.5	5.2
$\bar{X} = \dfrac{X_1 + X_2}{2}$	4.8	3.8	3.9	3.8	4.3	3.2	4.2	3.5	4.4	4.5	4.0	4.9	3.3	4.1	4.4
$R = \|X_1 - X_2\|$	0.8	1.5	2.8	0.1	0.1	0.2	0.7	0.6	0.2	1.6	0.8	0.2	0.6	1.2	1.5

Grand mean = $\bar{\bar{X}}$ = 4.08, Mean sample range = $\bar{R}$ = 0.86

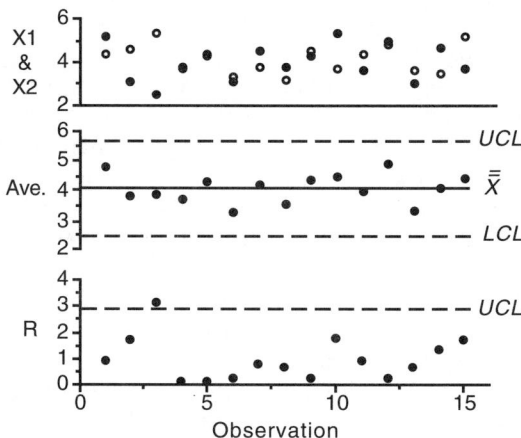

FIGURE 11.1 Three plots of the 15 pairs of quality control data with action and warning limits added to the charts for the average and range of X_1 and X_2.

show the typical variation of the measurement process. Objectivity is increased by setting *warning limits* and *action limits* to define an unusual condition so all viewers will react in the same way to the same signal in the data.

The two simplest control charts are the $\bar{X}$ (pronounced X-bar) chart and the Range (R) chart. The $\bar{X}$ chart (also called the Shewhart chart, after its inventor) provides a check on the process *level* and also gives some information about variation. The Range chart provides a check on *precision* (variability).

The acceptable variation in level and precision is defined by control limits that bound a specified percentage of all results expected as long as the process remains in control. A common specification is 99.7% of values within the control limits. Values falling outside these limits are unusual enough to activate a review of procedures because the process may have gone wrong. These control limits are valid only when the variation is random above and below the average level.

The equations for calculating the control limits are:

$\bar{X}$ chart   Central line = $\bar{\bar{X}}$

Control limits = $\bar{\bar{X}} \pm k_1 \bar{R}$

R chart   Central line = $\bar{R}$

Upper control limit (UCL) = $k_2 \bar{R}$

where $\bar{\bar{X}}$ is the grand mean of sample means (the average of the $\bar{X}$ values used to construct the chart), $\bar{R}$ is the mean sample range (the average of the ranges [R] used to construct the chart), and n is the number of replicates used to compute the average and the range at each sampling interval. R is the absolute difference between the largest and smallest values in the subset of n measured values at a particular sampling interval.

TABLE 11.2

Coefficients for Calculating Action Lines on $\bar{X}$ and Range Charts

n	k_1	k_2
2	1.880	3.267
3	1.023	2.575
4	0.729	2.282
5	0.577	2.115

Source: Johnson, R. A. (2000). *Probability and Statistics for Engineers,* 6th ed., Englewood Cliffs, NJ, Prentice-Hall.

The coefficients of k_1 and k_2 depend on the size of the subsample used to calculate $\bar{X}$ and R. A few values of k_1 and k_2 are given in Table 11.2. The term $k_1 \bar{R}$ is an unbiased estimate of the quantity $3\sigma/\sqrt{n}$, which is the half-length of a 99.7% confidence interval. Making more replicate measurements will reduce the width of the control lines.

The control charts in Figure 11.1 were constructed using values measured on two test specimens at each sampling time. The average of the two measurements, X_1 and X_2, is $\bar{X}$; and the range R is the absolute difference of the two values. The average of the 15 pairs of X values is $\bar{\bar{X}} = 4.08$. The average of the absolute range values is $\bar{R} = 0.86$. There are $n = 2$ observations used to calculate each $\bar{X}$ and R value. For the data in the Table 11.1 example, the action limits are:

$$\bar{X} \text{ action limits} = 4.08 \pm 1.880(0.86) = 4.08 \pm 1.61$$

$$\text{UCL} = 5.7 \qquad \text{LCL} = 2.5$$

The upper action limit for the range chart is:

$$R \text{ chart UCL} = 3.267(0.86) = 2.81$$

Usually, the value of $\bar{R}$ is not shown on the chart. We show no lower limits on a range chart because we are interested in detecting variability that is too large.

Using the Charts

Now examine the performance of a control chart for a simulated process that produced the data shown in Figure 11.2: the $\bar{X}$ chart and Range charts were constructed using duplicate measurements from the first 20 observation intervals when the process was in good control with $\bar{X} = 10.2$ and $\bar{R} = 0.54$. The $\bar{X}$ action limits are at 9.2 and 11.2. The $\bar{R}$ action limit is at 1.8. The action limits were calculated using the equations given in the previous section.

As new values become available, they are plotted on the control charts. At times 22 and 23 there are values above the upper $\bar{X}$ action limit. This signals a request to examine the measurement process to see if something has changed. (Values below the lower $\bar{X}$ action limit would also signal this need for action.) The R chart shows that process variability seems to remain in control although the level has shifted upward. These conditions of "high level" and "normal variability" continue until time 35 when the process level drops back to normal and the R chart shows increased variability.

The data in Figure 11.2 were simulated to illustrate the performance of the charts. From time 21 to 35, the level was increased by one unit while the variability was unchanged from the first 20-day period. From time 36 to 50, the level was at the original level (in control) and the variability was doubled. This example shows that control charts do not detect changes immediately and they do not detect every change that occurs.

Warning limits at $\bar{\bar{X}} \pm 2\sigma/\sqrt{n}$ could be added to the $\bar{X}$ chart. These would indicate a change sooner and more often that the action limits. The process will exceed warning limits approximately one time out of twenty when the process is in control. This means that one out of twenty indications will be a *false alarm*.

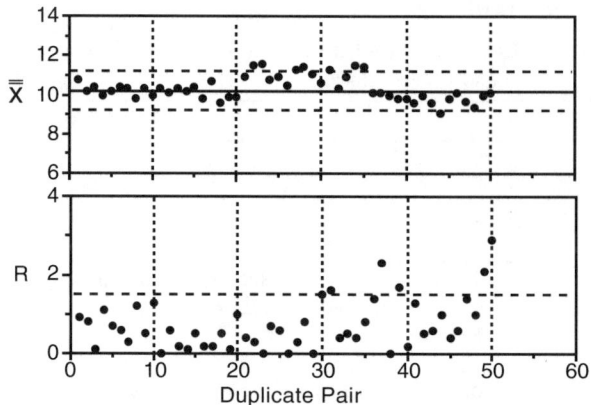

FIGURE 11.2 Using the quality control chart of duplicate pairs for process control. The level changes by one unit from time 21 to 35 while the variability is unchanged. From time 36 to 50, the level goes back to normal and the variability is doubled.

(A false alarm is an indication that the process is out of control when it really is not). The action limits give fewer false alarms (approximately 1 in 300). A compromise is to use both warning limits and action limits. A warning is not an order to start changing the process, but it could be a signal to run more quality control samples.

We could detect changes more reliably by making three replicate measurements instead of two. This will reduce the width of the action limits by about 20%.

Reacting to Unacceptable Conditions

The laboratory should maintain records of out-of-control events, identified causes of upsets, and corrective actions taken. The goal is to prevent repetition of problems, including problems that are not amenable to control charting (such as loss of sample, equipment malfunction, excessive holding time, and sample contamination).

Corrective action might include checking data for calculation or transcription errors, checking calibration standards, and checking work against standard operating procedures.

Comments

Quality assurance checks on measurement precision and bias are essential in engineering work. Do not do business with a laboratory that lacks a proper quality control program. A good laboratory will be able to show you the control charts, which should include $\bar{X}$ and Range charts on each analytical procedure. Charts are also kept on calibration standards, laboratory-fortified blanks, reagent blanks, and internal standards.

Do not trust quality control entirely to a laboratory's own efforts. Submit your own quality control specimens (known standards, split samples, or spiked samples). Submit these in a way that the laboratory cannot tell them from the routine test specimens in the work stream. If you send test specimens to several laboratories, consider Youden pairs (Chapter 9) as a way of checking for interlaboratory consistency. You pay for the extra analyses needed to do quality control, but it is a good investment. Shortcuts on quality do ruin reputations, but they do not save money.

The term "quality control" implies that we are content with a certain level of performance, the level that was declared "in control" in order to construct the control charts. A process that is in statistical control

Laboratory Quality Assurance

can be improved. Precision can be increased. Bias can be reduced. Lab throughput can be increased while precision and bias remain in control. Strive for quality assurance and *quality improvement*.

References

Johnson, R. A. (2000). *Probability and Statistics for Engineers,* 6th ed., Englewood Cliffs, NJ, Prentice-Hall.
Kateman, G. and L. Buydens (1993). *Quality Control in Analytical Chemistry,* 2nd ed., New York, John Wiley.
Miller, J. C. and J. N. Miller (1984). *Statistics for Analytical Chemistry,* Chichester, England, Ellis Horwood Ltd.
Tiao, George, et al., Eds. (2000). *Box on Quality and Discovery with Design, Control, and Robustness,* New York, John Wiley & Sons.

Exercises

11.1 Glucose BOD Standards. The data below are 15 paired measurements on a standard glucose/glutamate mixture that has a theoretical BOD of 200 mg/L. Use these data to construct a Range chart and an $\bar{X}$ chart.

1	2	3	4	5	6	7	8	9	10	11	12	13	14	15
203	213	223	205	209	200	200	196	201	206	192	206	185	199	201
206	196	214	189	205	201	226	207	214	210	207	188	199	198	200

11.2 BOD Range Chart. Use the Range chart developed in Exercise 11.1 to assess the precision of the paired BOD data given in Exercise 6.2.

12

Fundamentals of Process Control Charts

KEY WORDS *action limits, autocorrelation, control chart, control limits, cumulative sum, Cusum chart, drift, EWMA, identifiable variability, inherent variability, mean, moving average, noise, quality target, serial correlation, Shewhart chart, Six Sigma, specification limit, standard deviation, statistical control, warning limits, weighted average.*

Chapter 11 showed how to construct control charts to assure high precision and low bias in laboratory measurements. The measurements were assumed to be on independent specimens and to have normally distributed errors; the quality control specimens were managed to satisfy these conditions. The laboratory system can be imagined to be in a state of statistical control with random variations occurring about a fixed mean level, except when special problems intervene. A water or wastewater treatment process, or a river monitoring station will not have these ideal statistical properties. Neither do most industrial manufacturing systems. Except as a temporary approximation, random and normally distributed variation about a fixed mean level is a false representation. For these systems to remain in a fixed state that is affected only by small and purely random variations would be a contradiction of the second law of thermodynamics. A statistical scheme that goes against the second law of thermodynamics has no chance of success. One must expect a certain amount of drift in the treatment plant or the river, and there also may be more or less cyclic seasonal changes (diurnal, weekly, or annual). The statistical name for drift and seasonality is *serial correlation* or *autocorrelation*. Control charts can be devised for these more realistic conditions, but that is postponed until Chapter 13.

The industrial practitioners of *Six Sigma* programs[1] make an allowance of 1.5 standard deviations for process drift on either side of the target value. This drift, or long-term process instability, remains even after standard techniques of quality control have been applied. *Six Sigma* refers to the action limits on the control charts. One sigma (σ) is one standard deviation of the random, independent process variation. Six Sigma action limits are set at 6σ above and 6σ below the average or target level. Of the 6σ, 4.5σ are allocated to random variation and 1.5σ are allocated to process drift. This allocation is arbitrary, because the drift in a real process may be more than 1.5σ (or less), but making an allocation for drift is a large step in the right direction. This does not imply that standard quality control charts are useless, but it does mean that standard charts can fail to detect real changes *at the stated probability level* because they will see the drift as cause for alarm.

What follows is about *standard* control charts for *stable* processes. The assumptions are that variation is random about a fixed mean level and that changes in level are caused by some identifiable and removable factor. Process drift is not considered. This is instructive, if somewhat unrealistic.

Standard Control Chart Concepts

The greatest strength of a control chart is that it is a *chart*. It is a graphical guide to making process control decisions. The chart gives the process operator information about (1) how the process has been operating, (2) how the process is operating currently, and (3) provides an opportunity to infer from this information how the process may behave in the future. New observations are compared against a picture

---

[1] *Six Sigma* is the name for the statistical quality and productivity improvement programs used by such companies as Motorola, General Electric, Texas Instruments, Polaroid, and Allied Signal.

of typical performance. If typical performance were random variation about a fixed mean, the picture can be a classical control chart with warning limits and action limits drawn at some statistically defined distance above and below the mean (e.g., three standard deviations). Obviously, the symmetry of the action limits is based on assuming that the random fluctuations are normally distributed about the mean. A current observation outside control limits is presumptive evidence that the process has changed (is out of control), and the operator is expected to determine what has changed and what adjustment is needed to bring the process into acceptable performance.

This could be done without plotting the results on a chart. The operator could compare the current observation with two numbers that are posted on a bulletin board. A computer could log the data, make the comparison, and also ring an alarm or adjust the process. Eliminating the chart takes the human element out of the control scheme, and this virtually eliminates the elements of *quality improvement* and *productivity improvement*. The chart gives the human eye and brain a chance to recognize new patterns and stimulate new ideas.

A simple chart can incorporate rules for detecting changes other than "the current observations falls outside the control limits." If deviations from the fixed mean level have a normal distribution, and if each observation is independent and all measurements have the same precision (variance), the following are unusual occurrences:

1. One point beyond a 3σ control limit (odds of 3 in 1000)
2. Nine points in a row falling on one side of the central line (odds of 2 in 1000)
3. Six points in a row either steadily increasing or decreasing
4. Fourteen points in a row alternating up and down
5. Two out of three consecutive points more than 2σ from the central line
6. Four out of five points more than 1σ from the central line
7. Fifteen points in a row within 1σ of the central line both above and below
8. Eight points in a row on either side of the central line, none falling within 1σ of the central line

Variation and Statistical Control

Understanding variation is central to the theory and use of control charts. Every process varies. Sources of variation are numerous and each contributes an effect on the system. Variability will have two components; each component may have subcomponents.

1. Inherent variability results from common causes. It is characteristic of the process and can not be readily reduced without extensive change of the system. Sometimes this is called the *noise* of the system.
2. Identifiable variability is directly related to a specific cause or set of causes. These sometimes are called "assignable causes."

The purpose of control charts is to help identify periods of operation when assignable causes exist in the system so that they may be identified and eliminated. A process is in a *state of statistical control* when the assignable causes of variation have been detected, identified, and eliminated.

Given a process operating in a state of statistical control, we are interested in determining (1) when the process has changed in mean level, (2) when the process variation about that mean level has changed and (3) when the process has changed in both mean level and variation.

To make these judgments about the process, we must assume future observations (1) are generated by the process in the same manner as past observations, and (2) have the same statistical properties as past observations. These assumptions allow us to set control limits based on past performance and use these limits to assess future conditions.

There is a difference between "out of control" and "unacceptable process performance." A particular process may operate in a state of statistical control but fail to perform as desired by the operator. In this case, the system must be changed to improve the system performance. Using a control chart to bring it into statistical control solves the wrong problem. Alternatively, a process may operate in a way that is acceptable to the process operator, and yet from time to time be statistically out of control. A process is not necessarily in statistical control simply because it gives acceptable performance as defined by the process operator. Statistical control is defined by control limits. Acceptable performance is defined by *specification limits* or *quality targets*—the level of quality the process is supposed to deliver. Specification limits and control chart limits may be different.

Decision Errors

Control charts do not make perfect decisions. Two types of errors are possible:

1. Declare the process "out of control" when it is not.
2. Declare the process "in control" when it is not.

Charts can be designed to consider the relative importance of committing the two types of errors, but we cannot eliminate these two kinds of errors. We cannot simultaneously guard entirely against both kinds of errors. Guarding against one kind increases susceptibility to the other. Balancing these two errors is as much a matter of policy as of statistics.

Most control chart methods are designed to minimize falsely judging that an in-control process is out of control. This is because we do not want to spend time searching for nonexistent assignable causes or to make unneeded adjustments in the process.

Constructing a Control Chart

The first step is to describe the underlying statistical process of the system when it is in a state of statistical control. This description will be an equation. In the simplest possible case, like the ones studied so far, the process model is a straight horizontal line and the equation is:

$$\text{Observation} = \text{Fixed mean} + \text{Independent random error}$$

or

$$y_t = \eta + e_t$$

If the process exhibits some drift, the model needs to be expanded:

$$\text{Observation} = \text{Function of prior observations} + \text{Independent random error}$$

$$y_t = f(y_{t-1}, y_{t-2}, \ldots) + e_t$$

or

$$\text{Observation} = \text{Function of prior observations} + \text{Dependent error}$$

$$y_t = f(y_{t-1}, y_{t-2}, \ldots) + g(e_t, e_{t-1}, \ldots)$$

These are problems in time series analysis. Models of this kind are explained in Tiao et al. (2000) and Box and Luceno (1997). An exponentially weighted moving average will describe certain patterns of drift. Chapters 51 and 53 deal briefly with some relevant topics.

Once the typical underlying pattern (the inherent variability) has been described, the statistical properties of the deviations of observations from this typical pattern need to be characterized. If the deviations are random, independent, and have constant variance, we can construct a control chart that will examine these deviations. The average value of the deviations will be zero, and symmetrical control limits, calculated in the classical way, can be drawn above and below zero.

The general steps in constructing a control chart are these:

1. Sample the process at specific times (t, $t-1$, $t-2$,...) to obtain ... y_t, y_{t-1}, and y_{t-2}. These typically are averages of subgroups of n observations, but they may be single observations.
2. Calculate a quantity V_t, which is a function of the observations. The definition of V_t depends on the type of control chart.
3. Plot values V_t in a time sequence on the control chart.
4. Using appropriate control limits and rules, plot new observations and decide whether to take corrective action or to investigate.

Kinds of Control Charts

What has been said so far is true for control charts of all kinds. Now we look at the Shewhart[2] chart (1931), cumulative sum chart (Cusum), and moving average charts. Moving averages were used for smoothing in Chapter 4.

Shewhart Chart

The *Shewhart chart* is used to detect a change in the level of a process. It does not indicate a change in the variability. A Range chart (Chapter 11) is often used in conjunction with a Shewhart or other chart that monitors process level.

The quantity plotted on the Shewhart chart at each recording interval is an average, $\bar{y}_t$, of the subgroup of n observations y_t made at time t to calculate:

$$V_t = \bar{y}_t = \frac{1}{n}\sum_{i=1}^{n} y_t$$

If only one observation is made at time t, plot $V_t = y_t$. This is an *I*-chart (*I* for individual observation) instead of an $\bar{X}$ chart. Making only one observation at each sampling reduces the power of the chart to detect a shift in performance.

The central line on the control chart measures the general level of the process (i.e., the long-term average of the process). The upper control limit is drawn at $3s$ above the central control line; the lower limit is $3s$ below the central line. s is the standard error of averages of n observations used to calculate the average value at time t. This is determined from measurements made over a period of time when the process is in a state of stable operation.

Cumulative Sum Chart

The *cumulative sum*, or *Cusum, chart* is used to detect a change in the level of the process. It does not indicate a change in the variability. The Cusum chart will detect a change sooner (in fewer sampling intervals) than a Shewhart chart. It is the best chart for monitoring changes in process level.

---

[2] In Chapter 10, Shewhart charts were also called $\bar{X}$ (*X*-bar) charts and *X* was the notation used to indicate a measurement from a laboratory quality control setting. In all other parts of the book, we have used *y* to indicate the variable. Because the term *Y*-bar chart is not in common use and we wish to use *y* instead of *x*, in this chapter we will call these *X*-bar charts Shewhart charts.

Fundamentals of Process Control Charts

Cumulative deviations from T, the mean or target level of the process, are plotted on the chart. The target T is usually the average level of the process determined during some period when the process was in a stable operating condition. The deviation at time t is $y_t - T$. At time $t - 1$, the deviation is $y_{t-1} - T$, and so on. These are summed from time $t = 1$ to the current time t, giving the cumulative sum, or Cusum:

$$V_t = \sum_{t=1}^{t}(y_t - T)$$

If the process performance is stable, the deviations will vary randomly about zero. The sum of the deviations from the target level will average zero, and the cumulative sum of the deviations will drift around zero. There is no general trend either up or down.

If the mean process performance shifts upward, the deviations will include more positive values than before and the Cusum will increase. The values plotted on the chart will show an upward trend. Likewise, if the mean process performance shifts downward, the Cusum will trend downward.

The Cusum chart gives a lot of useful information even without control limits. The time when the change occurred is obvious. The amount by which the mean has shifted is the slope of the line after the change has occurred.

The control limits for a Cusum chart are not parallel lines as in the Shewhart chart. An unusual amount of change is judged using a V-Mask (Page, 1961). The V-Mask is placed on the control chart horizontally such that the apex is located a distance d from the current observation. If all previous points fall within the arms of the V-Mask, the process is in a state of statistical control.

Moving Average Chart

Moving average charts are useful when the single observations themselves are used. If the process has operated at a constant level with constant variance, the moving average gives essentially the same information as the average of several replicate observations at time t.

The moving average chart is based on the average of the k most recent observations. The quantity to be plotted is:

$$V_t = \frac{1}{k} \sum_{t-(k-1)}^{t} y_t$$

The central control line is the average for a period when the process performance is in stable control. The control limits are at distances $\pm 3 \frac{s}{\sqrt{k}}$, assuming single observations at each interval.

Exponentially Weighted Moving Average Chart

The exponentially weighted moving average (EWMA) chart is a plot of the weighted sum of all previous observations:

$$V_t = (1 - \lambda) \sum_{i=0}^{i} \lambda^i y_{t-i}$$

The EWMA control chart is started with $V_0 = T$, where T is the target or long-term average. A convenient updating equation is:

$$V_t = (1 - \lambda) y_t + \lambda V_{t-1}$$

The control limits are $\pm 3s \left(\frac{\lambda}{2 - \lambda}\right)^\lambda$.

The weight λ is a value less than 1.0, and often in the range 0.1 to 0.5. The weights decay exponentially from the current observation into the past. The current observation has weight $1 - \lambda$, the previous has weight $(1 - \lambda)\lambda$, the observation before that $(1 - \lambda)\lambda^2$, and so on. The value of λ determines the weight placed on the observations in the EWMA. A small value of λ gives a large weight to the current observation and the average does not remember very far into the past. A large value of λ gives a weighted average with a long memory. In practice, a weighted average with a long memory is dominated by the most recent four to six observations.

Comparison of the Charts

Shewhart, Cusum, Moving Average, and EWMA charts (Figures 12.1 to 12.3) differ in the way they weight previous observations. The Shewhart chart gives all weight to the current observation and no weight to all previous observations. The Cusum chart gives equal weight to all observations. The moving average chart gives equal weight to the k most recent observations and zero weight to all other observations. The EWMA chart gives the most weight to the most recent observation and progressively smaller weights to previous observations.

Figure 12.1 shows a Shewhart chart applied to duplicate observations at each interval. Figures 12.2 and 12.3 show Moving Average and EWMA, and Cusum charts applied to the data represented by open points in Figure 12.1. The Cusum chart gives the earliest and clearest signal of change.

The Shewhart chart needs no explanation. The first few calculations for the Cusum, MA(5), and EWMA charts are in Table 12.1. Columns 2 and 3 generate the Cusum using the target value of 12. Column 4 is the 5-day moving average. The EWMA (column 5) uses $\lambda = 0.5$ in the recursive updating formula starting from the target value of 12. The second row of the EWMA is $0.5(11.89) + 0.5(12.00) = 12.10$, the third row is $0.5(12.19) + 0.5(12.10) = 12.06$, etc.

No single chart is best for all situations. The Shewhart chart is good for checking the statistical control of a process. It is not effective unless the shift in level is relatively large compared with the variability.

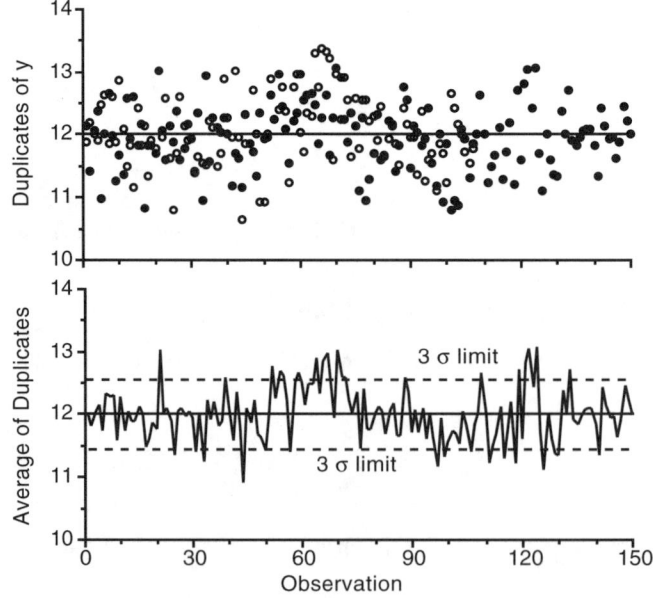

FIGURE 12.1 A Shewhart chart constructed using simulated duplicate observations (top panel) from a normal distribution with mean = 12 and standard deviation = 0.5. The mean level shifts up by 0.5 units from days 50–75, it is back to normal from days 76–92, it shifts down by 0.5 units from days 93–107, and is back to normal from day 108 onward.

Fundamentals of Process Control Charts 109

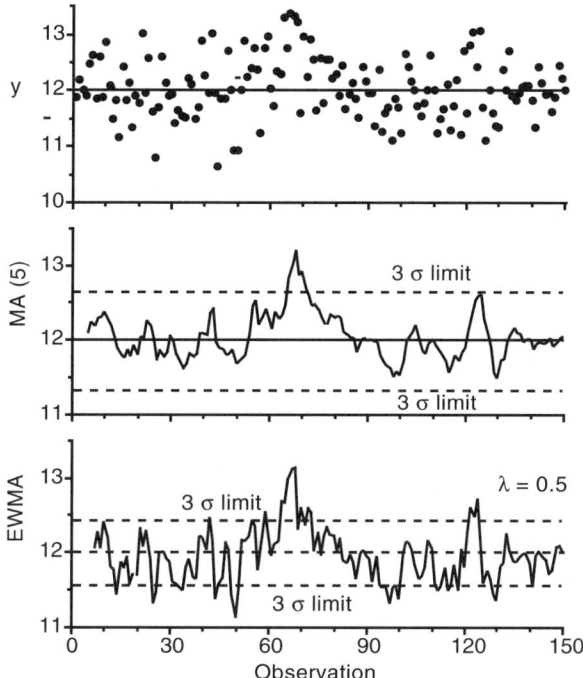

FIGURE 12.2 Moving average (5-day) and exponentially weighted moving average ($\lambda = 0.5$) charts for the single observations shown in the top panel. The mean level shifts up by 0.5 units from days 50–75, it is back to normal from days 76–92, it shifts down by 0.5 units from days 93–107, and is back to normal from day 108 onward.

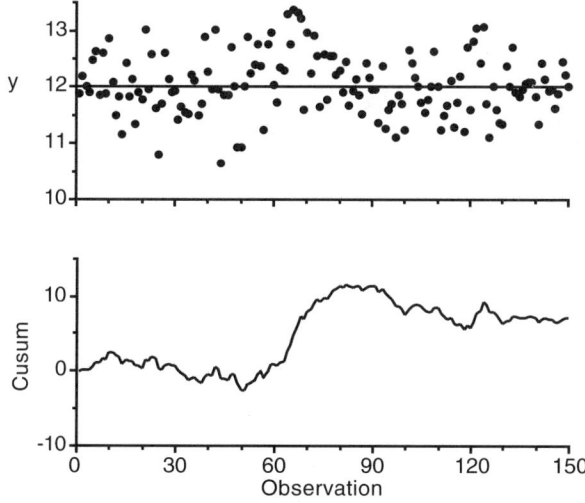

FIGURE 12.3 Cusum chart for the single observations in the top panel (also the top panel of Figure 12.2). The mean level shifts up by 0.5 units from day 50–75, it is back to normal from days 76–92, it shifts down by 0.5 units from days 93–107, and is back to normal from day 108 onward. The increase is shown by the upward trend that starts at day 50, the decrease is shown by the downward trend starting just after day 90. The periods of normal operation (days 1–50, 76–92, and 108–150) are shown by slightly drifting horizontal pieces.

TABLE 12.1

Calculations to Start the Control Charts for the Cusum, 5-Day Moving Average, and the Exponentially Weighted Moving Average ($\lambda = 0.5$)

(1) y_i	(2) $y_i - 12$	(3) Cusum	(4) MA(5)	(5) EWMA ($\lambda = 0.5$)
11.89	−0.11	−0.11		12.00
12.19	0.19	0.08		12.10
12.02	0.02	0.10		12.06
11.90	−0.10	0.00		11.98
12.47	0.47	0.47	12.09	12.22
12.64	0.64	1.11	12.24	12.43
11.86	−0.14	0.97	12.18	12.15
12.61	0.61	1.57	12.29	12.38
11.89	−0.11	1.47	12.29	12.13
12.87	0.87	2.33	12.37	12.50
12.09	0.09	2.42	12.26	12.30
11.50	−0.50	1.93	12.19	11.90
11.84	−0.16	1.76	12.04	11.87
11.17	−0.83	0.93	11.89	11.52

The Cusum chart detects small departures from the mean level faster than the other charts. The moving average chart is good when individual observations are being used (in comparison to the Shewhart chart in which the value plotted at time t is the average of a sample of size n taken at time t). The EWMA chart provides the ability to take into account serial correlation and drift in the time series of observations. This is a property of most environmental data and these charts are worthy of further study (Box and Luceno, 1997).

Comments

Control charts are simplified representations of process dynamics. They are not foolproof and come with the following caveats:

- Changes are not immediately obvious.
- Large changes are easier to detect than a small shift.
- False alarms do happen.
- Control limits in practice depend on the process data that is collected to construct the chart.
- Control limits can be updated and verified as more data become available.
- Making more than one measurement and averaging brings the control limits closer together and increases monitoring sensitivity.

The adjective "control" in the name control charts suggests that the best applications of control charts are on variables that can be changed by adjusting the process and on processes that are critical to saving money (energy, labor, or materials). This is somewhat misleading because some applications are simply monitoring without a direct link to control. Plotting the quality of a wastewater treatment effluent is a good idea, and showing some limits of typical or desirable performance is alright. But putting control limits on the chart does not add an important measure of process control because it provides no useful information about which factors to adjust, how much the factors should be changed, or how often they should be changed. In contrast, control charts on polymer use, mixed liquor suspended solids, bearing temperature, pump vibration, blower pressure, or fuel consumption may avoid breakdowns and upsets, and they may save money. Shewhart and Cusum charts are recommended for groundwater monitoring programs (ASTM, 1998).

The idea of using charts to assist operation is valid in all processes. Plotting the data in different forms — as time series, Cusums, moving averages — has great value and will reveal most of the important information to the thoughtful operator. Charts are not inferior or second-class statistical methods. They reflect the best of control chart philosophy without the statistical complications. They are statistically valid, easy to use, and not likely to lead to any serious misinterpretations.

Control charts, with formal action limits, are only dressed-up graphs. The control limits add a measure of objectivity, provided they are established without violating the underlying statistical conditions (independence, constant variance, and normally distributed variations). If you are not sure how to derive correct control limits, then use the charts without control limits, or construct an external reference distribution (Chapter 6) to develop approximate control limits. Take advantage of the human ability to recognize patterns and deviations from trends, and to reason sensibly.

Some special characteristics of environmental data include serial correlation, seasonality, nonnormal distributions, and changing variance. Nonnormal distribution and nonconstant variance can usually be handled with a transformation. Serial correlation and seasonality are problems because control charts are sensitive to these properties. One way to deal with this is the Six Sigma approach of arbitrarily widening the control limits to provide a margin for drift.

The next chapter deals with special control charts. Cumulative score charts are an extension of Cusum charts that can detect cyclic patterns and shifts in the parameters of models. Exponentially weighted moving average charts can deal with serial correlation and process drift.

References

ASTM (1998). Standard Guide for Developing Appropriate Statistical Approaches for Groundwater Detection Monitoring Programs, Washington, D.C., D 6312 , U.S. Government Printing Office.
Berthouex, P. M., W. G. Hunter, and L. Pallesen (1978). "Monitoring Sewage Treatment Plants: Some Quality Control Aspects," *J. Qual. Tech.,* 10(4).
Box, G. E. P. and A. Luceno (1997). *Statistical Control by Monitoring and Feedback Adjustment,* New York, Wiley Interscience.
Box, G. E. P. and L. Luceno (2000). "Six Sigma, Process Drift, Capability Indices, and Feedback Adjustment," *Qual. Engineer.,* 12(3), 297–302.
Page, E. S. (1961). "Continuous Inspection Schemes," *Biometrika,* 41, 100–115.
Page, E. S. (1961). "Cumulative Sum Charts," *Technometrics,* 3, 1–9.
Shewhart, W. A. (1931). *Economic Control of Quality of Manufacturing Product,* Princeton, NJ, Van Nostrand Reinhold.
Tiao, G. et al., Eds. (2000). *Box on Quality and Discovery with Design, Control, and Robustness,* New York, John Wiley & Sons.

Exercises

12.1 Diagnosing Upsets. Presented in the chapter are eight simple rules for defining an "unusual occurrence." Use the rules to examine the data in the accompanying chart. The average level is 24 and $\sigma = 1$.

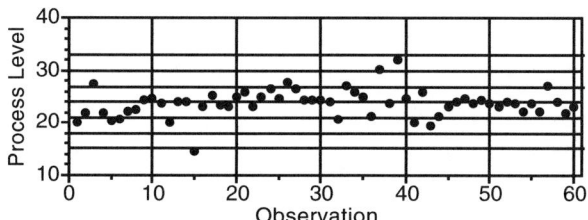

12.2 Charting. Use the first 20 duplicate observations in the data set below to construct Shewhart, Range, and Cusum charts. Plot the next ten observations and decide whether the process has remained in control. Compare the purpose and performance of the three charts.

Observ.	y_1	y_2	Observ.	y_1	y_2
1	5.88	5.61	16	5.70	5.96
2	5.64	5.63	17	4.90	5.65
3	5.09	5.12	18	5.40	6.71
4	6.04	5.36	19	5.32	5.67
5	4.66	5.24	20	4.86	4.34
6	5.58	4.50	21	6.01	5.57
7	6.07	5.41	22	5.55	5.55
8	5.31	6.30	23	5.44	6.40
9	5.48	5.83	24	5.05	5.72
10	6.63	5.23	25	6.04	4.62
11	5.28	5.91	26	5.63	4.62
12	5.97	5.81	27	5.67	5.70
13	5.82	5.19	28	6.33	6.58
14	5.74	5.41	29	5.94	5.94
15	5.97	6.60	30	6.68	6.09

12.3 Moving Averages. Use the first 20 duplicate observations in the Exercise 12.2 data set to construct an MA(4) moving average chart and an EWMA chart for $\lambda = 0.6$. Plot the next ten observations and decide whether the process has remained in control. Compare the purpose and performance of the charts.

13

Specialized Control Charts

KEY WORDS   *AR model, autocorrelation, bump disturbance, control chart, Cusum, Cuscore, cyclic variation, discrepancy vector, drift, EWMA, IMA model, linear model, moving average, process monitoring, random variation, rate of increase, serial correlation, Shewhart chart, sine wave disturbance, slope, spike, weighted average, white noise.*

Charts are used often for process monitoring and sometimes for process control. The charts used for these different objectives take different forms. This chapter deals with the situation where the object is not primarily to regulate but to monitor the process. The monitoring should verify the continuous stability of the process once the process has been brought into a state of statistical control. It should detect deviations from the stable state so the operator can start a search for the problem and take corrective actions.

The classical approach to this is the Shewhart chart. A nice feature of the Shewhart chart is that it is a direct plot of the actual data. Humans are skilled at extracting information from such charts and they can sometimes discover process changes of a totally unexpected kind. However, this characteristic also means that the Shewhart chart will not be as sensitive to some *specific* deviation from randomness as another specially chosen chart can be. When a specific kind of deviation is feared, a chart is needed that is especially sensitive to that kind of deviation. This chart should be used *in addition* to the Shewhart chart. The Page-Barnard cumulative sum (Cusum) chart is an example of a specialized control chart. It is especially sensitive to small changes in the mean level of a process, as indicated by the change in slope of the Cusum plot. The Cusum is one example of a *Cuscore statistic*.

The Cuscore Statistic

Consider a statistical model in which the y_t are observations, θ is some unknown parameter, and the x_t's are known independent variables. This can be written in the form:

$$y_t = f(x_t, \theta) + a_t \quad t = 1, 2, \ldots, n$$

Assume that when θ is the true value of the unknown parameter, the resulting a_t's are a sequence of independently, identically, normally distributed random variables with mean zero and variance $\sigma_a^2 = \sigma^2$. The series of a_t's is called a white noise sequence. The model is a way of reducing data to white noise:

$$a_t = y_t - f(x_t, \theta)$$

The Cusum chart is based on the simplest possible model, $y_t = \eta + a_t$. As long as the process is in control (varying randomly about the mean), subtracting the mean reduces the series of y_t to a series of white noise. The cumulative sum of the white noise series is the Cusum statistic and this is plotted on the Cusum chart. In a more general way, the Cusum is a *Cuscore* that relates how the residuals change with respect to changes in the mean (the parameter η).

Box and Ramirez (1992) defined the *Cuscore* associated with the parameter value $\theta = \theta_0$ as:

$$Q = \sum a_{t0} d_{t0}$$

where $d_{t0} = -\left.\frac{\partial a_t}{\partial \theta}\right|_{\theta=\theta_0}$ is a discrepancy vector related to how the residuals change with respect to changes in θ. The subscripts (0) indicate the reference condition for the in-control process. When the process is in control, the parameter has the value θ_0 and the process model generates residuals a_t. If the process shifts out of control, $\theta \neq \theta_0$ and the residuals are inflated as described by[1]:

$$a_{t0} = (\theta - \theta_0)d_{t0} + a_t$$

We see that if the value of the parameter θ is not equal to θ_0, an increment of the discrepancy vector d_{t0} is added to the vector a_t. The Cuscore statistic equation is thereby continuously searching for inflated residuals that are related to the presence of that particular discrepancy vector.

Plotting the Cuscore against t provides a very sensitive check for changes in θ, and such changes will be indicated by a change in the slope of the plot just as changes in the mean are indicated by a change in the slope of the Cusum plot. The specific times at which the change in slope occurs may be used to provide an estimate of the time of occurrence and this will help to identify specific problems.

The Linear Model

Consider this approach for the linear model:

$$y_t = \theta x_t + a_t$$

$$a_{t0} = y_t - \theta_0 x_t, \qquad d_{t0} = -\frac{\partial a_t}{\partial \theta} = x_t$$

and the Cuscore statistic is:

$$Q = \sum (y_t - \theta_0 x_t) x_t$$

In the particular case where $x_t = 1$ ($t = 1, 2, \ldots, n$), the test is for a change in mean, and the Cuscore is the familiar Cusum statistic $Q = \sum(y_t - \theta_0)$ with θ_0 as the reference (target) value.

Detecting a Spike

A spike is a short-term deviation that lasts only one sampling interval. Using the Cuscore to detect an individual spike in a background of white noise reproduces the Shewhart chart. Therefore, the Cuscore tells us to plot the individual deviations $y_t - T$, where T is the target value.

Detecting a Rectangular Bump

To detect a uniform change that lasts for b sampling intervals, what Box and Luceno (1997) call a *bump*, the Cuscore is the sum of the last b deviations:

$$Q = (y_t - T) + (y_{t-1} - T) + (y_{t-2} - T) + \cdots + (y_{t-b+1} - T)$$

Dividing this Q by b gives the *arithmetic moving average* of the last b deviations. An arithmetic moving average (AMA) chart is constructed by plotting the AMA on a Shewhart-like chart but with control lines at $\pm 2\sigma_b$ and $\pm 3\sigma_b$, where σ_b is the standard deviation of the moving average. If the original observations have standard deviation σ and are independent, a moving average of b successive values has standard deviation $\sigma_b = \sigma/\sqrt{b}$.

---

[1] This equation is exact if the model is linear in θ and approximate otherwise.

Detecting a Change in a Rate of Increase

Suppose that the normal rate of change in a process gives a straight line $y_t = \beta t + a_t$ with slope β and you want to monitor for the possibility that this rate has changed. The deviation from the line representing normal wear is $a_t = y_t - \beta t$ and the appropriate Cuscore is $Q = \Sigma(y_t - \beta t)t$.

Detecting a Sine Wave Buried in Noise

Figure 13.1 shows a Cuscore chart that identifies a small sinusoidal disturbance that is buried in random noise. The noise has $\sigma = 1$ (top panel) and the sine wave has amplitude $\theta = 0.5$ (second panel). The combined signal (noise plus sine wave) is shown in the third panel. The sinusoidal pattern is invisible although we know it is there. The bottom panel is the Cuscore chart that was designed to be especially sensitive to a sine disturbance, as explained below. The appearance and disappearance of the sine wave is now as clear as it was obscure in the original data.

If the sine wave has amplitude θ and period p, the model for the disturbance is:

$$y_t = T + \theta \sin(2\pi t/p) + a_t$$

and

$$a_t = y_t - T - \theta \sin(2\pi t/p)$$

When the process is operating correctly, there will be no sine component, $\theta = \theta_0 = 0$, and the residuals are $a_t = y_t - T$. Because $d_{t0} = -\frac{\partial a_t}{\partial \theta} = \sin(2\pi t/p)$, the Cuscore statistic that is looking for a recurrence of the

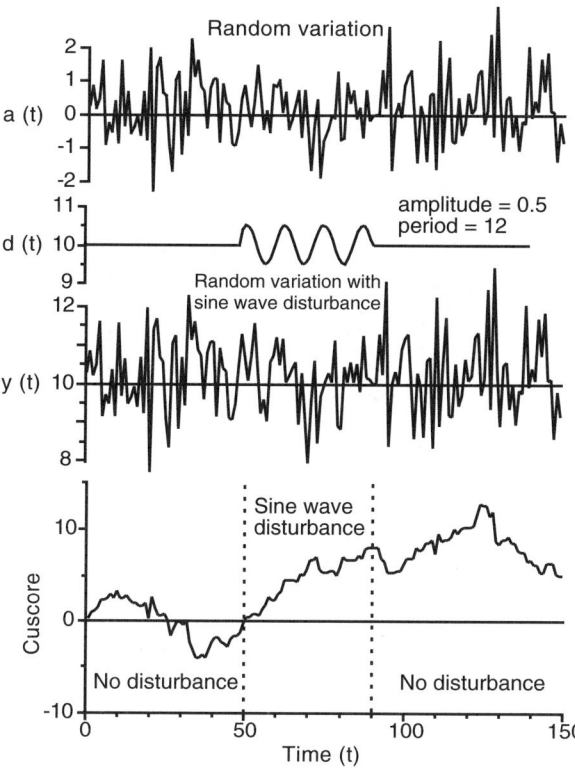

FIGURE 13.1 Cuscore chart to detect a sinusoidal disturbance buried in random noise.

sine component is:

$$Q = \sum (y_t - T) \sin(2\pi t/p)$$

For a sine wave of period $p = 12$, the plotted Cuscore is:

$$Q = \sum (y_t - T) \sin(2\pi t/12)$$

Notice that the amplitude of the sine wave is irrelevant.

Change in the Parameter of a Time Series Model

It was stated that process *drift* is a normal feature in most real processes. Drift causes serial correlation in subsequent observations. Time series models are used to describe this kind of data. To this point, serial correlation has been mentioned briefly, but not explained, and time series models have not been introduced. Nevertheless, to show the wonderful generality of the Cuscore principles, a short section about time series is inserted here. The material comes from Box and Luceno (1997). The essence of the idea is in the next two paragraphs and Figure 13.2, which require no special knowledge of time series analysis.

Figure 13.2 shows a time series that was generated with a simple model $y_t = 0.75 y_{t-1} + a_t$, which means that the process now (time t) remembers 75% of the level at the previous observation (time $t - 1$), with some random variation (the a_t) added. The 0.75 coefficient is a measure of the correlation between the observations at times t and $t - 1$. A process having this model drifts slowly up and down. Because of the correlation, when the level is high it tends to stay high for some time, but the random variation will sometimes bring it down and when low values occur they tend to be followed by low values. Hence, the drift in level.

Suppose that at some point in time, the process has a "memory failure" and it remembers only 50% of the previous value; the model changes to $y_t = 0.5 y_{t-1} + a_t$. The drift component is reduced and the random component is relatively more important. This is not a very big change in the model. The parameter θ is nothing like the slope parameter in the model of a straight line. In fact, our intuition tells us nothing helpful about how θ is related to the pattern of the time series. Even if it did, we could not see the change caused by this shift in the model. An appropriate Cuscore can detect this change, as Figure 13.2 shows.

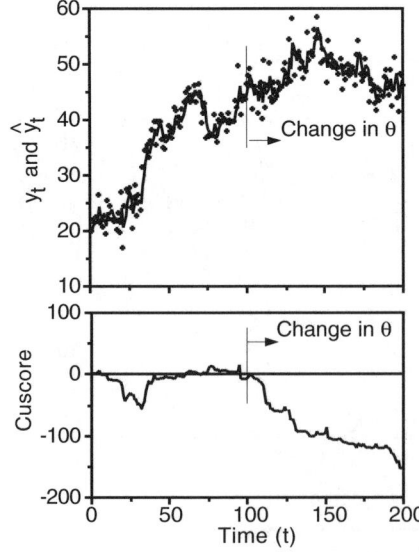

FIGURE 13.2 IMA time series $y_t = y_{t-1} + a_t - \theta a_{t-1}$. The value of θ was changed from 0.75 to 0.5 at $t = 100$. The Cuscore chart starts to show a change at $t = 125$ although it is difficult to see any difference in the time series.

Specialized Control Charts

At this point, readers who are not interested in time series analysis, or who have not studied this subject, may wish to skip to the Comments section. The next chapter brings us back to a comfortable topic.

The model for Figure 13.2 is an *integrated moving average* (IMA) time series model in its simplest form:

$$y_t - y_{t-1} = a_t - \theta a_{t-1}$$

where a_t is a random normal independent variable with mean zero and constant variance σ^2. This is the model for which the exponentially weighted moving average (EWMA) with smoothing constant θ is the minimum mean square error forecast of y_t from origin $t - 1$.

An IMA model describes a time series that is nonstationary; that is, it has no fixed mean level, except as an approximation over a short time interval. An IMA time series drifts. Many real processes do drift and the IMA can be a suitable model in these cases.

The disturbance model is $y_t - y_{t-1} = a_t - (\theta + \delta)a_{t-1}$, where δ is the change in the parameter. Tiao et al. (2000) has designed a Cuscore chart to detect this change.

The discrepancy vector is:

$$d_t = -\frac{\partial a_t}{\partial \theta} = \frac{-\hat{a}_t}{1 - \theta}$$

where $\hat{a}_t$ is the exponentially weighted moving average of previous a_t's with smoothing constant θ. This is upgraded using:

$$\hat{a}_t = (1 - \theta)a_{t-1} + \theta \hat{a}_{t-1}$$

The Cuscore statistic is:

$$Q = \frac{1}{1 - \theta} \sum a_{t0} \hat{a}_{t0}$$

If θ_0 is the true value of θ, the a_{t0}'s would be independent and the exponentially weighted average $\hat{a}_{t0}$ of previous a_{t0}'s would not be correlated with a_{t0}.

The first-order autoregressive model (AR1) is another common time series model. This model describes a stationary time series, that is, a series that has a fixed mean level:

$$z_t = \phi z_{t-1} + a_t$$

where z_t is the deviation $y_t - T$ from the target value. We wish to monitor for a possible change in the parameter from its normal value ϕ_0 to some new value $\phi_0 + \delta$. The null model is $a_{t0} = z_t - \phi_0 z_{t-1}$ and the discrepancy model is $a_{t0} = z_t - (\phi_0 + \delta)z_{t-1}$. Box and Luceno (1997) explain how this is done. We lack the space to provide more detailed explanation and example calculations. Details and applications are discussed in Box and Luceno (1997) and Tiao et al. (2000).

Comments

Control charts have developed well beyond the classical design of Shewhart. Automated data collection provides information that does not always fit a Shewhart chart. Individual observations are charted, and often there is no central limit property to rely on, so the reference distribution of the plotted values is not always the normal distribution. Autocorrelation and process drift are common, and data transformations are needed. Also, statistics other than the mean value are charted. Disturbances other than a simple shift in mean are important: a change in slope signals wear of a machine or tool, a sinusoidal disturbance signals vibration of a compressor shaft, a bump signals a possible harmful shock to the system, and so on.

The Cusum chart shown in Chapter 12 was one powerful development. It serves the same purpose as a Shewhart chart but has greater sensitivity. It will detect a smaller change than a Shewhart chart can detect. It gains this sensitivity by plotting the cumulative change from the target level. The accumulation of information about deviations provides useful knowledge about a process.

The power of the Cusum chart is generalized in the Cuscore charts, which also accumulate information about deviations from the normal pattern. They are designed for deviations of a special kind: a bump, a sine wave, a change in slope, or a change in a process parameter.

References

Box, G. E. P. and A. Luceno (1997). *Statistical Control by Monitoring and Feedback Adjustment*, New York, Wiley Interscience.
Box, G. E. P. and J. Ramirez (1992). "Cumulative Score Charts," *Qual. Rel. Eng. Int.*, 8, 17–27.
Tiao, George et al., Eds. (2000). *Box on Quality and Discovery with Design, Control, and Robustness*, New York, John Wiley & Sons.

Exercises

13.1 Find the Disturbance. Based on historical information, five machines vary randomly about an average level of 11.2 with a standard deviation of 1.0. The table below gives a series of operating data for each machine. Check each machine for change in average level, a "bump" of width = 3, a sinusoidal disturbance with period $p = 12$ and period $p = 24$, and a change in slope from 0 to 0.02.

		Machine						Machine			
Time	1	2	3	4	5	Time	1	2	3	4	5
1	11.8	11.8	11.8	11.8	11.8	26	11.4	12.3	14.9	11.4	11.9
2	11.7	11.7	11.7	11.7	11.7	27	13.0	14.0	16.5	13.0	13.5
3	13.0	13.0	13.0	13.0	13.0	28	10.6	11.5	14.1	10.6	11.2
4	11.1	11.1	11.1	11.1	11.1	29	11.3	11.8	14.8	11.3	11.9
5	10.5	10.5	10.5	10.5	10.5	30	10.9	10.9	14.4	10.9	11.5
6	10.5	10.5	10.5	10.5	10.5	31	13.2	12.7	16.7	13.2	13.8
7	11.3	11.3	11.3	15.6	11.3	32	13.4	12.6	16.9	13.4	14.1
8	10.5	10.5	13.0	10.5	10.5	33	13.4	12.4	13.4	13.4	14.0
9	12.3	12.3	14.8	12.3	12.3	34	12.8	12.0	12.8	12.8	13.5
10	11.6	11.6	14.1	11.6	11.6	35	12.3	11.8	12.3	12.3	13.0
11	11.7	11.7	14.2	11.7	11.7	36	11.1	11.1	11.1	11.1	11.1
12	11.0	11.0	13.5	13.3	11.2	37	11.8	11.8	11.8	11.8	11.8
13	10.5	10.5	10.5	10.5	10.8	38	12.7	12.7	12.7	12.7	12.7
14	10.9	11.4	10.9	10.9	11.2	39	12.8	12.8	12.8	12.8	12.8
15	10.5	11.4	10.5	7.4	10.8	40	10.9	10.9	10.9	10.9	10.9
16	11.5	12.4	11.5	11.5	11.9	41	11.7	11.7	11.7	11.7	11.7
17	10.8	11.3	10.8	10.8	11.1	42	12.0	12.0	12.0	12.0	12.0
18	12.8	12.8	12.8	12.8	13.1	43	11.7	11.7	11.7	11.7	11.7
19	9.9	9.4	9.9	9.9	10.3	44	10.4	10.4	10.4	10.4	10.4
20	11.4	10.6	11.4	11.4	11.8	45	9.8	9.8	9.8	9.8	9.8
21	13.6	12.6	13.6	13.6	14.0	46	10.2	10.2	10.2	10.2	10.2
22	12.7	11.8	12.7	12.7	13.1	47	11.4	11.4	11.4	11.4	11.4
23	12.1	11.6	12.1	12.1	12.5	48	12.3	12.3	12.3	12.3	12.3
24	10.6	10.6	14.1	10.6	11.1	49	11.8	11.8	11.8	11.8	11.8
25	9.1	9.6	12.6	9.1	9.6	50	10.9	10.9	10.9	10.9	10.9

13.2 Real World. Visit a nearby treatment plant or manufacturing plant to discover what kind of data they collect. Do not focus only on process or product quality. Pay attention to machine maintenance, chemical use, and energy use. Propose applications of Cuscore charts.

14

Limit of Detection

KEY WORDS *limit of detection, measurement error, method limit of detection, percentile, variance, standard deviation.*

The method limit of detection or method detection limit (MDL) defines the ability of a measurement method to determine an analyte in a sample matrix, regardless of its source of origin. Processing the specimen by dilution, extraction, drying, etc. introduces variability and it is essential that the MDL include this variability.

The MDL is often thought of as a chemical concept because it varies from substance to substance and it decreases as analytical methods improve. Nevertheless, the MDL is a statistic that is estimated from data. As such it has no scientific meaning until it is operationally defined in terms of a measurement process and a statistical method for analyzing the measurements that are produced. Without a precise statistical definition, one cannot determine a numerical value for the limit of detection, or expect different laboratories to be consistent in how they determine the limit of detection.

The many published definitions differ in detail, but all are defined in terms of a multiple of the standard deviation of measurements on blank specimens or, alternately, on specimens that have a very low concentration of the analyte of interest. All definitions exhibit the same difficulty with regard to how the standard deviation of blank specimens is estimated.

The U.S. EPA's definition and suggested method of estimation is reviewed and then we look at an alternative approach to understanding the precision of measurements at low concentrations.

Case Study: Lead Measurements

Lead is a toxic metal that is regulated in drinking water and in wastewater effluents. Five laboratories will share samples from time to time as a check on the quality of their work. They want to know whether the method detection limit (MDL) for lead is the same at each lab. As a first step in this evaluation, each laboratory analyzed 50 test specimens containing lead. They did not know how many levels of lead had been prepared (by spiking known amounts of lead into a large quantity of common solution), nor did they know what concentrations they were given, except that they were low, near the expected MDL. The background matrix of all the test specimens was filtered effluent from a well-operated activated sludge plant. A typical data set is shown in Table 14.1 (Berthouex, 1993).

Method Detection Limit: General Concepts

The method detection limit (MDL) is much more a statistical than a chemical concept. Without a precise statistical definition, one cannot determine a scientifically defensible value for the limit of detection, expect different laboratories to be consistent in how they determine the limit of detection, or be scientifically honest about declaring that a substance has (or has not) been detected. Beyond the statistical definition there must be a clear set of operational rules for how this measurement error is to be determined in the laboratory. Each step in processing a specimen introduces variability and the MDL should include this variability.

The EPA says, "The method detection limit (MDL) is defined as the minimum concentration of a substance that can be measured and reported with 99% confidence that the analyte concentration is

TABLE 14.1

Measurements on Filtered Activated Sludge Effluent to Which Known Amounts of Lead Have Been Added

Pb Added (μg/L)	Measured Lead Concentrations (μg/L)									
Zero	2.5	3.8	2.2	2.2	3.1	2.6				
1.25	2.8	2.7	3.4	2.4	3.0	3.7	4.6	3.1	3.6	3.1	4.3
	4.0	1.7	2.2	2.4	3.5	2.2	2.7	3.2	2.8		
2.50	4.5	3.7	3.8	4.4	5.4	3.9	4.1	3.7	3.0	4.5	4.8
	3.3	4.7	4.4								
5.0	3.9	5.0	5.4	4.9	6.2						
10	12.2	13.8	9.9	10.5	10.9						

Note: The "zero added" specimens are not expected to be lead-free.

greater than zero and is determined from analysis of a sample in a given matrix containing the analyte.... It is essential that all sample processing steps of the analytical method be included in the determination of the method detection limit."

The mention of "a sample in a given matrix" indicates that the MDL may vary as a function of specimen composition. The phrase "containing the analyte" may be confusing because the procedure was designed to apply to a wide range of samples, including reagent blanks that would not include the analyte.

The EPA points out that the variance of the analytical method may change with concentration and the MDL determined using their procedure may not truly reflect method variance at lower analyte concentrations. A minimum of seven aliquots of the prepared solution shall be used to calculate the MDL. If a blank measurement is required in the analytical procedure, obtain a separate blank measurement for each aliquot and subtract the average of the blank measurements from each specimen measurement.

Taylor (1987) pointed out that the term *limit of detection* is used in several different situations which should be clearly distinguished. One of these is the instrument limit of detection (IDL). This limit is based on the ability to detect the difference between *signal* and *noise*. If the measurement method is entirely instrumental (no chemical or procedural steps are used), the signal-to-noise ratio has some relation to MDL. Whenever other procedural steps are used, the IDL relates only to instrumental measurement error, which may have a small variance in comparison with variance contributed by other steps. In such a case, improving the signal-to-noise ratio by instrumental modifications would not necessarily lower the MDL. As a practical matter we are interested in the MDL. Processing the specimen by dilution, extraction, drying, etc. introduces variability and it is essential that the MDL reflect the variability contributed by these factors.

It should be remembered that the various limits of detection (MDL, IDL, etc.) are not unique constants of methodology (see APHA, AWWA, WEF, 1998, Section 1030C). They depend on the statistical definition and how measurement variability at low concentrations is estimated. They also depend on the expertise of the analyst, the quality control procedures used in the laboratory, and the sample matrix measured. Thus, two analysts in the same laboratory, using the same method, can show significant differences in precision and hence their MDLs will differ. From this it follows that published values for a MDL have no application in a specific case except possibly to provide a rough reference datum against which a laboratory or analyst could check a specifically derived MDL. The analyst's value and the published MDL could show poor agreement because the analyst's value includes specimen matrix effects and interferences that could substantially shift the MDL.

The EPA Approach to Estimating the MDL

The U.S. EPA defines the MDL as the minimum concentration of a substance that can be measured and reported with 99% confidence that the analyte concentration is greater than zero and is determined from analysis of a sample in a given matrix containing the analyte. Similar definitions have been given by Glaser et al. (1981), Hunt and Wilson (1986), the American Chemical Society (1983), Kaiser and Menzes (1968), Kaiser (1970), Holland and McElroy (1986), and Porter et al. (1988).

Limit of Detection

Measurements are made on a minimum of seven aliquots ($n \geq 7$) of a prepared solution that has a concentration near the expected limit of detection. These data are used to calculate the variance of the replicate measurements:

$$s^2 = \frac{1}{n-1}\sum_{i=1}^{n}(y_i - \bar{y})^2$$

where y_i ($i = 1$ to n) are the measured values. The MDL is:

$$\text{MDL} = s \times t_{v,\alpha=0.01}$$

where $t_{v,\alpha=0.01}$ is the Student's t value appropriate for a 99% confidence level and a standard deviation estimate α with $v = n - 1$ degrees of freedom. This is the t value that cuts off the upper 1% of the t distribution. For $n = 7$ and $v = 6$, $t_{6,0.01} = 3.143$ and the estimated MDL = $3.143s$.

The EPA points out that the variance of the analytical method may change with concentration. If this happens, the estimated MDL will also vary, depending on the concentration of the prepared solution on which the replicate measurements were made. The EPA suggests that the analyst check this by analyzing seven replicate aliquots at a slightly different concentration to "verify the reasonableness" of the estimate of the MDL. If variances at different concentrations are judged to be the same (based on the F statistic of their ratio), the variances are combined (pooled) to obtain a single estimate for s^2. The general relation for k pooled variances is:

$$s^2_{\text{pooled}} = \frac{(n_1-1)s_1^2 + (n_2-1)s_2^2 + \cdots + (n_k-1)s_k^2}{(n_1-1) + (n_2-1) + \cdots + (n_k-1)}$$

s^2_{pooled} has $v = (n_1 - 1) + (n_2 - 1) + \cdots + (n_k - 1)$ degrees of freedom. When the variances have identical degrees of freedom, pooling is just averaging the variances. In the case of two variances, s_1^2 and s_2^2, each determined from seven replicate measurements, the pooled variance is:

$$s^2_{\text{pooled}} = \frac{6s_1^2 + 6s_2^2}{6+6} = \frac{s_1^2 + s_2^2}{2}$$

The 6's are the degrees of freedom for the two sets of seven aliquot observations. s^2_{pooled} and s_{pooled} have 12 degrees of freedom, giving $t_{12,0.01} = 2.681$ and MDL = $2.681 s_{\text{pooled}}$. If more than seven aliquots were used, the appropriate Student's t value for $\alpha = 0.01$ would be taken from Table 14.2.

The EPA definition of the MDL is illustrated by Figure 14.1. Measurements on test specimens are pictured as being normally distributed about the true concentration. A small probability ($100\alpha\%$) exists that a measurement on a true blank will exceed the MDL, which is located a distance $st_{v,\alpha}$ to the right of the mean concentration. If the test specimens were true blanks (concentration equal zero), this definition of the MDL provides a $100(1 - \alpha)\%$ degree of protection against declaring that the analyte was detected in a specimen in which it was truly absent.

TABLE 14.2

Values of the t Statistic for EPA's Procedure to Calculate the MDL

For s^2	n	7	8	9	10
	df = $n - 1$	6	7	8	9
	$t_{n-1, 0.01}$	3.143	2.998	2.896	2.821
For s^2_{pooled}	$n_1 + n_2$	14	16	18	20
	df = $n_1 + n_2 - 2$	12	14	16	18
	$t_{n_1+n_2-2, 0.01}$	2.681	2.624	2.583	2.552

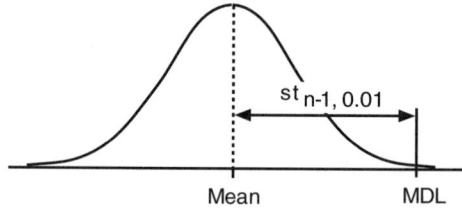

FIGURE 14.1 EPA definition of the MDL.

Example 14.1

Seven replicate specimens were measured to obtain the following results: 2.5, 2.7, 2.2, 2.2, 3.1, 2.6, 2.8. The estimated variance is $s^2 = 0.10$, $s = 0.32$, $t = 3.143$, and the MDL = 3.143(0.32) = 1.0. Another set of seven replicate specimens was analyzed to get a second estimate of the variance; the data were 1.6, 1.9, 1.3, 1.7, 2.1, 0.9, 1.8. These data give $s^2 = 0.16$, $s = 0.40$, and the MDL = 3.143(0.40) = 1.3. A statistical F test ($0.16/0.10 \leq F_{6,6} = 4.3$) shows that the two estimates of the variance are not significantly different. Therefore, the two samples are pooled to give:

$$s^2_{pooled} = \frac{0.10 + 0.16}{2} = 0.13$$

and

$$s_{pooled} = 0.36$$

s_{pooled} has $\nu = 12$ degrees of freedom. The improved (pooled) estimate of the MDL is:

$$\text{MDL} = 2.681 s_{pooled} = 2.681(0.36) = 0.96$$

This should be rounded to 1.0 μg/L to have a precision that is consistent with the measured values.

Example 14.2

Calculate, using the EPA method, the MDL for the lead data given in Table 14.1. The data come from a laboratory that did extra work to evaluate the limit of detection for lead measurements. Known amounts of lead were added to filtered effluent from a municipal activated sludge plant. Therefore, the aliquots with zero lead added are not believed to be entirely free of lead. By definition, a minimum of seven replicates containing the analyte is needed to compute the MDL. Therefore, we can use the 1.25 μg/L and 2.50 μg/L spiked specimens. The summary statistics are:

	1.25 μg/L	2.50 μg/L
$\bar{y}$	3.07	4.16
s^2	0.55	0.41
s	0.74	0.64
n	20	14

Pooling the sample variances yields:

$$s^2_{pooled} = \frac{19(0.55) + 13(0.41)}{19 + 13} = 0.49$$

Limit of Detection

with $v = 19 + 13 = 32$ degrees of freedom. The pooled standard deviation is:

$$s_{\text{pooled}} = 0.70\,\mu g/L$$

The appropriate t statistic is $t_{32,0.01} = 2.457$ and MDL $= 2.457 s_{\text{pooled}} = 2.457(0.70) = 1.7\,\mu g/L$.

An Alternate Model for the MDL

The purpose of giving an alternate model is to explain the sources of variation that affect the measurements used to calculate the MDL. Certified laboratories are obliged to follow EPA-approved methods, and this is not an officially accepted alternative.

Pallesen (1985) defined the limit of detection as "the smallest value of analyte that can be reliably detected above the random background noise present in the analysis of blanks." The MDL is defined in terms of the background noise of the analytical procedure for the matrix being analyzed. The variance of the analytical error and background noise are considered separately. Also, the variance of the analytical error can increase as the analyte concentration increases.

The error structure in the range where limit of detection considerations are relevant is assumed to be:

$$y_i = \eta + e_i = \eta + a_i + b_i$$

The a_i and b_i are two kinds of random error that affect the measurements. a_i is random analytical error and b_i is background noise. The total random error is the sum of these two component errors: $e_i = a_i + b_i$. Both errors are assumed to be randomly and normally distributed with mean zero.

Background noise (b_i) exists even in blank measurements and has constant variance, σ_b^2. The measurement error in the analytical signal (a_i) is assumed to be proportional to the measurement signal, η. That is, $\sigma_a = \kappa \eta$ and the variance $\sigma_a^2 = \kappa^2 \eta^2$. Under this assumption, the total error variance (σ_e^2) of any measurement is:

$$\sigma_e^2 = \sigma_b^2 + \sigma_a^2 = \sigma_b^2 + \kappa^2 \eta^2.$$

Both σ_e^2 and σ_a^2 decrease as the concentration decreases. For low η, the analytical error variance, $\sigma_a^2 = \kappa^2 \eta^2$ becomes small compared to the background variance, σ_b^2. When the specimen is blank ($\eta = 0$), σ_a^2 decreases to zero and $\sigma_e^2 = \sigma_b^2$, the variance of the blank. Figure 14.2 shows this relation in terms of the variances and the standard deviations.

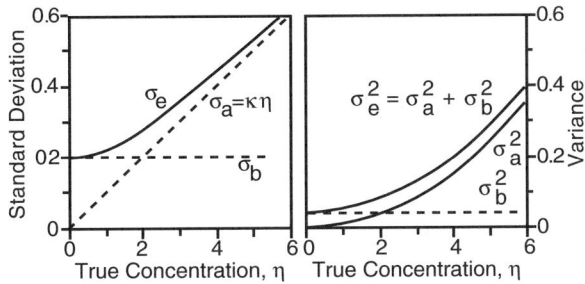

FIGURE 14.2 Error structure of the alternate model of the MDL. (From Pallesen, L. (1985). "The Interpretation of Analytical Measurements Made Near the Limit of Detection," Technical Report, IMSOR, Technical University of Denmark.)

The detection limit is the smallest value of y for which the hypothesis that $\eta = 0$ cannot be rejected with some stated level of confidence. If y > MDL, it is very improbable that $\eta = 0$ and it would be concluded that the analyte of interest has been detected. If y < MDL, we cannot say with confidence that the analyte of interest has been detected.

Assuming that y, in the absence of analyte, is normally distributed with mean zero and variance, the MDL as a multiple of the background noise, σ_b:

$$\text{MDL} = z_\alpha \sigma_b$$

where z_α is chosen such that the probability of y being greater than MDL is $(1 - \alpha)100\%$ (Pallesen, 1985). The probability that a blank specimen gives y > MDL is $\alpha 100\%$. Values of z_α can be found in a table of the normal distribution, where α is the area under the tail of the distribution that lies above z_α. For example:

$$\alpha = 0.05 \quad 0.023 \quad 0.01 \quad 0.0013$$
$$z_\alpha \quad 1.63 \quad 2.00 \quad 2.33 \quad 3.00$$

Using $z = 3.00$ ($\alpha = 0.0013$) means that observing y > MDL justifies a conclusion that η is not zero at a confidence level of 99.87%. Using this MDL repeatedly, just a little over one in one thousand (0.13%) of true blank specimens will be misjudged as positive determinations. (Note that the EPA definition is based on t_α instead of z_α, and chooses α so the confidence level is 99%.)

The replicate lead measurements in Table 14.1 will be used to estimate the parameters σ_b^2 and κ^2 in the model:

$$\sigma_e^2 = \sigma_b^2 + \sigma_a^2 = \sigma_b^2 + \kappa^2 \eta^2$$

The values of σ_e^2 and η^2 are estimated from the data by computing s_e^2 and $\bar{y}^2$. The average and variance for the laboratory are listed in Table 14.3. The parameters σ_b^2 and κ^2 are estimated by regression, as shown in Figure 14.3. The intercept is an estimate of σ_b^2 and the slope is estimate of κ^2. The estimated

TABLE 14.3

Estimated Averages ($\bar{y}$) and Variances (s_e^2) at Each Concentration Level for the Data from Table 14.1

Pb Added (μg/L)	$\bar{y}$	s_e^2
0	2.73	0.38
1.25	3.07	0.55
2.5	4.16	0.41
5.0	5.08	0.70
10.0	11.46	2.42

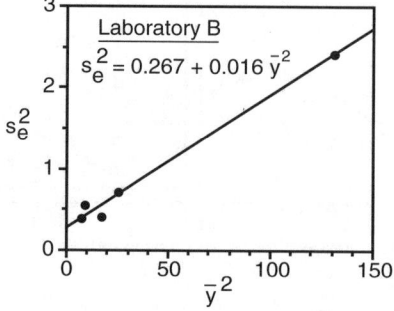

FIGURE 14.3 Plots of s_e^2 vs. $\bar{y}^2$ used to estimate the parameters κ^2 and σ_b^2.

parameter values are $\hat{\sigma}_b^2 = 0.267$ and $\hat{\kappa}^2 = 0.016$. These give $\hat{\sigma}_b = 0.52$ and $\hat{\kappa} = 0.13$. The MDL $= 3.0\hat{\sigma}_b = 3(0.52) = 1.56$. This should be rounded to 1.6 µg/L.

The EPA method gave MDL = 1.7 µg/L. Pallesen's method gave MDL = 1.6 µg/L. The difference has no practical importance, despite the following differences in the two methods.

1. The z or t statistic multipliers are different: $z_{\alpha=0.01} = 2.33$ for the EPA method and $z_{\alpha=0.0013} = 3.00$ for the Pallesen method. On purely statistical grounds, the EPA's specification of $\alpha = 0.01$ is no more justified than Pallesen's recommended multiplier of 3.00.

2. Users of the EPA method will often estimate the MDL using seven, or perhaps fourteen, replicate measurements because the procedure requires a minimum of seven replicates. In Example 14.2, the MDL (EPA method) for Laboratory B was based on 34 measurements. Pallesen's method requires replicated measurements at several concentration levels (how many levels or replicates are not specified). The case study used 50 measured values, divided among five concentration levels. Each of these values contributes to the estimation of the MDL.

3. The EPA method estimates the MDL from measurements in a given matrix containing the analyte and then assumes that the background noise observed is applicable to specimens that are blank (zero concentration). The alternate method extrapolates to zero analyte concentration to estimate the background noise (σ_b) that serves as the basis for computing the MDL. As shown in Table 14.3, the total error variance increases as the concentration increases and Pallesen's method accounts for this property of the data. If the data have this property, the EPA approach of pooling replicates from two different concentrations will tend to inflate the MDL over the estimate of the alternate method.

4. The EPA's MDL gives some information about the precision of measurements at low concentrations, but it says nothing about bias, which can be a major problem. Pallesen's model distinguishes between error contributions from the analytical method and from background noise. It also retains the average values, which could be compared with the known concentrations to make an assessment of bias.

Calibration Designs

Another way to estimate MDLs is to use a calibration design (Gibbons, 1994). A series of samples are spiked at known concentrations in the expected range of the MDL. Prediction limits (or tolerance limits in some methods) are determined for the very low concentrations and these define a limit of detection. Frequently, calibration data have nonconstant variance over a range of concentrations. If this is the case, then weighted least squares must be used to fit the calibration curve or else the variability is overestimated at low concentrations and the detection limit will be overestimated. Zorn et al. (1997) explain how this is done. Calibration is discussed in Chapter 37 and weighted least squares is discussed in Chapter 38.

Comments

The limit of detection is a troublesome concept. It causes difficulties for the chemist who must determine its value, for the analyst who must work with data that are censored by being reported as <MDL, and for the regulator and discharger who must make important decisions with incomplete information. Some statisticians and scientists think we would do better without it (Rhodes, 1981; Gilbert, 1987). Nevertheless, it is an idea that seems firmly fixed in environmental measurement methods, so we need to understand what it is and what it is not. It is an attempt to prevent deciding that a blank sample is contaminated with an analyte.

One weakness of the usual definitions of the MDL is that it is defined for a single analyst. ASTM D2777 (1988) is very clear that instead of repeated single-operator MDLs, the correct approach is to use the interlaboratory standard deviation from round-robin studies. This can change the MDL by a factor of 0.5 to 3.0 (Maddelone et al., 1988, 1991, 1993).

The MDL is based directly on an estimate of the variance of the measurement error at low concentrations. This quantity is not easy to estimate, which means that the MDL is not a well-known, fixed value. Different chemists and different laboratories working from the same specimens will estimate different MDLs. A result that is declared ≤MDL can come from a sample that has a concentration actually greater than the MDL.

What are the implications of the MDL on judging violations? Suppose that a standard is considered to be met as long as all samples analyzed are "non-detect." By the EPA's definition, if truly blank specimens were analyzed, the odds of a hit above the MDL on one particular specimen is 1%. As more specimens are analyzed, the collective odds of getting a hit increase. If, for example, a blank is analyzed every day for one month, the probability of having at least one blank declared above the MDL is about 25% and there is a 3% chance of having two above the MDL. Analyze blanks for 90 days and the odds of having at least one hit are 60% and the odds of having two hits is 16%. Koorse (1989, 1990) discusses some legal implications of the MDL.

References

American Chemical Society, Committee on Environmental Improvement (1983). "Principles of Environmental Analysis," *Anal. Chem.,* 55(14), 2210–2218.

APHA, AWWA, WEF (1998). *Standard Methods for the Examination of Water and Wastewater,* 20th ed., Clesceri, L. S., A. E. Greenberg, and A. D. Eaton, Eds.

ASTM (1988). *Standard Practice for Determination of Precision and Bias of Applicable Methods of Committee D-19 on Water,* D2777–86, ASTM Standards of Precision and Bias for Various Applications, 3rd ed.

Berthouex, P. M. (1993). "A Study of the Precision of Lead Measurements at Concentrations Near the Limit of Detection," *Water Envir. Res.,* 65(5), 620–629.

Currie, L. A. (1984). "Chemometrics and Analytical Chemistry," in *Chemometrics: Mathematics and Statistics in Chemistry,* NATO ASI Series C, 138, 115–146, Dordrecht, Germany, D. Reidel Publ. Co.

Gibbons, R. D. (1994). *Statistical Methods for Groundwater Monitoring,* New York, John Wiley.

Gilbert, R. O. (1987). *Statistical Methods for Environmental Pollution Monitoring,* New York, Van Nostrand Reinhold.

Glaser, J. A., D. L. Foerst, G. D. McKee, S. A. Quave, and W. L Budde (1981). "Trace Analyses for Wastewaters," *Environ. Sci. & Tech.,* 15(12), 1426–1435.

Holland, D. M. and F. F. McElroy (1986). "Analytical Method Comparison by Estimates of Precision and Lower Detection Limit," *Environ. Sci. & Tech.,* 20, 1157–1161.

Hunt, D. T. E. and A. L. Wilson (1986). *The Chemical Analysis of Water,* 2nd ed., London, The Royal Society of Chemistry.

Kaiser, H. H. and A. C. Menzes (1968). *Two Papers on the Limit of Detection of a Complete Analytical Procedure,* London, Adam Hilger Ltd.

Kaiser, H. H. (1970). "Quantitation in Elemental Analysis: Part II," *Anal. Chem.,* 42(4), 26A–59A.

Koorse, S. J. (1989). "False Positives, Detection Limits, and Other Laboratory Imperfections: The Regulatory Implications," *Environmental Law Reporter,* 19, 10211–10222.

Koorse, S. J. (1990). "MCL Noncompliance: Is the Laboratory at Fault," *J. AWWA,* Feb., pp. 53–58.

Maddelone, R. F., J. W. Scott, and J. Frank (1988). *Round-Robin Study of Methods for Trace Metal Analysis: Vols. 1 and 2: Atomic Absorption Spectroscopy — Parts 1 and 2,* EPRI CS-5910.

Maddelone, R. F., J. W. Scott, and N. T. Whiddon (1991). *Round-Robin Study of Methods for Trace Metal Analysis: Vol. 3: Inductively Coupled Plasma-Atomic Emission Spectroscopy,* EPRI CS-5910.

Maddelone, R. F., J. K. Rice, B. C. Edmondson, B. R. Nott, and J. W. Scott (1993). "Defining Detection and Quantitation Levels," *Water Env. & Tech.,* January, pp. 41–44.

Pallesen, L. (1985). "The Interpretation of Analytical Measurements Made Near the Limit of Detection," Technical Report, IMSOR, Technical University of Denmark.

Porter, P. S., R. C. Ward, and H. F. Bell (1988). "The Detection Limit," *Environ. Sci. & Tech.,* 22(8), 856–861.

Rhodes, R. C. (1981). "Much Ado About Next to Nothing, Or What to Do with Measurements Below the Detection Limit," *Environmetrics 81: Selected Papers,* pp. 174–175, SIAM–SIMS Conference Series No. 8, Philadelphia, Society for Industrial and Applied Mathematics, pp. 157–162.

Limit of Detection

Taylor, J. K. (1987). *Quality Assurance of Chemical Measurements,* Chelsea, MI, Lewis Publishers.
Zorn, M. E., R. D. Gibbons, and W. C. Sonzogni (1997). "Weighted Least Squares Approach to Calculating Limits of Detection and Quantification by Modeling Variability as a Function of Concentration," *Anal. Chem.,* 69(15), 3069–3075.

Exercises

14.1 Method Detection Limit. Determine the method limit of detection using the two sets of eight replicates.

Replicate Set 1	1.27	1.67	1.36	1.60	1.24	1.26	0.55	0.78
Replicate Set 2	2.56	1.43	1.20	1.21	1.93	2.31	2.11	2.33

14.2 MDL Update. An analyst has given you the concentrations of eight replicates: 1.3, 1.7, 1.4, 1.6, 1.2, 1.3, 0.6, and 0.8. (a) Calculate the limit of detection. (b) A few days later, the analyst brings an additional 13 replicate measurements (2.6, 1.4, 1.2, 1.2, 1.9, 2.3, 2.1, 2.3, 1.7, 1.9, 1.7, 2.9, and 1.6) and asks you to recalculate the limit of detection. How will you use these new data?

14.3 Alternate MDL. Use the data below to determine a limit of detection using Pallesen's method.

	Concentration of Standard Solution			
Replicate	**1 µg/L**	**2 µg/L**	**5 µg/L**	**8 µg/L**
1	1.3	1.8	5.6	8.0
2	1.1	1.9	5.7	7.9
3	1.2	2.6	5.2	7.7
4	1.1	2.0	5.4	7.7
5	1.1	1.8	5.2	8.9
6	0.8	1.8	6.2	8.7
7	0.9	2.5	5.1	8.9

15

Analyzing Censored Data

KEY WORDS *censored data, delta-lognormal distribution, median, limit of detection, trimmed mean, Winsorized mean, probability plot, rankit, regression, order statistics.*

Many important environmental problems focus on chemicals that are expected to exist at very low concentrations, or to be absent. Under these conditions, a set of data may include some observations that are reported as "not detected" or "below the limit of detection (MDL)." Such a data set is said to be *censored*.

Censored data are, in essence, missing values. Missing values in data records are common and they are not always a serious problem. If 50 specimens were collected and five of them, selected at random, were damaged or lost, we could do the analysis as though there were only 45 observations. If a few values are missing at random intervals from a time series, they can be filled in without seriously distorting the pattern of the series. The difficulty with censored data is that missing values are not selected at random. They are all missing at one end of the distribution. We cannot go ahead as if they never existed because this would bias the final results.

The odd feature of censored water quality data is that the censored values were not always missing. Some numerical value was measured, but the analytical chemist determined that the value was below the method limit of detection (MDL) and reported <MDL instead of the number. A better practice is to report all values along with a statement of their precision and let the data analyst decide what weight the very low values should carry in the final interpretation. Some laboratories do this, but there are historical data records that have been censored and there are new censored data being produced. Methods are needed to interpret these.

Unfortunately, there is no generally accepted scheme for replacing the censored observations with some arbitrary values. Replacing censored observations with zero or 0.5 MDL gives estimates of the mean that are biased low and estimates of the variance that are high. Replacing the censored values with the MDL, or omitting the censored observations, gives estimates of the mean that are high and variance that are low. The bias of both the mean and variance would increase as the fraction of observations censored increases, or the MDL increases (Berthouex and Brown, 1994).

The *median, trimmed mean*, and *Winsorized mean* are three unbiased estimates of the mean for normal or other symmetrical distributions. They are insensitive to information from the extremes of the distribution and can be used when the extent of censoring is moderate (i.e., not more than 15 to 25%). Graphical interpretation with probability plots is useful, especially when the degree of censoring is high. Cohen's maximum likelihood method for estimating the mean and variance is widely used when censoring is 25% or less.

The Median

The median is an unbiased estimate of the mean of any symmetric distribution (e.g., the normal distribution). The median is unaffected by the magnitude of observations on the tails of the distribution. It is also unaffected by censoring so long as more than half of the observations have been quantified.

The median is the middle value in a ranked data set if the number of observations is odd. If the number of observations is even, the two middle values are averaged to estimate the median. If more than half the values are censored, the median itself is below the MDL and cannot be estimated directly. Later a method is given to estimate the median from a probability plot.

Example 15.1

The sample of $n = 27$ observations given below has been censored of values below 6. The median is 7.6, the 14th largest of the 27 ranked observations shown below.

<MDL	<MDL	<MDL	<MDL	6.1	6.3	6.5	6.7	6.9
7.2	7.3	7.4	7.5	[7.6]	7.7	7.8	7.9	8.0
8.1	8.3	8.5	8.7	8.9	9.2	9.4	9.6	10.1

The confidence interval of the median is [6.9, 8.3]. This can be found using tables in Hahn and Meeker (1991) that give the nonparametric approximate 95% confidence interval for the median as falling between the 9th and 20th observation for sample size $n = 27$.

If the censored values (4.8, 5.2, 5.4, and 5.6) had been reported, we could have made parametric estimates of the mean and confidence interval. These are $\bar{y} = 7.51$, $s = 1.383$, and the confidence interval is [7.0, 8.0].

The Trimmed Mean

Censoring is, in effect, trimming. It trims away part of the lower tail of the distribution and creates an asymmetric data set, one with more known values above than below the median. Symmetry can be returned by trimming the upper tail of the distribution.

The trimmed mean and the Winsorized mean can be used to estimate the mean if the underlying distribution is symmetric (but not necessarily normal). In this case, they are unbiased estimators but they do not have minimum variance (Gilbert, 1987).

A 100 p% trimmed mean is computed from a set of n observations by trimming away (eliminating) the largest and the smallest np values. The average is computed using the remaining $n - 2np$ values. The degree of trimming (p) does not have to equal the percentage of observations that have been censored. It could be a higher, but not lower, percentage.

Hoaglin et al. (1983) suggest that a 25% trimmed mean is a good estimator of the mean for symmetric distributions. The 25% trimmed mean uses the middle 50% of the data. Hill and Dixon (1982) considered asymmetric distributions and found that a 15% trimmed mean was a safe estimator, in the sense that its performance did not vary markedly from one situation to another. If more than 15% of the observations in the sample have been censored, the 15% trimmed mean cannot be computed.

Example 15.2

Determine the trimmed mean of the data in Example 15.1. To create symmetry, trim away the largest four values, leaving the trimmed sample with $n = 19$ observations listed below.

6.1	6.3	6.5	6.7	6.9	7.2	7.3	7.4	7.5	7.6
7.7	7.8	7.9	8.0	8.1	8.3	8.5	8.7	8.9	

The trimmed mean is:

$$\bar{y}_t = \frac{\Sigma y_i}{n - 2np} = \frac{143.4}{27 - 2(4)} = 7.55 \ \mu g/L$$

The trimmed percent is $100(4/27) = 15\%$ and $\bar{y}_t = 7.55$ is a 15% trimmed mean.

The Winsorized Mean

Winsorization can be used to estimate the mean and standard deviation of a distribution although the data set has a few missing or unreliable values at either or both ends of the distribution.

Example 15.3

Again using the data in Example 15.1, replace the four censored values by the next largest value, which is 6.1. Replace the four largest values by the next smallest value, which is 8.9. The replaced values are shown in italics. This gives the Winsorized sample ($n = 27$) below:

6.1	*6.1*	*6.1*	*6.1*	6.1	6.3	6.5	6.7	6.9
7.2	7.3	7.4	7.5	7.6	7.7	7.8	7.9	8.0
8.1	8.3	8.5	8.7	8.9	*8.9*	*8.9*	*8.9*	*8.9*

Compute the sample mean ($\bar{y}$) and standard deviation (s) of the resulting Winsorized sample in the usual way:

$$\bar{y} = 7.53 \ \mu g/L \qquad s = 1.022 \ \mu g/L$$

The Winsorized mean ($\bar{y}_w$) is the mean of the Winsorized sample:

$$\bar{y}_w = \bar{y} = 7.53 \ \mu g/L$$

The Winsorized standard deviation (s_w) is an approximately unbiased estimator of s and is computed from the standard deviation of the Winsorized sample as:

$$s_w = \frac{s(n-1)}{\nu - 1} = \frac{1.02(27-1)}{19-1} = 1.48 \ \mu g/L$$

where n is the total number of observations and ν is the number of observations not replaced during Winsorization. $\nu = 27 - 4 - 4 = 19$ because the four "less-than" values and the four largest values have been replaced.

If the data are from a normal distribution, the upper and lower limits of a two-sided $100(1-\alpha)\%$ confidence interval of the mean η are:

$$\bar{y}_w \pm t_{\nu-1,\alpha/2} \frac{s_w}{\sqrt{n}}$$

where $t_{\nu-1,\alpha/2}$ cuts $100(\alpha/2)\%$ off each tail of the t distribution with $\nu - 1$ degrees of freedom. Note that the degrees of freedom are $\nu - 1$ instead of the usual $n - 1$, and s_w replaces s.

The trimmed mean and the Winsorized mean are best used on symmetric data sets that have either missing (censored) or unreliable data on the tail of the distribution. If the distribution is symmetric, they give unbiased estimates of the true mean.

Graphical Methods

The median, trimmed mean, and Winsorized mean give unbiased estimates of the mean only when the distribution is symmetrical. Many sets of environmental data are not symmetrical so other approaches are needed. Even if the distribution is symmetric, the proportion of censored observations may be so large that the trimmed or Winsorized estimates are not reliable. In these cases, we settle for doing something helpful without misleading.

Graphical methods, when properly used, reveal all the important features of the data. Graphical methods are especially important when one is unsure about a particular statistical method. It is better to display the data rather than to compute a summarizing statistic that hides the data structure and therefore may be misleading. Suitable plots very often enable a decision to be reached without further analysis.

Two useful graphs, the time series plot and the cumulative probability plot, are illustrated using a highly censored sample. Forty-five grab samples were taken on consecutive days from a large river. Table 15.1 gives the measured mercury concentrations and the days on which they were observed. Only 14 samples had a concentration above MDL of 0.2 $\mu g/L$ (today, the MDL for Hg is much lower).

With 31 out of 45 observations censored, it is impossible to compute any meaningful statistics for this data set. The median is below the MDL and, because more than half the data are censored, neither the trimmed or Winsorized mean can be computed. One approach is plot the data and see what can be learned. Figure 15.1 shows the measurable concentrations, which occur as seemingly random spikes.

Table 15.2 gives the plotting positions for the cumulative probability plot (Figure 15.2), which uses a logarithmic scale so the data will plot as a straight line. The plotting positions are $p = (i - 0.5)/n$. (See Chapter 5 for more about probability plots.)

TABLE 15.1

Measurable Mercury Concentrations

Time, Days	2	4	10	12	13	16	17	22	26	27	35	38	43	44
Hg ($\mu g/L$)	0.5	0.6	0.8	1.0	3.4	5.5	1.0	1.0	3.5	2.8	0.3	0.5	0.5	1.0

TABLE 15.2

Construction of the Cumulative Probability Distribution

Concentration y_i	No. of Obs. $\leq y_i$	Rank i	Probability Value $\leq y_i$ $p = (i - 0.5)/n$
<0.2	31	31	0.678
0.3	1	32	0.700
0.5	3	35	0.767
0.6	1	36	0.789
0.8	1	37	0.811
1.0	4	41	0.900
2.8	1	42	0.922
3.4	1	43	0.944
3.5	1	44	0.967
5.5	1	45	0.989

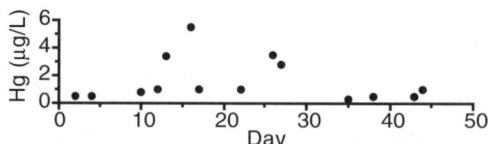

FIGURE 15.1 Time series plot of the data.

Analyzing Censored Data

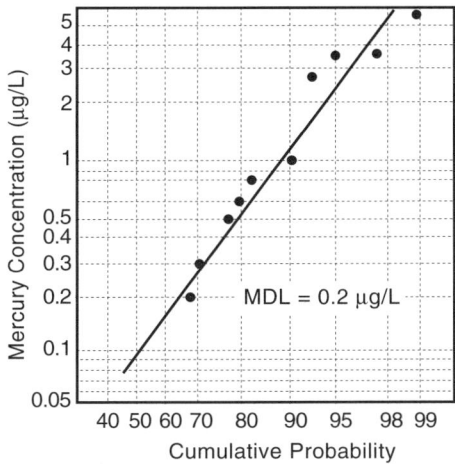

FIGURE 15.2 Cumulative probability plot.

The median is estimated as 0.1 μg/L by extrapolating the straight line below the MDL. We cannot be certain that data below the MDL will fall on the extrapolated straight line, but at least the available data do not preclude this possibility. The line was drawn by eye. In the next section, a regression method is used to fit the probability plot.

Regression on Rankits

It is possible to replace the probabilities with rankits (also called *normal order scores* or *order statistics*) and then to use regression to fit a line to the probability plot (Gilliom and Helsel, 1986; Hashimoto and Trussell, 1983; Travis and Land, 1990). This is equivalent to rescaling the graph in terms of standard deviations instead of probabilities.

If the data are normally distributed, or have been transformed to make them normal, the probabilities (p) are converted to rankits (normal order scores), $R_i = F^{-1}(p_i)$ where F^{-1} is the inverse cumulative normal probability distribution and p_i is the plotting position of the ith ranked observation. The rankits can be calculated (using the NORMSINV function in EXCEL) or looked up in standard statistical tables (for $n \leq 50$). The analysis could also be done using *probits*, which are obtained by adding five to the rankits to eliminate the negative values.

A straight line $y_i = b_0 + b_1 R_i$ is fitted to the rankits of the noncensored portion of the data. The rankits (R) are treated as the independent variable and the data are the dependent variable. For normally distributed data, the b_0 and b_1 are estimates of the mean and standard deviation of the noncensored distribution. The intercept b_0 estimates the median because rankits are symmetrical about zero and the 50th percentile corresponds to $R_i = 0$. For the normal distribution, the median equals the mean, so the 50th percentile also estimates the mean. The slope b_1 estimates the standard deviation because the rankits are scaled so that one rankit is one standard deviation on the original scale.

The method is demonstrated using the 45 values in Table 15.3 (these are random normal values). The concentrations have been ranked and the rankits are from the tables of Rohlf and Sokal (1981). Figure 15.3 shows the probability plots. The top panel shows the 45 normally distributed observations. The linear regression of concentration on rankits estimates a mean of 33.3 and a standard deviation of 5.4. The middle panel shows the result when the 10 smallest observations are censored (22% censoring). The bottom panel has the lowest 20 values censored (44% censoring) so the mean can still be estimated without extrapolation. The censoring hardly changed the estimates of the mean and standard deviation. All three data sets estimate a mean of about 33.5 and a standard deviation of about 5. It is apparent, however, that increased censoring gives more relative weight to values in the upper tail of the distribution.

TABLE 15.3

Censored Data Analysis Using Rankits (Normal Order Statistics)

Conc.	Rankit	Conc.	Rankit	Conc.	Rankit
23.8	−2.21	31.0	−0.40	36.4	0.46
24.4	−1.81	31.6	−0.34	37.0	0.52
25.0	−1.58	31.6	−0.28	37.0	0.59
25.0	−1.41	31.6	−0.22	37.0	0.65
25.6	−1.27	32.2	−0.17	37.0	0.72
25.6	−1.16	32.8	−0.11	37.0	0.80
28.0	−1.05	33.4	−0.06	37.6	0.88
28.0	−0.96	34.1	0.0	37.6	0.96
28.0	−0.88	34.6	0.06	38.2	1.05
28.6	−0.80	35.2	0.11	38.2	1.16
28.6	−0.72	35.2	0.17	39.4	1.27
29.2	−0.65	35.2	0.22	39.4	1.41
29.8	−0.59	35.8	0.28	40.6	1.58
29.8	−0.52	35.8	0.34	43.6	1.81
31.0	−0.46	35.8	0.40	47.8	2.21

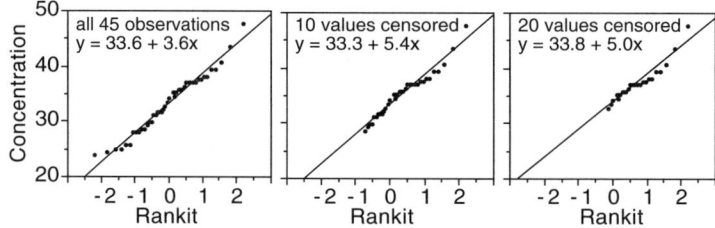

FIGURE 15.3 The left panel shows the rankit plot for a sample of 45 normally distributed values. In the middle panel, 10 values have been censored; the right panel has 20 censored values. Regression with rankits as the independent variables and concentration as the dependent variables gives the intercept (rankit = 0) as an estimated of the median = mean, and the slope as an estimate of the standard deviation.

If the data have been transformed to obtain normality, b_0 and b_1 estimate the mean and standard deviation of the transformed data, but not of the data on the original measurement scale. If normality was achieved by a logarithmic transformation $[z = \ln(y)]$, the 50th percentile (median of z) estimates the mean of z and is also the log of the geometric mean in the original metric. The slope will estimate the standard deviation of z. Exponentiation transforms the median on the log-scale transform to the median in the original metric. The back transformations to estimate the mean and standard deviation in the original metric are given in Chapter 7 and in Example 15.5.

Cohen's Maximum Likelihood Estimator Method

There are several methods to estimate the mean of a sample of censored data. Comparative studies show that none is always superior so we have chosen to present *Cohen's maximum likelihood method* (Cohen, 1959, 1961; Gilliom and Helsel, 1986; Haas and Scheff, 1990). It is easy to compute for samples from a normally distributed parent population or from a distribution that can be made normal by a log-arithmic transformation.

A sample of n observations has measured values of the variable only at $y \geq y_c$, where y_c is a known and fixed point of censoring. In our application, y_c is the MDL and it is assumed that the same MDL applies to each observation. Of the n total observations in the sample, n_c observations have $y \leq y_c$ and are censored. The number of observations with $y > y_c$ is $k = n - n_c$. The fraction of censored data is

Analyzing Censored Data

TABLE 15.4
Cohen's $\hat{\lambda}$ as a Function of $h = n/n_c$ and $\gamma = s^2/(\bar{y} - y_c)$

$\gamma \backslash h$	0.1	0.2	0.3	0.4	0.5
0.2	0.12469	0.27031	0.4422	0.6483	0.9012
0.3	0.13059	0.28193	0.4595	0.6713	0.9300
0.4	0.13595	0.29260	0.4755	0.6927	0.9570
0.5	0.14090	0.30253	0.4904	0.7129	0.9826
0.6	0.14552	0.31184	0.5045	0.7320	1.0070

Source: Cohen (1961).

$h = n_c/n$. Using the k fully measured observations, calculate crude estimates of the mean and variance of the noncensored data using:

$$\bar{y} = \frac{\sum y_i}{k} \quad \text{and} \quad s^2 = \frac{\sum (y_i - \bar{y})^2}{k}$$

Note that s^2 is calculated using a denominator of k instead of the usual $n - 1$.

For the case where the lower tail of the distribution is truncated, the estimated mean will be too high and the estimated variance will be too low. These estimates are adjusted using a factor λ. The maximum likelihood estimates for the mean η and the variance σ^2 are:

$$\hat{\eta} = \bar{y} - \hat{\lambda}(\bar{y} - y_c)$$

$$\hat{\sigma}^2 = s^2 + \hat{\lambda}(\bar{y} - y_c)^2$$

The hat (^) indicates that the value is an estimate ($\hat{\eta}$ is an estimate of the true mean η). Cohen (1961) provides tables of $\hat{\lambda}$ as a function of h and $\gamma = s^2/(\bar{y} - y_c)^2$. A portion of his table is reproduced in Table 15.4.

Example 15.4

Using Cohen's method, estimate the mean and variance of the censored sample from Example 15.1. The sample of $n = 27$ observations includes 23 measured and 4 censored values. The censoring limit is 6 µg/L. The 23 uncensored values are 6.9, 7.8, 8.9, 7.7, 9.6, 8.7, 6.7, 8.0, 8.5, 6.5, 9.2, 7.4, 6.3, 7.3, 8.3, 7.2, 7.5, 6.1, 9.4, 7.6, 8.1, 7.9, and 10.1. The average and variance of the $k = 23$ observations with measurable values are:

$$\bar{y} = \frac{\sum y_i}{k} = 7.9 \ \mu g/L$$

$$s^2 = \frac{\sum (y_i - \bar{y})^2}{k} = 1.1078 \ (\mu g/L)^2$$

The limit of censoring is at $y_c = 6$ and the proportion of censored data is:

$$h = 4/27 = 0.1481,$$

which gives:

$$\gamma = s^2/(\bar{y} - y_c)^2 = 1.1078/(7.9 - 6.0)^2 = 0.30688$$

The value of the adjustment factor $\hat{\lambda} = 0.20392$ is found by interpolating Table 15.4. The estimates of the mean and variance are:

$$\hat{\eta} = \bar{y} - \hat{\lambda}(\bar{y} - y_c) = 7.9 - 0.20392(7.9 - 6.0) = 7.51 \ \mu g/L$$

$$\hat{\sigma}^2 = s^2 + \hat{\lambda}(\bar{y} - y_c)^2 = 1.1078 + 0.20392(7.9 - 6.0)^2 = 1.844$$

$$\hat{\sigma} = 1.36 \ \mu g/L$$

The estimates $\hat{\eta}$ and $\hat{\sigma}^2$ cannot be used to compute confidence intervals in the usual way because both are affected by censoring. This means that these two statistics are not estimated independently (as they would be without censoring). Cohen (1961) provides formulas to correct for this dependency and explains how to compute confidence intervals.

The method described above can be used to analyze lognormally distributed data if the sample can be transformed to make it normal. The calculations are done on the transformed values. That is, if y is distributed lognormally, then $x = \ln(y)$ has a normal distribution. The computations are the same as illustrated above, except that they are done on the log-transformed values instead of on the original measured values.

Example 15.5

Use Cohen's method to estimate the mean and variance of a censored sample of $n = 30$ that is lognormally distributed. The 30 individual values ($y_1, y_2, \ldots, y_{30}$) are not given, but assume that a logarithmic transformation, $x = \ln(y)$, will make the distribution normal with mean η_x and variance σ_x^2.

Of the 30 measurements, 12 are censored at a limit of 18 $\mu g/L$, so the fraction censored is $h = 12/30 = 0.40$. The mean and variance computed from the logarithms of the noncensored values are:

$$\bar{x} = 3.2722 \quad \text{and} \quad s_x^2 = 0.03904$$

The limit of censoring is also transformed:

$$x_c = \ln(y_c) = \ln(18) = 2.8904$$

Using these values, we compute:

$$\gamma = s_x^2/(\bar{x} - x_c)^2 = 0.03904/(3.2722 - 2.8904)^2 = 0.2678.$$

which is used with $h = 0.4$ to interpolate $\hat{\lambda} = 0.664$ in Table 15.4. The estimated mean and variance of the log-transformed values are:

$$\hat{\eta}_x = 3.2722 - 0.664(3.2722 - 2.8904) = 3.0187$$

$$\hat{\sigma}_x^2 = 0.03904 + 0.664(3.2722 - 2.8904)^2 = 0.1358$$

The transformation equations to convert these into estimates of the mean and variance of the untransformed y's are:

$$\hat{\eta}_y = \exp(\hat{\eta}_x + 0.5\hat{\sigma}_x^2)$$

$$\hat{\sigma}_y^2 = \hat{\eta}_y^2 [\exp(\hat{\sigma}_x^2) - 1]$$

Substituting the parameter estimates $\hat{\eta}_x$ and $\hat{\sigma}_x^2$ gives:

$$\hat{\eta}_y = \exp[3.0187 + 0.5(0.1358)] = \exp(3.0866) = 21.9 \ \mu g/L$$

$$\hat{\sigma}_y^2 = (21.90)^2[\exp(0.1358) - 1] = 69.76 (\mu g/L)^2$$

$$\hat{\sigma}_y = 8.35 \ \mu g/L$$

The Delta-Lognormal Distribution

The *delta-lognormal method* estimates the mean of a sample of size n as a weighted average of n_c replaced censored values and $n - n_c$ uncensored lognormally distributed values. The Aitchison method (1955, 1969) assumes that all censored values are replaced by zeros ($D = 0$) and the noncensored values have a lognormal distribution. Another approach is to replace censored values by the detection limit ($D = $ MDL) or by some value between zero and the MDL (U.S. EPA, 1989; Owen and DeRouen, 1980).

The estimated mean is a weighted average of the mean of n_c values that are assigned value D and the mean of the $n - n_c$ fully measured values that are assumed to have a lognormal distribution with mean η_x and variance σ_x^2.

$$\hat{\eta}_y = D\frac{n_c}{n} + \left(1 - \frac{n_c}{n}\right)\exp(\hat{\eta}_x - 0.5\hat{\sigma}_x^2)$$

where $\hat{\eta}_x$ and $\hat{\sigma}_x$ are the estimated mean and variance of the log-transformed noncensored values.

This method gives results that agree well with Cohen's method, but it is not consistently better than Cohen's method. One reason is that the user is required to assume that all censored values are located at a single value, which may be zero, the limit of detection, or something in between.

Comments

The problem of censored data starts when an analyst decides not to report a numerical value and instead reports "not detected." It would be better to have numbers, even if the measurement error is large relative to the value itself, as long as a statement of the measurement's precision is provided. Even if this were practiced universally in the future, there remain many important data sets that have already been censored and must be analyzed.

Simply replacing and deleting censored values gives biased estimates of both the mean and the variance. The median, trimmed mean, and Winsorized mean provide unbiased estimates of the mean when the distribution is symmetric. The trimmed mean is useful for up to 25% censoring, and the Winsorized mean for up to 15% censoring. These methods fail when more than half the observations are censored. In such cases, the best approach is to display the data graphically. Simple time series plots and probability plots will reveal a great deal about the data and will never mislead, whereas presenting any single numerical value may be misleading.

The time series plot gives a good impression about variability and randomness. The probability plot shows how frequently any particular value has occurred. The probability plot can be used to estimate the median value. If the median is above the MDL, draw a smooth curve through the plotted points and estimate the median directly. If the median is below the MDL, extrapolation will often be justified on the basis of experience with similar data sets. If the data are distributed normally, the median is also the arithmetic mean. If the distribution is lognormal, the median is the geometric mean.

The precision of the estimated mean and variances becomes progressively worse as the fraction of observations censored increases. Comparative studies (Gilliom and Helsel, 1986; Haas and Scheff, 1990; Newman et al., 1989) on simulated data show that Cohen's method works quite well for up to 20% censoring. Of the methods studied, none was always superior, but Cohen's was always one of the best. As the extent of censoring reaches 20 to 50%, the estimates suffer increased bias and variability.

Historical records of environmental data often consist of information combined from several different studies that may be censored at different detection limits. Older data may be censored at 1 mg/L while the most recent are censored at 10 μg/L. Cohen (1963), Helsel and Cohen (1988), and NCASI (1995) provide methods for estimating the mean and variance of progressively censored data sets.

The Cohen method is easy to use for data that have a normal or lognormal distribution. Many sets of environmental samples are lognormal, at least approximately, and a log transformation can be used. Failing to transform the data when they are skewed causes serious bias in the estimates of the mean.

The normal and lognormal distributions have been used often because we have faith in these familiar models and it is not easy to verify any other true distribution for a small sample ($n = 20$ to 50), which is the size of many data sets. Hahn and Shapiro (1967) showed this graphically and Shumway et al. (1989) have shown it using simulated data sets. They have also shown that when we are unsure of the correct distribution, making the log transformation is usually beneficial or, at worst, harmless.

References

Aitchison, J. (1955). "On the Distribution of a Positive Random Variable Having a Discrete Probability Mass at the Origin," *J. Am. Stat. Assoc.,* 50, 901–908.

Aitchison, J. and J. A. Brown (1969). *The Lognormal Distribution,* Cambridge, England, Cambridge University Press.

Berthouex, P. M. and L. C. Brown (1994). *Statistics for Environmental Engineers,* Boca Raton, FL, Lewis Publishers.

Blom, G. (1958). *Statistical Estimates and Transformed Beta Variables,* New York, John Wiley.

Cohen, A. C., Jr. (1959). "Simplified Estimators for the Normal Distribution when Samples are Singly Censored or Truncated," *Technometrics,* 1, 217–237.

Cohen, A. C., Jr. (1961). "Tables for Maximum Likelihood Estimates: Singly Truncated and Singly Censored Samples," *Technometrics,* 3, 535–551.

Cohen, A. C. (1979). "Progressively Censored Sampling in the Three Parameter Log-Normal Distribution," *Technometrics,* 18, 99–103.

Cohen, A. C., Jr. (1963). "Progressively Censored Samples in Life Testing," *Technometrics,* 5(3), 327–339.

Gibbons, R. D. (1994). *Statistical Methods for Groundwater Monitoring,* New York, John Wiley.

Gilbert, R. O. (1987). *Statistical Methods for Environmental Pollution Monitoring,* New York, Van Nostrand Reinhold.

Gilliom, R. J. and D. R. Helsel (1986). "Estimation of Distribution Parameters for Censored Trace Level Water Quality Data. 1. Estimation Techniques," *Water Resources Res.,* 22, 135–146.

Hashimoto, L. K. and R. R. Trussell (1983). *Proc. Annual Conf. of the American Water Works Association,* p. 1021.

Haas, C N. and P. A. Scheff (1990). "Estimation of Averages in Truncated Samples," *Environ. Sci. Tech.,* 24, 912–919.

Hahn, G. A. and W. Q. Meeker (1991). *Statistical Intervals: A Guide for Practitioners,* New York, John Wiley.

Hahn, G. A. and S. S. Shapiro (1967). *Statistical Methods for Engineers,* New York, John Wiley.

Helsel, D. R. and T. A. Cohen (1988). "Estimation of Descriptive Statistics for Multiply Censored Water Quality Data," *Water Resources Res.,* 24(12), 1997–2004.

Helsel, D. R. and R. J. Gilliom (1986). "Estimation of Distribution Parameters for Censored Trace Level Water Quality Data: 2. Verification and Applications," *Water Resources Res.,* 22, 146–55.

Hill, M. and W. J. Dixon (1982). "Robustness in Real Life: A Study of Clinical Laboratory Data," *Biometrics,* 38, 377–396.

Hoaglin, D. C., F. Mosteller, and J. W. Tukey (1983). *Understanding Robust and Exploratory Data Analysis,* New York, Wiley.

Mandel, J. (1964). *The Statistical Analysis of Experimental Data,* New York, Interscience Publishers.

NCASI (1991). "Estimating the Mean of Data Sets that Include Measurements Below the Limit of Detection," *Tech. Bull. No. 621,*

NCASI (1995). "Statistical Method and Computer Program for Estimating the Mean and Variance of Multi-Level Left-Censored Data Sets," *NCASI Tech. Bull.* 703. Research Triangle Park, NC.

Newman, M. C. and P. M. Dixon (1990). "UNCENSOR: A Program to Estimate Means and Standard Deviations for Data Sets with Below Detection Limit Observations," *Anal. Chem.,* 26(4), 26–30.

Newman, M. C., P. M. Dixon, B. B. Looney, and J. E. Pinder (1989). "Estimating Means and Variance for Environmental Samples with Below Detection Limit Observations," *Water Resources Bull.,* 25(4), 905–916.

Owen, W. J. and T. A. DeRouen (1980). "Estimation of the Mean for Lognormal Data Containing Zeros and Left-Censored Values, with Applications to the Measurement of Worker Exposure to Air Contaminants," *Biometrics,* 36, 707–719.

Rohlf, F. J. and R. R. Sokal (1981). *Statistical Tables,* 2nd ed., San Francisco, W. H. Freeman and Co.

Shumway, R. H., A. S. Azari, and P. Johnson (1989). "Estimating Mean Concentrations under Transformation for Environmental Data with Detection Limits," *Technometrics,* 31(3), 347–356.

Travis, C. C. and M. L. Land (1990). "The Log-Probit Method of Analyzing Censored Data," *Envir. Sci. Tech.,* 24(7), 961–962.

U.S. EPA (1989). *Methods for Evaluating the Attainment of Cleanup Standards, Vol. 1: Soils and Solid Media,* Washington, D.C.

Exercises

15.1 Chlorophenol. The sample of $n = 20$ observations of chlorophenol was reported with the four values below 50 g/L, shown in brackets, reported as "not detected" (ND).

63	78	89	[32]	77	96	87	67	[28]	80
100	85	[45]	92	74	63	[42]	73	83	87

(a) Estimate the average and variance of the sample by (i) replacing the censored values with 50, (ii) replacing the censored values with 0, (iii) replacing the censored values with half the detection limit (25) and (iv) by omitting the censored values. Comment on the bias introduced by these four replacement methods.

(b) Estimate the median and the trimmed mean.

(c) Estimate the population mean and standard deviation by computing the Winsorized mean and standard deviation.

15.2 Lead in Tap Water. The data below are lead measurements on tap water in an apartment complex. Of the total $n = 140$ apartments sampled, 93 had a lead concentration below the limit of detection of 5 μg/L. Estimate the median lead concentration in the 140 apartments. Estimate the mean lead concentration.

Pb (μg/L)	0–4.9	5.0–9.9	10–14.9	15–19.9	20–29.9	30–39.9	40–49.9	50–59.9	60–69.9	70–79.9
Number	93	26	6	4	7	1	1	1	0	1

15.3 Lead in Drinking Water. The data below are measurements of lead in tap water that were sampled early in the morning after the tap was allowed to run for one minute. The analytical limit of detection was 5 μg/L, but the laboratory has reported values that are lower than this. Do the values below 5 μg/L fit the pattern of the other data? Estimate the median and the 90th percentile concentrations.

Pb (µg/L)	Number	%	Cum. %
0–0.9	20	0.143	0.143
1–1.9	16	0.114	0.257
2–2.9	32	0.229	0.486
3–3.9	11	0.079	0.564
4–4.9	13	0.093	0.657
5–9.9	27	0.193	0.850
10–14.9	7	0.050	0.900
15–19.9	4	0.029	0.929
20–29.9	6	0.043	0.971
30–39.9	1	0.007	0.979
40–49.9	1	0.007	0.986
50–59.9	1	0.007	0.993
60–69.9	0	0.000	0.993
70–79.9	1	0.007	1.000

Source: Prof. David Jenkins, University of California-Berkeley.

15.4 Rankit Regression. The table below gives eight ranked observations of a lognormally distributed variable y, the log-transformed values x, and their rankits.

(a) Make conventional probability plots of the x and y values. (b) Make plots of x and y versus the rankits. (c) Estimate the mean and standard deviation. ND = not detected (<MDL).

y	ND	ND	11.6	19.4	22.9	24.6	26.8	119.4
$x = \ln(y)$	—	—	2.451	2.965	3.131	3.203	3.288	4.782
Rankit	−1.424	−0.852	−0.473	−0.153	0.153	0.473	0.852	1.424

15.5 Cohen's Method — Normal. Use Cohen's method to estimate the mean and standard deviation of the $n = 26$ observations that have been censored at $y_c = 7$.

ND	ND	ND	ND	ND	ND	ND	ND	7.8	8.9	7.7	9.6	8.7
8.0	8.5	9.2	7.4	7.3	8.3	7.2	7.5	9.4	7.6	8.1	7.9	10.1

15.6 Cohen's Method — Lognormal. Use Cohen's method to estimate the mean and standard deviation of the following lognormally distributed data, which has been censored at 10 mg/L.

14	15	16	ND	72	ND	12	ND	ND	20	52	16	25	33	ND	62

15.7 PCB in Sludge. Seven of the sixteen measurements of PCB in a biological sludge are below the MDL of 5 mg/kg. Do the data appear better described by a normal or lognormal distribution? Use Cohen's method to obtain MLE estimates of the population mean and standard deviation.

ND	ND	ND	ND	ND	ND	ND	6	10	12	16	16	17
19	37	41										

16
Comparing a Mean with a Standard

KEY WORDS t-*test, hypothesis test, confidence interval, dissolved oxygen, standard.*

A common and fundamental problem is making inferences about mean values. This chapter is about problems where there is only one mean and it is to be compared with a known value. The following chapters are about comparing two or more means.

Often we want to compare the mean of experimental data with a known value. There are four such situations:

1. In laboratory quality control checks, the analyst measures the concentration of test specimens that have been prepared or calibrated so precisely that any error in the quantity is negligible. The specimens are tested according to a prescribed analytical method and a comparison is made to determine whether the measured values and the known concentration of the standard specimens are in agreement.
2. The desired quality of a product is known, by specification or requirement, and measurements on the process are made at intervals to see if the specification is accomplished.
3. A vendor claims to provide material of a certain quality and the buyer makes measurements to see whether the claim is met.
4. A decision must be made regarding compliance or noncompliance with a regulatory standard at a hazardous waste site (ASTM, 1998).

In these situations there is a single known or specified numerical value that we set as a standard against which to judge the average of the measured values. Testing the magnitude of the difference between the measured value and the standard must make allowance for random measurement error. The statistical method can be to (1) calculate a confidence interval and see whether the known (standard) value falls within the interval, or (2) formulate and test a hypothesis. The objective is to decide whether we can confidently declare the difference to be positive or negative, or whether the difference is so small that we are uncertain about the direction of the difference.

Case Study: Interlaboratory Study of DO Measurements

This example is loosely based on a study by Wilcock et al. (1981). Fourteen laboratories were sent standardized solutions that were prepared to contain 1.2 mg/L dissolved oxygen (DO). They were asked to measure the DO concentration using the Winkler titration method. The concentrations, as mg/L DO, reported by the participating laboratories were:

    1.2 1.4 1.4 1.3 1.2 1.35 1.4 2.0 1.95 1.1 1.75 1.05 1.05 1.4

Do the laboratories, on average, measure 1.2 mg/L, or is there some bias?

Theory: *t*-Test to Assess Agreement with a Standard

The known or specified value is defined as η_0. The true, but unknown, mean value of the tested specimens is η, which is estimated from the available data by calculating the average $\bar{y}$.

We do not expect to observe that $\bar{y} = \eta_0$, even if $\eta = \eta_0$. However, if $\bar{y}$ is near η_0, it can reasonably be concluded that $\eta = \eta_0$ and that the measured value agrees with the specified value. Therefore, some statement is needed as to how close we can reasonably expect the estimate to be. If the process is on-standard or on-specification, the distance $\bar{y} - \eta_0$ will fall within bounds that are a multiple of the standard deviation of the measurements.

We make use of the fact that for $n < 30$,

$$t = \frac{\bar{y} - \eta}{s/\sqrt{n}}$$

is a random variable which has a t distribution with $\nu = n - 1$ degrees of freedom. s is the sample standard deviation. Consequently, we can assert, with probability $1 - \alpha$, that the inequality:

$$-t_{\nu, \alpha/2} \leq \frac{\bar{y} - \eta}{s/\sqrt{n}} \leq t_{\nu, \alpha/2}$$

will be satisfied. This means that the maximum value of the error $|\bar{y} - \eta|$ is:

$$|\bar{y} - \eta| = t_{\nu, \alpha/2} \frac{s}{\sqrt{n}}$$

with probability $1 - \alpha$. In other words, we can assert with probability $1 - \alpha$ that the error in using $\bar{y}$ to estimate η will be at most $t_{\nu, \alpha/2} \frac{s}{\sqrt{n}}$.

From here, the comparison of the estimated mean with the standard value can be done as a hypothesis test or by computing a confidence interval. The two approaches are equivalent and will lead to the same conclusion. The confidence interval approach is more direct and often appeals to engineers.

Testing the Null Hypothesis

The comparison between $\bar{y}$ and η_0 can be stated as a null hypothesis:

$$H_0: \eta - \eta_0 = 0$$

which is read "the expected difference between η and η_0 is zero." The "null" is the zero. The extent to which $\bar{y}$ differs from η will be due to only random measurement error and not to bias. The extent to which $\bar{y}$ differs from η_0 will be due to both random error and bias. We hypothesize the bias ($\eta - \eta_0$) to be zero, and test for evidence to the contrary.

The sample average is:

$$\bar{y} = \frac{\sum y_i}{n}$$

The sample variance is:

$$s^2 = \frac{\sum(y_i - \bar{y})}{n - 1}$$

and the standard error of the mean is:

$$s_{\bar{y}} = \frac{s}{\sqrt{n}}$$

The t statistic is constructed assuming the null hypothesis to be true (i.e., $\eta = \eta_0$):

$$t_0 = \frac{\bar{y} - \eta_0}{s_{\bar{y}}} = \frac{\bar{y} - \eta_0}{s/\sqrt{n}}$$

On the assumption of random sampling from a normal distribution, t_0 will have a t-distribution with $v = n - 1$ degrees of freedom. Notice that t_0 may be positive or negative, depending upon whether $\bar{y}$ is greater or less than η_0.

For a one-sided test that $\eta > \eta_0$ (or $\eta < \eta_0$), the null hypothesis is rejected if the absolute value of the calculated t_0 is greater than $t_{v,\alpha}$ where α is the selected probability point of the t distribution with $v = n - 1$ degrees of freedom.

For a two-sided test ($\eta > \eta_0$ or $\eta < \eta_0$), the null hypothesis is rejected if the absolute value of the calculated t_0 is greater than $t_{v,\alpha/2}$, where α/z is the selected probability point of the t distribution with $v = n - 1$ degrees of freedom. Notice that the one-sided test uses t_α and the two-sided test uses $t_{\alpha/2}$, where the probability α is divided equally between the two tails of the t distribution.

Constructing the Confidence Interval

The $(1 - \alpha)100\%$ confidence interval for the difference $\bar{y} - \eta$ is constructed using t distribution as follows:

$$-t_{v,a/2} s_{\bar{y}} < \bar{y} - \eta < +t_{v,a/2} s_{\bar{y}}$$

If this confidence interval does not include $(\bar{y} - \eta_0)$, the difference between the known and measured values is so large that it is unlikely to arise from chance. It is concluded that there is a difference between the estimated mean and the known value η_0.

A similar confidence interval can be defined for the true population mean:

$$\bar{y} - t_{v,a/2} s_{\bar{y}} < \eta < \bar{y} + t_{v,a/2} s_{\bar{y}}$$

If the standard η_0 falls outside this interval, it is declared to be different from the true population mean η, as estimated by $\bar{y}$, which is declared to be different from η_0.

Case Study Solution

The concentration of the standard specimens that were analyzed by the participating laboratories was 1.2 mg/L. This value was known with such accuracy that it was considered to be the standard: $\eta_0 = 1.2$ mg/L. The average of the 14 measured DO concentrations is $\bar{y} = 1.4$ mg/L, the standard deviation is $s = 0.31$ mg/L, and the standard error is $s_{\bar{y}} = 0.083$ mg/L. The difference between the known and measured average concentrations is $1.4 - 1.2 = 0.2$ mg/L. A t-test can be used to assess whether 0.2 mg/L is so large as to be unlikely to occur through chance. This must be judged relative to the variation in the measured values.

The test t statistic is $t_0 = (1.4 - 1.2)/0.083 = 2.35$. This is compared with the t distribution with $v = 13$ degrees of freedom, which is shown in Figure 16.1a. The values $t = -2.16$ and $t = +2.16$ that cut off 5% of the area under the curve are shaded in Figure 16.1. Notice that the $\alpha = 5\%$ is split between 2.5% on the upper tail plus 2.5% on the lower tail of the distribution. The test value of $t_0 = 2.35$, located by the arrow, falls outside this range and therefore is considered to be exceptionally large. We conclude that it is highly unlikely (less than 5% chance) that such a difference would occur by chance. The estimate of the true mean concentration, $\bar{y} = 1.4$, is larger than the standard value, $\eta_0 = 1.2$, by an amount that cannot be attributed to random experimental error. There must be *bias error* to explain such a large difference.

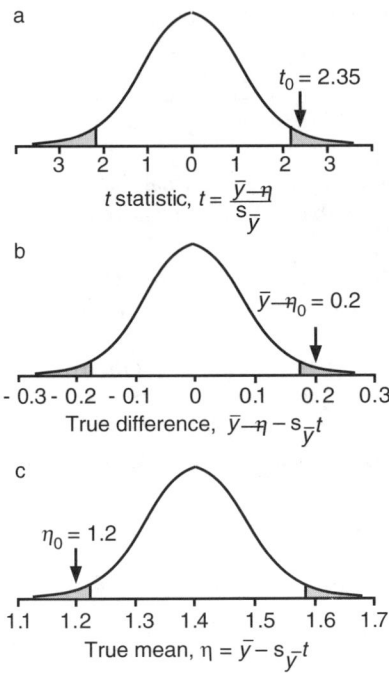

FIGURE 16.1 Three equivalent reference distributions scaled to compare the observed average with the known value on the basis of the distribution of the (a) t statistic, (b) difference between the observed average and the known level, and (c) true mean. The distributions were constructed using $\eta_0 = 1.2$ mg/L, $\bar{y} = 1.4$ mg/L, $t_{\nu=13,\,\alpha/2=0.025} = 2.16$, and $S_{\bar{y}} = 0.083$ mg/L.

In statistical jargon this means "the null hypothesis is rejected." In engineering terms this means "there is strong evidence that the measurement method used in these laboratories gives results that are too high."

Now we look at the equivalent interpretation using a 95% confidence interval for the difference $\bar{y} - \eta$. This is constructed using $t = 2.16$ for $\alpha/2 = 0.025$ and $\nu = 13$. The difference has expected value zero under the null hypothesis, and will vary over the interval $\pm t_{13,0.025} s_{\bar{y}} = \pm 2.16(0.083) = \pm 0.18$ mg/L. The portion of the reference distribution for the difference that falls outside this range is shaded in Figure 16.1b. The difference between the observed and the standard, $\bar{y} - \eta_0 = 0.2$ mg/L, falls beyond the 95% confidence limits. We conclude that the difference is so large that it is unlikely to occur due to random variation in the measurement process. "Unlikely" means "a probability of 5% that a difference this large could occur due to random measurement variation."

Figure 16.1c is the reference distribution that shows the expected variation of the true mean (η) about the average. It also shows the 95% confidence interval for the mean of the concentration measurements. The true mean is expected to fall within the range of $1.4 \pm 2.16(0.083) = 1.4 \pm 0.18$. The lower bound of the 95% confidence interval is 1.22 and the upper bound is 1.58. The standard value of 1.2 mg/L does not fall within the 95% confidence interval, which leads us to conclude that the true mean of the measured concentration is higher than 1.2 mg/L.

The shapes of the three reference distributions are identical. The only difference is the scaling of the horizontal axis, whether we choose to consider the difference in terms of the t statistic, the difference, or the concentration scale. Many engineers will prefer to make this judgment on the basis of a value scaled as the measured values are scaled (e.g., as mg/L instead of on the dimensionless scale of the t statistic). This is done by computing the confidence intervals either for the difference ($\bar{y} - \eta$) or for the mean η.

The conclusion that the average of the measured concentrations is higher than the known concentration of 1.2 mg/L could be viewed in two ways. The high average could happen because the measurement method is biased: only three labs measured less than 1.2 mg/L. Or it could result from the high concentrations (1.75 mg/L and 1.95 mg/L) measured by two laboratories. To discover which is the case, send out more standard specimens and ask the labs to try again. (This may not answer the question. What often happens when labs get feedback from quality control checks is that they improve their performance. This is actually the desired result because the objective is to attain uniformly excellent performance and not to single out poor performers.)

On the other hand, the measurement method might be all right and the true concentration might be higher than 1.2 mg/L. This experiment does not tell us which interpretation is correct. It is not a simple matter to make a standard solution for DO; dissolved oxygen can be consumed in a variety of reactions. Also, its concentration can change upon exposure to air when the specimen bottle is opened in the laboratory. In contrast, a substance like chloride or zinc will not be lost from the standard specimen, so the concentration actually delivered to the chemist who makes the measurements is the same concentration in the specimen that was shipped. In the case of oxygen at low levels, such as 1.2 mg/L, it is not likely that oxygen would be lost from the specimen during handling in the laboratory. If there is a change, the oxygen concentration is more likely to be increased by dissolution of oxygen from the air. We cannot rule out this causing the difference between 1.4 mg/L measured and 1.2 mg/L in the original standard specimens. Nevertheless, the chemists who arranged the test believed they had found a way to prepare stable test specimens, and they were experienced in preparing standards for interlaboratory tests. We have no reason to doubt them. More checking of the laboratories seems a reasonable line of action.

Comments

The classical *null hypothesis* is that "The difference is zero." No scientist or engineer ever believes this hypothesis to be strictly true. There will always be a difference, at some decimal point. Why propose a hypothesis that we believe is not true? The answer is a philosophical one. We cannot prove equality, but we may collect data that shows a difference so large that it is unlikely to arise from chance. The null hypothesis therefore is an artifice for letting us conclude, at some stated level of confidence, that there is a difference. If no difference is evident, we state, "The evidence at hand does not permit me to state with a high degree of confidence that the measurements and the standard are different." The null hypothesis is tested using a *t*-test.

The alternate, but equivalent, approach to testing the null hypothesis is to compute the interval in which the difference is expected to fall if the experiment were repeated many, many times. This interval is a *confidence interval*. Suppose that the value of a primary standard is 7.0 and the average of several measurements is 7.2, giving a difference of 0.20. Suppose further that the 95% confidence interval shows that the true difference is between 0.12 to 0.28. This is what we want to know: the true difference is not zero.

A confidence interval is more direct and often less confusing than null hypotheses and significance tests. In this book we prefer to compute confidence intervals instead of making significance tests.

References

ASTM (1998). *Standard Practice for Derivation of Decision Point and Confidence Limit Testing of Mean Concentrations in Waste Management Decisions,* D 6250, Washington, D.C., U.S. Government Printing Office.

Wilcock, R. J., C. D. Stevenson, and C. A. Roberts (1981). "An Interlaboratory Study of Dissolved Oxygen in Water," *Water Res.,* 15, 321–325.

Exercises

16.1 Boiler Scale. A company advertises that a chemical is 90% effective in cleaning boiler scale and cites as proof a sample of ten random applications in which an average of 81% of boiler scale was removed. The government says this is false advertising because 81% does not equal 90%. The company says the statistical sample is 81% but the true effectiveness may easily be 90%. The data, in percentages, are 92, 60, 77, 92, 100, 90, 91, 82, 75, 50. Who is correct and why?

16.2 Fermentation. Gas produced from a biological fermentation is offered for sale with the assurance that the average methane content is 72%. A random sample of $n = 7$ gas specimens gave methane contents (as %) of 64, 65, 75, 67, 65, 74, and 75. (a) Conduct hypothesis tests at significance levels of 0.10, 0.05, and 0.01 to determine whether it is fair to claim an average of 72%. (b) Calculate 90%, 95%, and 99% confidence intervals to evaluate the claim of an average of 72%.

16.3 TOC Standards. A laboratory quality assurance protocol calls for standard solutions having 50 mg/L TOC to be randomly inserted into the work stream. Analysts are blind to these standards. Estimate the bias and precision of the 16 most recent observations on such standards. Is the TOC measurement process in control?

50.3 51.2 50.5 50.2 49.9 50.2 50.3 50.5 49.3 50.0 50.4 50.1 51.0 49.8 50.7 50.6

16.4 Discharge Permit. The discharge permit for an industry requires the monthly average COD concentration to be less than 50 mg/L. The industry wants this to be interpreted as "50 mg/L falls within the confidence interval of the mean, which will be estimated from 20 observations per month." For the following 20 observations, would the industry be in compliance according to this interpretation of the standard?

57 60 49 50 51 60 49 53 49 56 64 60 49 52 69 40 44 38 53 66

17

Paired t-Test for Assessing the Average of Differences

KEY WORDS *confidence intervals, paired t-test, interlaboratory tests, null hypothesis, t-test, dissolved oxygen, pooled variance.*

A common question is: "Do two different methods of doing A give different results?" For example, two methods for making a chemical analysis are compared to see if the new one is equivalent to the older standard method; algae are grown under different conditions to study a factor that is thought to stimulate growth; or two waste treatment processes are tested at different levels of stress caused by a toxic input. In the strict sense, we do not believe that the two analytical methods or the two treatment processes are identical. There will always be some difference. What we are really asking is: "Can we be highly confident that the difference is positive or negative?" or "How large might the difference be?"

A key idea is that the *design of the experiment* determines the way we compare the two treatments. One experimental design is to make a series of tests using treatment A and then to independently make a series of tests using method B. Because the data on methods A and B are independent of each other, they are compared by computing the average for each treatment and using an *independent* t-*test* to assess the difference of the two averages.

A second way of designing the experiment is to pair the samples according to time, technician, batch of material, or other factors that might contribute to a difference between the two measurements. Now the test results on methods A and B are produced in pairs that are not independent of each other, so the analysis is done by averaging the differences for each pair of test results. Then a *paired* t-*test* is used to assess whether the average of these difference is different from zero. The paired *t*-test is explained here; the independent *t*-test is explained in Chapter 18.

Two samples are said to be paired when each data point in the first sample is matched and related to a unique data point in the second sample. Paired experiments are used when it is difficult to control all factors that might influence the outcome. If these factors cannot be controlled, the experiment is arranged so they are equally likely to influence both of the paired observations.

Paired experiments could be used, for example, to compare two analytical methods for measuring influent quality at a wastewater treatment plant. The influent quality will change from moment to moment. To eliminate variation in influent quality as a factor in the comparative experiment, paired measurements could be made using both analytical methods on the same specimen of wastewater. The alternative approach of using method A on wastewater collected on day one and then using method B on wastewater collected at some later time would be inferior because the difference due to analytical method would be overwhelmed by day-to-day differences in wastewater quality. This difference between paired same-day tests is not influenced by day-to-day variation. Paired data are evaluated using the paired *t*-test, which assesses the average of the differences of the pairs.

To summarize, the test statistic that is used to compare two treatments is as follows: when assessing the *difference of two averages*, we use the independent *t*-test; when assessing the *average of paired differences*, we use the paired *t*-test. Which method is used depends on the design of the experiment. We know which method will be used before the data are collected.

Once the appropriate difference has been computed, it is examined to decide whether we can confidently declare the difference to be positive, or negative, or whether the difference is so small that we are uncertain about the direction of the difference. The standard procedure for making such comparisons is to construct a null hypothesis that is tested statistically using a *t*-test. The classical null hypothesis is: "The difference between the two methods is zero." We do not expect two methods to give exactly the same results, so it may seem strange to investigate a hypothesis that is certainly wrong. The philosophy is the same as in law where the accused is presumed innocent until proven guilty. We cannot prove a person innocent, which is why the verdict is worded "not guilty" when the evidence is insufficient to convict. In a statistical comparison, we cannot prove that two methods are the same, but we can collect evidence that shows them to be different. The null hypothesis is therefore a philosophical device for letting us avoid saying that two things are equal. Instead we conclude, at some stated level of confidence, that "there is a difference" or that "the evidence does not permit me to confidently state that the two methods are different."

An alternate, but equivalent, approach to constructing a null hypothesis is to compute the difference and the interval in which the difference is expected to fall if the experiment were repeated many, many times. This interval is called the *confidence interval*. For example, we may determine that "A − B = 0.20 and that the true difference falls in the interval 0.12 to 0.28, this statement being made at a 95% level of confidence." This tells us all that is important. We are highly confident that A gives a result that is, on average, higher than B. And it tells all this without the sometimes confusing notions of null hypothesis and significance tests.

Case Study: Interlaboratory Study of Dissolved Oxygen

An important procedure in certifying the quality of work done in laboratories is the analysis of standard specimens that contain known amounts of a substance. These specimens are usually introduced into the laboratory routine in a way that keeps the analysts blind to the identity of the sample. Often the analyst is blind to the fact that quality assurance samples are included in the assigned work. In this example, the analysts were asked to measure the dissolved oxygen (DO) concentration of the same specimen using two different methods.

Fourteen laboratories were sent a test solution that was prepared to have a low dissolved oxygen concentration (1.2 mg/L). Each laboratory made the measurements using the Winkler method (a titration) and the electrode method. The question is whether the two methods predict different DO concentrations. Table 17.1 shows the data (Wilcock et al., 1981). The observations for each method may be assumed random and independent as a result of the way the test was designed. The differences plotted in Figure 17.1 suggest that the Winkler method may give DO measurements that are slightly lower than the electrode method.

TABLE 17.1

Dissolved Oxygen Data from the Interlaboratory Study

Laboratory	1	2	3	4	5	6	7	8	9	10	11	12	13	14
Winkler	1.2	1.4	1.4	1.3	1.2	1.3	1.4	2.0	1.9	1.1	1.8	1.0	1.1	1.4
Electrode	1.6	1.4	1.9	2.3	1.7	1.3	2.2	1.4	1.3	1.7	1.9	1.8	1.8	1.8
Diff. (W − E)	−0.4	0.0	−0.5	−1.0	−0.5	0.0	−0.8	0.6	0.6	−0.6	−0.1	−0.8	−0.7	−0.4

Source: Wilcock, R. J., C. D. Stevenson, and C. A. Roberts (1981). *Water Res.,* 15, 321–325.

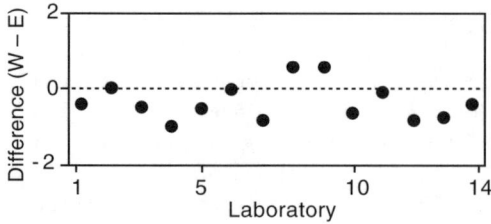

FIGURE 17.1 The DO data and the differences of the paired values.

Theory: The Paired t-Test Analysis

Define δ as the true mean of differences between random variables y_1 and y_2 that were observed as matched pairs under identical experimental conditions. δ will be zero if the means of the populations from which y_1 and y_2 are drawn are equal. The estimate of δ is the average of differences between n paired observations:

$$\bar{d} = \frac{\Sigma d_i}{n} = \frac{1}{n}\Sigma(y_{1,i} - y_{2,i})$$

Because of measurement error, the value of $\bar{d}$ is not likely to be zero, although it will tend toward zero if δ is zero.

The sample variance of the differences is:

$$s_d^2 = \frac{\Sigma(d_i - \bar{d})^2}{n-1}$$

The standard error of the average difference $\bar{d}$ is:

$$s_{\bar{d}} = \frac{s_d}{\sqrt{n}}$$

This is used to establish the $1 - \alpha$ confidence limits for δ, which are $\bar{d} \pm s_{\bar{d}} t_{n-1,\alpha/2}$. The correctness of this confidence interval depends on the data being independent and coming from distributions that are approximately normal with the same variance.

Case Study Solution

The differences were calculated by subtracting the electrode measurements from the Winkler measurements. The average of the paired differences is:

$$\bar{d} = \frac{(-0.4) + 0 + (-0.5) + (-1.0) + \cdots + (-0.4)}{14} = -0.329 \text{ mg/L}$$

and the variance of the paired differences is:

$$s_d^2 = \frac{[-0.4 - (-0.329)]^2 + [0 - (-0.329)]^2 + \cdots + [-0.4 - (-0.329)]^2}{14 - 1} = 0.244 \ (\text{mg/L})^2$$

giving $s_d = 0.494$ mg/L. The standard error of the average of the paired differences is:

$$s_{\bar{d}} = \frac{0.494}{\sqrt{14}} = 0.132 \text{ mg/L}$$

The $(1 - \alpha)100\%$ confidence interval is computed using the t distribution with $\nu = 13$ degrees of freedom at the $\alpha/2$ probability point. For $(1 - \alpha) = 0.95$, $t_{13,0.025} = 2.160$, and the 95% confidence interval of the true difference δ is:

$$\bar{d} - s_{\bar{d}} t_{13,0.025} < \delta < \bar{d} + s_{\bar{d}} t_{13,0.025}$$

For the particular values of this example:

$$-0.326 - 0.132(2.160) < \delta < -0.326 + 0.132(2.160)$$

$$-0.61 \text{ mg/L} < \delta < -0.04 \text{ mg/L}$$

We are highly confident that the difference between the two methods is not zero because the confidence interval does not include the difference of zero. The methods give different results and, furthermore, the electrode method has given higher readings than the Winkler method.

If the confidence interval had included zero, the interpretation would be that we cannot say with a high degree of confidence that the methods are different. We should be reluctant to report that the methods are the same or that the difference between the methods is zero because what we know about chemical measurements makes it unlikely that these statements are strictly correct. We may decide that the difference is small enough to have no practical importance. Or the range of the confidence interval might be large enough that the difference, if real, would be important, in which case additional tests should be done to resolve the matter.

An alternate but equivalent evaluation of the results is to test the null hypothesis that *the difference between the two averages is zero*. The way of stating the conclusion when the 95% confidence interval does not include zero is to say that "the difference was significant at the 95% confidence level." Significant, in this context, has a purely statistical meaning. It conveys nothing about how interesting or important the difference is to an engineer or chemist. Rather than reporting that the difference was significant (or not), communicate the conclusion more simply and directly by giving the confidence interval. Some reasons for preferring to look at the confidence interval instead of doing a significance test are given at the end of this chapter.

Why Pairing Eliminates Uncontrolled Disturbances

Paired experiments are used when it is difficult to control all the factors that might influence the outcome. A paired experimental design ensures that the uncontrolled factors contribute equally to both of the paired observations. The difference between the paired values is unaffected by the uncontrolled disturbances, whereas the differences of unpaired tests would reflect the additional component of experimental error. The following example shows how a large seasonal effect can be *blocked out* by the paired design. Block out means that the effect of seasonal and day-to-day variations are removed from the comparison.

Blocking works like this. Suppose we wish to test for differences in two specimens, A and B, that are to be collected on Monday, Wednesday, and Friday (M, W, F). It happens, perhaps because of differences in production rate, that Wednesday is always two (2) units higher than Monday, and Friday is always three (3) units higher than Monday. The data are:

| | Method | | |
Day	A	B	Difference
M	5	3	2
W	7	5	2
F	8	6	2
Averages	6.67	4.67	2
Variances	2.3	2.3	0

This day-to-day variation is blocked out if the analysis is done on $(A - B)_M$, $(A - B)_W$, and $(A - B)_F$ instead of the alternate $(A_M + A_W + A_F)/3 = \overline{A}$ and $(B_M + B_W + B_F)/3 = \overline{B}$. The difference between A and B is two (2) units. This is true whether we calculate the average of the differences $[(2 + 2 + 2)/3 = 2]$ or the difference of the averages $[6.67 - 4.67 = 2]$. The variance of the differences is zero, so it is clear that the difference between A and B is 2.0.

Paired t-Test for Assessing the Average of Differences

A *t*-test on the difference of the averages would conclude that A and B are not different. The reason is that the variance of the averages over M, W, and F is inflated by the day-to-day variation. This day-to-day variation overwhelms the analysis; pairing removes the problem.

The experimenter who does not think of pairing (blocking) the experiment works at a tremendous handicap and will make many wrong decisions. Imagine that the collecting for A was done on M, W, F, of one week and collection for B was done in another week. Now the paired analysis cannot be done and the difference will not be detected. This is why we speak of a *paired design* as well as of a *paired t-test analysis*. The crucial step is making the correct design. Pairing is always recommended.

Case Study to Emphasize the Benefits of a Paired Design

A once-through cooling system at a power plant is suspected of reducing the population of certain aquatic organisms. The copepod population density (organisms per cubic meter) were measured at the inlet and outlet of the cooling system on 17 different days (Simpson and Dudaitis, 1981). On each sampling day, water specimens were collected within a short time interval, first at the inlet and then at the outlet. The sampling plan represents a thoughtful effort to block out the effect of day-to-day and month-to-month variations in population counts. It pairs the inlet and outlet measurements. Of course, it is impossible to sample the same parcel of water at the inlet and outlet (i.e., the pairing is not exact), but any variation caused by this will be reflected as a component of the random measurement error.

The data are plotted in Figure 17.2. The plot gives the impression that the cooling system may not affect the copepods. The outlet counts are higher than inlets counts on 10 of the 17 days. There are some big differences, but these are on days when the count was very high and we expect that the measurement error in counting will be proportional to the population. (If you count 10 pennies you will get the right answer, but if you count 1000, you are certain to have some error; the more pennies the more counting error.) Before doing the calculations, consider once more why the paired comparison should be done.

Specimens 1 through 6 were taken in November 1977, specimens 7 through 12 in February 1978, and specimens 13 through 17 in August 1978. A large seasonal variation is apparent. If we were to compute the variances of the inlet and outlet counts, it would be huge and it would consist largely of variation due to seasonal differences. Because we are not trying to evaluate seasonal differences, this would be a poor way to analyze the data. The paired comparison operates on the differences of the daily inlet and outlet counts, and these differences do not reflect the seasonal variation (except, as we shall see in a moment, to the extent that the differences are proportional to the population density).

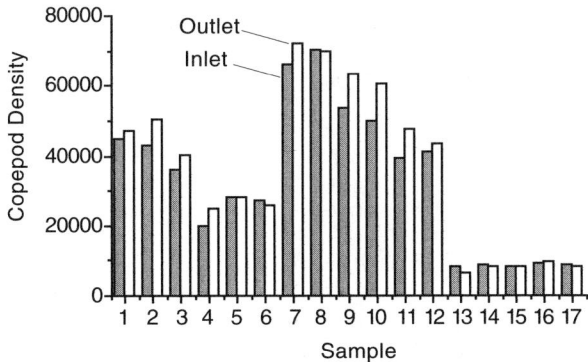

FIGURE 17.2 Copepod population density (organisms/m^3).

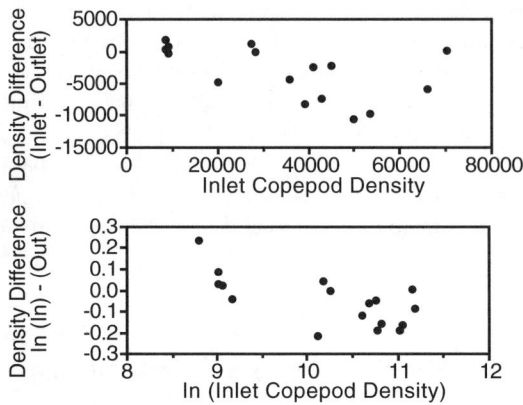

FIGURE 17.3 The difference in copepod inlet and outlet population density is larger when the population is large, indicating nonconstant variance at different population levels.

It is tempting to tell ourselves that "I would not be foolish enough not to do a paired comparison on data such as these." Of course we would not when the variation due to the nuisance factor (season) is both huge and obvious. But almost every experiment is at risk of being influenced by one or more nuisance factors, which may be known or unknown to the experimenter. Even the most careful experimental technique cannot guarantee that these will not alter the outcome. The paired experimental design will prevent this and it is recommended whenever the experiment can be so arranged.

Biological counts usually need to be transformed to make the variance uniform over the observed range of values. The paired analysis will be done on the differences between inlet and outlet, so it is the variance of these differences that should be examined. The differences are plotted in Figure 17.3. Clearly, the differences are larger when the counts are larger, which means that the variance is not constant over the range of population counts observed. Constant variance is one condition of the t-test because we want each observation to contribute in equal weight to the analysis. Any statistics computed from these data would be dominated by the large differences of the high population counts and it would be misleading to construct a confidence interval or test a null hypothesis using the data in their original form.

A transformation is needed to make the variance constant over the ten-fold range of the counts in the sample. A square-root transformation is often used on biological counts (Sokal and Rohlf, 1969), but for these data a log transformation seemed to be better. The bottom section of Figure 17.3 shows that the differences of the log-transformed data are reasonably uniform over the range of the transformed values.

Table 17.2 shows the data, the transformed data $[z = \ln(y)]$, and the paired differences. The average difference of $\ln(\text{in}) - \ln(\text{out})$ is $\bar{d} = \Sigma d_{\text{in}}/17 = -0.051$. The variance of the differences is $s^2 = \Sigma(d_i - \bar{d})^2/16 = 0.014$ and the standard error of average difference $s_{\bar{d}} = s/\sqrt{17} = 0.029$.

The 95% confidence interval is constructed using $t_{16, 0.025} = 2.12$. It can be stated with 95% confidence that the true difference falls in the region:

$$\bar{d}_{\ln} - s_{\bar{d}} t_{16, 0.025} < \delta_{\ln} < \bar{d}_{\ln} + s_{\bar{d}} t_{16, 0.025}$$

$$-0.051 - 2.12(0.029) < \delta_{\ln} < -0.051 + 2.12(0.029)$$

$$-0.112 < \delta_{\ln} < 0.010$$

This confidence interval includes zero so we can state with a high degree of confidence that outlet counts are not less than inlet counts.

TABLE 17.2

Outline of Computations for a Paired t-Test on the Copepod Data after a Logarithmic Transformation

Sample	Original Counts (no./m^3)			Transformed Data, $z = \ln(y)$		
	y_{in}	y_{out}	$d = y_{in} - y_{out}$	z_{in}	z_{out}	$d_{ln} = z_{in} - z_{out}$
1	44909	47069	−2160	10.712	10.759	−0.047
2	42858	50301	−7443	10.666	10.826	−0.160
3	35976	40431	−4455	10.491	10.607	−0.117
4	20048	24887	−4839	9.906	10.122	-0.216
5	28273	28385	−112	10.250	10.254	−0.004
6	27261	26122	1139	10.213	10.171	0.043
7	66149	72039	−5890	11.100	11.185	−0.085
8	70190	70039	151	11.159	11.157	0.002
9	53611	63228	−9617	10.890	11.055	−0.165
10	49978	60585	−10607	10.819	11.012	−0.192
11	39186	47455	−8269	10.576	10.768	−0.191
12	41074	43584	−2510	10.623	10.682	−0.059
13	8424	6640	1784	9.039	8.801	0.238
14	8995	8244	751	9.104	9.017	0.087
15	8436	8204	232	9.040	9.012	0.028
16	9195	9579	−384	9.126	9.167	−0.041
17	8729	8547	182	9.074	9.053	0.021
Average	33135	36196	−3062	10.164	10.215	−0.051
Std. deviation	20476	23013	4059	0.785	0.861	0.119
Std. error	4967	5582	984	0.190	0.209	0.029

Comments

The paired t-test examines the average of the differences between paired observations. This is not equivalent to comparing the difference of the average of two samples that are not paired. Pairing blocks out the variation due to uncontrolled or unknown experimental factors. As a result, the paired experimental design should be able to detect a smaller difference than an unpaired design. We do not have free choice of which t-test to use for a particular set of data. The appropriate test is determined by the experimental design.

We never really believe the null hypothesis. It is too much to expect that the difference between any two methods is truly zero. Tukey (1991) states this bluntly:

> Statisticians classically asked the wrong question — and were willing to answer with a lie ... They asked "Are the effects of A and B different?" and they were willing to answer "no."
>
> All we know about the world teaches us that A and B are always different — in some decimal place. Thus asking "Are the effects different?" is foolish.
>
> What we should be answering first is "Can we be confident about the direction from method A to method B? Is it up, down, or uncertain?"

If uncertain whether the direction is up or down, it is better to answer "we are uncertain about the direction" than to say "we reject the null hypothesis." If the answer was "direction certain," the follow-up question is how big the difference might be. This question is answered by computing confidence intervals.

Most engineers and scientists will like Tukey's view of this problem. Instead of accepting or rejecting a null hypothesis, compute and interpret the confidence interval of the difference. We want to know the confidence interval anyway, so this saves work while relieving us of having to remember exactly what it means to "fail to reject the null hypothesis." And it lets us avoid using the words *statistically significant*.

To further emphasize this, Hooke (1963) identified these inadequacies of significance tests.

1. The test is qualitative rather than quantitative. In dealing with quantitative variables, it is often wasteful to point an entire experiment toward determining the existence of an effect when the effect could also be measured at no extra cost. A confidence statement, when it can be made, contains all the information that a significance statement does, and more.
2. The word "significance" often creates misunderstandings, owing to the common habit of omitting the modifier "statistical." Statistical significance merely indicates that evidence of an effect is present, but provides no evidence in deciding whether the effect is large enough to be important. In a given experiment, *statistical significance is neither necessary nor sufficient for scientific or practical importance.* (emphasis added)
3. Since statistical significance means only that an effect can be seen in spite of the experimental error (a signal is heard above the noise), it is clear that the outcome of an experiment depends very strongly on the sample size. Large samples tend to produce significant results, while small samples fail to do so.

Now, having declared that we prefer not to state results as being significant or nonsignificant, we pass on two tips from Chatfield (1983) that are well worth remembering:

1. A nonsignificant difference is not necessarily the same thing as no difference.
2. A significant difference is not necessarily the same thing as an interesting difference.

References

Chatfield, C. (1983). *Statistics for Technology,* 3rd ed., London, Chapman & Hall.
Hooke, R. (1963). *Introduction to Scientific Inference,* San Francisco, CA, Holden-Day.
Simpson, R. D. and A. Dudaitis (1981). "Changes in the Density of Zooplankton Passing Through the Cooling System of a Power-Generating Plant," *Water Res.,* 15, 133–138.
Sokal, R. R. and F. J. Rohlf (1969). *Biometry: The Principles and Practice of Statistics in Biological Research,* New York, W. H. Freeman and Co.
Tukey, J. W. (1991). "The Philosophy of Multiple Comparisons," *Stat. Sci.,* 6(6), 100–116.
Wilcock, R. J., C. D. Stevenson, and C. A. Roberts (1981). "An Interlaboratory Study of Dissolved Oxygen in Water," *Water Res.,* 15, 321–325.

Exercises

17.1 Antimony. Antimony in fish was measured in three paired samples by an official standard method and a new method. Do the two methods differ significantly?

Sample No.	1	2	3
New method	2.964	3.030	2.994
Standard method	2.913	3.000	3.024

17.2 Nitrite Measurement. The following data were obtained from paired measurements of nitrite in water and wastewater by direct ion-selective electrode (ISE) and a colorimetric method. Are the two methods giving consistent results?

Method	Nitrite Measurements							
ISE	0.32	0.36	0.24	0.11	0.11	0.44	2.79	2.99	3.47
Colorimetric	0.36	0.37	0.21	0.09	0.11	0.42	2.77	2.91	3.52

17.3 BOD Tests. The data below are paired comparisons of BOD tests done in standard 300-mL bottles and experimental 60-mL bottles. Estimate the difference and the confidence interval of the difference between the results for the two bottle sizes.

300 mL	7.2	4.5	4.1	4.1	5.6	7.1	7.3	7.7	32	29	22	23	27
60 mL	4.8	4.0	4.7	3.7	6.3	8.0	8.5	4.4	30	28	19	26	28

Source: McCutcheon, S. C., *J. Env. Engr.* ASCE, 110, 697–701.

17.4 Leachate Tests. Paired leaching tests on a solid waste material were conducted for contact times of 30 and 75 minutes. Based on the following data, is the same amount of tin leached from the material at the two leaching times?

Leaching Time (min)	Tin Leached (mg/kg)							
30	51	60	48	52	46	58	56	51
75	57	57	55	56	56	55	56	55

17.5 Stream Monitoring. An industry voluntarily monitors a stream to determine whether its goal of raising the level of pollution by 4 mg/L or less is met. The observations below for September and April were made every fourth working day. Is the industry's standard being met?

September		April	
Upstream	Downstream	Upstream	Downstream
7.5	12.5	4.6	15.9
8.2	12.5	8.5	25.9
8.3	12.5	9.8	15.9
8.2	12.2	9.0	13.1
7.6	11.8	5.2	10.2
8.9	11.9	7.3	11.0
7.8	11.8	5.8	9.9
8.3	12.6	10.4	18.1
8.5	12.7	12.1	18.3
8.1	12.3	8.6	14.1

18

Independent t-Test for Assessing the Difference of Two Averages

KEY WORDS   *confidence interval, independent t-test, mercury.*

Two methods, treatments, or conditions are to be compared. Chapter 17 dealt with the experimental design that produces measurements from two treatments that were paired. Sometimes it is not possible to pair the tests, and then the averages of the two treatments must be compared using the *independent* t-*test*.

Case Study: Mercury in Domestic Wastewater

Extremely low limits now exist for mercury in wastewater effluent limits. It is often thought that whenever the concentration of heavy metals is too high, the problem can be corrected by forcing industries to stop discharging the offending substance. It is possible, however, for target effluent concentrations to be so low that they might be exceeded by the concentration in domestic sewage. Specimens of drinking water were collected from two residential neighborhoods, one served by the city water supply and the other served by private wells. The observed mercury concentrations are listed in Table 18.1. For future studies on mercury concentrations in residential areas, it would be convenient to be able to sample in either neighborhood without having to worry about the water supply affecting the outcome. Is there any difference in the mercury content of the two residential areas?

The sample collection cannot be paired. Even if water specimens were collected on the same day, there will be differences in storage time, distribution time, water use patterns, and other factors. Therefore, the data analysis will be done using the independent *t*-test.

t-Test to Compare the Averages of Two Samples

Two independently distributed random variables y_1 and y_2 have, respectively, mean values η_1 and η_2 and variances σ_1^2 and σ_2^2. The usual statement of the problem is in terms of testing the null hypothesis that the difference in the means is zero: $\eta_1 - \eta_2 = 0$, but we prefer viewing the problem in terms of the confidence interval of the difference.

The expected value of the difference between the averages of the two treatments is:

$$E(\bar{y}_1 - \bar{y}_2) = \eta_1 - \eta_2$$

If the data are from random samples, the variances of the averages $\bar{y}_1$ and $\bar{y}_2$ are:

$$V(\bar{y}_1) = \sigma_1^2/n_1 \quad \text{and} \quad V(\bar{y}_2) = \sigma_2^2/n_2$$

where n_1 and n_2 are the sample sizes. The variance of the difference is:

$$V(\bar{y}_1 - \bar{y}_2) = \frac{\sigma_1^2}{n_1} + \frac{\sigma_2^2}{n_2}$$

TABLE 18.1
Mercury Concentrations in Wastewater Originating in an Area Served by the City Water Supply (*c*) and an Area Served by Private Wells (*p*)

Source	Mercury Concentrations (μg/L)											
City (n_c = 13)	0.34	0.18	0.13	0.09	0.16	0.09	0.16	0.10	0.14	0.26	0.06	0.26	0.07
Private (n_p = 10)	0.26	0.06	0.16	0.19	0.32	0.16	0.08	0.05	0.10	0.13		

Data provided by Greg Zelinka, Madison Metropolitan Sewerage District.

Usually the variances σ_1^2 and σ_2^2 are unknown and must be estimated from the sample data by computing:

$$s_1^2 = \frac{\Sigma(y_{1i} - \bar{y}_1)^2}{n_1 - 1} \quad \text{and} \quad s_2^2 = \frac{\Sigma(y_{2i} - \bar{y}_2)^2}{n_2 - 1}$$

These can be pooled if they are of equal magnitude. Assuming this to be true, the pooled estimate of the variance is:

$$s_{\text{pool}}^2 = \frac{(n_1 - 1)s_1^2 + (n_2 - 1)s_2^2}{n_1 + n_2 - 2}$$

This is the weighted average of the variances, where the weights are the degrees of freedom of each variance. The number of observations used to compute each average and variance need not be equal.

The estimated variance of the difference is:

$$V(\bar{y}_1 - \bar{y}_2) = \frac{s_{\text{pool}}^2}{n_1} + \frac{s_{\text{pool}}^2}{n_2} = s_{\text{pool}}^2 \left(\frac{1}{n_1} + \frac{1}{n_2}\right)$$

and the standard error is the square root:

$$s_{\bar{y}_1 - \bar{y}_2} = \sqrt{\frac{s_{\text{pool}}^2}{n_1} + \frac{s_{\text{pool}}^2}{n_2}} = s_{\text{pool}} \sqrt{\frac{1}{n_1} + \frac{1}{n_2}}$$

Student's *t* distribution is used to compute the level confidence interval. To construct the $(1 - \alpha)100\%$ percent confidence interval use the *t* statistic for $\alpha/2$ and $\nu = n_1 + n_2 - 2$ degrees of freedom.

The correctness of this confidence interval depends on the data being independent and coming from distributions that are approximately normal with the same variance. If the variances are very different in magnitude, they cannot be pooled unless uniform variance can be achieved by means of a transformation. This procedure is robust to moderate nonnormality because the central limit effect will tend to make the distributions of the averages and their difference normal even when the parent distributions of y_1 and y_2 are not normal.

Case Solution: Mercury Data

Water specimens collected from a residential area that is served by the city water supply are indicated by subscript *c*; *p* indicates specimens taken from a residential area that is served by private wells. The averages, variances, standard deviations, and standard errors are:

City (n_c = 13)	$\bar{y}_c = 0.157$ μg/L	$s_c^2 = 0.0071$	$s_c = 0.084$	$s_{\bar{y}_c} = 0.023$
Private (n_p = 10)	$\bar{y}_p = 0.151$ μg/L	$s_p^2 = 0.0076$	$s_p = 0.087$	$s_{\bar{y}_p} = 0.028$

The difference in the averages of the measurements is $\bar{y}_c - \bar{y}_p = 0.157 - 0.151 = 0.006$ µg/L. The variances s_c^2 and s_p^2 of the city and private samples are nearly equal, so they can be pooled by weighting in proportion to their degrees of freedom:

$$s_{pool}^2 = \frac{12(0.0071) + 9(0.0076)}{12 + 9} = 0.00734 \; (\mu g/L)^2$$

The estimated variance of the difference between averages is:

$$V(\bar{y}_c - \bar{y}_p) = s_{\bar{y}_c}^2 + s_{\bar{y}_p}^2 = s_{pool}^2 \left(\frac{1}{n_c} + \frac{1}{n_p}\right) = 0.00734\left(\frac{1}{13} + \frac{1}{10}\right) = 0.0013 \; (\mu g/L)^2$$

and the standard error of $\bar{y}_c - \bar{y}_p = 0.006$ µg/L is $s_{\bar{y}_c - \bar{y}_p} = \sqrt{0.0013} = 0.036$ µg/L.

The variance of the difference is estimated with $v = 12 + 9 = 21$ degrees of freedom. The 95% confidence interval is calculated using $\alpha/2 = 0.025$ and $t_{21, 0.025} = 2.080$:

$$(\bar{y}_c - \bar{y}_p) \pm t_{21, 0.025} s_{\bar{y}_c - \bar{y}_p} = 0.006 \pm 2.080(0.036) = 0.006 \pm 0.075 \; \mu g/L$$

It can be stated with 95% confidence that the true difference between the city and private water supplies falls in the interval of −0.069 µg/L and 0.081 µg/L. This confidence interval includes zero so there is no persuasive evidence in these data that the mercury contents are different in the two residential areas. Future sampling can be done in either area without worrying that the water supply will affect the outcome.

Comments

The case study example showed that one could be highly confident that there is no statistical difference between the average mercury concentrations in the two residential neighborhoods. In planning future sampling, therefore, one might proceed as though the neighborhoods are identical, although we understand that this cannot be strictly true.

Sometimes a difference is statistically significant but small enough that, in practical terms, we do not care. It is statistically significant, but unimportant. Suppose that the mercury concentrations in the city and private waters had been 0.15 mg/L and 0.17 mg/L (not µg/L) and that the difference of 0.02 mg/L was statistically significant. We would be concerned about the dangerously high mercury levels in both neighborhoods. The difference of 0.02 mg/L and its statistical significance would be unimportant. This reminds us that *significance* in the statistical sense and *important* in the practical sense are two different concepts.

In this chapter the test statistic used to compare two treatments was the difference of two averages and the comparison was made using an independent *t*-test. Independent, in this context, means that all sources of uncontrollable random variation will equally affect each treatment. For example, specimens tested on different days will reflect variation due to any daily difference in materials or procedures in addition to the random variations that always exist in the measurement process. In contrast, Chapter 17 explains how a paired *t*-test will block out some possible sources of variation. Randomization is also effective for producing independent observations.

Exercises

18.1 Biosolids. Biosolids from an industrial wastewater treatment plant were applied to 10 plots that were randomly selected from a total of 20 test plots of farmland. Corn was grown on the treated (T) and untreated (UT) plots, with the following yields (bushels/acre).

UT	126	122	90	135	95	180	68	99	122	113
T	144	122	135	122	77	149	122	117	131	149

Calculate a 95% confidence limit for the difference in means.

18.2 **Lead Measurements.** Below are measurements of lead in solutions that are identical except for the amount of lead that has been added. Fourteen specimens had an addition of 1.25 µg/L and 14 had an addition of 2.5 µg/L. Is the difference in the measured values consistent with the known difference of 1.25 µg/L?

Addition = 1.25 µg/L	1.1	2.0	1.3	1.0	1.1	0.8	0.8	0.9	0.8	1.6	1.1	1.2	1.3	1.2
Addition = 2.5 µg/L	2.8	3.5	2.3	2.7	2.3	3.1	2.5	2.5	2.5	2.7	2.5	2.5	2.6	2.7

18.3 **Bacterial Densities.** The data below are the natural logarithms of bacterial counts as measured by two analysts on identical aliquots of river water. Are the analysts getting the same result?

Analyst A	1.60	1.74	1.72	1.85	1.76	1.72	1.78
Analyst B	1.72	1.75	1.55	1.67	2.05	1.51	1.70

18.4 **Highway TPH Contamination.** Use a t-test analysis of the data in Exercise 3.6 to compare the TPH concentrations on the eastbound and westbound lanes of the highway.

18.5 **Water Quality.** A small lake is fed by streams from a watershed that has a high density of commercial land use, and a watershed that is mainly residential. The historical data below were collected at random intervals over a period of four years. Are the chloride and alkalinity of the two streams different?

Commercial Land Use		Residential Land Use	
Chloride (mg/L)	Alkalinity (mg/L)	Chloride (mg/L)	Alkalinity (mg/L)
140	49	120	40
135	45	114	38
130	28	142	38
132	40	100	45
135	38	100	43
145	43	92	51
118	36	122	33
157	48	97	45
		145	51
		130	55

19

Assessing the Difference of Proportions

KEY WORDS *bioassay, binomial distribution, binomial model, censored data, effluent testing, normal distribution, normal approximation, percentages, proportions, ratio, toxicity, t-test.*

Ratios and proportions arise in biological, epidemiological, and public health studies. We may want to study the proportion of people infected at a given dose of virus, the proportion of rats showing tumors after exposure to a carcinogen, the incidence rate of leukemia near a contaminated well, or the proportion of fish affected in bioassay tests on effluents. Engineers would study such problems only with help from specialists, but they still need to understand the issues and some of the relevant statistical methods.

A situation where engineers will use ratios and proportions is when samples have been censored by a limit of detection. A data set on an up-gradient groundwater monitoring well has 90% of all observations censored and a down-gradient well has only 75% censored. Does this difference indicate that contamination has occurred in the groundwater flowing between the two wells?

Case Study

Biological assays are a means of determining the toxicity of an effluent. There are many ways such tests might be organized: species of test organism, number of test organisms, how many dilutions of effluent to test, specification of response, physical conditions, etc. Most of these are biological issues. Here we consider some statistical issues in a simple bioassay.

Organisms will be put into (1) an aquarium containing effluent or (2) a control aquarium containing clean water. Equal numbers of organisms are assigned randomly to the control and effluent groups. The experimental response is a binary measure: presence or absence of some characteristic. In an acute bioassay, the binary characteristic is survival or death of the organism. In a chronic bioassay, the organisms are exposed to nonlethal conditions and the measured response might be loss of equilibrium, breathing rate, loss of reproductive capacity, rate of weight gain, formation of neoplasms, etc.

In our example, 80 organisms ($n_1 = n_2 = 80$) were exposed to each treatment condition (control and effluent) and toxicity was measured in terms of survival. The data shown in Table 19.1 were observed. Are the survival proportions in the two groups so different that we can state with a high degree of confidence that the two treatments truly differ in toxicity?

The Binomial Model

The data from a binomial process consist of two discrete outcomes (binary). A test organism is either dead or alive after a given period of time. An effluent is either in compliance or it is not. In a given year, a river floods or it does not flood. The binomial probability distribution gives the probability of observing an event x times in a set of n trials (experiment). If the event is observed, the trial is said to be successful. Success in this statistical sense does not mean that the outcome is desirable. A success may be the death of an organism, failure of a machine, or violation of a regulation. It means success in

161

TABLE 19.1

Data from a Bioassay on Wastewater Treatment Plant Effluent

Group	Number			%	
	Surviving	Not Surviving	Totals	Surviving	Not Surviving
Control	72	8	80	90	10
Effluent	64	16	80	80	20
Totals	136	24	160	Avg. = 85	15

TABLE 19.2

Cumulative Binomial Probability for x Successes in 20 Trials, where p is the True Random Probability of Success in a Single Trial

x	$p = 0.05$	0.10	0.15	0.20	0.25	0.50
0	0.36	0.12	0.04	0.01	0.00	0.00
1	0.74	0.39	0.18	0.07	0.02	0.00
2	0.92	0.68	0.40	0.21	0.09	0.00
3	0.98	0.87	0.65	0.41	0.23	0.00
4	1.00	0.96	0.83	0.63	0.41	0.01
5	1.00	0.99	0.93	0.80	0.62	0.02
6		1.00	0.98	0.91	0.79	0.06
7		1.00	0.99	0.97	0.90	0.13
8			1.00	0.99	0.96	0.25
9			1.00	1.00	0.99	0.41
10				1.00	1.00	0.59
11					1.00	0.75
12						0.87
13						0.94
14						0.98
15						0.99
16						1.00

observing the behavior of interest. The true probability of the event of interest occurring in a given trial is p, and $1 - p$ is the probability of the event not occurring. In most environmental problems, the desired outcome is for the event to occur infrequently, which means that we are interested in cases where both x and p are small.

The *binomial probability* that x will occur for given values of n and p is:

$$f(x: n, p) = \frac{n!}{x!(n-x)!} p^x (1-p)^{n-x} \quad x = 0, 1, 2, \ldots, n$$

The terms with the factorials indicates the number of ways that x successes can occur in a sample of size n. These terms are known as the *binomial coefficients*. The probability of a success in a single trial is p (the corresponding probability of failure in a single trial is $(1 - p)$). The expected number of occurrences in n trials is the mean of x, which is $\mu_x = np$. The variance is $\sigma_x^2 = np(1 - p)$. This will be correct when p is constant and outcomes are independent from trial to trial.

The probability of r or fewer success in n independent trials for a probability of success p in a single trial is given by the cumulative binomial distribution:

$$\Pr(x \le r) = F(r: n, p) = \sum_0^r \frac{n!}{x!(n-x)!} p^x (1-p)^{n-x}$$

Table 19.2, calculated from this equation, gives the cumulative probability of x successes in $n = 20$ trials for several values of p. Table 19.3, which gives the probability of *exactly* x occurrences for the same

TABLE 19.3

Probability of Observing Exactly x Successes in 20 Trials, where p is the True Random Probability of Success in a Single Trial

x	$p = 0.05$	0.10	0.15	0.20	0.25	0.50
0	0.36	0.12	0.04	0.01	—	—
1	0.38	0.27	0.14	0.06	0.02	—
2	0.18	0.29	0.22	0.14	0.07	—
3	0.06	0.19	0.25	0.20	0.14	—
4	0.02	0.09	0.18	0.22	0.18	0.01
5		0.03	0.10	0.17	0.21	0.01
6		0.01	0.05	0.11	0.17	0.04
7			0.01	0.06	0.11	0.07
8			0.01	0.02	0.06	0.12
9				0.01	0.02	0.16
10					0.01	0.18
11						0.16
12						0.12
13						0.07
14						0.04
15						0.01
16						0.01

conditions as Table 19.2, can be constructed by differencing the entries in Table 19.2. For example, if $p = 0.05$, we expect an average of $\eta_x = 0.05(20) = 1$ success in a 20 trials. In one particular experiment, however, the result might be no successes, or one, or two, or three.

Some examples based on the binomial acute toxicity bioassay will demonstrate how these functions are used. Each test organism is a trial and the event of interest is its death within the specified test period. It is assumed that (1) theoretical probability of death is equal for all organisms subjected to the same treatment and (2) the fate of each organism is independent of the fate of other organisms. If n organisms are exposed to a test condition, the probability of observing any specific number of dead organisms, given a true underlying random probability of death for an individual organism, is computed from the binomial distribution.

From Table 19.2, the probability of getting one or no deaths is $\Pr(x \leq 1) = 0.74$. From Table 19.3, the probability of observing exactly zero deaths is $\Pr(x = 0) = 0.36$. These two values are used to compute the probability of exactly one death:

$$\Pr(x = 1) = \Pr(x \leq 1) - \Pr(x \leq 0) = 0.74 - 0.36 = 0.38$$

This is the value given in Table 19.3 for $x = 1$ and $p = 0.05$. Likewise, the probability of two or less deaths is $\Pr(x \leq 2) = 0.92$ and the probability of exactly two deaths is:

$$\Pr(x = 2) = \Pr(x \leq 2) - \Pr(x \leq 1) = 0.92 - 0.74 = 0.18$$

Table 19.3 shows that the range of possible outcomes in a binomial process can be quite large and that the variability about the expected value increases as p increases. This characteristic of the variance needs to be considered in designing bioassay experiments.

Most binomial tables only provide for up to $n = 20$. Fortunately, under certain circumstances, the normal distribution provides a good approximation to the binomial distribution. The binomial probability distribution is symmetric when $p = 0.5$. The distribution approaches symmetry as n becomes large, the approach being more rapid when p is close to 0.5. For values of $p \leq 0.5$, it is skewed to the right; for $p > 0.5$, it is skewed left. For large n, however, the skewness is not great unless p is near to 0 or 1.

As n increases, the binomial distribution can be approximated by a normal distribution with the same mean and variance [$\eta_x = np$ and $\sigma_x^2 = np(1 - p)$]. The normal approximation is reasonable if $np > 5$ and $n(1 - p) > 5$. Table 19.3 and Figure 19.1 show that the distribution is exactly symmetric for $n = 20$ and $p = 0.5$. For $n = 20$ and $p = 0.2$, where $np(1 - p)$ is only 3.2, the distribution is approaching symmetry.

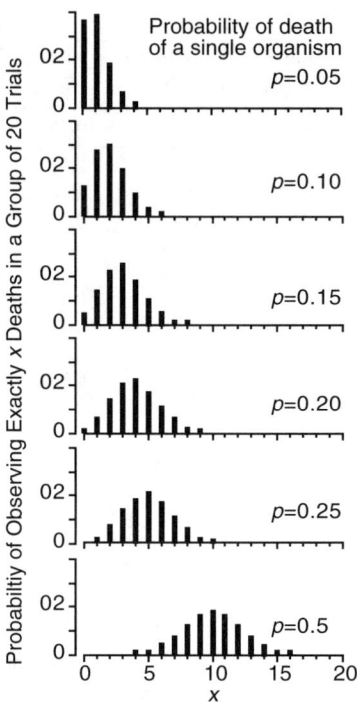

FIGURE 19.1 The binomial distribution with $n = 20$ for several values of p.

Assessing the Difference Between Two Proportions

The binomial distribution expresses the number of occurrences of an event x in n trials, where p is the probability of occurrence in a single trial. Usually the population probability p in a binomial process is unknown, so it is often more useful to examine the proportion of occurrences rather than their absolute number, x. Contrary to our guidelines on notation (Chapter 2), the population parameter p is not denoted with a Greek letter symbol. A hat (^) is used to distinguish the population parameter p and the *sample proportion*, which will be called $\hat{p} = x/n$. The hat (^) is a reminder that the sample proportion is a statistic computed from the data and that it estimates the population proportion. The sample proportion (x/n) is an unbiased estimator of the underlying population probability (p) in a binomial process.

The *sample variance* of p is:

$$s^2_{x/n} = p(1-p)/n$$

Two independent test groups of size n_1 and n_2 are to be compared. Suppose that group 1 represents a control (no exposure) and group 2 is the treatment group (i.e., exposed to effluent). The number of surviving organisms is x_1 in the control and x_2 in the treatment, giving observed sample proportions of $\hat{p}_1 = x_1/n_1$ and $\hat{p}_2 = x_2/n_2$. If we assume, as in a null hypothesis, that the control and treatment populations have the same true underlying population probability p (i.e., $p = p_1 = p_2$), then $\hat{p}_1$ and $\hat{p}_2$ will be normally distributed with mean p and variances $p(1-p)/n_1$ and $p(1-p)/n_2$, respectively.

The difference between the two sample proportions $(\hat{p}_1 - \hat{p}_2)$ will be normally distributed with mean zero and variance:

$$\frac{p(1-p)}{n_1} + \frac{p(1-p)}{n_2} = p(1-p)\left(\frac{1}{n_1} + \frac{1}{n_2}\right)$$

Assessing the Difference of Proportions

The standardized difference also is normally distributed:

$$z = \frac{(\hat{p}_1 - \hat{p}_2) - 0}{\sqrt{p(1-p)\left(\frac{1}{n_1} + \frac{1}{n_2}\right)}}$$

The difficulty with using the standardized difference is that p is unknown and the denominator cannot be computed until some estimate is found. The best estimate of p is the weighted average of the sample proportions (Rosner, 1990):

$$p = \frac{n_1 \hat{p}_1 + n_2 \hat{p}_2}{n_1 + n_2}$$

Use this to compute z as given above under the null hypothesis that the two population proportions are equal ($p_1 = p_2$). This value of z is compared with the tabulated standard normal variate, z_α, for a specified significance level α. Usually only the case where $p_1 > p_2$ (or $p_2 > p_1$) is of interest, so a one-sided test is used. If $z > z_\alpha$, the difference is too large to accept the hypothesis that the population proportions are the same.

The $(1 - \alpha)100\%$ confidence limits for $p_1 - p_2$ for a large sample under the normal approximation is:

$$\hat{p}_1 - \hat{p}_2 \pm z_{\alpha/2} \sqrt{\frac{\hat{p}_1(1-\hat{p}_1)}{n_1} + \frac{\hat{p}_2(1-\hat{p}_2)}{n_2}}$$

To set a one-sided upper bound with a $(1 - \alpha)100\%$ confidence level, use the above equation with z_α. To set a one-sided lower bound, use $-z_\alpha$.

When the two test populations are the same size ($n = n_1 = n_2$), the standardized difference simplifies to:

$$z = \frac{(\hat{p}_1 - \hat{p}_2) - 0}{\sqrt{\frac{2p(1-p)}{n}}}$$

where $p = (\hat{p}_1 + \hat{p}_2)/2$.

Fleiss (1981) suggests that when n is small (e.g., $n < 20$), the computation of z should be modified to:

$$z = \frac{(\hat{p}_1 - \hat{p}_2) - \frac{1}{n}}{\sqrt{\frac{2p(1-p)}{n}}}$$

The $(-1/n)$ in the numerator is the Yates continuity correction, which takes into account the fact that a continuous distribution (the normal) is being used to represent the discrete binomial distribution of sample proportions. For reasonably small n (e.g., $n = 20$), the correction is $-1/n = -0.05$, which can be substantial relative to the differences usually observed (e.g., $p_1 - p_2 = 0.15$). Not everyone uses this correction (Rosner, 1990). If it is omitted, there is a reduced probability of detecting a difference in rates. As n becomes large, this correction factor becomes negligible because the normal distribution very closely approximates the binomial distribution.

Case Study Solution

Eighty organisms ($n_1 = n_1 = 80$) were exposed to each treatment condition (control and effluent) and the sample survival proportions are observed to be:

$$\hat{p}_1 = 72/80 = 0.90 \quad \text{and} \quad \hat{p}_2 = 64/80 = 0.80$$

We would like to use the normal approximation to the binomial distribution. Checking to see whether this is appropriate gives:

$$n\hat{p}_1 = 80(0.90) = 72 \quad \text{and} \quad n(1-\hat{p}_1) = 80(0.10) = 8$$
$$n\hat{p}_2 = 80(0.80) = 64 \quad \text{and} \quad n(1-\hat{p}_2) = 80(0.20) = 16$$

Because these are greater than 5, the normal approximation can be used. The sample size is large enough that the Yates continuity correction is not needed.

Might the true difference of the two proportions be zero? Could we observe a difference as large as 0.1 just as a result of random variation? This can be checked by examining the lower 95% confidence limit. From the observed proportions, we estimate $p = (0.9 + 0.8)/2 = 0.85$ and $1 - p = 0.15$. For a one-sided test at the 95% level, $z_{\alpha=0.05} = 1.645$ and the lower 95% confidence bound is:

$$(0.9 - 0.8) - 1.645\sqrt{\frac{2(0.15)(0.85)}{80}} = 0.1 - 1.645(0.056) = 0.007$$

The lower limit is greater than zero, so it is concluded that the observed difference is larger than is expected to occur by chance and that $p_2 = 0.8$ is less than $p_1 = 0.9$.

Alternately, we could have compared a computed sample z-statistic:

$$z = \frac{0.90 - 0.80}{\sqrt{\frac{2(0.15)(0.85)}{80}}} = 1.77$$

with the tabulated value of $z_{0.05} = 1.645$. Finding $z_{sample} = 1.77 > z_{0.05} = 1.645$ means that the observed difference is quite far onto the tail of the distribution. It is concluded that the difference between the proportions is large enough to say that the two treatments are different.

Comments

The example problem found a small difference in proportions to be statistically significant because the number of test organisms was large. Many bioassays use about 20 test organisms, in which case the observed difference in proportions will have to be quite large before we can have a high degree of confidence that a difference did not arise merely from chance.

Testing proportions is straightforward if the sample size and proportions are large enough for the normal approximation to be used. However, when the proportions approach one or zero, a transformation is needed (Mowery et al., 1985). When the sample size is small, the normal approximation becomes invalid; consult Fleiss (1981) or Rosner (1990).

References

Fleiss, J. L. (1981). *Statistical Methods for Rates and Proportions,* New York, John Wiley.
Mowery, P. D., J. A. Fava, and L. W. Clatlin (1985). "A Statistical Test Procedure for Effluent Toxicity Screening," *Aquatic Toxicol. Haz. Assess., 7th Symp.,* ASTM STP 854.
Rosner, B. (1990). *Fundamentals of Biostatistics,* 3rd ed., Boston, PWS-Kent Publishing Co.

Exercises

19.1 **Well Monitoring.** Twenty monitoring wells that are down-gradient from a hazardous waste landfill are monitoring groundwater quality every three months. The standard is that 19 of the 20 wells must have a concentration less than 10 μg/L or special studies must be undertaken. To allow for sampling error, the monitoring agency is allowed to judge compliance based on a 5% probability that a well truly in compliance may register above the 10 μg/L limit. Assuming that all wells are truly in compliance: (a) What is the probability that 2 of 20 wells will be observed above the limit? (b) What is the probability that 3 of 20 are above? (c) That none are above?

19.2 Control Group. Two groups, A and B, consist of 100 healthy organisms each. Group A, a control group, is put into a pond that is fed by a clean stream. Group B is put into a pond that receives a portion of wastewater effluent in the same stream water. After some time it is found that in groups A and B, 25 and 35 organisms, respectively, have an unusual condition. At significance levels of (a) 0.01, (b) 0.05, and (c) 0.10, test the hypothesis that something in the wastewater is harming the organisms.

19.3 Censored Data. Test specimens of water from monitoring wells sometimes contain a concentration of the target pollutant that is too low to be detected. These are reported as ND, to signify not detected. Thirty of forty wells beneath a landfill had ND levels of the target pollutant, while nine of ten wells some distance from the landfill had ND levels. How strong is the evidence that the landfill is polluting the groundwater?

19.4 Microbiology. A screening survey of 128 New England drinking-water wells that were potentially vulnerable to fecal contamination were sampled four times during a one-year period to assess microbiological quality. A portion of the results are shown below. "Present" means that class of organism was found in the collected well water during the year; "Absent" means it was not observed in any of the four test specimens. (a) Estimate the upper 95% confidence limit for the proportion of wells testing positive for each biological class of organism. (b) Is there a significant difference between the proportion of wells testing positive for total coliforms as measured by the membrane filter (MF) and Colilert methods? (c) Is the proportion of positives different for somatic coliphage and total coliform? (d) Is the proportion of positives different for total culturable viruses and total coliform?

Class of Mircobial Contaminant	Present	Absent	No Data
Total coliform (MF method)	20	104	4
Total coliform (Colilert method)	15	109	4
Fecal coliform (EC)	3	121	4
E. coli (Colilert method)	0	124	4
Enterococci	20	104	4
Male-specific coliphage	4	75	49
Somatic coliphage	1	78	49
Total culturable viruses	2	122	4

Source: K. M. Doherty (2001). M. S. Project Report, CEE Dept., Tufts University.

19.5 Operations. The maintenance division says that at least 40% of alarms at pump stations are urgent and uses this claim to support a request for increased staffing. Management collected data on the alarms and found that only 49 of 150 were urgent and that staffing was adequate to service these. Can the null hypothesis $p \geq 0.40$ be rejected at the 99% level of probability?

20

Multiple Paired Comparisons of k Averages

KEY WORDS *data snooping, data dredging, Dunnett's procedure, multiple comparisons, sliding reference distribution, studentized range,* t*-tests, Tukey's procedure.*

The problem of comparing several averages arises in many contexts: compare five bioassay treatments against a control, compare four new polymers for sludge conditioning, or compare eight new combinations of media for treating odorous ventilation air. One multiple paired comparison problem is to compare all possible pairs of k treatments. Another is to compare $k - 1$ treatments with a control.

Knowing how to do a *t*-test may tempt us to compare several combinations of treatments using a series of paired *t*-tests. If there are k treatments, the number of pair-wise comparisons that could be made is $k(k-1)/2$. For $k = 4$, there are 6 possible combinations, for $k = 5$ there are 10, for $k = 10$ there are 45, and for $k = 15$ there are 105. Checking 5, 10, 45, or even 105 combinations is manageable but not recommended. Statisticians call this *data snooping* (Sokal and Rohlf, 1969) or *data dredging* (Tukey, 1991). We need to understand why data snooping is dangerous.

Suppose, to take a not too extreme example, that we have 15 different treatments. The number of possible pair-wise comparisons that could be made is $15(15-1)/2 = 105$. If, before the results are known, we make one selected comparison using a *t*-test with a $100\alpha\% = 5\%$ error rate, there is a 5% chance of reaching the wrong decision each time we repeat the data collection experiment for those two treatments. If, however, several pairs of treatments are tested for possible differences using this procedure, the error rate will be larger than the expected 5% rate. Imagine that a two-sample *t*-test is used to compare the largest of the 15 average values against the smallest. The null hypothesis that this difference, the largest of all the 105 possible pair-wise differences, is likely to be rejected almost every time the experiment is repeated, instead of just at the 5% rate that would apply to making one pair-wise comparison selected at random from among the 105 possible comparisons.

The number of comparisons does not have to be large for problems to arise. If there are just three treatment methods and of the three averages, A is larger than B and C is slightly larger than A ($\bar{y}_C > \bar{y}_A > \bar{y}_B$), it is possible for the three possible *t*-tests to indicate that A gives higher results than B ($\eta_A > \eta_B$), A is not different from C ($\eta_A = \eta_C$), and B is not different from C ($\eta_B = \eta_C$). This apparent contradiction can happen because different variances are used to make the different comparisons. Analysis of variance (Chapter 21) eliminates this problem by using a common variance to make a single test of significance (using the F statistic).

The multiple comparison test is similar to a *t*-test *but* an allowance is made in the error rate to keep the collective error rate at the stated level. This collective rate can be defined in two ways. Returning to the example of 15 treatments and 105 possible pair-wise comparisons, the probability of getting the wrong conclusion for a single randomly selected comparison is the *individual error rate*. The *family error rate* (also called the *Bonferroni error rate*) is the chance of getting one or more of the 105 comparisons wrong in each repetition of data collection for all 15 treatments. The family error rate counts an error for each wrong comparison in each repetition of data collection for all 15 treatments. Thus, to make valid statistical comparisons, the individual per comparison error rate must be shrunk to keep the simultaneous family error rate at the desired level.

TABLE 20.1

Ten Measurements of Lead Concentration (μg/L) Measured on Identical Wastewater Specimens by Five Laboratories

	Lab 1	Lab 2	Lab 3	Lab 4	Lab 5
	3.4	4.5	5.3	3.2	3.3
	3.0	3.7	4.7	3.4	2.4
	3.4	3.8	3.6	3.1	2.7
	5.0	3.9	5.0	3.0	3.2
	5.1	4.3	3.6	3.9	3.3
	5.5	3.9	4.5	2.0	2.9
	5.4	4.1	4.6	1.9	4.4
	4.2	4.0	5.3	2.7	3.4
	3.8	3.0	3.9	3.8	4.8
	4.2	4.5	4.1	4.2	3.0
Mean $\bar{y}_i =$	4.30	3.97	4.46	3.12	3.34
Variance $s_i^2 =$	0.82	0.19	0.41	0.58	0.54

TABLE 20.2

Ten Possible Differences of Means ($\bar{y}_i - \bar{y}_j$) Between Five Laboratories

	Laboratory i (Average = $\bar{y}_i$)				
	1	2	3	4	5
Laboratory j	(4.30)	(3.97)	(4.46)	(3.12)	(3.34)
1	—	—	—	—	
2	0.33	—	—	—	
3	−0.16	−0.49	—	—	
4	1.18	0.85	1.34	—	
5	0.96	0.63	1.12	−0.22	—

Case Study: Measurements of Lead by Five Laboratories

Five laboratories each made measurements of lead on ten replicate wastewater specimens. The data are given in Table 20.1 along with the means and variance for each laboratory. The ten possible comparisons of mean lead concentrations are given in Table 20.2. Laboratory 3 has the highest mean (4.46 μg/L) and laboratory 4 has the lowest (3.12 μg/L). Are the differences consistent with what one might expect from random sampling and measurement error, or can the differences be attributed to real differences in the performance of the laboratories?

We will illustrate *Tukey's multiple t-test* and *Dunnett's method of multiple comparisons with a control*, with a minimal explanation of statistical theory.

Tukey's Paired Comparison Method

A $(1 - \alpha)100\%$ confidence interval for the true difference between the means of two treatments, say treatments i and j, is:

$$(\bar{y}_i - \bar{y}_j) \pm t_{\nu, \alpha/2} s_{\text{pool}} \sqrt{\frac{1}{n_i} + \frac{1}{n_j}}$$

TABLE 20.3

Values of the Studentized Range Statistic $q_{k,v,\alpha/2}$ for $k(k-1)/2$ Two-Sided Comparisons for a Joint 95% Confidence Interval Where There are a Total of k Treatments

	k						
v	2	3	4	5	6	8	10
5	4.47	5.56	6.26	6.78	7.19	7.82	8.29
10	3.73	4.47	4.94	5.29	5.56	5.97	6.29
15	3.52	4.18	4.59	4.89	5.12	5.47	5.74
20	3.43	4.05	4.43	4.70	4.91	5.24	5.48
30	3.34	3.92	4.27	4.52	4.72	5.02	5.24
60	3.25	3.80	4.12	4.36	4.54	4.81	5.01
∞	3.17	3.68	3.98	4.20	4.36	4.61	4.78

Note: Family error rate = 5%; $\alpha/2 = 0.05/2 = 0.025$.

Source: Harter, H. L. (1960). *Annals Math. Stat.,* 31, 1122–1147.

where it is assumed that the two treatments have the same variance, which is estimated by pooling the two sample variances:

$$s^2_{pool} = \frac{(n_i - 1)s_i^2 + (n_j - 1)s_j^2}{n_i + n_j - 2}$$

The chance that the interval includes the true value for any *single comparison* is exactly $1 - \alpha$. But the chance that all possible $k(k-1)/2$ intervals will simultaneously contain their true values is less than $1 - \alpha$.

Tukey (1949) showed that the confidence interval for the difference in two means (η_i and η_j), taking into account that all possible comparisons of k treatments may be made, is given by:

$$\bar{y}_i - \bar{y}_j \pm \frac{q_{k,v,\alpha/2}}{\sqrt{2}} s_{pool} \sqrt{\frac{1}{n_i} + \frac{1}{n_j}}$$

where $q_{k,v,\alpha/2}$ is the upper significance level of the *studentized range* for k means and v degrees of freedom in the estimate s^2_{pool} of the variance σ^2. This formula is exact if the numbers of observations in all the averages are equal, and approximate if the k treatments have different numbers of observations. The value of s^2_{pool} is obtained by pooling sample variances over all k treatments:

$$s^2_{pool} = \frac{(n_1 - 1)s_1^2 + \cdots + (n_k - 1)s_k^2}{n_1 + \cdots + n_k - k}$$

The size of the confidence interval is larger when $q_{k,v,\alpha/2}$ is used than for the t statistic. This is because the studentized range allows for the possibility that any one of the $k(k-1)/2$ possible pair-wise comparisons might be selected for the test. Critical values of $q_{k,v,\alpha/2}$ have been tabulated by Harter (1960) and may be found in the statistical tables of Rohlf and Sokal (1981) and Pearson and Hartley (1966). Table 20.3 gives a few values for computing the two-sided 95% confidence interval.

Solution: Tukey's Method

For this example, $k = 5$, $s^2_{pool} = 0.51$, $s_{pool} = 0.71$, $v = 50 - 5 = 45$, and $q_{5,40,0.05/2} = 4.49$. This gives the 95% confidence limits of:

$$(\bar{y}_i - \bar{y}_j) \pm \frac{4.49}{\sqrt{2}} (0.71) \sqrt{\frac{1}{10} + \frac{1}{10}}$$

TABLE 20.4

Table of $t_{k-1,v,0.05/2}$ for $k-1$ Two-Sided Comparisons for a Joint 95% Confidence Level Where There are a Total of k Treatments, One of Which is a Control

v	\multicolumn{6}{c}{$k-1$ = Number of Treatments Excluding the Control}					
	2	3	4	5	6	8	10
5	3.03	3.29	3.48	3.62	3.73	3.90	4.03
10	2.57	2.76	2.89	2.99	3.07	3.19	3.29
15	2.44	2.61	2.73	2.82	2.89	3.00	3.08
20	2.38	2.54	2.65	2.73	2.80	2.90	2.98
30	2.32	2.47	2.58	2.66	2.72	2.82	2.89
60	2.27	2.41	2.51	2.58	2.64	2.73	2.80
∞	2.21	2.35	2.44	2.51	2.57	2.65	2.72

Source: Dunnett, C. W. (1964). *Biometrics,* 20, 482–491.

and the difference in the true means is, with 95% confidence, within the interval:

$$-1.01 \leq \bar{y}_i - \bar{y}_j \leq 1.01$$

We can say, with a high degree of confidence, that any observed difference larger than 1.01 µg/L or smaller than −1.01 µg/L is not likely to be zero. We conclude that laboratories 3 and 1 are higher than 4 and that laboratory 3 is also different from laboratory 5. We cannot say which laboratory is correct, or which one is best, without knowing the true concentration of the test specimens.

Dunnett's Method for Multiple Comparisons with a Control

In many experiments and monitoring programs, one experimental condition (treatment, location, etc.) is a standard or a control treatment. In bioassays, there is always an unexposed group of organisms that serve as a control. In river monitoring, one location above a waste outfall may serve as a control or reference station. Now, instead of k treatments to compare, there are only $k-1$. And there is a strong likelihood that the control will be different from at least one of the other treatments.

The quantities to be tested are the differences $\bar{y}_i - \bar{y}_c$, where $\bar{y}_c$ is the observed average response for the control treatment. The $(1-\alpha)100\%$ confidence intervals for all $k-1$ comparisons with the control are given by:

$$(\bar{y}_i - \bar{y}_c) \pm t_{k-1,v,\alpha/2} s_{\text{pool}} \sqrt{\frac{1}{n_i} + \frac{1}{n_c}}$$

This expression is similar to Tukey's as used in the previous section except the quantity $q_{k,v,\alpha/2}/\sqrt{2}$ is replaced with Dunnett's $t_{k-1,v,\alpha/2}$. The value of s_{pool} is obtained by pooling over all treatments. An abbreviated table for 95% confidence intervals is reproduced in Table 20.4. More extensive tables for one- and two-sided tests are found in Dunnett (1964).

Solution: Dunnet's Method

Rather than create a new example we reconsider the data in Table 20.1 supposing that laboratory 2 is a reference (control) laboratory. Pooling sample variances over all five laboratories gives the estimated within-laboratory variance, $s_{\text{pool}}^2 = 0.51$ and $s_{\text{pool}} = 0.71$. For $k-1 = 4$ treatments to be compared with the control and $v = 45$ degrees of freedom, the value of $t_{4,45,0.05/2} = 2.55$ is found in Table 20.4. The 95%

TABLE 20.5

Comparing Four Laboratories with a Reference Laboratory

Laboratory	Control	1	3	4	5
Average	3.97	4.30	4.46	3.12	3.34
Difference ($\bar{y}_i - \bar{y}_c$)	—	0.33	0.49	−0.85	−0.63

confidence limits are:

$$\bar{y}_i - \bar{y}_c \pm 2.55(0.71)\sqrt{\frac{1}{10} + \frac{1}{10}}$$

$$-0.81 \leq \bar{y}_i - \bar{y}_c \leq 0.81$$

We can say with 95% confidence that any observed difference greater than 0.81 or smaller than −0.81 is unlikely to be zero. The four comparisons with laboratory 2 shown in Table 20.5 indicate that the measurements from laboratory 4 are smaller than those of the control laboratory.

Comments

Box et al. (1978) describe yet another way of making multiple comparisons. The simple idea is that if k treatment averages had the same mean, they would appear to be k observations from the same, nearly normal distribution with standard deviation $\sigma/\sqrt{n}$. The plausibility of this outcome is examined graphically by constructing such a normal reference distribution and superimposing upon it a dot diagram of the k average values. The reference distribution is then moved along the horizontal axis to see if there is a way to locate it so that all the observed averages appear to be typical random values selected from it. This sliding reference distribution is a "…rough method for making what are called multiple comparisons." The Tukey and Dunnett methods are more formal ways of making these comparisons.

Dunnett (1955) discussed the allocation of observations between the control group and the other $p = k - 1$ treatment groups. For practical purposes, if the experimenter is working with a joint confidence level in the neighborhood of 95% or greater, then the experiment should be designed so that $n_c/n = \sqrt{p}$ approximately, where n_c is the number of observations on the control and n is the number on each of the p noncontrol treatments. Thus, for an experiment that compares four treatments to a control, $p = 4$ and n_c is approximately $2n$.

References

Box, G. E. P., W. G. Hunter, and J. S. Hunter (1978). *Statistics for Experimenters: An Introduction to Design, Data Analysis, and Model Building,* New York, Wiley Interscience.

Dunnett, C. W. (1955). "Multiple Comparison Procedure for Comparing Several Treatments with a Control," *J. Am. Stat. Assoc.,* 50, 1096–1121.

Dunnett, C. W. (1964). "New Tables for Multiple Comparisons with a Control," *Biometrics,* 20, 482–491.

Harter, H. L. (1960). "Tables of Range and Studentized Range," *Annals Math. Stat.,* 31, 1122–1147.

Pearson, E. S. and H. O. Hartley (1966). *Biometrika Tables for Statisticians,* Vol. 1, 3rd ed., Cambridge, England, Cambridge University Press.

Rohlf, F. J. and R. R. Sokal (1981). *Statistical Tables,* 2nd ed., New York, W. H. Freeman & Co.

Sokal, R. R. and F. J. Rohlf (1969). *Biometry: The Principles and Practice of Statistics in Biological Research,* New York, W. H. Freeman and Co.

Tukey, J. W. (1949). "Comparing Individual Means in the Analysis of Variance," *Biometrics,* 5, 99.

Tukey, J. W. (1991). "The Philosophy of Multiple Comparisons," *Stat. Sci.,* 6(6), 100–116.

Exercises

20.1 Storage of Soil Samples. The concentration of benzene (μg/g) in soil was measured after being stored in sealed glass ampules for different times, as shown in the data below. Quantities given are average ± standard deviation, based on $n = 3$. Do the results indicate that storage time must be limited to avoid biodegradation?

Day 0	Day 5	Day 11	Day 20
6.1 ± 0.7	5.9 ± 0.2	6.2 ± 0.2	5.7 ± 0.2

Source: Hewitt, A. D. et al. (1995). *Am. Anal. Lab.*, Feb., p. 26.

20.2 Biomonitoring. The data below come from a biological monitoring test for chronic toxicity on fish larvae. The control is clean (tap) water. The other four conditions are tap water mixed with the indicated percentages of treated wastewater effluent. The lowest observed effect level (LOEL) is the lowest percentage of effluent that is statistically different from the control. What is the LOEL?

| Replicate | Control | Percentage Effluent | | | |
		1.0	3.2	10.0	32.0
1	1.017	1.157	0.998	0.837	0.715
2	0.745	0.914	0.793	0.935	0.907
3	0.862	0.992	1.021	0.839	1.044
Mean	0.875	1.021	0.937	0.882	0.889
Variance	0.0186	0.0154	0.0158	0.0031	0.0273

20.3 Biological Treatment. The data below show the results of applying four treatment conditions to remove a recalcitrant pollutant from contaminated groundwater. All treatments were replicated three times. The "Controls" were done using microorganisms that have been inhibited with respect to biodegrading the contaminant. The "Bioreactor" uses organisms that are expected to actively degrade the contaminant. If the contaminant is not biodegraded, it could be removed by chemical degradation, volatilization, sorption, etc. Is biodegradation a significant factor in removing the contaminant?

Condition	T_0	T_1	T_2	T_3
Control	1220	1090	695	575
	1300	854	780	580
	1380	1056	688	495
Bioreactor	1327	982	550	325
	1320	865	674	310
	1253	803	666	465

Source: Dobbins, D. C. (1994). *J. Air & Waste Mgmt. Assoc.*, 44, 1226–1229.

21

Tolerance Intervals and Prediction Intervals

KEY WORDS *confidence interval, coverage, groundwater monitoring, interval estimate, lognormal distribution, mean, normal distribution, point estimate, precision, prediction interval, random sampling, random variation, spare parts inventory, standard deviation, tolerance coefficient, tolerance interval, transformation, variance, water quality monitoring.*

Often we are interested more in an interval estimate of a parameter than in a point estimate. When told that the average efficiency of a sample of eight pumps was 88.3%, an engineer might say, "The *point estimate* of 88.3% is a concise summary of the results, but it provides no information about their precision." The estimate based on the sample of 8 pumps may be quite different from the results if a different sample of 8 pumps were tested, or if 50 pumps were tested. Is the estimate $88.3 \pm 1\%$, or $88.3 \pm 5\%$? How good is 88.3% as an estimate of the efficiency of the next pump that will be delivered? Can we be reasonably confident that it will be within 1% or 10% of 88.3%?

Understanding this uncertainty is as important as making the point estimate. The main goal of statistical analysis is to quantify these kinds of uncertainties, which are expressed as intervals.

The choice of a statistical interval depends on the application and the needs of the problem. One must decide whether the main interest is in *describing the population or process* from which the sample has been selected or in *predicting the results of a future sample* from the same population. *Confidence intervals* enclose the population mean and *tolerance intervals* contain a specified proportion of a population. In contrast, intervals for a future sample mean and intervals to include all of *m* future observations are called *prediction intervals* because they deal with predicting (or containing) the results of a future sample from a previously sampled population (Hahn and Meeker, 1991).

Confidence intervals were discussed in previous chapters. This chapter briefly considers tolerance intervals and prediction intervals.

Tolerance Intervals

A *tolerance interval* contains a specified proportion (p) of the units from the sampled population or process. For example, based upon a past sample of copper concentration measurements in sludge, we might wish to compute an interval to contain, with a specified degree of confidence, the concentration of at least 90% of the copper concentrations from the sampled process. The tolerance interval is constructed from the data using two coefficients, the coverage and the tolerance coefficient. The *coverage* is the proportion of the population (p) that an interval is supposed to contain. The *tolerance coefficient* is the degree of confidence with which the interval reaches the specified coverage. A tolerance interval with coverage of 95% and a tolerance coefficient of 90% will contain 95% of the population distribution with a confidence of 90%.

The form of a *two-sided tolerance interval* is the same as a confidence interval:

$$\bar{y} \pm K_{1-\alpha, p, n} s$$

where the factor $K_{1-\alpha, p, n}$ has a $100(1-\alpha)\%$ confidence level and depends on n, the number of observations in the given sample. Table 21.1 gives the factors ($t_{n-1, \alpha/2}/\sqrt{n}$) for two-sided 95% confidence intervals

TABLE 21.1

Factors for Two-Sided 95% Confidence Intervals and Tolerance Intervals for the Mean of a Normal Distribution

n	Confidence Intervals $(t_{n-1,\,\alpha/2}/\sqrt{n})$	$K_{1-\alpha,p,n}$ for Tolerance Intervals		
		$p = 0.90$	$p = 0.95$	$p = 0.99$
4	1.59	5.37	6.34	8.22
5	1.24	4.29	5.08	6.60
6	1.05	3.73	4.42	5.76
7	0.92	3.39	4.02	5.24
8	0.84	3.16	3.75	4.89
9	0.77	2.99	3.55	4.63
10	0.72	2.86	3.39	4.44
12	0.64	2.67	3.17	4.16
15	0.50	2.49	2.96	3.89
20	0.47	2.32	2.76	3.62
25	0.41	2.22	2.64	3.46
30	0.37	2.15	2.55	3.35
40	0.32	2.06	2.45	3.22
60	0.26	1.96	2.34	3.07
∞	0.00	1.64	1.96	2.58

Source: Hahn, G. J. (1970). *J. Qual. Tech.*, 3, 18–22.

for the population mean η and values of $K_{1-\alpha,p,n}$ for two-sided tolerance intervals to contain at least a specified proportion (coverage) of $p = 0.90$, 0.95, or 0.99 of the population at a $100(1 - \alpha)\% = 95\%$ confidence level. Complete tables for one-sided and two-sided confidence intervals, tolerance intervals, and prediction intervals are given by Hahn and Meeker (1991) and Gibbons (1994).

The factors in these tables were calculated assuming that the data are a random sample. *Simple random sampling* gives every possible sample of n units from the population the same probability of being selected. The assumption of random sampling is critical because the statistical intervals reflect only the randomness introduced by the sampling process. They do not take into account bias that might be introduced by a nonrandom sample.

The use of these tables is illustrated by example.

Example 21.1

A random sample of $n = 5$ observations yields the values $\bar{y} = 28.4$ µg/L and $s = 1.18$ µg/L. The second row of Table 21.1 gives the needed factors for $n = 5$

1. The two-sided 95% confidence interval for the mean η of the population is:

$$28.4 \pm 1.24(1.18) = [26.9, 29.9]$$

The coefficient $1.24 = t_{4,0.025}/\sqrt{5}$. We are 95% confident that the interval 26.9 to 29.9 µg/L contains the true (but unknown) mean concentration of the population. The 95% confidence describes the percentage of time that a claim of this type is correct. That is, 95% of intervals so constructed will contain the true mean concentration.

2. The two-sided 95% tolerance interval to contain at least 99% of the sampled population is:

$$28.4 \pm 6.60(1.18) = [20.6, 36.2]$$

The factor 6.60 is for $n = 5$, $p = 0.99$, and $1 - \alpha = 0.95$. We are 95% confident that the interval 20.6 to 36.2 µg/L contains at least 99% of the population. This is called a 95% tolerance interval for 99% of the population.

Prediction Intervals

A *prediction interval* contains the expected results of a *future sample* to be obtained from a previously sampled population or process. Based upon a past sample of measurements, we might wish to construct a prediction interval to contain, with a specified degree of confidence: (1) the concentration of a randomly selected single future unit from the sampled population, (2) the concentrations for five future specimens, or (3) the average concentration of five future units.

The form of a *two-sided prediction interval* is the same as a confidence interval or a tolerance interval:

$$\bar{y} \pm K_{1-\alpha, n} s$$

The factor $K_{1-\alpha,n}$ has a 100(1 − α)% confidence level and depends on n, the number of observations in the given sample, and also on whether the prediction interval is to contain a single future value, several future values, or a future mean. Table 21.2 gives the factors to calculate (1) two-sided simultaneous prediction intervals to contain all of m future observations from the previously sampled normal population for m = 1, 2, 10, 20 and $m = n$; and (2) two-sided prediction intervals to contain the mean of $m = n$ future observations. The confidence level associated with these intervals is 95%.

The two-sided (1 − α)100% prediction limit for the next single measurement of a normally distributed random variable is:

$$\bar{y} \pm t_{n-1, \alpha/2} s \sqrt{1 + \frac{1}{n}}$$

where the t statistic is for n − 1 degrees of freedom, based on the sample of n measurements used to estimate the mean and standard deviation. For the one-sided upper (1 − α)100% confidence prediction limit use $\bar{y} + t_{n-1,\alpha} s \sqrt{1 + \frac{1}{n}}$.

TABLE 21.2

Factors for Two-Sided 95% Prediction Intervals for a Normal Distribution

	Simultaneous Prediction Intervals to Contain All m Future Observations					Prediction Intervals to Contain the Mean of n Future Observations
n	$m = 1$	$m = 2$	$m = 5$	$m = 10$	$m = \infty$	
4	3.56	4.41	5.56	6.41	5.29	2.25
5	3.04	3.70	4.58	5.23	4.58	1.76
6	2.78	3.33	4.08	4.63	4.22	1.48
7	2.62	3.11	3.77	4.26	4.01	1.31
8	2.51	2.97	3.57	4.02	3.88	1.18
9	2.43	2.86	3.43	3.85	3.78	1.09
10	2.37	2.79	3.32	3.72	3.72	1.01
12	2.29	2.68	3.17	3.53	3.63	0.90
15	2.22	2.57	3.03	3.36	3.56	0.78
20	2.14	2.48	2.90	3.21	3.50	0.66
30	2.08	2.39	2.78	3.06	3.48	0.53
40	2.05	2.35	2.73	2.99	3.49	0.45
60	2.02	2.31	2.67	2.93	3.53	0.37
∞	1.96	2.24	2.57	2.80	∞	0.00

Source: Hahn, G. J. (1970). *J. Qual. Tech.*, 3, 18–22.

Example 21.2

A random sample of $n = 5$ observations yields the values $\bar{y} = 28.4$ μg/L and $s = 1.18$ μg/L. An additional $m = 10$ specimens are to be taken at random from the same population.

1. Construct a two-sided (simultaneous) 95% prediction interval to contain the concentrations of all 10 additional specimens. For $n = 5$, $m = 10$, and $\alpha = 0.05$, the factor is 5.23 from the second row of Table 21.2. The prediction interval is:

$$28.4 \pm 5.23(1.18) = [22.2, 34.6]$$

We are 95% confident that the concentration of all 10 specimens will be contained within the interval 22.2 to 34.6 μg/L.

2. Construct a two-sided prediction interval to contain the mean of the concentration readings of five additional specimens randomly selected from the same population. For $n = 5$, $m = 5$, and $1 - \alpha = 0.95$, the factor is 1.76 and the interval is:

$$28.4 \pm 1.76(1.18) = [26.3, 30.5]$$

We are 95% confident that the mean of the readings of five additional concentrations will be in the interval 26.3 to 30.5 μg/L.

There are two sources of imprecision in statistical prediction. First, because the given data are limited, there is uncertainty with respect to the parameters of the previously sampled population. Second, there is random variation in the future sample. Say, for example, that the results of an initial sample of size n from a normal population with unknown mean η and unknown standard deviation σ are used to predict the value of a single future randomly selected observation from the same population. The mean $\bar{y}$ of the initial sample is used to predict the future observation. Now $\bar{y} = \eta + e$, where e, the random variation associated with the mean of the initial sample, is normally distributed with mean 0 and variance σ^2/n. The future observation to be predicted is $y_f = \eta + e_f$, where e_f is the random variation associated with the future observation, normally distributed with mean 0 and variance σ^2. Thus, the prediction error is $y_f - \bar{y} = e_f - e$, which has variance $\sigma^2 + (\sigma^2/n)$. The length of the prediction interval to contain y_f will be proportional to $\sqrt{\sigma^2 + (\sigma^2/n)}$. Increasing the initial sample will reduce the imprecision associated with the sample mean $\bar{y}$ (i.e., σ^2/n), but it will not reduce the sampling error in the estimate of the variation (σ^2) associated with the future observations. Thus, an increase in the size of the initial sample size beyond the point where the inherent variation in the future sample tends to dominate will not materially reduce the length of the prediction interval.

A confidence interval to contain a population parameter converges to a point as the sample size increases. A prediction interval converges to an interval. Thus, it is not possible to obtain a prediction interval consistently shorter than some limiting interval, no matter how large an initial sample is taken (Hahn and Meeker, 1991).

Statistical Interval for the Standard Deviation of a Normal Distribution

Confidence and prediction intervals for the standard deviation of a normal distribution can be calculated using factors from Table 21.3. The factors are based on the χ^2 distribution and are asymmetric. They are multipliers and the intervals have the form:

$$[k_1 s, k_2 s]$$

TABLE 21.3

Factors for Two-Sided 95% Statistical Intervals for a Standard Deviation of a Normal Distribution

n	Confidence Intervals		Simultaneous Prediction Intervals to Contain All $m = n$ Future Observations	
	k_1	k_2	k_1	k_2
4	0.57	3.73	0.25	3.93
5	0.60	2.87	0.32	3.10
6	0.62	2.45	0.37	2.67
7	0.64	2.20	0.41	2.41
8	0.66	2.04	0.45	2.23
10	0.69	1.83	0.50	2.01
15	0.73	1.58	0.58	1.73
20	0.76	1.46	0.63	1.59
40	0.82	1.28	0.73	1.38
60	0.85	1.22	0.77	1.29
∞	1.00	1.00	1.00	1.00

Source: Hahn, G. J. and W. Q. Meeker (1991). *Statistical Intervals: A Guide for Practitioners,* New York, John Wiley.

Example 21.3

A random sample of $n = 5$ observations yields the values $\bar{y} = 28.4$ μg/L and $\sigma = 1.18$ μg/L.

1. Using the factors in Table 21.3, find a two-sided confidence interval for the standard deviation σ of the population. For $n = 5$, $k_1 = 0.6$ and $k_2 = 2.87$. The 95% confidence interval is:

$$[0.60(1.18), 2.87(1.18)] = [0.7, 3.4]$$

We are 95% confident that the interval 0.7 to 3.4 μg/L contains the unknown standard deviation s of the population of concentration readings.

2. Construct a two-sided 95% prediction interval to contain the standard deviation of five additional concentrations randomly sampled from the same population. For $n = m = 5$, $k_1 = 0.32$, $k_2 = 3.10$, and the prediction interval is:

$$[0.32(1.18), 3.1(1.18)] = [0.4, 3.7]$$

We are 95% confident that the standard deviation of the five additional concentration readings will be in the interval 0.4 to 3.7 μg/L.

Notice how wide the intervals are compared with confidence intervals and tolerance intervals for the mean.

Case Study: Spare Parts Inventory

Village water supply projects in Africa have installed thousands of small pumps that use bearings from a company that will soon discontinue the manufacture of bearings. The company has agreed to create an inventory of bearings that will meet, with 95% confidence, the demand for replacement bearings for at least 8 years. The number of replacement bearings required in each of the past 6 years were:

$$282, 380, 318, 298, 368, \text{ and } 348$$

For the given data, $\bar{y} = 332.3$ and $s = 39.3$. Assume that the number of units sold per year follows a normal distribution with a mean and standard deviation that are constant from year to year, and will continue to be so, and that the number of units sold in one year is independent of the number sold in any other year.

Under the stated assumptions, $\bar{y} = 332.3$ provides a prediction for the average yearly demand. The demand for replacement bearings over 8 years is thus $8(332.3) = 2658$. However, because of statistical variability in both the past and future yearly demands, the actual total would be expected to differ from this prediction. A one-sided upper 95% prediction bound ($\bar{y}_U$) for the mean of the yearly sales for the next $m = 8$ years is:

$$\bar{y}_U = \bar{y} + t_{n-1,\alpha} s \left(\frac{1}{m} + \frac{1}{n} \right)^{1/2} = 332.3 + 2.015(39.3)\left(\frac{1}{8} + \frac{1}{6} \right)^{1/2} = 375.1$$

Thus, an upper 95% prediction bound for the total 8-year demand is $8(375.1) = 3001$ bearings.

We are 95% confident that the total demand for the next 8 years will not exceed 3001 bearings. At the same time, if the manufacturer actually built 3001 bearings, we would predict that the inventory would most likely last for $3001/332.3 \cong 9$ years. A one-sided *lower* prediction bound for the total 8-year demand is only 2316 bearings.

Case Study: Groundwater Monitoring

A hazardous waste landfill operator is required to take quarterly groundwater samples from $m = 25$ monitoring wells and analyze each sample for $n = 20$ constituents. The total number of quarterly comparisons of measurements with published regulatory limits is $nm = 25(20) = 500$. There is a virtual certainty that some comparisons will exceed the limits even if all wells truly are in compliance for all constituents. The regulations make no provision for the "chance" failures, but substantial savings would be possible if they would allow for a small (i.e., 1%) chance failure rate. This could be accomplished using a two-stage monitoring plan based on tolerance intervals for screening and prediction intervals for resampling verification (Gibbons, 1994; ASTM, 1998).

The one-sided tolerance interval is of the form $\bar{y} + ks$, where s, the standard deviation for each constituent, has been estimated from an available sample of n_b background measurements. Values of k are tabulated in Gibbons (1994, Table 4.2).

Suppose that we want a tolerance interval that has 95% confidence ($\alpha = 0.05$) of including 99% of all values in the interval (coverage $p = 0.99$). This is $n_b = 20$, $1 - \alpha = 0.95$, and $p = 0.99$, $k = 3.295$ and:

$$\bar{y} + 3.295s$$

For the failure rate of $(1 - 0.99) = 1\%$, we expect that $k = 0.01(500) = 5$ comparisons might exceed the published standards. If there are more than the expected five exceedances, the offending wells should be resampled, but only for those constituents that failed.

The resampling data should be compared to a 95% prediction interval for the expected number of exceedances and not the number that happens to be observed. The one-sided prediction interval is:

$$\bar{y} + ks\sqrt{1 + \frac{1}{n_b}}$$

If n_b is reasonably large (i.e., $n_b \geq 40$), the quantity under the square root is approximately 1.0 and can be ignored. Assuming this to be true, this case study uses $k = 2.43$, which is from Gibbons (1994, Table 1.2) for $n_b = 40$, $1 - \alpha = 0.95$, and $p = 0.99$. Thus, the prediction interval is $\bar{y} + 2.43s$.

Case Study: Water Quality Compliance

A company is required to meet a water quality limit of 300 ppm in a river. This has been monitored by collecting specimens of river water during the first week of each of the past 27 quarters. The data are from Hahn and Meeker (1991).

48	94	112	44	93	198	43	52	35
170	25	22	44	16	139	92	26	116
91	113	14	50	75	66	43	10	83

There have been no violations so far, but the company wants to use the past data to estimate the probability that a future quarterly reading will exceed the regulatory limit of $L = 300$.

The data are a time series and should be evaluated for trend, cycles, or correlations among the observations. Figure 21.1 shows considerable variability but gives no clear indication of a trend or cyclical pattern. Additional checking (see Chapters 32 and 53) indicates that the data may be treated as random.

Figure 21.2 shows histograms of the original data and their logarithms. The data are not normally distributed and the analysis will be made using the (natural) logarithms of the data. The sample mean and standard deviation of the log-transformed readings are $\bar{x} = 4.01$ and $s = 0.773$.

A point estimate for the probability that $y \geq 300$ [or $x \geq \ln(300)$], assuming the logarithm of chemical concentration readings follow a normal distribution, is:

$$\hat{p} = 1 - \Phi\left[\frac{\ln(L) - \bar{x}}{s}\right]$$

where $\Phi[x]$ is the percentage point on the cumulative lognormal distribution that corresponds to $x \geq \ln(300)$. For our example:

$$\hat{p} = 1 - \Phi\left[\frac{\ln(300) - 4.01}{0.773}\right] = 1 - \Phi\left[\frac{5.7 - 4.01}{0.773}\right] = 1 - \Phi(2.19) = 0.0143$$

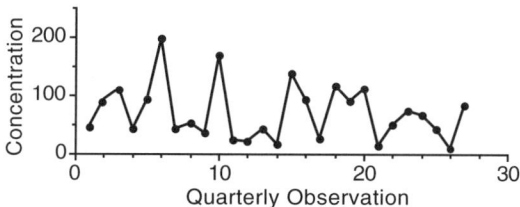

FIGURE 21.1 Chemical concentration data for the water quality compliance case study. (From Hahn G. J. and W. Q. Meeker (1991). *Statistical Methods for Groundwater Monitoring*, New York, John Wiley.)

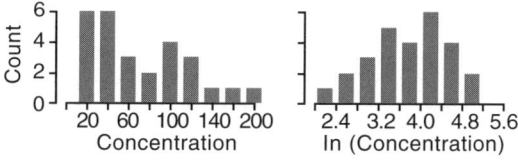

FIGURE 21.2 Histograms of the chemical concentrations and their logarithms show that the data are not normally distributed.

The value 0.0143 can be looked up in a table of the standard normal distribution. It is the area under the tail that lies beyond $z = 2.19$.

A two-sided confidence interval for $p = \text{Prob}(y \leq L)$ has the form:

$$[h_{1-\alpha/2;-K,n},\ h_{1-\alpha/2;K,n}]$$

where $K = (\bar{y} - L)/s$. A one-sided upper confidence bound is $[h_{1-\alpha;K,n}]$. The h factors are found in Table 7 of Odeh and Owen (1980).

For $1 - \alpha = 0.95$, $n = 27$, and $K = [4.01 - \ln(300)]/0.773 = -2.2$, the factor is $h = 0.94380$ and the upper 95% confidence bound for p is $1 - 0.9438 = 0.0562$. Thus, we are 95% confident that the probability of a reading exceeding 300 ppm is less than 5.6%. This 5.6% probability of getting a future value above $L = 300$ may be disappointing given that all of the previous 27 observations have been below the limit.

Had the normal distribution been incorrectly assumed, the upper 95% confidence limit obtained would have been 0.015%, the contrast between 0.00015 and 0.05620 (a ratio of about 375). This shows that confidence bounds on probabilities in the tail of a distribution are badly wrong when the incorrect distribution is assumed.

Comments

In summary, a confidence interval contains the unknown value of a parameter (a mean), a tolerance interval contains a proportion of the population, and a prediction interval contains one or more future observations from a previously sampled population.

The lognormal distribution is frequently used in environmental assessments. The logarithm of a variable with a lognormal distribution has a normal distribution. Thus, the methods for computing statistical intervals for the normal distribution can be used for the lognormal distribution. Tolerance limits, confidence limits for distribution percentiles, and prediction limits are calculated on the logarithms of the data, and then are converted back to the scale of the original data.

Intervals based on the Poisson distribution can be determined for the number of occurrences. Intervals based on the binomial distribution can be determined for proportions and percentages.

All the examples in this chapter were based on assuming the normal or lognormal distribution. Tolerance and prediction intervals can be computed by distribution-free methods (nonparametric methods). Using the distribution gives a more precise bound on the desired probability than the distribution-free methods (Hahn and Meeker, 1991).

References

ASTM (1998). *Standard Practice for Derivation of Decision Point and Confidence Limit Testing of Mean Concentrations in Waste Management Decisions,* D 6250, Washington, D.C., U.S. Government Printing Office.

Hahn, G. J. (1970). "Statistical Intervals for a Normal Population. Part I. Tables, Examples, and Applications," *J. Qual. Tech.,* 3, 18–22.

Hahn, G. J. and W. Q. Meeker (1991). *Statistical Intervals: A Guide for Practitioners,* New York, John Wiley.

Gibbons, R. D. (1994). *Statistical Methods for Groundwater Monitoring,* New York, John Wiley.

Johnson, R. A. (2000). *Probability and Statistics for Engineers,* 6th ed., Englewood Cliffs, NJ, Prentice-Hall.

Odeh, R. E. and D. B. Owen (1980). *Tables for Normal Tolerance Limits, Sampling Plans, and Screening,* New York, Marcel Dekker.

Owen, D. B. (1962). *Handbook of Statistical Tables,* Palo Alto, CA, Addison-Wesley.

Tolerance Intervals and Prediction Intervals

Exercises

21.1 Phosphorus in Biosolids. A random sample of $n = 5$ observations yields the values $\bar{y} = 39.0$ µg/L and $s = 2.2$ µg/L for total phosphorus in biosolids from a municipal wastewater treatment plant.

1. Calculate the two-sided 95% confidence interval for the mean and the two-sided 95% tolerance interval to contain at least 99% of the population.
2. Calculate the two-sided 95% prediction interval to contain all of the next five observations.
3. Calculate the two-sided 95% confidence interval for the standard deviation of the population.

21.2 TOC in Groundwater. Two years of quarterly measurements of TOC from a monitoring well are 10.0, 11.5, 11.0, 10.6, 10.9, 12.0, 11.3, and 10.7.

1. Calculate the two-sided 95% confidence interval for the mean and the two-sided 95% tolerance interval to contain at least 95% of the population.
2. Calculate the two-sided 95% confidence interval for the standard deviation of the population.
3. Determine the upper 95% prediction limit for the next quarterly TOC measurement.

21.3 Spare Parts. An international agency offers to fund a warehouse for spare parts needed in small water supply projects in South Asian countries. They will provide funds to create an inventory that should last for 5 years. A particular kind of pump impeller needs frequent replacement. The number of impellers purchased in each of the past 5 years is 2770, 3710, 3570, 3080, and 3270. How many impellers should be stocked if the spare parts inventory is created? How long will this inventory be expected to last?

22

Experimental Design

KEY WORDS *blocking, Box-Behnken, composite design, direct comparison, empirical models, emulsion breaking, experimental design, factorial design, field studies, interaction, iterative design, mechanistic models, one-factor-at-a-time experiment, OFAT, oil removal, precision, Plakett-Burman, randomization, repeats, replication, response surface, screening experiments, standard error, t-test.*

> "It is widely held by nonstatisticians that if you do good experiments statistics are not necessary. They are quite right.... The snag, of course, is that doing good experiments is difficult. Most people need all the help they can get to prevent them making fools of themselves by claiming that their favorite theory is substantiated by observations that do nothing of the sort...." (Coloquhon, 1971).

We can all cite a few definitive experiments in which the results were intuitively clear without statistical analysis. This can only happen when there is an excellent experimental design, usually one that involves direct comparisons and replication. Direct comparison means that nuisance factors have been removed. Replication means that credibility has been increased by showing that the favorable result was not just luck. (If you do not believe me, I will do it again.) On the other hand, we have seen experiments where the results were unclear even after laborious statistical analysis was applied to the data. Some of these are the result of an inefficient experimental design.

Statistical experimental design refers to the work plan for manipulating the settings of the independent variables that are to be studied. Another kind of experimental design deals with building and operating the experimental apparatus. The more difficult and expensive the operational manipulations, the more statistical design offers gains in efficiency.

This chapter is a descriptive introduction to experimental design. There are many kinds of experimental designs. Some of these are one-factor-at-a-time, paired comparison, two-level factorials, fractional factorials, Latin squares, Graeco-Latin squares, Box-Behnken, Plackett-Burman, and Taguchi designs. An efficient design gives a lot of information for a little work. A "botched" design gives very little information for a lot of work. This chapter has the goal of convincing you that one-factor-at-a-time designs are poor (so poor they often may be considered botched designs) and that it is possible to get a lot of information with very few experimental runs. Of special interest are two-level factorial and fractional factorial experimental designs. Data interpretation follows in Chapters 23 through 48.

What Needs to be Learned?

Start your experimental design with a clear statement of the question to be investigated and what you know about it. Here are three pairs of questions that lead to different experimental designs:

1. a. If I observe the system without interference, what function best predicts the output y?
   b. What happens to y when I change the inputs to the process?
2. a. What is the value of θ in the mechanistic model $y = x^\theta$?
   b. What smooth polynomial will describe the process over the range $[x_1, x_2]$?

TABLE 22.1

Five Classes of Experimental Problems Defined in Terms of What is Unknown in the Model, $\eta = f(\mathbf{X}, \theta)$, Which is a Function of One or More Independent Variables $\mathbf{X}$ and One or More Parameters θ

Unknown	Class of Problem	Design Approach	Chapter
$f, \mathbf{X}, \theta$	Determine a subset of important variables from a given larger set of potentially important variables	Screening variables	23, 29
f, θ	Determine empirical "effects" of known input variables $\mathbf{X}$	Empirical model building	27, 38
f, θ	Determine a local interpolation or approximation function, $f(\mathbf{X}, \theta)$	Empirical model building	36, 37, 38, 40, 43
f, θ	Determine a function based on mechanistic understanding of the system	Mechanistic model building	46, 47
θ	Determine values for the parameters	Model fitting	35, 44

Source: Box, G. E. P. (1965). Experimemtal Strategy, Madison WI, Department of Statistics, Wisconsin Tech. Report #111, University of Wisconsin-Madison.

   3.a. Which of seven potentially active factors are important?

     b. What is the magnitude of the effect caused by changing two factors that have been shown important in preliminary tests?

A clear statement of the experimental objectives will answer questions such as the following:

1. What factors (variables) do you think are important? Are there other factors that might be important, or that need to be controlled? Is the experiment intended to show which variables are important or to estimate the effect of variables that are known to be important?

2. Can the experimental factors be set precisely at levels and times of your choice? Are there important factors that are beyond your control but which can be measured?

3. What kind of a model will be fitted to the data? Is an empirical model (a smoothing polynomial) sufficient, or is a mechanistic model to be used? How many parameters must be estimated to fit the model? Will there be interactions between some variables?

4. How large is the expected random experimental error compared with the expected size of the effects? Does my experimental design provide a good estimate of the random experimental error? Have I done all that is possible to eliminate bias in measurements, and to improve precision?

5. How many experiments does my budget allow? Shall I make an initial commitment of the full budget, or shall I do some preliminary experiments and use what I learn to refine the work plan?

Table 22.1 lists five general classes of experimental problems that have been defined by Box (1965). The model $\eta = f(\mathbf{X}, \theta)$ describes a response η that is a function of one or more independent variables $\mathbf{X}$ and one or more parameters θ. When an experiment is planned, the functional form of the model may be known or unknown; the active independent variables may be known or unknown. Usually, the parameters are unknown. The experimental strategy depends on what is unknown. A well-designed experiment will make the unknown known with a minimum of work.

Principles of Experimental Design

Four basic principles of good experimental design are direct comparison, replication, randomization, and blocking.

Comparative Designs

If we add substance X to a process and the output improves, it is tempting to attribute the improvement to the addition of X. But this observation may be entirely wrong. X may have no importance in the process.

Its addition may have been coincidental with a change in some other factor. The way to avoid a false conclusion about X is to do a comparative experiment. Run parallel trials, one with X added and one with X not added. All other things being equal, a change in output can be attributed to the presence of X. Paired t-tests (Chapter 17) and factorial experiments (Chapter 27) are good examples of comparative experiments.

Likewise, if we passively observe a process and we see that the air temperature drops and output quality decreases, we are not entitled to conclude that we can cause the output to improve if we raise the temperature. Passive observation or the equivalent, dredging through historical records, is less reliable than direct comparison. If we want to know what happens to the process when we change something, we must observe the process when the factor is actively being changed (Box, 1966; Joiner, 1981).

Unfortunately, there are situations when we need to understand a system that cannot be manipulated at will. Except in rare cases (TVA, 1962), we cannot control the flow and temperature in a river. Nevertheless, a fundamental principle is that we should, whenever possible, do designed and controlled experiments. By this we mean that we would like to be able to establish specified experimental conditions (temperature, amount of X added, flow rate, etc.). Furthermore, we would like to be able to run the several combinations of factors in an order that we decide and control.

Replication

Replication provides an internal estimate of random experimental error. The influence of error in the effect of a factor is estimated by calculating the standard error. All other things being equal, the standard error will decrease as the number of observations and replicates increases. This means that the precision of a comparison (e.g., difference in two means) can be increased by increasing the number of experimental runs. Increased precision leads to a greater likelihood of correctly detecting small differences between treatments. It is sometimes better to increase the number of runs by replicating observations instead of adding observations at new settings.

Genuine repeat runs are needed to estimate the random experimental error. "Repeats" means that the settings of the x's are the same in two or more runs. "Genuine repeats" means that the runs with identical settings of the x's capture all the variation that affects each measurement (Chapter 9). Such replication will enable us to estimate the standard error against which differences among treatments are judged. If the difference is large relative to the standard error, confidence increases that the observed difference did not arise merely by chance.

Randomization

To assure validity of the estimate of experimental error, we rely on the principle of randomization. It leads to an unbiased estimate of variance as well as an unbiased estimate of treatment differences. Unbiased means free of systemic influences from otherwise uncontrolled variation.

Suppose that an industrial experiment will compare two slightly different manufacturing processes, A and B, on the same machinery, in which A is always used in the morning and B is always used in the afternoon. No matter how many manufacturing lots are processed, there is no way to separate the difference between the machinery or the operators from morning or afternoon operation. A good experiment does not assume that such systematic changes are absent. When they affect the experimental results, the bias cannot be removed by statistical manipulation of the data. Random assignment of treatments to experimental units will prevent systematic error from biasing the conclusions.

Randomization also helps to eliminate the corrupting effect of serially correlated errors (i.e., process or instrument drift), nuisance correlations due to lurking variables, and inconsistent data (i.e., different operators, samplers, instruments).

Figure 22.1 shows some possibilities for arranging the observations in an experiment to fit a straight line. Both replication and randomization (run order) can be used to improve the experiment.

Must we randomize? In some experiments, a great deal of expense and inconvenience must be tolerated in order to randomize; in other experiments, it is impossible. Here is some good advice from Box (1990).

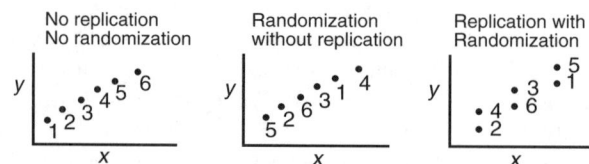

FIGURE 22.1 The experimental designs for fitting a straight line improve from left to right as replication and randomization are used. Numbers indicate order of observation.

1. In those cases where randomization only slightly complicates the experiment, always randomize.
2. In those cases where randomization would make the experiment impossible or extremely difficult to do, but you can make an honest judgment about existence of nuisance factors, run the experiment without randomization. Keep in mind that wishful thinking is not the same as good judgment.
3. If you believe the process is so unstable that without randomization the results would be useless and misleading, and randomization will make the experiment impossible or extremely difficult to do, then do not run the experiment. Work instead on stabilizing the process or getting the information some other way.

Blocking

The paired t-test (Chapter 17) introduced the concept of blocking. Blocking is a means of reducing experimental error. The basic idea is to partition the total set of experimental units into subsets (blocks) that are as homogeneous as possible. In this way the effects of nuisance factors that contribute systematic variation to the difference can be eliminated. This will lead to a more sensitive analysis because, loosely speaking, the experimental error will be evaluated in each block and then pooled over the entire experiment.

Figure 22.2 illustrates blocking in three situations. In (a), three treatments are to be compared but they cannot be observed simultaneously. Running A, followed by B, followed by C would introduce possible bias due to changes over time. Doing the experiment in three blocks, each containing treatment A, B, and C, in random order, eliminates this possibility. In (b), four treatments are to be compared using four cars. Because the cars will not be identical, the preferred design is to treat each car as a block and balance the four treatments among the four blocks, with randomization. Part (c) shows a field study area with contour lines to indicate variations in soil type (or concentration). Assigning treatment A to only the top of the field would bias the results with respect to treatments B and C. The better design is to create three blocks, each containing treatment A, B, and C, with random assignments.

Attributes of a Good Experimental Design

A good design is simple. A simple experimental design leads to simple methods of data analysis. The simplest designs provide estimates of the main differences between treatments with calculations that amount to little more than simple averaging. Table 22.2 lists some additional attributes of a good experimental design.

If an experiment is done by unskilled people, it may be difficult to guarantee adherence to a complicated schedule of changes in experimental conditions. If an industrial experiment is performed under production conditions, it is important to disturb production as little as possible.

In scientific work, especially in the preliminary stages of an investigation, it may be important to retain flexibility. The initial part of the experiment may suggest a much more promising line of investigation, so that it would be a bad thing if a large experiment has to be completed before any worthwhile results are obtained. Start with a simple design that can be augmented as additional information becomes available.

Experimental Design

TABLE 22.2

Attributes of a Good Experiment

A good experimental design should:

1. Adhere to the basic principles of randomization, replication, and blocking.
2. Be simple:
   a. Require a minimum number of experimental points
   b. Require a minimum number of predictor variable levels
   c. Provide data patterns that allow visual interpretation
   d. Ensure simplicity of calculation
3. Be flexible:
   a. Allow experiments to be performed in blocks
   b. Allow designs of increasing order to be built up sequentially
4. Be robust:
   a. Behave well when errors occur in the settings of the x's
   b. Be insensitive to wild observations
   c. Be tolerant to violation of the usual normal theory assumptions
5. Provide checks on goodness of fit of model:
   a. Produce balanced information over the experimental region
   b. Ensure that the fitted value will be as close as possible to the true value
   c. Provide an internal estimate of the random experimental error
   d. Provide a check on the assumption of constant variance

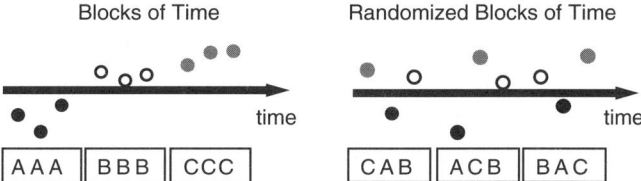

(a) Good and bad designs for comparing treatments A, B, and C

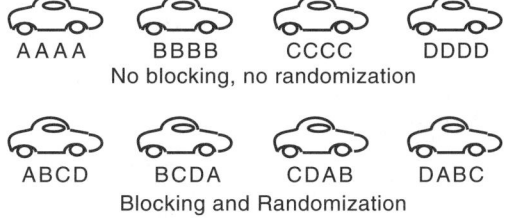

(b) Good and bad designs for comparing treatments A, B, C, and D for pollution reduction in automobiles

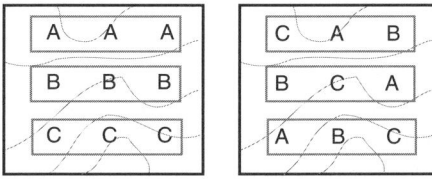

(b) Good and bad designs for comparing treatments A, B, and C in a field of non-uniform soil type.

FIGURE 22.2 Successful strategies for blocking and randomization in three experimental situations.

One-Factor-At-a-Time (OFAT) Experiments

Most experimental problems investigate two or more factors (independent variables). The most inefficient approach to experimental design is, "Let's just vary one factor at a time so we don't get confused." If this approach does find the best operating level for all factors, it will require more work than experimental designs that simultaneously vary two or more factors at once.

These are some advantages of a good multifactor experimental design compared to a one-factor-at-a-time (OFAT) design:

- It requires less resources (time, material, experimental runs, etc.) for the amount of information obtained. This is important because experiments are usually expensive.
- The estimates of the effects of each experimental factor are more precise. This happens because a good design multiplies the contribution of each observation.
- The interaction between factors can be estimated systematically. Interactions cannot be estimated from OFAT experiments.
- There is more information in a larger region of the factor space. This improves the prediction of the response in the factor space by reducing the variability of the estimates of the response. It also makes the process optimization more efficient because the optimal solution is searched for over the entire factor space.

Suppose that jar tests are done to find the best operating conditions for breaking an oil–water emulsion with a combination of ferric chloride and sulfuric acid so that free oil can be removed by flotation. The initial oil concentration is 5000 mg/L. The first set of experiments was done at five levels of ferric chloride with the sulfuric acid dose fixed at 0.1 g/L. The test conditions and residual oil concentration (oil remaining after chemical coagulation and gravity flotation) are given below.

$FeCl_3$ (g/L)	1.0	1.1	1.2	1.3	1.4
H_2SO_4 (g/L)	0.1	0.1	0.1	0.1	0.1
Residual oil (mg/L)	4200	2400	1700	175	650

The dose of 1.3 g/L of $FeCl_3$ is much better than the other doses that were tested. A second series of jar tests was run with the $FeCl_3$ level fixed at the apparent optimum of 1.3 g/L to obtain:

$FeCl_3$ (g/L)	1.3	1.3	1.3
H_2SO_4 (g/L)	0	0.1	0.2
Oil (mg/L)	1600	175	500

This test seems to confirm that the best combination is 1.3 g/L of $FeCl_3$ and 0.1 g/L of H_2SO_4.

Unfortunately, this experiment, involving eight runs, leads to a wrong conclusion. The response of oil removal efficiency as a function of acid and iron dose is a valley, as shown in Figure 22.3. The first one-at-a-time experiment cut across the valley in one direction, and the second cut it in the perpendicular direction. What appeared to be an optimum condition is false. A valley (or a ridge) describes the response surface of many real processes. The consequence is that one-factor-at-a-time experiments may find a false optimum. Another weakness is that they fail to discover that a region of higher removal efficiency lies in the direction of higher acid dose and lower ferric chloride dose.

We need an experimental strategy that (1) will not terminate at a false optimum, and (2) will point the way toward regions of improved efficiency. Factorial experimental designs have these advantages. They are simple and tremendously productive and every engineer who does experiments of any kind should learn their basic properties.

We will illustrate two-level, two-factor designs using data from the emulsion breaking example. A two-factor design has two independent variables. If each variable is investigated at two levels (high and

Experimental Design

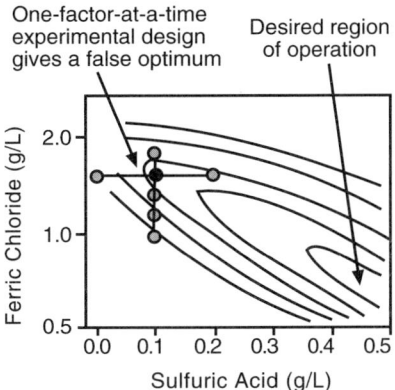

FIGURE 22.3 Response surface of residual oil as a function of ferric chloride and sulfuric acid dose, showing a valley-shaped region of effective conditions. Changing one factor at a time fails to locate the best operating conditions for emulsion breaking and oil removal.

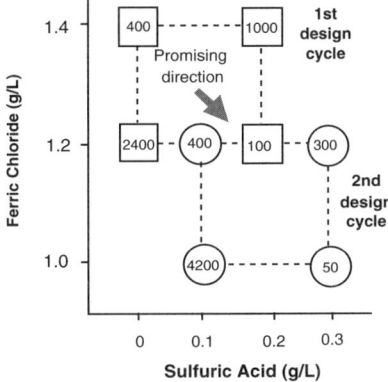

FIGURE 22.4 Two cycles (a total of eight runs) of two-level, two-factor experimental design efficiently locate an optimal region for emulsion breaking and oil removal.

low, in general terms), the experiment is a two-level design. The total number of experimental runs needed to investigate two levels of two factors is $n = 2^2 = 4$. The 2^2 experimental design for jar tests on breaking the oil emulsion is:

Acid (g/L)	FeCl$_3$ (g/L)	Oil (mg/L)
0	1.2	2400
0	1.4	400
0.2	1.2	100
0.2	1.4	1000

These four experimental runs define a small section of the response surface and it is convenient to arrange the data in a graphical display like Figure 22.4, where the residual oil concentrations are shown in the squares. It is immediately clear that the best of the tested conditions is high acid dose and low FeCl$_3$ dose. It is also clear that there might be a payoff from doing more tests at even higher acid doses and even lower iron doses, as indicated by the arrow. The follow-up experiment is shown by the circles in Figure 22.4.

The eight observations used in the two-level, two-factor designs come from the 28 actual observations made by Pushkarev et al. (1983) that are given in Table 22.3. The factorial design provides information

TABLE 22.3

Residual Oil (mg/L) after Treatment by Chemical Emulsion Breaking and Flotation

FeCl$_3$ Dose (g/L)	Sulfuric Acid Dose (g/L H$_2$SO$_4$)				
	0	0.1	0.2	0.3	0.4
0.6	—	—	—	—	600
0.7	—	—	—	—	50
0.8	—	—	—	4200	50
0.9	—	—	2500	50	150
1.0	—	4200	150	50	200
1.1	—	2400	50	100	400
1.2	2400	1700	100	300	700
1.3	1600	175	500	—	—
1.4	400	650	1000	—	—
1.5	350	—	—	—	—
1.6	1600	—	—	—	—

Source: Pushkarev et al. 1983. *Treatment of Oil-Containing Wastewater*, New York, Allerton Press.

that allows the experimenter to iteratively and quickly move toward better operating conditions if they exist, and provides information about the interaction of acid and iron on oil removal.

More about Interactions

Figure 22.5 shows two experiments that could be used to investigate the effect of pressure and temperature. The one-factor-at-a-time experiment (shown on the left) has experimental runs at these conditions:

Test Condition	Yield
(1) Standard pressure and standard temperature	10
(2) Standard pressure and new temperature	7
(3) New pressure and standard temperature	11

Imagine a total of $n = 12$ runs, 4 at each condition. Because we had four replicates at each test condition, we are highly confident that changing the temperature at standard pressure decreased the yield by 3 units. Also, we are highly confidence that raising the temperature at standard pressure increased the yield by 1 unit.

Will changing the temperature at the new pressure also decrease the yield by 3 units? The data provide no answer. The effect of temperature on the response at the new temperature cannot be estimated.

Suppose that the 12 experimental runs are divided equally to investigate four conditions as in the two-level, two-factor experiment shown on the right side of Figure 22.5.

Test Condition	Yield
(1) Standard pressure and standard temperature	10
(2) Standard pressure and new temperature	7
(3) New pressure and standard temperature	11
(4) New pressure and new temperature	12

At the standard pressure, the effect of change in the temperature is a decrease of 3 units. At the new pressure, the effect of change in temperature is an increase of 1 unit. The effect of a change in temperature depends on the pressure. There is an *interaction* between temperature and pressure. The experimental effort was the same (12 runs) but this experimental design has produced new and useful information (Czitrom, 1999).

Experimental Design

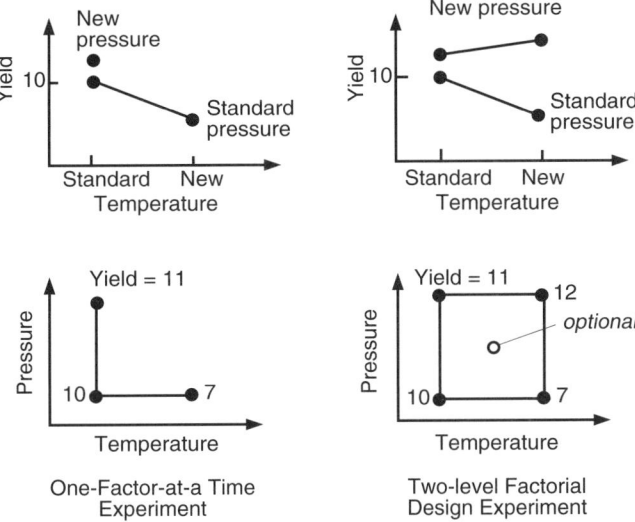

FIGURE 22.5 Graphical demonstration of why one-factor-at-a-time (OFAT) experiments cannot estimate the two-factor interaction between temperature and pressure that is revealed by the two-level, two-factor design.

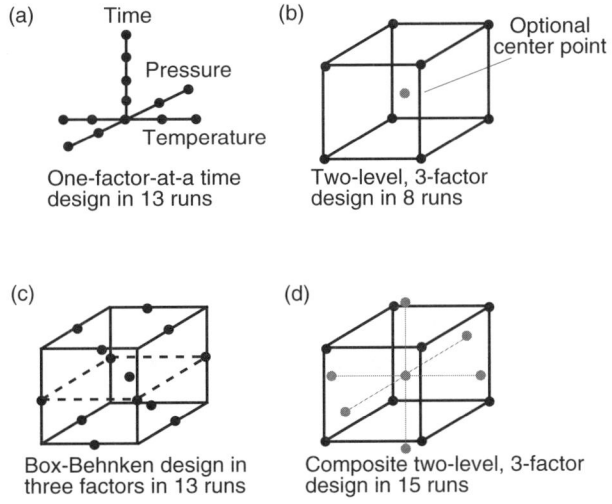

FIGURE 22.6 Four possible experimental designs for studying three factors. The worst is (a), the one-factor-at-a-time design (top left). (b) is a two-level, three-factor design in eight runs and can describe a smooth nonplanar surface. The Box-Behnken design (c) and the composite two-level, three-factor design (d) can describe quadratic effects (maxima and minima). The Box-Behnken design uses 12 observations located on the face of the cube plus a center point. The composite design has eight runs located at the corner of the cube, plus six "star" points, plus a center point. The corner and star points are equidistant from the center (i.e., located on a sphere having a diameter equal to the distance from the center to a corner).

It is generally true that (1) the factorial design gives better precision than the OFAT design if the factors *do* act additively; and (2) if the factors *do not* act additively, the factorial design can detect and estimate interactions that measure the nonadditivity.

As the number of factors increases, the benefits of investigating several factors simultaneously increases. Figure 22.6 illustrates some designs that could be used to investigate three factors. The one-factor-at-a time design (Figure 22.6a) in 13 runs is the worst. It provides no information about interactions and no information about curvature of the response surface. Designs (b), (c), and (d) do provide estimates

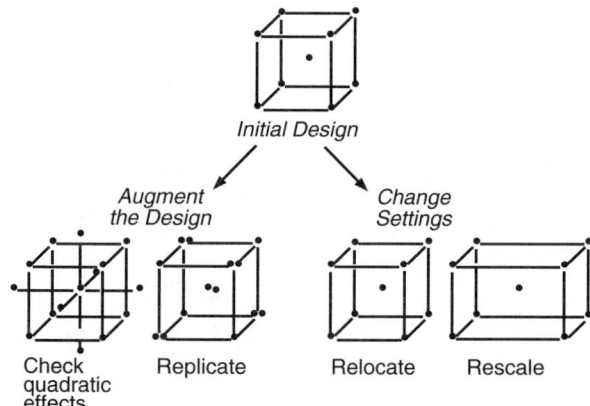

FIGURE 22.7 Some of the modifications that are possible with a two-level factorial experimental design. It can be stretched (rescaled), replicated, relocated, or augmented.

of interactions as well as the effects of changing the three factors. Figure 22.6b is a two-level, three-factor design in eight runs that can describe a smooth nonplanar surface. The *Box-Behnken design* (c) and the composite two-level, three-factor design (d) can describe quadratic effects (maxima and minima). The Box-Behnken design uses 12 observations located on the face of the cube plus a center point. The *composite design* has eight runs located at the corner of the cube, plus six "star" points, plus a center point. There are advantages to setting the corner and star points equidistant from the center (i.e., on a sphere having a diameter equal to the distance from the center to a corner).

Designs (b), (c), and (d) can be replicated, stretched, moved to new experimental regions, and expanded to include more factors. They are ideal for iterative experimentation (Chapters 43 and 44).

Iterative Design

Whatever our experimental budget may be, we never want to commit everything at the beginning. Some preliminary experiments will lead to new ideas, better settings of the factor levels, and to adding or dropping factors from the experiment. The oil emulsion-breaking example showed this. The importance of iterative experimentation is discussed again in Chapters 43 and 44. Figure 22.7 suggests some of the iterative modifications that might be used with two-level factorial experiments.

Comments

A good experimental design is simple to execute, requires no complicated calculations to analyze the data, and will allow several variables to be investigated simultaneously in few experimental runs.

Factorial designs are efficient because they are balanced and the settings of the independent variables are completely uncorrelated with each other (orthogonal designs). Orthogonal designs allow each effect to be estimated independently of other effects.

We like factorial experimental designs, especially for treatment process research, but they do not solve all problems. They are not helpful in most field investigations because the factors cannot be set as we wish. A professional statistician will know other designs that are better. Whatever the final design, it should include replication, randomization, and blocking.

Chapter 23 deals with selecting the sample size in some selected experimental situations. Chapters 24 to 26 explain the analysis of data from factorial experiments. Chapters 27 to 30 are about two-level factorial and fractional factorial experiments. They deal mainly with identifying the important subset of experimental factors. Chapters 33 to 48 deal with fitting linear and nonlinear models.

References

Berthouex, P. M. and D. R. Gan (1991). "Fate of PCBs in Soil Treated with Contaminated Municipal Sludge," *J. Envir. Engr. Div.,* ASCE, 116(1), 1–18.

Box, G. E. P. (1965). Experimental Strategy, Madison, WI, Department of Statistics, Wisconsin Tech. Report #111, University of Wisconsin–Madison.

Box, G. E. P. (1966). "The Use and Abuse of Regression," *Technometrics,* 8, 625–629.

Box, G. E. P. (1982). "Choice of Response Surface Design and Alphabetic Optimiality," *Utilitas Mathematica,* 21B, 11–55.

Box, G. E. P. (1990). "Must We Randomize?," *Qual. Eng.,* 2, 497–502.

Box, G. E. P., W. G. Hunter, and J. S. Hunter (1978). *Statistics for Experimenters: An Introduction to Design, Data Analysis, and Model Building,* New York, Wiley Interscience.

Colquhoun, D. (1971). *Lectures in Biostatistics,* Oxford, England, Clarendon Press.

Czitrom, Veronica (1999). "One-Factor-at-a Time Versus Designed Experiments," *Am. Stat.,* 53(2), 126–131.

Joiner, B. L. (1981). "Lurking Variables: Some Examples," *Am. Stat.,* 35, 227–233.

Pushkarev et al. (1983). *Treatment of Oil-Containing Wastewater,* New York, Allerton Press.

Tennessee Valley Authority (1962). *The Prediction of Stream Reaeration Rates,* Chattanooga, TN.

Tiao, George, S. Bisgarrd, W. J. Hill, D. Pena, and S. M. Stigler, Eds. (2000). *Box on Quality and Discovery with Design, Control, and Robustness,* New York, John Wiley & Sons.

Exercises

22.1 Straight Line. You expect that the data from an experiment will describe a straight line. The range of x is from 5 to 50. If your budget will allow 12 runs, how will you allocate the runs over the range of x? In what order will you execute the runs?

22.2 OFAT. The instructions to high school science fair contestants states that experiments should only vary one factor at a time. Write a letter to the contest officials explaining why this is bad advice.

22.3 Planning. Select one of the following experimental problems and (a) list the experimental factors, (b) list the responses, and (c) explain how you would arrange an experiment. Consider this a brainstorming activity, which means there are no wrong answers. Note that in 3, 4, and 5 some experimental factors and responses have been suggested, but these should not limit your investigation.

1. Set up a bicycle for long-distance riding.
2. Set up a bicycle for mountain biking.
3. Investigate how clarification of water by filtration will be affected by such factors as pH, which will be controlled by addition of hydrated lime, and the rate of flow through the filter.
4. Investigate how the dewatering of paper mill sludge would be affected by such factors as temperature, solids concentration, solids composition (fibrous vs. granular material), and the addition of polymer.
5. Investigate how the rate of disappearance of oil from soil depends on such factors as soil moisture, soil temperature, wind velocity, and land use (tilled for crops vs. pasture, for example).
6. Do this for an experiment that you have done, or one that you would like to do.

22.4 Soil Sampling. The budget of a project to explore the extent of soil contamination in a storage area will cover the collection and analysis of 20 soil specimens, or the collection of 12 specimens with duplicate analyses of each, or the collection of 15 specimens with duplicate analyses of 6 of these specimens selected at random. Discuss the merits of each plan.

22.5 Personal Work. Consider an experiment that you have performed. It may be a series of analytical measurements, an instrument calibration, or a process experiment. Describe how the principles of direct comparison, replication, randomization, and blocking were incorporated into the experiment. If they were not practiced, explain why they were not needed, or why they were not used. Or, suggest how the experiment could have been improved by using them.

22.6 Trees. It is proposed to study the growth of two species of trees on land that is irrigated with treated industrial wastewater effluent. Ten trees of each species will be planted and their growth will be monitored over a number of years. The figure shows two possible schemes. In one (left panel) the two kinds of trees are allocated randomly to 20 test plots of land. In the other (right panel) the species A is restricted to half the available land and species B is planted on the other. The investigator who favors the randomized design plans to analyze the data using an independent t-test. The investigator who favors the unrandomized design plans to analyze the data using a paired t-test, with the average of 1a and 1b being paired with 1c and 1d. Evaluate these two plans. Suggest other possible arrangements. Optional: Design the experiment if there are four species of tress and 20 experimental plots.

	a	b	c	d			a	b	c	d
1	A	B	B	A		1	A	A	B	B
2	A	B	A	B		2	A	A	B	B
3	A	A	B	A		3	A	A	B	B
4	B	B	A	B		4	A	A	B	B
5	B	A	A	B		5	A	A	B	B

Randomized | Unrandomized

22.7 Solar Energy. The production of hot water is studied by installing ten units of solar collector A and ten units of solar collector B on homes in a Wisconsin town. Propose some experimental designs and discuss their advantages and disadvantages.

22.8 River Sampling. A river and one of its tributary streams were monitored for pollution and the following data were obtained:

River	16	12	14	11	
Tributary	9	10	8	6	5

It was claimed that this proves the tributary is cleaner than the river. The statistician who was asked to confirm this impression asked a series of questions. *When were the data taken? All in one day? On different days? Were the data taken during the same time period for the two streams? Were the temperatures of the two streams the same? Where in the streams were the data taken? Why were these points chosen? Are they representative?*

Why do you think the statistician asked these questions? Are there other questions that should have been asked? Is there any set of answers to these questions that would justify the use of a t-test to draw conclusions about pollution levels?

23

Sizing the Experiment

KEY WORDS arcsin, binomial, bioassay, census, composite sample, confidence limit, equivalence of means, interaction, power, proportions, random sampling, range, replication, sample size, standard deviation, standard error, stratified sampling, t-test, t distribution, transformation, type I error, type II error, uniform distribution, variance.

Perhaps the most frequently asked question in planning experiments is: "How large a sample do I need?" When asked the purpose of the project, the question becomes more specific:

What size sample is needed to estimate the average within X units of the true value?

What size sample is needed to detect a change of X units in the level?

What size sample is needed to estimate the standard deviation within 20% of the true value?

How do I arrange the sampling when the contaminate is spotty, or different in two areas?

How do I size the experiment when the results will be proportions of percentages?

There is no single or simple answer. It depends on the experimental design, how many effects or parameters you want to estimate, how large the effects are expected to be, and the standard error of the effects. The value of the standard error depends on the intrinsic variability of the experiment, the precision of the measurements, and the sample size.

In most situations where statistical design is useful, only limited improvement is possible by modifying the experimental material or increasing the precision of the measuring devices. For example, if we change the experimental material from sewage to a synthetic mixture, we remove a good deal of intrinsic variability. This is the "lab-bench" effect. We are able to predict better, but what we can predict is not real.

Replication and Experimental Design

Statistical experimental design, as discussed in the previous chapter, relies on blocking and randomization to balance variability and make it possible to estimate its magnitude. After refining the experimental equipment and technique to minimize variance from nuisance factors, we are left with replication to improve the informative power of the experiment.

The standard error is the measure of the magnitude of the experimental error of an estimated statistic (mean, effect, etc.). For the sample mean, the standard error is $\sigma/\sqrt{n}$, compared with the standard deviation σ. The standard deviation (or variance) refers to the intrinsic variation of observations within individual experimental units, whereas the standard error refers to the random variation of an estimate from the whole experiment.

Replication will not reduce the standard deviation but it will reduce the standard error. The standard error can be made arbitrarily small by increased replication. All things being equal, the standard error is halved by a fourfold increase in the number of experimental runs; a 100-fold increase is needed to divide the standard error by 10. This means that our goal is a standard error small enough to make convincing conclusions, but not too small. If the standard error is large, the experiment is worthless, but resources have been wasted if it is smaller than necessary.

In a paired *t*-test, each pair is a block that is not affected by nuisance factors that change during the time between runs. Each pair provides one estimate of the difference between the treatments being compared. If we have only one pair, we can estimate the average difference but we can say nothing about the precision of the estimate because we have no degrees of freedom with which to estimate the experimental error. Making two replicates (two pairs) is an improvement, and going to four pairs is a big improvement. Suppose the variance of each difference is σ^2. If we run two replicates (two pairs), the approximate 95% confidence interval would be $\pm 2\sigma/\sqrt{2} = \pm 1.4\sigma$. Four replicates would reduce the confidence interval to $\pm 2\sigma/\sqrt{4} = \pm \sigma$. Each quadrupling of the sample size reduces the standard error and the confidence interval by half.

Two-level factorial experiments, mentioned in the previous chapter as an efficient way to investigate several factors at one time, incorporate the effect of replication. Suppose that we investigate three factors by setting each at two levels and running all eight possible combinations, giving an experiment with $n = 8$ runs. From these eight runs we get four independent estimates of the effect of *each* factor. This is like having a paired experiment repeated four times for factor A, four times for factor B, and four times for factor C. Each measurement is doing triple duty. In short, we gain a benefit similar to what we gain from replication, but without actually repeating any tests. It is better, of course, to actually repeat some (or all) runs because this will reduce the standard error of the estimated effects and allow us to detect smaller differences. If each test condition were repeated twice, the $n = 16$ run experiment would be highly informative.

Halving the standard error is a big gain. If the true difference between two treatments is one standard error, there is only about a 17% chance that it will be detected at a confidence level of 95%. If the true difference is two standard errors, there is slightly better than a 50/50 chance that it will be identified as statistically significant at the 95% confidence level.

We now see the dilemma for the engineer and the statistical consultant. The engineer wants to detect a small difference without doing many replicates. The statistician, not being a magician, is constrained to certain mathematical realities. The consultant will be most helpful at the planning stages of an experiment when replication, randomization, blocking, and experimental design (factorial, paired test, etc.) can be integrated.

What follows are recipes for a few simple situations in single-factor experiments. The theory has been mostly covered in previous chapters.

Confidence Interval for a Mean

The $(1 - \alpha)100\%$ confidence interval for the mean η has the form $\bar{y} \pm E$, where E is the half-length $E = z_{\alpha/2}\sigma/\sqrt{n}$. The sample size n that will produce this interval half-length is:

$$n = \left(\frac{z_{\alpha/2}\sigma}{E}\right)^2$$

The value obtained is rounded to the next highest integer. This assumes random sampling. It also assumes that n is large enough that the normal distribution can be used to define the confidence interval. (For smaller sample sizes, the t distribution is used.)

To use this equation we must specify E, α or $1 - \alpha$, and σ. Values of $1 - \alpha$ that might be used are:

$1 - \alpha = 0.997$	$1 - \alpha = 0.99$	$1 - \alpha = 0.955$	$1 - \alpha = 0.95$	$1 - \alpha = 0.90$
$z = 3.0$	$z = 2.56$	$z = 2.0$	$z = 1.96$	$z = 1.64$

The most widely used value of $1 - \alpha$ is 0.95 and the corresponding value of $z = 1.96$. For an *approximate* 95% confidence interval, use $z = 2$ instead of 1.96 to get $n = 4\sigma^2/E^2$. This corresponds to $1 - \alpha = 0.955$.

The remaining problem is that the true value of σ is unknown, so an estimate is substituted based on prior data of a similar kind or, if necessary, a good guess. If the estimate of σ is based on prior data,

TABLE 23.1

Reduction in the Width of the 95% Confidence Interval ($\alpha = 0.05$) as the Sample Size is Increased, Assuming $E = t_{\alpha/2} s/\sqrt{n}$

n	2	3	4	5	8	10	15	20	25
$t_{\alpha/2}$	4.30	3.18	2.78	2.57	2.31	2.23	2.13	2.09	2.06
$\sqrt{n}$	1.41	1.73	2.00	2.2	2.8	3.2	3.9	4.5	5.0
$E = t_{\alpha/2} s/\sqrt{n}$	3.0s	1.8s	1.4s	1.2s	0.8s	0.7s	0.55s	0.47s	0.41s

we assume that the system will not change during the next phase of sampling. This can be checked as data are collected and the sampling plan can be revised if necessary.

For smaller sample sizes, say $n < 30$, and assuming that the distribution of the sample mean is approximately normal, the confidence interval half-width is $E = t_{\alpha/2} s/\sqrt{n}$ and we can assert with $(1 - \alpha)$ 100% confidence that E is the maximum error made in using $\bar{y}$ to estimate η.

The value of t decreases as n increases, but there is little change once n exceeds 5, as shown in Table 23.1. The greatest gain in narrowing the confidence interval comes from the decrease in $1/\sqrt{n}$ and not in the decrease in t. Doubling n decreases the size of confidence interval by a factor of $1/\sqrt{2}$ when the sample is large ($n > 30$). For small samples the gain is more impressive. For a stated level of confidence, doubling the size from 5 to 10 reduces the half-width of the confidence by about one-third. Increasing the sample size from 5 to 20 reduces the half-width by almost two-thirds.

An exact solution of the sample size for small n requires an iterative solution, but a good approximate solution is obtained by using a rounded value of $t = 2.1$ or 2.2, which covers a good working range of $n = 10$ to $n = 25$. When analyzing data we carry three decimal places in the value of t, but that kind of accuracy is misplaced when sizing the sample. The greatest uncertainty lies in the value of the specified s, so we can conveniently round off t to one decimal place.

Another reason not to be unreasonably precise about this calculation is that the sample size you calculate will usually be rounded up, not just to the next higher integer, but to some even larger convenient number. If you calculate a sample size of $n = 26$, you might well decide to collect 30 or 35 specimens to allow for breakage or other loss of information. If you find after analysis that your sample size was too small, it is expensive to go back to collect more experimental material, and you will find that conditions have shifted and the overall variability will be increased. In other words, the calculated n is guidance and not a limitation.

Example 23.1

We wish to estimate the mean of a process to within ten units of the true value, with 95% confidence. Assuming that a large sample is needed, use:

$$n = \left(\frac{z_{\alpha/2} \sigma}{E}\right)^2$$

Ten random preliminary measurements [233, 266, 283, 233, 201, 149, 219, 179, 220, and 214] give $\bar{y} = 220$ and $s = 38.8$. Using s as an estimate of σ and $E = 10$:

$$n = \left(\frac{1.96(38.8)}{10}\right)^2 \approx 58$$

Example 23.2

A monitoring study is intended to estimate the mean concentration of a pollutant at a sewer monitoring station. A preliminary survey consisting of ten representative observations gave [291, 320, 140, 223, 219, 195, 248, 251, 163, and 292]. The average is $\bar{y} = 234.2$ and the sample standard deviation is $s = 58.0$. The 95% confidence interval of this estimate is calculated using $t_{9,0.025} = 2.228$:

$$\bar{y} \pm t_{9,0.025} \frac{s}{\sqrt{n}} = 234.2 \pm 2.228 \frac{58.0}{\sqrt{10}} = 234.2 \pm 40.8$$

The true mean lies in the interval 193 to 275.

What sample size is needed to estimate the true mean with ±20 units? Assume the needed sample size will be large and use $z = 1.96$. The solution is:

$$n = \left(\frac{z_{\alpha/2}\sigma}{E}\right)^2 = \left(\frac{1.96(58)}{20}\right)^2 = 32$$

Ten of the recommended 32 observations have been made, so 22 more are needed. The recommended sample size is based on anticipation of $\sigma = 58$. The σ value actually realized may be more or less than 58, so $n = 32$ observations may give an estimation error more or less than the target of ±20 units.

The approximation using $z = 2.0$ leads to $n = 34$.

The number of samples in Example 23.2 might be adjusted to obtain balance in the experimental design. Suppose that a study period of about 4 to 6 weeks is desirable. Taking $n = 32$ and collecting specimens on 32 consecutive days would mean that four days of the week are sampled five times and the other three days are sampled four times. Sampling for 35 days (or perhaps 28 days) would be a more attractive design because each day of the week would be sampled five times (or four times).

In Examples 23.1 and 23.2, σ was estimated by calculating the standard deviation from prior data. Another approach is to estimate σ from the range of the data. If the data come from a normal distribution, the standard deviation can be estimated as a multiple of the range. If $n > 15$, the factor is 0.25 (estimated σ = range/4). For $n < 15$, use the factors in Table 23.2. These factors change with sample size because the range is expected to increase as more observations are made.

If you are stuck without data and have no information except an approximate range of the expected data (smaller than a garage but larger than a refrigerator), assume a uniform distribution over this range. The standard deviation of a uniform distribution with range R is $s_U = \sqrt{R^2/12} = 0.29R$. This helps to set a reasonable planning value for σ.

TABLE 23.2

Factors for Estimating the Standard Deviation from the Range of a Sample from a Normal Distribution

$n =$	2	3	4	5	6	7	8	9	10
Factor	0.886	0.591	0.486	0.430	0.395	0.370	0.351	0.337	0.325
$n =$	11	12	13	14	15	>15			
Factor	0.315	0.307	0.300	0.294	0.288	0.250			

The following example illustrates that it is not always possible to achieve a stated objective by increasing the sample size. This happens when the stated objective is inconsistent with statistical reality.

Example 23.3

A system has been changed with the expectation that the intervention would reduce the pollution level by 25 units. That is, we wish to detect whether the pre-intervention mean η_1 and the post-intervention mean η_2 differ by 25 units. The pre-intervention estimates are $\bar{y}_1 = 234.3$ and $s_1 = 58$, based on a survey with $n_1 = 10$. The project managers would like to determine, with a 95% confidence level, whether a reduction of 25 units has been accomplished.

The observed mean after the intervention will be estimated by $\bar{y}_2$ and the estimate of the change $\eta_1 - \eta_2$ will be $\bar{y}_1 - \bar{y}_2$. Because we are interested in a change in one direction (a decrease), the test condition is a one-sided 95% confidence interval such that:

$$t_{\nu,0.05} s_{\bar{y}_1 - \bar{y}_2} \leq 25$$

Sizing the Experiment

If this condition is satisfied, the confidence interval for $\eta_1 - \eta_2$ does not include zero and we do not reject the hypothesis that the true change could be as large as 25 units.

The standard error of the difference is:

$$s_{\bar{y}_1 - \bar{y}_2} = s_{\text{pool}} \sqrt{\frac{1}{n_1} + \frac{1}{n_2}}$$

which is estimated with $v = n_1 + n_2 - 2$ degrees of freedom. Assuming the variances before and after the intervention are the same, $s_1 = s_2 = 58$ and therefore $s_{\text{pool}} = 58$.

For $\alpha = 0.05$, $n_1 = 10$, and assuming $v = 10 + n_2 - 2 > 30$, $t_{0.05} = 1.70$. The sample size n_2 must be large enough that:

$$1.7(58)\sqrt{\frac{1}{10} + \frac{1}{n_2}} \leq 25$$

This condition is impossible to satisfy. Even with $n_2 = \infty$, the left-hand side of the expression gets only as small as 31.2.

The managers should have asked the sampling design question *before* the pre-change survey was made, and when a larger pre-change sample could be taken. A sample of $n_1 = n_2 \approx 32$ would be about right.

What about Type II Error?

So far we have mentioned only the error that is controlled by selecting α. That is the so-called *type I error*, which is the error of declaring an effect is real when it is in fact zero. Setting $\alpha = 0.05$ controls this kind of error to a probability of 5%, when all the assumptions of the test are satisfied.

Protecting only against type I error is not totally adequate, however, because a type I error probably never occurs in practice. Two treatments are never likely to be truly equal; inevitably they will differ in some respect. No matter how small the difference is, provided it is non-zero, samples of a sufficiently large size can virtually guarantee statistical significance. Assuming we want to detect only differences that are of practical importance, we should impose an additional safeguard against a type I error by not using sample sizes larger than are needed to guard against the second kind of error.

The *type II error* is failing to declare an effect is significant when the effect is real. Such a failure is not necessarily bad when the treatments differ only trivially. It becomes serious only when the difference is important. Type II error is not made small by making α small. The first step in controlling type II error is specifying just what difference is important to detect. The second step is specifying the probability of actually detecting it. This probability $(1 - \beta)$ is called the *power* of the test. The quantity β is the probability of failing to detect the specified difference to be statistically significant.

Figure 23.1 shows the situation. The normal distribution on the left represents the two-sided condition when the true difference between population means is zero ($\delta = 0$). We may, nevertheless, with a probability of $\alpha/2$, observe a difference d that is quite far above zero. This is the type I error. The normal distribution on the right represents the condition where the true difference is larger than d. We may, with probability β, collect a random sample that gives a difference much lower than d and wrongly conclude that the true difference is zero. This is the type II error.

The experimental design problem is to find the sample size necessary to assure that (1) any smaller sample will reduce the chance below $1 - \beta$ of detecting the specified difference and (2) any larger sample may increase the chance well above α of declaring a trivially small difference to be significant (Fleiss, 1981). The required sample size for detecting a difference in the mean of two treatments is:

$$n = \frac{2\sigma^2(z_{\alpha/2} + z_\beta)}{\Delta^2}$$

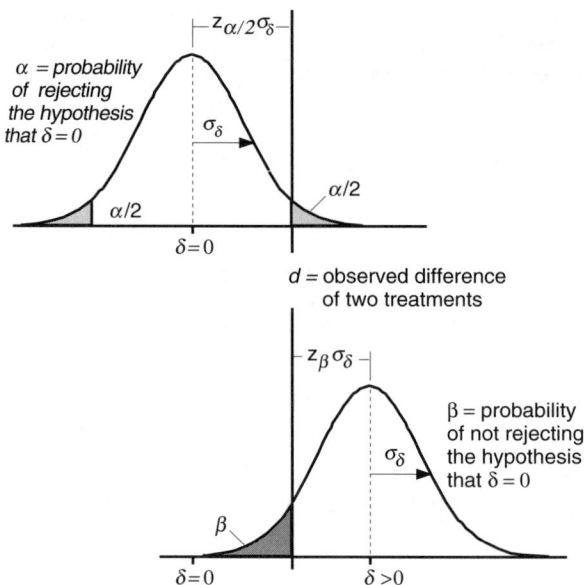

FIGURE 23.1 Definition of type I and type II errors for a one-sided test of the difference between two means.

where $\Delta = |\eta_1 - \eta_2|$ and α and β are probabilities of type I and type II errors. If the variance σ^2 is not known, it is replaced with the sample variance s^2.

The sample size for a *one-sided test* on whether a mean is above a fixed standard level (i.e., a regulatory threshold) is:

$$n = \frac{2\sigma^2(z_\alpha + z_\beta)}{\Delta^2} + \frac{1}{2}z_\alpha^2$$

This is an approximate, but very accurate, sample size estimate (U.S.EPA, 1994).

Example 23.4

A decision rule is being developed for directing loads of contaminated material to a sanitary landfill or a hazardous waste landfill. Each load is fairly homogeneous but there is considerable variation between loads. The standard deviation of contaminant in a given load is 0.06 mg/kg. The decision rule conditions are (1) a probability of 95% of declaring a load hazardous when the true mean concentration is 1.0 mg/kg and (2) a probability of 10% of declaring a load hazardous when the true mean concentration is 0.75 mg/kg. What size sample should be analyzed from each load, assuming samples can be collected at random?

For the stated conditions, $\alpha = 0.05$ and $\beta = 0.10$, giving $z_{0.05/2} = 1.96$ and $z_{0.10} = 1.28$. With $\sigma = 0.06$ and $\Delta = 1.00 - 0.75$:

$$n = \frac{2\sigma^2(z_{\alpha/2} + z_\beta)}{\Delta^2} = \frac{2(0.06)^2 1.96 + (1.280)^2}{0.25^2} = 1.21 \approx 2$$

Setting the probability of the type I and type II errors may be difficult. Typically, α is specified first. If declaring the two treatments to differ significantly will lead to a decision to conduct further expensive research or to initiate a new and expensive form of treatment, then a type I error is serious and it should be kept small ($\alpha = 0.01$ or 0.02). On the other hand, if additional confirmatory testing is to be done in any case, as in routine monitoring of an effluent, the type I error is less serious and α can be larger.

Also, if the experiment was intended primarily to add to the body of published literature, it should be acceptable to increase α to 0.05 or 0.10.

Having specified α, the investigator needs to specify β, or $1-\beta$. Cohen (1969) suggests that in the context of medical treatments, a type I error is roughly four times as serious as a type II error. This implies that one should use approximately $\beta = 4\alpha$ so that the power of the test is $1 - \beta = 1 - 4\alpha$. Thus, when $\alpha = 0.05$, set $1 - \beta = 0.80$, or perhaps less.

Sample Size for Assessing the Equivalence of Two Means

The previous sections dealt with selecting a sample size that is large enough to detect a *difference* between two processes. In some cases we wish to establish that two processes are not different, or at least are close enough to be considered equivalent. Showing a difference and showing equivalence are not the same problem.

One statistical definition of equivalence is the classical null hypothesis H_0: $\eta_1 - \eta_2 = 0$ versus the alternate hypothesis H_1: $\eta_1 - \eta_2 \neq 0$. If we use this problem formulation to determine the sample size for a two-sided test of no difference, as shown in the previous section, the answer is likely to be a sample size that is impracticably large when Δ is very small.

Stein and Dogansky (1999) present an alternate formulation of this classical problem that is often used in bioequivalence studies. Here the hypothesis is formed to demonstrate a *difference* rather than *equivalence*. This is sometimes called the *interval testing approach*. The interval hypothesis (H_1) requires the difference between two means to lie with an equivalence interval $[\theta_L, \theta_U]$ so that the rejection of the null hypothesis, H_0 at a nominal level of significance (α), is a declaration of equivalence. The interval determines how close we require the two means to be to declare them equivalent as a practical matter:

$$H_0: \eta_1 - \eta_2 \leq \theta_L \quad \text{or} \quad \eta_1 - \eta_2 \geq \theta_U$$

versus

$$H_1: \theta_L < \eta_1 - \eta_2 < \theta_U$$

This is decomposed into two one-sided hypotheses:

$$H_{01}: \eta_1 - \eta_2 \leq \theta_L \quad \text{and} \quad H_{02}: \eta_1 - \eta_2 \geq \theta_U$$
$$H_{11}: \eta_1 - \eta_2 > \theta_L \qquad\qquad H_{12}: \eta_1 - \eta_2 < \theta_U$$

where each test is conducted at a nominal level of significance, α. If H_{01} and H_{02} are both rejected, we conclude that $\theta_L < \eta_1 - \eta_2 < \theta_U$ and declare that the two treatments are equivalent.

We can specify the equivalence interval such that $\theta = \theta_U = -\theta_L$. When the common variance σ^2 is known, the rule is to reject H_0 in favor of H_1 if:

$$-\theta + z_\alpha \sigma_{\bar{y}_1 - \bar{y}_2} \leq \bar{y}_1 - \bar{y}_2 \leq \theta - z_\alpha \sigma_{\bar{y}_1 - \bar{y}_2}$$

The approximate sample size for the case where $n_1 = n_2 = n$ is:

$$n = \frac{2\sigma^2 (z_\alpha + z_\beta)^2}{(\theta - \Delta)^2} + 1$$

θ defines (*a priori*) the practical equivalence limits, or how close the true treatment means are required to be before they are declared equivalent. Δ is the true difference between the two treatment means under which the comparison is made.

Stein and Dogansky (1999) give an iterative solution for the case where a different sample size will be taken for each treatment. This is desirable when data from the standard process is already available.

In the interval hypothesis, the type I error rate (α) denotes the probability of falsely declaring equivalence. It is often set to $\alpha = 0.05$. The power of the hypothesis test $(1 - \beta)$ is the probability of correctly declaring equivalence. Note that the type I and type II errors have the reverse interpretation from the classical hypothesis formulation.

Example 23.5

A standard process is to be compared with a new process. The comparison will be based on taking a sample of size n from each process. We will consider the two process means equivalent if they differ by no more than 3 units ($\theta = 3.0$), and we wish to determine this with risk levels $\alpha = 0.05$, $\beta = 0.10$, $\sigma = 1.8$, when the true difference is at most 1 unit ($\Delta = 1.0$). The sample size from each process is to be equal. For these conditions, $z_{0.05} = 1.645$ and $z_{0.10} = 1.28$, and:

$$n = \frac{2(1.8)^2(1.645 + 1.28)^2}{(3.0 - 1.0)^2} + 1 \approx 15$$

Confidence Interval for an Interaction

Here we insert an example that does not involve a *t*-test. The statistic to be estimated measures a change that occurs between two locations and over a span of time. A control area and a potentially affected area are to be monitored before and after a construction project. This is shown by Figure 23.2. The dots in the squares indicate multiple specimens collected at each monitoring site. The figure shows four replicates, but this is only for illustration; there could be more or less than four per area.

The averages of pre-construction and post-construction control areas are $\bar{y}_{B1}$ and $\bar{y}_{B2}$. The averages of pre-construction and post-construction affected areas are $\bar{y}_{A1}$ and $\bar{y}_{A2}$. In an ideal world, if the construction caused a change, we would find $\bar{y}_{B1} = \bar{y}_{B2} = \bar{y}_{A1}$ and $\bar{y}_{A2}$ would be different. In the real world, $\bar{y}_{B1}$ and $\bar{y}_{B2}$ may be different because of their location, and $\bar{y}_{B1}$ and $\bar{y}_{A1}$ might be different because they are monitored at different times. The effect that should be evaluated is the interaction effect (I) over time and space, and that is:

$$I = (\bar{y}_{A2} - \bar{y}_{A1}) - (\bar{y}_{B2} - \bar{y}_{B1})$$

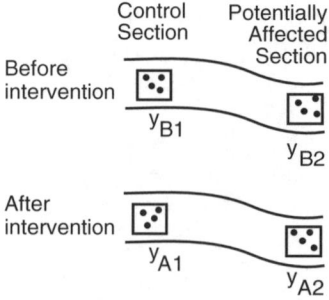

FIGURE 23.2 The arrangement of before and after monitoring at control (upstream) and possibly affected (downstream) sites. The dots in the monitoring areas (boxes) indicate that multiple specimens will be collected for analysis.

The variance of the interaction effect is:

$$\text{Var}(I) = \text{Var}(\bar{y}_{A2}) + \text{Var}(\bar{y}_{A1}) + \text{Var}(\bar{y}_{B2}) + \text{Var}(\bar{y}_{B1})$$

Assume that the variance of each average is σ^2/r, where r is the number of replicate specimens collected from each area. This gives:

$$\text{Var}(I) = \frac{\sigma^2}{r} + \frac{\sigma^2}{r} + \frac{\sigma^2}{r} + \frac{\sigma^2}{r} = \frac{4\sigma^2}{r}$$

The approximate 95% confidence interval of the interaction I is:

$$I \pm 2\sqrt{\frac{4\sigma^2}{r}} \quad \text{or} \quad I \pm \frac{4\sigma}{\sqrt{r}}$$

If only one specimen were collected per area, the confidence interval would be 4σ. Four specimens per area gives a confidence interval of 2σ, 16 specimens gives 1σ, etc. in the same pattern we saw earlier. Each quadrupling of the sample size reduces the confidence interval by half.

The number of replicates from each area needed to estimate the interaction I with a maximum error of $E = 2(4\sigma/\sqrt{r}) = 8\sigma/\sqrt{r}$ is:

$$r = 64\sigma^2/E^2$$

The total sample size is $4r$.

One-Way Analysis of Variance

The next chapter deals with comparing k mean values using analysis of variance (ANOVA). Here we somewhat prematurely consider the sample size requirements. Kastenbaum et al. (1970) give tables for sample size requirements when the means of k groups, each containing n observations, are being compared at α and β levels of risk. Figure 23.3 is a plot of selected values from the tables for $k = 5$ and $\alpha = 0.05$,

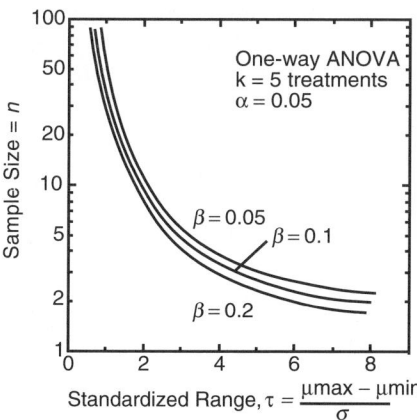

FIGURE 23.3 Sample size requirements to analyze the means of five treatments by one-way analysis of variance. α is the type I error risk, β is the type II error risk, σ is the planning value for the standard deviation. μ_{max} and μ_{min} are the maximum and minimum expected mean values in the five groups. (From data in tables of Kastenbaum et al., 1970.)

with curves for $\beta = 0.05$, 0.1, and 0.2. The abscissa is the standardized range, $\tau = (\mu_{max} - \mu_{min})/\sigma$, where μ_{max} and μ_{min} are the planning values for the largest and smallest mean values, and σ is the planning standard deviation.

Example 23.6

How small a difference can be detected between five groups of contaminated soil with a sample size of $n = 10$, assuming for planning purposes that $\sigma = 2.0$, and for risk levels $\alpha = 0.05$ and $\beta = 0.1$? Read from the graph ($\tau = 1.85$) and calculate:

$$\mu_{max} - \mu_{min} = \tau\sigma = 1.85(2.0) = 3.7$$

Sample Size to Estimate a Binomial Population

A binomial population consists of binary individuals. The horses are black and white, the pump is faulty or fault-free, an organism is sick or healthy, the air stinks or it does not. The problem is to determine how many individuals to examine in order to estimate the true proportion p for each binary category.

An approximate expression for sample size of a binomial population is:

$$n = p^*(1 - p^*)\left(\frac{z_{\alpha/2}}{E}\right)^2$$

where p^* is the *a priori* estimate of the proportion (i.e., the planning value). If no information is available from prior data, we can use $p^* = 1/2$, which will give the largest possible n, which is:

$$n = \frac{1}{4}\left(\frac{z_{\alpha/2}}{E}\right)^2$$

This sample size will give a $(1 - \alpha)100\%$ confidence interval for p with half-length E. This is based on a normal approximation and is generally satisfactory if both np and $n(1 - p)$ exceed 10 (Johnson, 2000).

Example 23.7

Some unknown proportion p of a large number of installed devices (i.e., flow meters or UV lamp ballasts) were assembled incorrectly and have to be repaired. To assess the magnitude of the problem, the manufacturer wishes to estimate the proportion (p) of installed faulty devices. How many units must be examined at random so that the estimate ($\hat{p}$) will be within ± 0.08 of the true proportion p, with 95% confidence? Based on consumer complaints, the proportion of faulty devices is thought to be less than 20%.

In this example, the planning value is $p^* = 0.2$. Also, $E = 0.08$, $\alpha = 0.05$, and $z_{0.025} = 1.96$, giving:

$$n = 0.2(1 - 0.2)\left(\frac{1.96}{0.08}\right)^2 \approx 96$$

If fewer than 96 units have been installed, the manufacturer will have to check all of them. (A sample of an entire population is called a *census*.)

The test on proportions can be developed to consider type I and type II errors. There is typically large inherent variability in biological tests, so bioassays are designed to protect against the two kinds of decision errors. This will be illustrated in the context of bioassay testing where raw data are usually converted into proportions.

The proportion of organisms affected is compared with the proportion of affected organisms in an unexposed control group. For simplicity of discussion, assume that the response of interest is survival

Sizing the Experiment

of the organism. A further simplification is that we will consider only two groups of organisms, whereas many bioassay tests will have several groups.

The true difference in survival proportions (p) that is to be detected with a given degree of confidence must be specified. That difference ($\delta = p_e - p_c$) should be an amount that is deemed scientifically or environmentally important. The subscript e indicates the exposed group and c indicates the control group.

The variance of a binomial response is $\text{Var}(p) = p(1-p)/n$. In the experimental design problem, the variances of the two groups are not equal. For example, using $n = 20$, $p_c = 0.95$ and $p_e = 0.8$, gives:

$$\text{Var}(p_e) = p_e(1 - p_e)/n = 0.8(1 - 0.8)/20 = 0.008$$

and

$$\text{Var}(p_c) = p_c(1 - p_c)/n = 0.95(1 - 0.95)/20 = 0.0024$$

As the difference increases, the variances become more unequal (for $p = 0.99$, $\text{Var}(p) = 0.0005$). This distortion must be expected in the bioassay problem because the survival proportion in the control group should approach 1.00. If it does not, the bioassay is probably invalid on biological grounds.

The transformation $x = \arcsin\sqrt{p}$ will "stretch" the scale near $p = 1.00$ and make the variances more nearly equal (Mowery et al., 1985). In the following equations, x is the transformed survival proportion and the difference to be detected is:

$$\delta = x_c - x_e = \arcsin\sqrt{p_c} - \arcsin\sqrt{p_e}$$

For a binomial process, δ is approximately normally distributed. The difference of the two proportions is also normally distributed. When x is measured in radians, $\text{Var}(x) = 1/4n$. Thus, $\text{Var}(\delta) = \text{Var}(x_1 - x_2) = 1/4n + 1/4n = 1/2n$. These results are used below.

Figure 23.1 describes this experiment, with one small change. Here we are doing a one-sided test, so the left-hand normal distribution will have the entire probability α assigned to the upper tail, where α is the probability of rejecting the null hypothesis and inferring that an effect is real when it is not. The true difference must be the distance ($z_\alpha + z_\beta$) in order to have probability β of detecting a real effect at significance level α. Algebraically this is:

$$z_\alpha + z_\beta = \frac{\delta}{\sqrt{0.5n}}$$

The denominator is the standard error of δ. Rearranging this gives:

$$n = \frac{1}{2}\left(\frac{z_\alpha + z_\beta}{\delta}\right)^2$$

Table 23.3 gives some selected values of α and β that are useful in designing the experiment.

TABLE 23.3

Selected Values of $z_\alpha + z_\beta$ for One-Sided Tests in a Bioassay Experiment to Compare Two Groups

p	$x = \arcsin\sqrt{p}$	α or β	$z_\alpha + z_\beta$
1.00	1.571	0.01	2.236
0.98	1.429	0.02	2.054
0.95	1.345	0.05	1.645
0.90	1.249	0.10	1.282
0.85	1.173	0.20	0.842
0.80	1.107	0.25	0.674
0.75	1.047	0.30	0.524

Example 23.8

We expect the control survival proportion to be $p_c^* = 0.95$ and we wish to detect effluent toxicity corresponding to an effluent survival proportion of $p_e^* = 0.75$. The probability of detecting a real effect is to be $1 - \beta = 0.9$ ($\beta = 0.1$) with confidence level $\alpha = 0.05$. The transformed proportions are $x_c = \arcsin \sqrt{0.95} = 1.345$ and $x_e = \arcsin \sqrt{0.8} = 1.047$, giving $\delta = 1.345 - 1.047 = 0.298$. Using $z_{0.05} = 1.645$ and $z_{0.1} = 1.282$ gives:

$$n = 0.5 \left(\frac{1.645 + 1.282}{1.345 - 1.047} \right)^2 = 48.2$$

This would probably be adjusted to $n = 50$ organisms for each test condition.

This may be surprisingly large although the design conditions seem reasonable. If so, it may indicate an unrealistic degree of confidence in the widely used design of $n = 20$ organisms. The number of organisms can be decreased by increasing α or β, or by decreasing δ.

This approach has been used by Cohen (1969) and Mowery et al. (1985). An alternate approach is given by Fleiss (1981). Two important conclusions are (1) there is great statistical benefit in having the control proportion high (this is also important in terms of biological validity), and (2) small sample sizes ($n < 20$) are useful only for detecting very large differences.

Stratified Sampling

Figure 23.4 shows three ways that sampling might be arranged in a area. Random sampling and systematic sampling do not take account of any special features of the site, such as different soil type of different levels of contamination. Stratified sampling is used when the study area exists in two or more distinct strata, classes, or conditions (Gilbert, 1987; Mendenhall et al., 1971). Often, each class or stratum has a different inherent variability. In Figure 23.4, samples are proportionally more numerous in stratum 2 than in stratum 1 because of some known difference between the two strata.

We might want to do stratified sampling of an oil company's properties to assess compliance with a stack monitoring protocol. If there were 3 large, 30 medium-sized, and 720 small properties, these three sizes define three strata. One could sample these three strata proportionately; that is, one third of each, which would be 1 large, 10 medium, and 240 small facilities. One could examine all the large facilities, half of the medium facilities, and a random sample of 50 small ones. Obviously, there are many possible sampling plans, each having a different precision and a different cost. We seek a plan that is low in cost and high in information.

The overall population mean $\bar{y}$ is estimated as a weighted average of the estimated means for the strata:

$$\bar{y} = w_1 \bar{y}_1 + w_2 \bar{y}_2 + \cdots + w_{n_s} \bar{y}_{n_s}$$

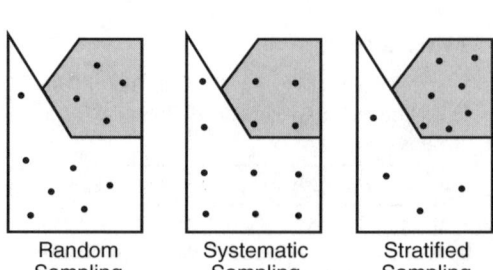

FIGURE 23.4 Comparison of random, systematic, and stratified random sampling of a contaminated site. The shaded area is known to be more highly contaminated than the unshaded area.

Sizing the Experiment

TABLE 23.4
Data for the Stratified Sample for Examples 23.9, 23.10, and 23.11

	Observations n_i	Mean $\bar{y}_i$	Variance s_i^2	Size of Stratum	Weight w_i
Stratum 1	20	34	35.4	1500	0.5
Stratum 2	8	25	180	750	0.25
Stratum 3	12	19	12	750	0.25

where n_s is the number of strata and the w_i are weights that indicate the proportion of the population included in stratum i. The estimated variance of $\bar{y}$ is:

$$s_{\bar{y}}^2 = w_1^2 \frac{s_1^2}{n_1} + w_1^2 \frac{s_1^2}{n_1} + \cdots + w_{n_s}^2 \frac{s_{n_s}^2}{n_{n_s}}$$

Example 23.9

Suppose we have the data in Table 23.4 from sampling a contaminated site that was known to have three distinct areas. There were a total of 3000 parcels (acres, cubic meters, barrels, etc.) that could have been sampled. A total of $n = 40$ observations were collected from randomly selected parcels within each stratum. The allocation was 20 observations in stratum 1, 8 in stratum 2, and 12 in stratum 3. Notice that one-half of the 40 observations were in stratum 1, which is also one-half of the population of 3000 sampling units, but the observations in strata 2 and 3 are not proportional to their populations. This allocation might have been made because of the relative cost of collecting the data, or because of some expected characteristic of the site that we do not know about. Or, it might just be an inefficient design. We will check that later.

The overall mean is estimated as a weighted average:

$$\bar{y} = 0.5(34) + 0.25(25) + 0.25(19) = 28$$

The estimated variance of the overall average is the sum of the variances of the three strata weighted with respect to their populations:

$$s_{\bar{y}}^2 = 0.5^2 \left(\frac{35.4}{20}\right) + 0.25^2 \left(\frac{180}{8}\right) + 0.25^2 \left(\frac{12}{12}\right) = 1.9$$

The confidence interval of the mean is $\bar{y} \pm 1.96 \sqrt{s_{\bar{y}}^2}$, or 28 ± 2.7.

The confidence intervals for the randomly sampled individual strata are interpreted using familiar equations. The 95% confidence interval for stratum 2 is $\bar{y}_2 \pm 1.96 \sqrt{s_2^2/n_2}$ and $25 \pm 1.96 \sqrt{180/8} = 25 \pm 9.3$. This confidence interval is large because the variance is large and the sample size is small. If this had been known, or suspected, before the sampling was done, a better allocation of the $n = 40$ samples could have been made.

Samples should be allocated to strata according to the size of the strata, its variance, and the cost of sampling. The cost of the sampling plan is:

$$\text{Cost} = \text{Fixed cost} + (c_1 n_1 + c_2 n_2 + \cdots + c_{n_s} n_{n_s})$$

The c_i are the costs to collect and analyze each specimen. The optimal sample size per stratum is:

$$n_i = n \frac{w_i s_i / \sqrt{c_i}}{w_i s_i / \sqrt{c_i} + \cdots + w_{n_s} s_{n_s} / \sqrt{c_{n_s}}}$$

This says that the sample size in stratum i will be large if the stratum is large, the variance is large, or the cost is low. If sampling costs are equal in all strata, then:

$$n_i = n \frac{w_i s_i}{w_i s_i + \cdots + w_{n_s} s_{n_s}}$$

Using these equations requires knowing the total sample size, n. This might be constrained by budget, or it might be determined to meet an estimation error criterion for the population mean, or to have a specified variance (Gilbert, 1987).

The sample size needed to estimate the overall mean with a specified margin of error (E) and an approximate probability $(1 - \alpha)100\% = 95\%$ of exceeding that error is:

$$n = \frac{4}{E^2} \sum w_i s_i^2$$

Example 23.10

Using the data from Example 23.9 (Table 23.4), design a stratified sampling plan to estimate the mean with a margin of error of 1.0 unit with 95% confidence. There are three strata, with variances $s_1^2 = 35.4$, $s_2^2 = 180$, and $s_3^2 = 12$, and having weights $w_1 = 0.5$, $w_2 = 0.25$, and $w_3 = 0.25$. Assume equal sampling costs in the three strata. The total sample size required is:

$$n = \frac{4}{1^2}\left[0.5(35.4) + 0.25(180) + 0.25(12)\right] = 263$$

The allocation among strata is:

$$n_1 = 263\left[\frac{w_i s_i}{0.5(35.4) + 0.25(180) + 0.25(12)}\right] = 4 w_i s_i$$

giving

$$n_1 = 4(0.5)(35.4) = 71$$
$$n_2 = 4(0.25)(180) = 180$$
$$n_3 = 4(0.25)(12) = 12$$

This large sample size results from the small margin of error (1 unit) and the large variance in stratum 2.

Example 23.11

The allocation of the $n = 40$ samples in Example 23.9 gave a 95% confidence interval of ± 2.7. The allocation according to Example 23.10 is $n_i = 40(w_i s_i^2 / 65.7) = 0.61 w_i s_i^2$, which gives $n_1 = 11$, $n_2 = 28$, and $n_3 = 2$. Because of rounding, this adds to $n = 41$ instead of 40.

If this allocation leads to the same sample variances listed in Table 23.4, the variance of the mean would be $s_{\bar{y}}^2 = 1.58$ and the 95% confidence interval would be $\pm 1.96\sqrt{1.53} = \pm 2.5$. This is a reduction of about 10% without doing any additional work, except to make a better sampling plan.

Comments

This chapter omits many useful sampling strategies. Fortunately there is a rich collection in other books. Cochran (1977) and Cochran and Cox (1957) are the classic texts on sampling and experimental design.

Gilbert (1987) is the best book on environmental sampling and monitoring. He gives examples of random sampling, stratified sampling, two- and three-stage sampling, systematic sampling, double sampling, and locating hot spots. Two-stage sampling is also called subsampling. It is used when large field samples are collected and one or more aliquots are selected from the field sample for analysis. This is common with particulate materials. The subsampling introduces additional uncertainty into estimates of means and variances. Three-stage sampling involves compositing field samples and then subsampling for the composite for analysis. The compositing does some beneficial averaging.

Hahn and Meeker (1991) give examples and many useful tables for deriving sample size requirements for confidence intervals for means, a normal distribution standard deviation, binomial proportions, Poisson occurrence rates, tolerance bounds and intervals, and prediction intervals. This is a readable and immensely practical and helpful book. Mendenhall et al. (1971) explain random sampling, stratified sampling, ratio estimation, cluster sampling, systematic sampling, and two-stage cluster sampling.

This chapter was not written to provide a simple answer to the question, "What sample size do I need?" There is no simple answer. The objective was to encourage asking a different question, "How can the experiment be arranged so the standard error is reduced and better decisions will be made with minimum work?"

Many kinds of sampling plans and experimental designs have been developed to give maximum efficiency in special cases. A statistician can help us recognize what is special about a sampling problem and then select the right sampling design. But the statistician must be asked to help *before* the data are collected. Then all the tools of experimental design — replication, blocking, and randomization — can be organized in your favor.

References

Cochran, W. G. (1977). *Sampling Techniques,* 3rd ed., New York, John Wiley.

Cochran, W. G. and G. M, Cox (1957). *Experimental Designs,* 2nd ed., New York, John Wiley.

Cohen, J. (1969). S*tatistical Power Analysis for the Behavioral Sciences,* New York, New York Academy Press.

Fleiss, J. L. (1981). *Statistical Methods for Rates and Proportions,* New York, John Wiley.

Gibbons, R. D. (1994). *Statistical Methods for Groundwater Monitoring,* New York, John Wiley.

Gilbert, R.O. (1987). *Statistical Methods for Environmental Pollution Monitoring,* New York, Van Nostrand Reinhold.

Hahn, G. A. and W. Q. Meeker (1991). *Statistical Intervals: A Guide for Practitioners,* New York, John Wiley.

Johnson, R. A. (2000). *Probability and Statistics for Engineers,* 6th ed., Englewood Cliffs, NJ, Prentice-Hall.

Kastenbaum, M. A., D. G. Hoel, and K. O. Bowman (1970). "Sample Size Requirements: One-Way Analysis of Variance," *Biometrika,* 57, 421–430.

Mendenhall, William, L. Ott, and R. L. Schaeffer (1971). *Elementary Survey Sampling,* Duxbury Press.

Mowery, P. D., J. A. Fava, and L. W. Catlin (1985). "A Statistical Test Procedure for Effluent Toxicity Screening," *Aquatic Toxicol. Haz. Assess., 7th Symp.,* ASTM STP 854.

Stein, J. and N. Dogansky (1999). "Sample Size Considerations for Assessing the Equivalence of Two Process Means," *Qual. Eng.,* 12(1), 105–110.

U.S. EPA (1994). *Guidance for the Data Quality Objectives Process,* (EPA QA/G-4), Washington, D.C., Quality Assurance Management Staff.

Exercises

23.1 Heavy Metals. Preliminary analysis of some spotty but representative historical data shows standard deviations of 14, 13, and 40 µg/L for three heavy metals in a groundwater monitoring well. What sample size is required to estimate the mean concentration of each metal, with 95% confidence, to within 5, 5, and 10 µg/L, respectively. Assume that all three metals can be measured on the same water specimen.

23.2 Blood Lead Levels. The concentration of lead in the bloodstream was measured for a sample of $n = 50$ children from a large school near a main highway. The sample mean was 11.02 ng/mL and the standard deviation was 0.60 ng/mL. Calculate the 95% confidence interval for the mean lead concentration for all children in the school. What sample size would be needed to reduce the half-length of the confidence interval to 0.1 ng/mL?

23.3 Sewer Monitoring. Redo Example 23.2 to estimate the mean with a maximum error of $E = 25$. Start by assuming $n > 30$ and do an iterative solution. Explain how you will arrange the sample collection to have a balanced design.

23.4 Range. Determine the sample size to estimate a mean value within ±10 units when there is no prior data but you are confident the range of data values will be 60 to 90.

23.5 Difference of Two Means. We wish to be able to distinguish between true mean concentrations of mercury that differ by at least 0.05 µg/L for $\alpha = 0.05$ (the probability of falsely rejecting that the two means are equal) and $\beta = 0.20$ (the probability of failing to reject they are equal) if $\Delta = |\eta_1 - \eta_2| = 0.02$. From the case study in Chapter 18, we use $s_p = 0.0857$ µg/L as an estimate of σ. Determine the required sample size, assuming $n_p = n_c$.

23.6 Equivalence of Two Means. We wish to determine whether the mercury concentrations in wastewater from areas served by the city water supply and by private wells are *equivalent*. We wish to distinguish true mean concentrations that differ by at least $\Delta = 0.02$ µg/L, but are willing to establish, on a practical basis, an equivalence interval of $\theta = \pm 0.05$ µg/L for the difference. Assuming $n_p = n_c$, use $\alpha = 0.05$, $\beta = 0.20$, and $s_p = 0.0857$ µg/L to determine the required sample size.

23.7 Phosphorus Inventory. To estimate the average and the total amount of phosphorus in a reservoir, the reservoir was divided into three strata (surface layer, bottom layer, and intermediate layer). Portions of reservoir water were collected at random (with respect to vertical and horizontal location) within each strata. The data are:

Strata	Volume	w_h	n_h	Measurements (µg/L)									
1	4,000,000	0.40	10	21	12	15	25	30	14	11	19	17	31
2	3,500,000	0.35	10	40	36	41	29	48	33	37	30	44	45
3	2,500,000	0.25	8	60	58	34	51	39	53	57	69		

Find the total amount of phosphorus in the reservoir, its average concentration, and its standard error.

23.8 Precision Guidelines. Some years ago the National Council for Air and Stream Improvement (NCASI) proposed the following guidelines for monitoring pulp and paper mill waste streams.

Monitoring Objective	Maximum Error (%)
Exploratory survey to locate areas of heavy pollution	±40
Evaluate unit process efficiency	±5
Estimate design loads for wastewater treatment	±10
Evaluate pollution prevention projects	±5–10
Monitoring for effluent compliance	±5
Monitoring for stream standards	±5–10
Monitoring to assess sewer charges	±5

Use the following 12 observations to estimate the standard deviation and determine the required sample size for (a) exploratory survey, (b) design load estimation, and (c) sewer charge assessment.

3160, 2140, 1210, 1400, 2990, 1420, 2000, 2610, 2720, 2700, 2900, 3390

23.9 Coliforms. Use the data from Exercise 19.4 to determine the sample size necessary to estimate the proportion of wells that contain coliform bacteria (MF method) with a 95% confidence interval no larger than ±0.05. Use a planning value of $p = 0.15$. Would you use the same sample size to estimate the proportion of wells that contain coliforms as indicated by the Colilert method?

24

Analysis of Variance to Compare k Averages

KEY WORDS ANOVA, ANOVA table, analysis of variance, average, between treatment variance, grand average, F test, F distribution, one-way ANOVA, sum of squares, within-treatment variance.

Analysis of variance (ANOVA) is a method for testing two or more treatments to determine whether their sample means could have been obtained from populations with the same true mean. This is done by estimating the amount of variation within treatments and comparing it to the variance between treatments. If the treatments are alike (from populations with the same mean), the variation within each treatment will be about the same as the variation between treatments. If the treatments come from populations with different means, the variance between treatments will be inflated. The "within variance" and the "between variance" are compared using the F statistic, which is a measure of the variability in estimated variances in the same way that the t statistic is a measure of the variability in estimated means.

Analysis of variance is a rich and widely used field of statistics. "…the analysis of variance is more than a technique for statistical analysis. Once understood, analysis of variance provides an insight into the nature of variation of natural events, into Nature in short, which is possibly of even greater value than the knowledge of the method as such. If one can speak of beauty in a statistical method, analysis of variance possesses it more than any other" (Sokal and Rohlf, 1969).

Naturally, full treatment of such a powerful subject has been the subject of entire books and only a brief introduction will be attempted here. We seek to illustrate the key ideas of the method and to show it as an alternative to the multiple paired comparisons that were discussed in Chapter 20.

Case Study: Comparison of Five Laboratories

The data shown in Table 24.1 (and in Table 20.1) were obtained by dividing a large quantity of prepared material into 50 identical aliquots and having five different laboratories each analyze 10 randomly selected specimens. By design of the experiment there is no real difference in specimen concentrations, but the laboratories have produced different mean values and different variances. In Chapter 20 these data were analyzed using a multiple t-test to compare the mean levels. Here we will use a *one-way ANOVA*, which focuses on comparing the variation within laboratories with the variation between laboratories. The analysis is one-way because there is one factor (laboratories) to be assessed.

One-Way Analysis of Variance

Consider an experiment that has k treatments (techniques, methods, etc.) with n_t replicate observations made under each treatment, giving a total of $N = \sum k n_t$ observations, where $t = 1, 2,…,k$. If the variation within each treatment is due only to random measurement error, this *within-treatment variance* is a good estimate of the pure random experimental error. If the k treatments are different, the variation *between* the k treatments will be greater than might be expected in light of the variation that occurs within the treatment.

TABLE 24.1

Ten Measurements of Lead Concentration (μg/L) on Identical Specimens from Five Laboratories

Lab 1	Lab 2	Lab 3	Lab 4	Lab 5
3.4	4.5	5.3	3.2	3.3
3.0	3.7	4.7	3.4	2.4
3.4	3.8	3.6	3.1	2.7
5.0	3.9	5.0	3.0	3.2
5.1	4.3	3.6	3.9	3.3
5.5	3.9	4.5	2.0	2.9
5.4	4.1	4.6	1.9	4.4
4.2	4.0	5.3	2.7	3.4
3.8	3.0	3.9	3.8	4.8
4.2	4.5	4.1	4.2	3.0
$\bar{y}_i = 4.30$	3.97	4.46	3.12	3.34
$s_i^2 = 0.82$	0.19	0.41	0.58	0.54

The within-treatment sum of squares is calculated from the residuals of the observations within a treatment and the average for that treatment. The variance within each treatment is:

$$s_t^2 = \sum_{i=1}^{n_t} \frac{(y_{ti} - \bar{y}_t)}{n_t - 1}$$

where y_{ti} are the n_t observations under each treatment.

Assuming that all treatments have the same population variance, we can pool the k sample variances to estimate the within-treatment variance (s_w^2):

$$s_w^2 = \frac{\sum_{t=1}^{k}(n_t - 1)s_t^2}{\sum_{t=1}^{k}(n_t - 1)}$$

The between-treatment variance (s_b^2) is calculated using the treatment averages $\bar{y}_t$ and the grand average, $\bar{\bar{y}}$:

$$s_b^2 = \frac{\sum_{t=1}^{k} n_t (\bar{y}_t - \bar{\bar{y}})^2}{k - 1}$$

If there are an equal number of observations in each treatment the equations for s_w^2 and s_b^2 simplify to:

$$s_w^2 = \frac{(n_t - 1)\sum_{t=1}^{k} s_t^2}{N - k}$$

and

$$s_b^2 = \frac{n_t \sum_{t=1}^{k} (\bar{y}_t - \bar{\bar{y}})^2}{k - 1}$$

The logic of the comparison between for s_w^2 and s_b^2 goes like this:

1. The pooled variance within treatments (s_w^2) is based on $N - k$ degrees of freedom. It will be unaffected by real differences between the means of the k treatments. Assuming no hidden factors are affecting the results, s_w^2 estimates the pure measurement error variance σ^2.
2. If there are no real differences between the treatment averages other than what would be expected to occur by chance, the variance between treatments (s_b^2) also reflects only random measurement error. As such, it would be nearly the same magnitude as s_w^2 and would give a second estimate of σ^2.

Analysis of Variance to Compare k Averages

3. If the true means do vary from treatment to treatment, s_b^2 will be inflated and it will tend to be larger than s_w^2.
4. The null hypothesis is that no difference exists between the k means. It is tested by checking to see whether the two estimates of σ^2 (s_b^2 and s_w^2) are the same. Strict equality ($s_w^2 = s_b^2$) of these two variances is not expected because of random variation; but if the null hypothesis is true, they will be of the same magnitude. Roughly speaking, the same magnitude means that the ratio s_b^2/s_w^2 will be no larger than about 2.5 to 5.0. More precisely, this ratio is compared with the F statistic having $k - 1$ degrees of freedom in the numerator and $N - k$ degrees of freedom in the denominator (i.e., the degree of freedom are the same as s_b^2 and s_w^2). If s_b^2 and s_w^2 are of the same magnitude, there is no strong evidence to support a conclusion that the means are different. On the other hand, an indication that s_b^2 is inflated (large s_b^2/s_w^2) supports a conclusion that there is a difference between treatments.

The sample variances have a χ^2 distribution. Ratios of sample variances are distributed according to the F distribution. The χ^2 and F are skewed distributions whose exact shape depends on the degrees of freedom involved. The distributions are related to each other in much the same way that the normal and t distributions are related in the t-test. The two estimates of σ^2 are compared using the analysis of variance (ANOVA) table and the F test.

An Example Calculation

The computations for the one-way ANOVA are simpler than the above equations may suggest. Suppose that an experiment comparing treatments A, B, and C yields the data shown below.

	A	B	C
	12	13	18
	10	17	16
	13	20	21
	9	14	17
Treatment average	$\bar{y}_A = 11.0$	$\bar{y}_B = 16.0$	$\bar{y}_C = 18.0$
Treatment variance	$s_A^2 = 3.33$	$s_B^2 = 10.0$	$s_C^2 = 4.67$
Grand average	$\bar{y} = 15$		

The order of the experimental runs was randomized within and between treatments. The *grand average* of all 12 observed values is $\bar{y} = 15$. The averages for each treatment are $\bar{y}_A = 11$, $\bar{y}_B = 16$, and $\bar{y}_C = 18$. The within-treatment variances are:

$$s_A^2 = \frac{(12-11)^2 + (10-11)^2 + (13-11)^2 + (9-11)^2}{4-1} = \frac{10}{3} = 3.33$$

$$s_B^2 = \frac{(13-16)^2 + (17-16)^2 + (20-16)^2 + (14-16)^2}{4-1} = \frac{30}{3} = 10.00$$

$$s_C^2 = \frac{(18-18)^2 + (16-18)^2 + (21-18)^2 + (17-18)^2}{4-1} = \frac{14}{3} = 4.67$$

The pooled *within-treatment variance* is:

$$s_w^2 = \frac{(4-1)(3.33 + 10 + 4.67)}{12 - 3} = 6.0$$

The *between-treatment variance* is computed from the mean for each treatment and the grand mean, as follows:

$$s_b^2 = \frac{4(11-15)^2 + 4(16-15)^2 + 4(18-15)^2}{3-1} = 52$$

TABLE 24.2

Analysis of Variance Table for Comparing Treatments A, B, and C

Source of Variation	Sum of Squares	Degrees of Freedom	Mean Square	F Ratio
Between treatments	104	2	52	8.7
Within treatments	54	9	6	
Total	158	11		

Is the between-treatment variance larger than the within-treatment variance? This is judged by comparing the ratio of the between variance and the within variance. The ratios of sample variances are distributed according to the F distribution. The tabulation of F values is arranged according to the degrees of freedom in the variances used to compute the ratio. The numerator is the mean square of the "between-treatments" variance, which has v_1 degrees of freedom. The denominator is always the estimate of the pure random error variance, in this case the "within-treatments" variance, which has v_2 degrees of freedom. An F value with these degrees of freedom is denoted by $F_{v_1,v_2,\alpha}$, where α is the upper percentage point at which the test is being made. Usually $\alpha = 0.05$ (5%) or $\alpha = 0.01$ (1%). Geometrically, α is the area under the F_{v_1,v_2} distribution that lies on the upper tail beyond the value $F_{v_1,v_2,\alpha}$.

The test will be made at the 5% level with degrees of freedom $v_1 = k - 1 = 3 - 1 = 2$ and $v_2 = N - k = 12 - 3 = 9$. The relevant value is $F_{2,9,0.05} = 4.26$. The ratio computed for our experiment, $F = 52/6 = 8.67$ is greater than $F_{2,9,\alpha=0.05} = 4.26$, so we conclude that $\sigma_b^2 > \sigma_w^2$. This provides sufficient evidence to conclude at the 95% confidence level that the means of the three treatments are not equal. We are entitled only to conclude that $\eta_A \neq \eta_B \neq \eta_C$. This analysis does not tell us whether one treatment is different from the other two (i.e., $\eta_A \neq \eta_B$ but $\eta_B = \eta_C$), or whether all three are different. To determine which are different requires the kind of analysis described in Chapter 20.

When ANOVA is done by a commercial computer program, the results are presented in a special ANOVA table that needs some explanation. For the example problem just presented, this table would be as given in Table 24.2. The "sum of squares" in Table 24.2 is the sum of the squared deviations in the numerator of each variance estimate. The "mean square" in Table 24.2 is the sum of squares divided by the degrees of freedom of that sum of squares. The mean square values estimate the within-treatment variance (s_w^2) and the between-treatment variance (s_b^2). Note that the mean square for variation between treatments is 52, which is the *between-treatment variance* computed above. Also, note that the within treatment mean square of 6 is the *within-treatment variance* computed above. The F ratio is the ratio of these two mean squares and is the same as the F ratio of the two estimated variances computed above.

Case Study Solution

Figure 24.1 is a dot diagram showing the location and spread of the data from each laboratory. It appears that the variability in the results is about the same in each lab, but laboratories 4 and 5 may be giving low readings. The data are replotted in Figure 24.2 as deviations about their respective means. An analysis of variance will tell us if the means of these laboratories are statistically different.

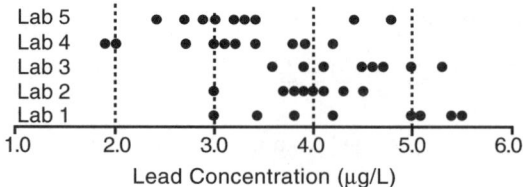

FIGURE 24.1 Dot plots comparing the results from five laboratories.

TABLE 24.3

ANOVA Table for the Comparison of Five Laboratories

Source of Variation	Sum of Squares	Degrees of Freedom	Mean Square	F Ratio
Between laboratories	13.88	4	3.47	6.81
Within laboratories	22.94	45	0.51	
Total	36.82	49		

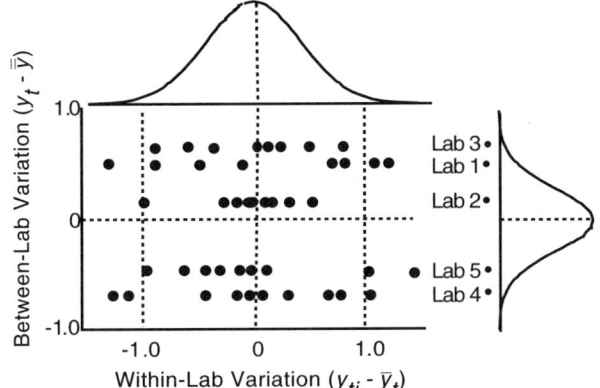

FIGURE 24.2 Plot of the data from the five laboratories and the distributions of within-laboratory and between-laboratory variation.

The variance within a single laboratory should be due to random errors arising in handling and analyzing the specimens. The variation between laboratories might be due to a real difference in performance, or it might also be due to random variation. If the variation between laboratories is random, the five observed laboratory means will vary randomly about the grand mean of all 50 measured concentrations ($\bar{\bar{y}} = 3.84\ \mu g/L$) and, furthermore, the variance of the five laboratories' means with respect to the grand mean will be the same as the variance within laboratories.

The ANOVA table for the laboratory data is given in Table 24.3. The F ratio is compared with the critical value $F_{4,45,0.05} = 2.59$. The value of $F = 6.81$ found for this experiment is much larger than $F_{4,45,0.05} = 2.59$ so we conclude that the variation between laboratories has been inflated by real differences in the mean level of performance of the labs.

Knowing this result of the analysis of variance, a plausible conclusion would be that laboratory 4, having the lowest average, is different from the others. But laboratories 4 and 5 may both be different from the other three. Or laboratory 3 may also be different, but on the high side. Unfortunately, ANOVA does not tell us how many or which laboratories are different; we only know that they are not giving the same results.

Comments

When the ANOVA indicates differences between laboratories, additional questions arise.

1. Which laboratories are different and which are the same? Making multiple pair-wise comparisons to answer this was discussed in Chapter 20.
2. Which laboratories, if any, are giving correct results? Without knowing the true concentration of the samples analyzed, there is no answer to this question. We remind ourselves, however, that the performer who is different may be the champion!

3. Is a special program needed to bring all laboratories into conformance? Maybe the only needed step is to show the participating laboratories the results. Simply knowing there is a possible problem usually will stimulate improvement. Quality improvement depends more on collecting data and communicating the results than on fixing blame or identifying poor performers. This means that we do not always need to find out which laboratories are different. (Recall point 2.)

A common question is: "Are the differences between labs large enough to have important consequences in practice?" Importance and statistical significance are different concepts. Importance depends on the actual use to which the measurements are put. Statistically significant differences are not always important. We can change significance to nonsignificance by changing the probability level of the test (or by using a different statistical procedure altogether). This obviously would not change the practical importance of a real difference in performance. Furthermore, the importance of a difference will exist whether we have data to detect it or not.

Analysis of variance can be applied to problems having many factors. One such example, a four-way ANOVA, is discussed in Chapter 26. Chapter 25 discusses the use of ANOVA to discover the relative magnitude of several sources of variability in a sampling and measurement procedure. Box et al. (1978) provide an interesting geometric interpretation of the analysis of variance.

References

Box, G. E. P., W. G. Hunter, and J. S. Hunter (1978). *Statistics for Experimenters: An Introduction to Design, Data Analysis, and Model Building,* New York, Wiley Interscience.

Johnson, R. A. and D. W. Wichern (1992). *Applied Multivariate Statistical Analysis,* Englewood Cliffs, NJ, Prentice-Hall.

Sokal, R. R. and F. J. Rohlf (1969). *Biometry: The Principles and Practice of Statistics in Biological Research,* New York, W. H. Freeman and Co.

Exercises

24.1 Chromium Measurements. A large portion of chromium contaminated water was divided into 32 identical aliquots. Eight aliquots were sent to each of four laboratories and the following data were produced. Are the laboratories making consistent measurements?

Lab 1	26.1	21.5	22.0	22.6	24.9	22.6	23.8	23.2
Lab 2	18.3	19.7	18.0	17.4	22.6	11.6	11.0	15.7
Lab 3	19.1	13.9	15.7	18.6	19.1	16.8	25.5	19.7
Lab 4	30.7	27.3	20.9	29.0	20.9	26.1	26.7	30.7

24.2 Aerated Lagoon. Conductivity measurements (μmho/cm) were taken at four different locations in the aerated lagoon of a pulp and paper mill. The lagoon is supposed to be mixed by aerators so the contents are homogeneous. Is the lagoon homogeneously mixed?

Location A	Location B	Location C	Location D
620	630	680	560
600	670	660	620
630	710	710	600
590	640	670	610
	650	680	630
	660	680	640
			630
			590

24.3 Adsorption. Six identical column adsorption units were fed identical influent for 20 weeks, with the exception that the influent to three units had a surfactant added. The data below are weekly average effluent concentrations. Can you discover from analysis of variance which units received surfactant? Has the surfactant affected the performance of the adsorption units?

Reactor\Week	1	2	3	4	5	6	7	8	9	10	11	12	13	14	15	16	17	18	19	20
1	7	4	4	4	4	5	6	4	6	5	3	4	6	11	10	17	8	7	8	8
2	5	8	10	7	7	9	7	6	6	11	10	15	18	19	22	23	10	20	17	6
3	6	7	9	8	7	15	12	7	8	11	8	14	18	19	6	10	4	4	4	3
4	4	4	4	7	8	9	10	7	8	10	11	12	11	13	8	13	5	11	11	10
5	7	10	6	3	4	5	5	6	6	4	5	7	6	6	5	9	6	5	7	5
6	3	4	3	9	12	12	14	11	9	10	11	11	13	14	17	19	7	11	14	6

25

Components of Variance

KEY WORDS *analysis of variance, ANOVA, components of variance, composite sample, foundry waste, nested design, sampling cost, sample size, system sand.*

A common problem arises when extreme variation is noted in routine measurements of a material. What are the important sources of variability? How much of the observed variability is caused by real differences in the material, how much is related to sampling, and how much is physical or chemical measurement error? A well-planned experiment and statistical analysis can answer these questions by breaking down the total observed variance into components that are attributed to specific sources.

Multiple Sources of Variation

When we wish to compare things it is important to collect the right data and use the appropriate estimate of error. An estimate of the appropriate error standard deviation can be obtained only from a design in which *the operation we wish to test* is repeated.

The kind of mistake that is easily made is illustrated by the following example. Three comparisons we might wish to make are (1) two testing methods, (2) two sampling methods, or (3) two processing methods. Table 25.1 lists some data that might have been collected to make these comparisons.

Two Testing Methods. Suppose we wish to compare two testing methods, A and B. Then we must compare, for example, ten *determinations* using test method A with ten determinations using test method B with all tests made on the *same sample specimen*.

Two Sampling Methods. To compare two different sampling methods, we might similarly compare determinations made on ten *different specimens* using sampling method A with those made on ten specimens using method B with all samples coming from the *same batch*.

Two Processing Methods. By the same principle, to compare standard and modified methods of processing, we might compare determinations from ten *batches* from the standard process with determinations from ten batches made by the modified process.

The total deviation $y - \eta$ of an observation from the process mean is due, in part, to three sources: (1) variation resulting only from the testing method, which has error e_t and standard deviation σ_t; (2) variation due only to sampling, which has error e_s and standard deviation σ_s; and (3) variation due only to the process, which has error e_p and standard deviation σ_p.

Figure 25.1 shows the dot plots of the data from the three testing schemes along with the sample statistics from Table 25.1. The estimated variance of 0.50 obtained from making ten tests on one specimen provides an estimate of σ_t^2, the testing variance. The variance $V_s = 5.21$ obtained from ten specimens each tested once does not estimate σ_s^2 alone, but rather $\sigma_s^2 + \sigma_t^2$. This is because each specimen value includes not only the deviation e_s due to the specimen but also the deviation e_t due to the test. The variance of $e_s + e_t$ is $\sigma_s^2 + \sigma_t^2$ because the variance of a sum is the sum of the variances for independent sources of variation. In a similar way, the variance $V_p = 35.79$ obtained from ten batches each sampled and tested once is an estimate of $\sigma_p^2 + \sigma_s^2 + \sigma_t^2$.

Using a "hat" notation ($\hat{\sigma}^2$) to indicate an estimate:

$$\hat{\sigma}_t^2 = 0.50$$

TABLE 25.1
Three Sets of Ten Measurements that Reflect Different Sources of Variation

Ten Tests on One Specimen	Ten Specimens Each Tested Once	Ten Batches Each Sampled and Tested Once
36.6	34.5	39.2
36.4	40.2	33.8
38.3	38.4	41.5
37.4	41.9	47.3
36.7	34.9	31.1
37.6	39.5	31.3
35.8	36.7	41.2
36.7	38.3	36.7
37.4	37.5	39.4
37.1	37.2	48.4
$\bar{y}_t = 37.0$	$\bar{y}_s = 37.9$	$\bar{y}_p = 39.0$
$V_t = 0.50$	$V_s = 5.21$	$V_p = 35.79$

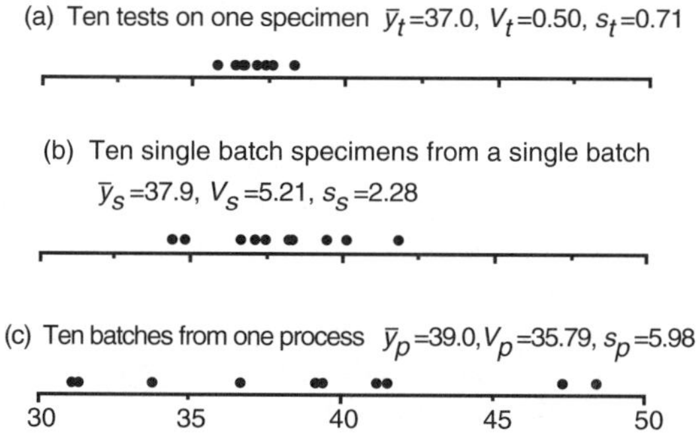

(a) Ten tests on one specimen $\bar{y}_t = 37.0$, $V_t = 0.50$, $s_t = 0.71$

(b) Ten single batch specimens from a single batch
$\bar{y}_s = 37.9$, $V_s = 5.21$, $s_s = 2.28$

(c) Ten batches from one process $\bar{y}_p = 39.0$, $V_p = 35.79$, $s_p = 5.98$

FIGURE 25.1 Ten observations made in three ways to illustrate the differences in variance (a) in single specimens, (b) in batches, and (c) in the total process.

$$\hat{\sigma}_s^2 = 5.21 - 0.50 = 4.71$$

$$\hat{\sigma}_p^2 = 35.79 - 5.21 - 0.50 = 30.08$$

Taking square roots gives the estimated standard deviations:

$$\hat{\sigma}_t = 0.71 \qquad \hat{\sigma}_s = 2.17 \qquad \hat{\sigma}_p = 5.48$$

It is seen from the dot plots (Figure 25.1) and from the calculations that the testing variation is relatively small and the process variation is large. For another process, sampling could be the largest source of variation, or testing could be. Usually, however, in environmental processes and systems, the laboratory analytical variation is smaller than the other sources of variation.

Case Study: Foundry Wastes

Foundries produce castings by pouring molten metal into molds made of molding sands and core sands. Molding sand and core sand can be recycled several times, but eventually are discarded as a mixture called "system sand." Molding sand is a mixture of silica sand, clay, carbon, and water. Core sand is composed of silica sand with a small amount of chemical binder, which may be natural substances (e.g., vegetable or petroleum oils, sodium silicate, ground corn flour and oil, ground hardwood cellulose) and synthetic binders (e.g., phenol formaldehyde, phenol isocyanate, alkyl isocyanate). The typical solid waste of a foundry is two thirds or more system sand, 2 to 20% core sand and core butts, and up to 11% dust collected in the baghouse that is used for air pollution control.

These wastes are generally put into landfills. Most of the waste is inert (sand) but certain components have the potential of being leached from the landfill and entering subsurface soils and groundwater. Studies of leachate from foundry landfills have shown large variations in chemical composition. The variation may arise from the nonuniformity of the waste materials deposited in the landfill, but it also may have other causes. This raises questions about how large a sample should be collected in the field, and whether this sample should be a composite of many small portions. There are also questions about how to partition a large field specimen into smaller portions that are suitable for laboratory analysis. Finally, the laboratory work itself cannot be overlooked as a possible source of variability.

These considerations point to the need for replicate specimens and replicate measurements. An efficient protocol for sampling and analytical replication cannot be designed until the sources of variation are identified and quantified. The major source of variation might be between field specimens, from subspecimens prepared in the laboratory, or from the analytical procedure itself.

Krueger (1985) studied three kinds of solid waste (system sand, core butts, or baghouse dust) from a foundry. One of his objectives was to assess the magnitude of the variance components due to (1) batch of material sampled (system sand, core butts, or baghouse dust); (2); specimen preparation in the laboratory (which includes a leaching extraction performed on that specimen); and (3) the analytical test for a specific substance. One of the substances measured was copper. Table 25.2 is an abridgment of Krueger's data on copper concentrations in a leaching extract from the three batches of solid waste.

TABLE 25.2

Copper Concentrations in the Leachate Extract of Foundry Solid Waste

Batch of Solid Waste	Specimen Number	Copper Concentrations (mg/L)		
		Specimen y_{bst}	Specimen Ave. $\bar{y}_{bs}$	Batch Ave. $\bar{y}_b$
Baghouse dust	1	0.082		
		0.084	0.0830	
	2	0.108		
		0.109	0.1085	
	3	0.074		
		0.070	0.0720	
	4	0.074		
		0.071	0.0725	0.0840
Core butts	1	0.054		
		0.051	0.0525	
	2	0.050		
		0.050	0.0500	
	3	0.047		
		0.050	0.0485	
	4	0.092		
		0.091	0.0915	0.0606
System sand	1	0.052		
		0.050	0.0510	
	2	0.084		
		0.080	0.0820	
	3	0.044		
		0.041	0.0425	
	4	0.050		
		0.044	0.0470	0.0556

Source: Krueger, R. C. (1985). *Characterization of Three Types of Foundry Waste and Estimation of Variance Components,* Research Report, Madison, WI, Department of Civil and Environmental Engineering, The University of Wisconsin–Madison.

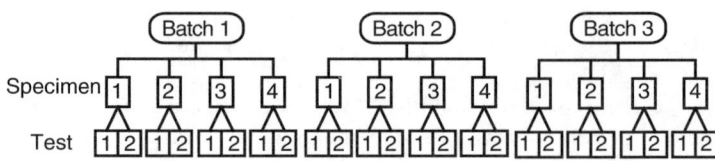

FIGURE 25.2 Nested design to estimate components of variance.

Variance Components Analysis

Variance components analysis is a method for learning what fraction of the total variance in a measurement process is caused by different components (factors) that contribute random variation into the sampling and testing process. If we have n measurements, denoted by y_i, $i = 1,...,n$, the sample variance for the entire data set is:

$$s_y^2 = \frac{\Sigma(y_i - \bar{y})^2}{n-1}$$

One design that allows the variance of each of these factors to be estimated independently of the other factors is the nested (or hierarchical) design shown in Figure 25.2.

The analysis can be generalized for k components, but it is more convenient to explain it specifically for a three-factor analysis (note that the case study involves three factors). In general, there are n_b batches, n_s specimens, and n_t chemical tests, for a total of $n = n_b n_s n_t$ observations. The nested experimental design shown in Figure 25.2 consists of three batches, with two chemical tests on each of the four specimens from each batch, giving a total of $n = (3)(4)(2) = 24$ observations.

The overall error of any particular measurement y_i will be $e_i = y_i - \eta$, where η is the true mean of the population of specimens. In practice, we estimate this mean by computing the average of all the measurements in the variance components experiment. A measurement on any one of the 24 test specimens produced by the design shown in Figure 25.2 will reflect variability contributed by each component, so:

$$e_i = e_b + e_s + e_t$$

where e_b, e_s, and e_t are the error contributions from the batch, specimen, and test, respectively. Assuming these errors are random and independent, their variances will add to give the total population variance σ_y^2:

$$\sigma_y^2 = \sigma_b^2 + \sigma_s^2 + \sigma_t^2$$

where the subscripts b, s, and t identify the variance components of the batches, specimens, and chemical tests, respectively. The aggregation of the error and variance components is diagrammed in Figure 25.3.

The variation among replicate chemical tests on each specimen provides an estimate of σ_t^2. The variation among specimen averages reflects both test and specimen variance and provides an estimate of the quantity $n_t\sigma_s^2 + \sigma_t^2$. The variation among batches embodies all three sources of variance and provides an estimate of the quantity $n_s n_t \sigma_b^2 + n_t\sigma_s^2 + \sigma_t^2$. The case study provides an opportunity to demonstrate the calculations to estimate these variances. A similar example with additional explanation is given by Box et al. (1978).

These calculations can be organized as an analysis of variance table. Table 25.3 shows the algebra of the analysis of variance for the three-factor nested design discussed here. Some additional nomenclature

Components of Variance

TABLE 25.3
General Analysis of Variance Table for Estimating the Variance Components from a Three-Factor Nested Experimental Design

Source of Variation	Sum of Squares	Degrees of Freedom	Mean Square	Mean Square Estimates
Average	$SS_{ave} = n_b n_s n_t \bar{y}^2$	1		
Batches	$SS_b = n_s n_t \sum_{b=1}^{n_b}(\bar{y}_b - \bar{y})^2$	$n_b - 1$	$MS_b = \dfrac{SS_b}{n_b - 1}$	$n_s n_t \sigma_b^2 + n_t \sigma_s^2 + \sigma_t^2$
Specimens	$SS_s = n_t \sum_{b=1}^{n_b}\sum_{s=1}^{n_s}(\bar{y}_{bs} - \bar{y}_s)^2$	$n_b(n_s - 1)$	$MS_s = \dfrac{SS_s}{n_b(n_s - 1)}$	$n_t \sigma_s^2 + \sigma_t^2$
Tests	$SS_t = \sum_{b=1}^{n_b}\sum_{s=1}^{n_s}\sum_{t=1}^{n_t}(y_{bst} - \bar{y}_{bs})^2$	$n_b n_s(n_t - 1)$	$MS_t = \dfrac{SS_t}{n_b n_s(n_t - 1)}$	σ_t^2
Total	$SS_T = \sum_{b=1}^{n_b}\sum_{s=1}^{n_s}\sum_{t=1}^{n_t} y_{bst}^2$			

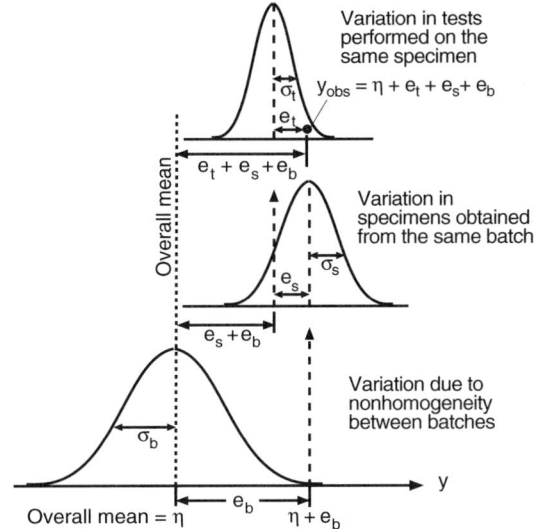

FIGURE 25.3 Error components contributing to variance in the measured quantity.

is also given. In practice, the calculations usually are done by computer and many software packages have this capability.

Case Study: Sampling Waste Foundry Sand

The data in Table 25.2 showed that three batches of foundry solid waste (baghouse dust, core butts, system sand) were collected. Each batch was subdivided in the laboratory into four specimens, and each specimen was analyzed in duplicate. This gives $n_b = 3$, $n_s = 4$, and $n_t = 2$, for a total of $3 \times 4 \times 2 = 24$ observations.

The average of the 24 observations is $\bar{y} = 0.06673$ mg/L. Table 25.2 also gives the averages for the duplicate measurements on each specimen, $\bar{y}_{bs}$, and the average of the eight measurements made on each batch of waste ($\bar{y}_b$).

TABLE 25.4
Analysis of Variance Table for the Foundry Waste Example

Source of Variation	Sum of Squares	df	MS	F
Average	0.10693	1		
Batches	0.00367	2	0.001835	367
Specimens	0.00624	9	0.000693	139
Tests	0.00006	12	0.000005	
Total	0.11690	24		

The sum of squares for each variance component is:

$$\text{Total SS}_t = \sum_{b=1}^{n_b}\sum_{s=1}^{n_s}\sum_{t=1}^{n_t} y_{bst}^2 = 0.082 + 0.084^2 + \cdots + 0.044^2 = 0.11690$$

$$\text{Average SS}_{ave} = n_b n_s n_t \bar{y}^2 = 3(4)(2)(0.06675)^2 = 0.106934$$

$$\text{Batch SS}_b = n_s n_t \sum_{b=1}^{n_b} (\bar{y}_b - \bar{y})^2$$

$$= (4)(2)[(0.0840 - 0.06675)^2 + (0.0606 - 0.06675)^2 + (0.0556 - 0.06675)^2]$$

$$= 0.003671$$

$$\text{Specimen SS}_s = n_t \sum_{b=1}^{n_b} \sum_{s=1}^{n_s} (\bar{y}_{bs} - \bar{y}_b)^2$$

$$= 2[(0.083 - 0.0840)^2 + (0.1085 - 0.0840)^2 + \cdots + (0.047 - 0.0556)^2]$$

$$= 0.000624$$

$$\text{Test SS}_t = \sum_{b=1}^{n_b}\sum_{s=1}^{n_s}\sum_{t=1}^{n_t} (y_{bst} - \bar{y}_{bs})^2$$

$$= (0.082 - 0.083)^2 + (0.084 - 0.083)^2 + \cdots + (0.044 - 0.0470)^2 = 0.000057$$

Table 25.4 gives the full analysis of variance table, with sums of squares, degrees of freedom, and mean square values. The mean squares (the sums of squares divided by the respective degrees of freedom) are $MS_b = 0.001835$, $MS_s = 0.000693$, and $MS_t = 0.000005$. The mean squares are used to estimate the variances. Table 25.3 shows that MS_t estimates σ_t^2, MS_s estimates $n_t\sigma_s^2 + \sigma_t^2$, and MS_b estimates $n_s n_t \sigma_b^2 + n_t \sigma_s^2 + \sigma_t^2$. Using these relations with the computed mean square values gives the following estimates:

$$\hat{\sigma}_t^2 = MS_t = 0.000005$$

$$\hat{\sigma}_s^2 = \frac{MS_s - MS_t}{n_t} = \frac{0.000693 - 0.000005}{2} = 0.000344$$

$$\hat{\sigma}_b^2 = \frac{MS_b - MS_s}{n_s n_t} = \frac{0.001836 - 0.000693}{4 \times 2} = 0.000143$$

The analysis shows that the variance between specimens is about twice the variance of between batches and about 60 times the variance of the chemical analysis of copper. Variation in the actual copper measurements is small relative to other variance components. Extensive replication of copper measurements scarcely helps in reducing the total variance of the solid waste characterization program. This suggests making only enough replicate copper measurements to maintain quality control and putting the major effort into reducing other variance components.

TABLE 25.5

Calculation of the Individual Errors e_t, e_s, and e_b that are Squared and Added to Compute the Sums of Squares

Experimental Design			Copper Conc. (mg/L)	Deviations of Tests		Deviations of Specimens		Deviations of Batches
Batch	Specimen	Test	y (1)	$\bar{y}_{bs}$ (2)	e_t (3)	$\bar{y}_b$ (4)	e_s (5)	e_b (6)
Baghouse	1	1	0.082	0.0830	−0.0010	0.0840	−0.0010	0.0173
Baghouse	1	2	0.084	0.0830	0.0010	0.0840	−0.0010	0.0173
Baghouse	2	1	0.108	0.1085	−0.0005	0.0840	0.0245	0.0173
Baghouse	2	2	0.109	0.1085	0.0005	0.0840	0.0245	0.0173
Baghouse	3	1	0.074	0.0720	0.0020	0.0840	−0.0120	0.0173
Baghouse	3	2	0.070	0.0720	−0.0020	0.0840	−0.0120	0.0173
Baghouse	4	1	0.074	0.0725	0.0015	0.0840	−0.0115	0.0173
Baghouse	4	2	0.071	0.0725	−0.0015	0.0840	−0.0115	0.0173
Core butts	1	1	0.054	0.0525	0.0015	0.0606	−0.0081	−0.0061
Core butts	1	2	0.051	0.0525	−0.0015	0.0606	−0.0081	−0.0061
Core butts	2	1	0.050	0.0500	0.0000	0.0606	−0.0106	−0.0061
Core butts	2	2	0.050	0.0500	0.0000	0.0606	−0.0106	−0.0061
Core butts	3	1	0.047	0.0485	−0.0015	0.0606	−0.0121	−0.0061
Core butts	3	2	0.050	0.0485	0.0015	0.0606	−0.0121	−0.0061
Core butts	4	1	0.092	0.0915	0.0005	0.0606	0.0309	−0.0061
Core butts	4	2	0.091	0.0915	−0.0005	0.0606	0.0309	−0.0061
System sand	1	1	0.052	0.0510	0.0010	0.0556	−0.0046	−0.0111
System sand	1	2	0.050	0.0510	−0.0010	0.0556	−0.0046	−0.0111
System sand	2	1	0.084	0.0820	0.0020	0.0556	0.0264	−0.0111
System sand	2	2	0.080	0.0820	−0.0020	0.0556	0.0264	−0.0111
System sand	3	1	0.044	0.0425	0.0015	0.0556	−0.0131	−0.0111
System sand	3	2	0.041	0.0425	−0.0015	0.0556	−0.0131	−0.0111
System sand	4	1	0.050	0.0470	0.0030	0.0556	−0.0086	−0.0111
System sand	4	2	0.044	0.0470	−0.0030	0.0556	−0.0086	−0.0111
Sums of squares:	From column (3)		$SS_t = 0.00006$					
	From column (5)		$SS_s = 0.00624$					
	From column (6)		$SS_b = 0.00367$					

Knowing the source of the three batches of foundry solid waste, it is reasonable that they should be different; the variation between batches should be large. One goal of sampling is to assess the *difference* between batches of waste material taken from the foundry, and perhaps samples taken from the landfill as well. This assessment will be complicated by the difficulty in dividing a large field specimen into smaller representative portions for laboratory analysis. Work should focus on the variability between batches and how they can be subdivided. One approach would be to prepare more specimens for laboratory analysis and enhance the statistical discrimination by averaging.

How Much Testing and Sampling are Needed?

The variance of an average of n independent observations is obtained by dividing the variance of the data by n. Also, the variances of summed independent variances are additive. If we take n_s specimens from a batch and do n_t tests on each specimen, the total number of measurements used to calculate the batch average is $n_s n_t$ and the variance of the batch average is:

$$\sigma^2 = \frac{\sigma_s^2}{n_s} + \frac{\sigma_t^2}{n_s n_t}$$

The reason σ_t^2 is divided by $n_s n_t$ and not by n_t is that the total of $n_s n_t$ measurements would be averaged.

TABLE 25.6

Some Possible Sampling and Testing Schemes with Their Variance and Cost

Scheme	Number of Specimens (n_s)	Number of Tests per Specimens (n_t)	Variance	Total Cost ($)
1	1	1	17.3	21
2	5	1	4.5	25
3	8	1	3.3	28
4	16	1	2.3	36
5	32	2	1.1	72

Computed using $\sigma_t^2 = 1.3$, $\sigma_s^2 = 16.0$, $C_s = \$1$, and $C_t = \$20$.

Let us experiment with this formula using $\sigma_t^2 = 0.5$ and $\sigma_s^2 = 5.0$. The variance of the average of results coming from n_t tests on each of n_s specimens is $\sigma^2 = \frac{5.0}{n_s} + \frac{0.5}{n_s n_t}$. If we take one specimen and test it once, then $n_s = 1$ and $n_t = 1$ and $\sigma^2 = \frac{5.0}{1} + \frac{0.5}{1} = 5.5$ and $\sigma = 2.34$. If instead we test this one specimen ten times, then $n_s = 1$ and $n_t = 10$, giving $\sigma^2 = \frac{5.0}{1} + \frac{0.5}{10} = 5.05$ and $\sigma = 2.24$. The nine extra tests reduced the standard deviation by only 4%. This is hardly a worthwhile investment.

Now see the result if we tested ten samples once, instead of one sample ten times. The variance is $\sigma^2 = \frac{5.0}{10} + \frac{0.5}{1 \times 10} = 0.55$ and $\sigma = 0.74$. The impressive reduction of nearly 70% is achieved.

In the case of environmental sampling, you could take n_s specimens, mix them together, and then do only a *single* test on the mixture. The mixing does the averaging for us. Then by making a *single* test on the mixture of ten samples, you would get $\sigma^2 = \frac{5.0}{10} + \frac{0.5}{1} = 0.55$ and $\sigma = 0.74$.

This mixture of specimens is a *composite sample*. Composite sampling is widely used. For additional information about composite sampling, see Edland and van Belle (1994), Mack and Robinson (1985), Elder (1980), Finucan (1964), Rajagopal and Williams (1989), Kussmaul and Anderson (1967), and Edelman (1974).

For any given application you can manipulate the formula to get a number of alternative schemes. If you know the cost of taking a sample and the cost of making an analysis, you can calculate the cost of each scheme and choose the one that is the best bargain.

The cost and variance can be considered when designing the experiment. Table 25.6 shows some possibilities that were calculated by the following method.

Let C_s be the cost of taking one specimen and C_t be the cost of making one test. The cost of the sampling scheme is:

$$C_{\text{total}} = C_s n_s + C_t n_t$$

where n_s and n_t are the number of specimens and tests per specimen included in the scheme. The variance of the scheme is $\sigma^2 = \sigma_s^2/n_s + \sigma_t^2/n_s n_t$. The minimum cost for any desired value of the variance can be achieved by choosing the ratio $n_s/n_t = \sqrt{(\sigma_s^2/\sigma_t^2) \times (C_t/C_s)}$. If we assume, for example, that $\sigma_s^2 = 16.0$ and $\sigma_t^2 = 1.3$, and the costs are $20 per test and $1 per sample, we obtain:

$$\frac{n_s}{n_t} = \sqrt{\frac{16.0}{1.3} \times \frac{20}{1}} = 15.7$$

There is no need to make more than one test per specimen unless more than 16 specimens are to be taken. The total cost of the scheme will be $16(\$1) + 1(\$20) = \$36$. The variance of the compositing scheme is $\sigma^2 = \frac{16}{16} + \frac{1.3}{1(16)} = 1.08$. If we did not composite, the cost would be $16(\$1) + 16(\$20) = \$336$.

Change the costs so sampling and testing are each $20 and we get:

$$\frac{n_s}{n_t} = \sqrt{\frac{16.0}{1.3} \times \frac{20}{20}} = 3.5$$

Making one test on each specimen would require four specimens. This scheme would have a variance of $\sigma^2 = \frac{16}{4} + \frac{1.3}{1(4)} = 4.3$, $\sigma = 2.07$, and a cost of $4(\$20) + 4(\$20) = \$160$. If this variance is too large, we might try $n_s = 8$ and $n_t = 2$, which gives $\sigma^2 = \frac{16}{8} + \frac{1.3}{2 \times 8} = 2.08$ and $\sigma = 1.44$, at a cost of 8($20) + 2(8)($20) = $480.

Comments

The variance components analysis is an effective method for quantifying the sources of variation in an experimental program. Obviously, this kind of study will be most useful if it is done *at the beginning* of a monitoring program. The information gained from the analysis can be used to plan a cost-effective program for collecting and analyzing samples. Cost-effective means that both the cost of the measurement program and the variance of the measurements produced are minimized.

It can happen in practice that some estimated variance components are negative. Such a nonsense result often can be interpreted as a sign of the variance component being zero or having some insignificant positive value. It may result from the lack of normality in the residual errors (Leone and Nelson, 1966). In any case, it should indicate that a review of the data structure and the basic assumptions of the components of the variance test should be made.

References

Box, G. E. P. (1998). "Multiple Sources of Variation: Variance Components," *Qual. Eng.*, 11(1), 171–174.

Box, G. E. P., W. G. Hunter, and J. S. Hunter (1978). *Statistics for Experimenters: An Introduction to Design, Data Analysis, and Model Building*, New York, Wiley Interscience.

Davies, O. L. and P. L. Goldsmith (1972). *Statistical Methods in Research and Production*, 4th ed., rev., New York, Hafner Publishing Co.

Edelman, D. A. (1974). "Three-Stage Nested Designs with Composited Samples," *Technometrics*, 16(3), 409–417.

Edland, Steven D. and Gerald van Belle (1994). "Decreased Sampling Costs and Improved Accuracy with Composite Sampling," in *Environmental Statistics, Assessment, and Forecasting*, C. R. Cothern and N. P. Ross (Eds.), Boca Raton, FL, CRC Press.

Elder, R. S. (1980). "Properties of Composite Sampling Procedures," *Technometrics*, 22(2), 179–186.

Finucan, H. M. (1964). "The Blood-Testing Problem," *Appl. Stat.*, 13, 43–50.

Hahn, G. J. (1977). "Random Samplings: Estimating Sources of Variability," *Chemtech*, Sept., pp. 580–582.

Krueger, R. C. (1985). *Characterization of Three Types of Foundry Waste and Estimation of Variance Components*, Research Report, Madison, WI, Department of Civil and Environmental Engineering, The University of Wisconsin–Madison.

Kussmaul, K. and R. L. Anderson (1967). "Estimation of Variance Components in Two-Stage Nested Design with Composite Samples," *Technometrics*, 9(3), 373–389.

Leone, F. C. and L. S. Nelson (1966). "Sampling Distributions of Variance Components. I. Empirical Studies of Balanced Nested Designs," *Technometrics*, 8, 457–468.

Mack, G. A. and P. E. Robinson (1985). "Use of Composited Samples to Increase the Precision and Probability of Detection of Toxic Chemicals," *Am. Chem. Soc. Symp. on Environmental Applications of Chemometrics*, J. J. Breen and P. E. Robinson (Eds.), Philadelphia, PA, 174–183.

Rajagopal, R. and L. R. Williams (1989). "Economics of Compositing as a Screening Tool in Ground Water Monitoring," *GWMR*, Winter, pp. 186–192.

Exercises

25.1 Fish Liver. The livers of four fish were analyzed for a bioaccumulative chemical. Three replicate measurements were made on each liver. Using the following results, verify that the variance between fish is significantly greater than the analytical variance.

Fish	Replicate Measurement		
1	0.25	0.22	0.23
2	0.22	0.20	0.19
3	0.19	0.21	0.20
4	0.24	0.22	0.22

25.2 Industrial Waste. The data below represent six deliveries of industrial waste to a municipal wastewater treatment plant. Two batches were collected at random from each delivery, and four tests were done on each batch. Analyze the data to determine the amount of variability due to delivery (D), batch (B), or test (C).

Delivery 1		Delivery 2		Delivery 3		Delivery 4		Delivery 5		Delivery 6	
B1	B2	B1	B2	B1	B2	B1	B2	B1	B2	B1	B2
35.6	38.6	30.7	31.7	30.0	27.9	34.3	38.7	33.2	35.8	39.5	38.7
33.6	41.6	30.5	30.0	35.0	27.7	36.4	38.5	35.2	37.1	42.1	36.1
34.1	40.7	27.2	33.8	35.0	29.0	33.4	43.3	37.8	37.1	38.5	35.9
34.5	39.9	26.8	29.6	32.6	32.8	33.4	36.7	35.4	40.2	40.2	42.8

25.3 Automobile Pollution. Triplicate measurements of a chemical contaminant (TPH) were made on soil specimens collected at four locations beside the eastbound and westbound lanes of an interstate highway. Perform a components of variance analysis to assess location, direction (east vs. west), and specimen (soil collection and analysis).

Location	Direction	Specimen	TPH
Mile 0–10	East	1	40
Mile 0–10	East	2	23
Mile 0–10	East	3	29
Mile 0–10	West	1	19
Mile 0–10	West	2	48
Mile 0–10	West	3	118
Mile 10–20	East	1	19
Mile 10–20	East	2	18
Mile 10–20	East	3	20
Mile 10–20	West	1	91
Mile 10–20	West	2	16
Mile 10–20	West	3	44
Mile 20–30	East	1	53
Mile 20–30	East	2	157
Mile 20–30	East	3	103
Mile 20–30	West	1	101
Mile 20–30	West	2	119
Mile 20–30	West	3	129
Mile 30–40	East	1	76
Mile 30–40	East	2	208
Mile 30–40	East	3	230
Mile 30–40	West	1	30
Mile 30–40	West	2	272
Mile 30–40	West	3	242

26
Multiple Factor Analysis of Variance

KEY WORDS air pollution, dioxin, furan, incineration, samplers, ANOVA, analysis of variance, factorial experiment, sampling error.

Environmental monitoring is expensive and complicated. Many factors may contribute variation to measured values. An obvious source of variation is the sampling method. An important question is: "Do two samplers give the same result?" This question may arise because a new sampler has come on the market, or because a monitoring program needs to be expanded and there are not enough samplers of one kind available.

It might seem natural to compare the two (or more) available sampling methods under a fixed set of conditions. This kind of experiment would estimate random error under only that specific combination of conditions. The samplers, however, will be used under a variety of conditions. A sampler that is effective under one condition may be weak under others. The error of one or both samplers might depend on plant operation, weather, concentration level being measured, or other factors. The variance due to laboratory measurements may be a significant part of the total variance. Interactions between sampling methods and other possible sources of variation should be checked. The experimental design should take into account all these factors.

Comparing two samplers under fixed conditions pursues the wrong goal. A better plan would be to assess performance under a variety of conditions. It is important to learn whether variation between samplers is large or small in comparison with variation due to laboratory analysis, operating conditions, etc. A good experiment would provide an analysis of variance of all factors that might be important in planning a sampling program.

It is incorrect to imagine that one data point provides one piece of information and therefore the information content of a data set is determined entirely by the number of measurements. The amount of information available from a fixed number of measurements increases dramatically if each observation contributes to estimating more than one parameter (mean, factor effect, variance, etc.). An exciting application of statistical experimental design is to make each observation do double duty or even triple or heavier duty. However, any valid statistical analysis can only extract the information existing in the data at hand. This content is largely determined by the experimental design and cannot be altered by the statistical analysis.

This chapter discusses an experimental design that was used to efficiently evaluate four factors that were expected to be important in an air quality monitoring program. The experiment is based on a factorial design (but not the two-level design discussed in Chapter 27). The method of computing the results is not discussed because this can be done by commercial computer programs. Instead, discussion focuses on how the *four-factor analysis of variance* is interpreted. References are given for the reader who wishes to know how such experiments are designed and how the calculations are done (Scheffe, 1959).

Case Study: Sampling Dioxin and Furan Emissions from an Incinerator

Emission of dioxins and furans from waste incinerators has been under investigation in many countries. It is important to learn whether different samplers (perhaps used at different incinerators or in different cities or countries) affect the amount of dioxin or furan measured. It is also important to assess whether differences, if any, are independent of other factors (such as incinerator loading rate and feed materials which change from one sampling period to another).

TABLE 26.1

Dioxin and Furan Data from a Designed Factorial Experiment

Sample Period	1		2		3		4	
Sampler	A	B	A	B	A	B	A	B
Dioxins								
Sum TetraCDD	0.4	1.9	0.5	1.7	0.3	0.7	1.0	2.0
Sum PentaCDD	1.8	28	3.0	7.3	2.7	5.5	7.0	11
Sum HexaCDD	2.5	24	2.6	7.3	3.8	5.1	4.7	6.0
Sum HeptaCDD	17	155	16	62	29	45	30	40
OctoCDD	7.4	55	7.3	28	14	21	12	17
Furans								
Sum TetraCDF	4.9	26	7.8	18	5.8	9.0	13	13
Sum PentaCDF	4.2	31	11	22	7.0	12	17	24
Sum HexaCDF	3.5	31	11	28	8.0	14	18	19
Sum HeptaCDF	9.1	103	32	80	32	41	47	62
OctoCDF	3.8	19	6.4	18	6.6	7.0	6.7	6.7

Note: Values shown are concentrations in ng/m^3 normal dry gas at actual CO_2 percentage.

The data in Table 26.1 were collected at a municipal incinerator by the Danish Environmental Agency (Pallesen, 1987). Two different kinds of samplers were used to take simultaneous samples during four 3.5-hour sampling periods, spread over a three-day period. Operating load, temperature, pressure, etc. were variable. Each sample was analyzed for five dioxin groups (TetraCDD, PentaCDD, HexaCDD, HeptaCDD, and OctoCDD) and five furan groups (TetraCDF, PentaCDF, HexaCDF, HeptaCDF, and OctoCDF). The species within each group are chlorinated to different degrees (4, 5, 6, 7, and 8 chlorine atoms per molecule). All analyses were done in one laboratory.

There are four factors being evaluated in this experiment: two kinds of samplers (S), four sampling periods (P), two dioxin and furan groups (DF), five levels of chlorination within each group (CL). This gives a total of $n = 2 \times 4 \times 2 \times 5 = 80$ measurements. The data set is completely balanced; all conditions were measured once with no repeats. If there are any missing values in an experiment of this kind, or if some conditions are measured more often than others, the analysis becomes more difficult (Milliken and Johnson, 1992).

When the experiment was designed, the two samplers were expected to perform similarly but that variation over sampling periods would be large. It was also expected that the levels of dioxins and furans, and the amounts of each chlorinated species, would be different. There was no prior expectation regarding interactions. A four-factor analysis of variance (ANOVA) was done to assess the importance of each factor and their interactions.

Method: Analysis of Variance

Analysis of variance addresses the problem of identifying which factors contribute significant amounts of variance to measurements. The general idea is to partition the total variation in the data and assign portions to each of the four factors studied in the experiment and to their interactions.

Total variance is measured by the total residual sum of squares:

$$\text{Total SS} = \sum_{\text{all obs}}^{n} (y_{\text{obs}} - \bar{\bar{y}})^2$$

where the residuals are the deviations of each observation from the grand mean

$$\bar{\bar{y}} = \frac{1}{n} \sum_{\text{all obs}}^{n} y_i$$

of the $n = 80$ observations. This is also called the total adjusted sum of squares (corrected for the mean). Each of the n observations provides one degree of freedom. One of them is consumed in computing the grand average, leaving $n - 1$ degrees of freedom available to assign to each of the factors that contribute variability. The Total SS and its $n - 1$ degrees of freedom are separated into contributions from the factors controlled in the experimental design. For the dioxin/furan emissions experiment, these sums of squares (SS) are:

$$\text{Total SS} = \text{Periods SS} + \text{Samplers SS} + \text{Dioxin/Furan SS} + \text{Chlorination SS} + \text{Interaction(s) SS} + \text{Error SS}$$

Another approach is to specify a general model to describe the data. It might be simple, such as:

$$y_{ijkl} = \bar{\bar{y}} + \alpha_i + \beta_j + \gamma_k + \lambda_l + \text{(interaction terms)} + e_i$$

where the Greek letters indicate the true response due to the four factors and e_i is the random residual error of the ith observation. The residual errors are assumed to be independent and normally distributed with mean zero and constant variance σ^2 (Rao, 1965; Box et al., 1978).

The assumptions of independence, normality, and constant variance are not equally important to the ANOVA. Scheffe (1959) states, "In practice, the statistical inferences based on the above model are not seriously invalidated by violation of the normality assumption, nor,... by violation of the assumption of equality of cell variances. However, there is no such comforting consideration concerning violation of the assumption of statistical independence, except for experiments in which randomization has been incorporated into the experimental procedure."

If measurements had been replicated, it would be possible to make a direct estimate of the error sum of squares (σ^2). In the absence of replication, the usual practice is to use the higher-order interactions as estimates of σ^2. This is justified by assuming, for example, that the fourth-order interaction has no meaningful physical interpretation. It is also common that third-order interactions have no physical significance. If sums of squares of third-order interactions are of the same magnitude as the fourth-order interaction, they can be pooled to obtain an estimate of σ^2 that has more degrees of freedom.

Because no one is likely to manually do the computations for a four-factor analysis of variance, we assume that results are available from some commercial statistical software package. The analysis that follows emphasizes variance decomposition and interpretation rather than model specification.

The first requirement for using available statistical software is recognizing whether the problem to be solved is one-way ANOVA, two-way ANOVA, etc. This is determined by the number of factors that are considered. In the example problem there are four factors: S, P, DF, and CL. It is therefore a four-way ANOVA.

In practice, such a complex experiment would be designed in consultation with a statistician, in which case the method of data analysis is determined by the experimental design. The investigator will have no need to guess which method of analysis, or which computer program, will suit the data. As a corollary, we also recommend that happenstance data (data from unplanned experiments) should not be subjected to analysis of variance because, in such data sets, randomization will almost certainly have not been incorporated.

Dioxin Case Study Results

The ANOVA calculations were done on the natural logarithm of the concentrations because this transformation tended to strengthen the assumption of constant variance.

The results shown in Table 26.2 are the complete variance decomposition, specifying all sum of squares (SS) and degrees of freedom (df) for the main effects of the four factors and all interactions between the four factors. These are produced by any computer program capable of handling a four-way ANOVA

TABLE 26.2

Variance Decomposition of the Dioxin/Furan Incinerator Emission Data

Source of Variation	SS	df	MS	F
S	18.3423	1	18.3423	573
CL	54.5564	4	13.6391	426
DF	11.1309	1	11.1305	348
DF × CL	22.7618	4	5.6905	178
S × P	9.7071	3	3.2357	101
P	1.9847	3	0.6616	21
DF × P	1.1749	3	0.3916	12.2
DF × S	0.2408	1	0.2408	7.5
P × CL	1.4142	12	0.1179	3.7
DF × P × CL	0.8545	12	0.0712	2.2
S × P × CL	0.6229	12	0.0519	a
S × CL	0.0895	4	0.0224	0.7
DF × S × CL	0.0826	4	0.0206	0.6
DF × S × P × CL	0.2305	12	0.0192	a
DF × S × P	0.0112	3	0.0037	a

[a] F calculated using $\sigma^2 = 0.032$, which is estimated with 27 degrees of freedom.

(e.g., SAS, 1982). The main effects and interactions are listed in descending order with respect to the mean sums of squares (MS = SS/df).

The individual terms in the sums of squares column measure the variability due to each factor plus some random measurement error. The expected contribution of variance due to random error is the random error variance (σ^2) multiplied by the degrees of freedom of the individual factor. If the true effect of the factor is small, its variance will be of the same magnitude as the random error variance. Whether this is the case is determined by comparing the individual variance contributions with σ^2, which is estimated below.

There was no replication in the experiment so no independent estimate of σ^2 can be computed. Assuming that the high-order interactions reflect only random measurement error, we can take the fourth-order interaction, DF × S × P × CL, as an estimate of the error sum of squares, giving $\hat{\sigma}^2 = 0.2305/12 = 0.0192$. We note that several other interactions have mean squares of about the same magnitude as the DF × S × P × CL interaction and it is tempting to pool these. There are, however, no hard and fast rules about which terms may be pooled. It depends on the data analyst's concept of a model for the data. Pooling more and more degrees of freedom into the random error term will tend to make $\hat{\sigma}^2$ smaller. This carries risks of distorting the decision regarding significance and we will follow Pallesen (1987) who pooled only the fourth-order and two third-order interactions (S × P × CL and of S × P × DF) to estimate $\hat{\sigma}^2 = (0.2305 + 0.6229 + 0.0112)/(12 + 12 + 3) = 0.8646/27 = 0.032$.

The estimated error variance ($\hat{\sigma}^2 = 0.032 = 0.18^2$) on the logarithmic scale can be interpreted as a measurement error with a standard deviation of about 18% in terms of the original concentration scale.

The main effects of all four factors are all significant at the 0.05% level. The largest source of variation is due to differences between the two samplers. Clearly, it is not acceptable to consider the samplers as equivalent. Presumably sampler B gives higher concentrations (Table 26.1), implying greater efficiency of contaminant recovery. The differences between samplers is much greater than differences between sampling periods, although "periods" represents a variety of operating conditions.

The interaction of the sampler with dioxin/furan groups (S × DF) was small, but statistically significant. The interpretation is that the difference between the samplers changes, depending on whether the contaminant is dioxin or furan. The S × P interaction is also significant, indicating that the difference between samplers was not constant over the four sampling periods.

The *a priori* expectation was that the dioxin and furan groups (DF) would have different levels and that the amounts of the various chlorinated species (CL) with chemical groups would not be equal. The large mean squares for DF and CL supports this.

Comments

When the experiment was planned, variation between sampling periods was expected to be large and differences between samplers were expected to be small. The data showed both expectations to be wrong. The major source of variation was between the two samplers. Variation between periods was small, although statistically significant.

Several interactions were statistically significant. These, however, have no particular practical importance until the matter of which sampler to use is settled. Presumably, after further research, one of the samplers will be accepted and the other rejected, or one will be modified. If one of the samplers were modified to make it perform more like the other, this analysis of variance would not represent the performance of the modified equipment.

Analysis of variance is a useful tool for breaking down the total variability of designed experiments into interpretable components. For well-designed (complete and fully balanced) experiments, this partitioning is unique and allows clear conclusions to be drawn from the data. If the design contains missing data, the partition of the variation is not unique and the interpretation depends on the number of missing values, their location in the table, and the relative magnitude of the variance components (Cohen and Cohen, 1983).

References

Box, G. E. P., W. G. Hunter, and J. S. Hunter (1978). *Statistics for Experimenters: An Introduction to Design, Data Analysis, and Model Building,* New York, Wiley Interscience.

Cohen, J. and P. Cohen (1983). *Applied Multiple Regression & Correlation Analysis for the Behavioral Sciences,* 2nd ed., New York, Lawrence Erlbann Assoc.

Milliken, G. A. and D. E. Johnson (1992). *Analysis of Messy Data, Vol. I: Designed Experiments,* New York, Van Nostrand Reinhold.

Milliken, G. A. and D. E. Johnson (1989). *Analysis of Messy Data, Vol. II: Nonreplicated Experiments,* New York, Van Nostrand Reinhold.

Pallesen, L. (1987). "Statistical Assessment of PCDD and PCDF Emission Data," *Waste Manage. Res.,* 5, 367–379.

Rao, C. R. (1965). *Linear Statistical Inference and Its Applications,* New York, John Wiley.

SAS Institute Inc. (1982). *SAS User's Guide: Statistics,* Cary, NC.

Scheffe, H. (1959). *The Analysis of Variance,* New York, John Wiley.

Exercises

26.1 Dioxin and Furan Sampling. Reinterpret the Pallesen example in the text after pooling the higher-order interactions to estimate the error variance according to your own judgment.

26.2 Ammonia Analysis. The data below are the percent recovery of 2 mg/L of ammonia (as NH_3-N) added to wastewater final effluent and tap water. Is there any effect of pH before distillation or water type?

pH Before Distillation	Final Effluent (initial conc. = 13.8 mg/L)			Tap Water (initial conc. $\leq$ 0.1 mg/L)		
9.5[a]	98	98	100	96	97	95
6.0	100	88	101	98	96	96
6.5	102	99	98	98	93	94
7.0	98	99	99	95	95	97
7.5	105	103	101	97	94	98
8.0	102	101	99	95	98	94

[a] Buffered.

Source: Dhaliwal, B. S., *J. WPCF,* 57, 1036–1039.

27

Factorial Experimental Designs

KEY WORDS *additivity, cube plot, density, design matrix, effect, factor, fly ash, factorial design, interaction, main effect, model matrix, normal order scores, normal plot, orthogonal, permeability, randomization, rankits, two-level design.*

Experiments are performed to (1) screen a set of factors (independent variables) and learn which produce an effect, (2) estimate the magnitude of effects produced by changing the experimental factors, (3) develop an empirical model, and (4) develop a mechanistic model. Factorial experimental designs are efficient tools for meeting the first two objectives. Many times, they are also excellent for objective three and, at times, they can provide a useful strategy for building mechanistic models.

Factorial designs allow a large number of variables to be investigated in few experimental runs. They have the additional advantage that no complicated calculations are needed to analyze the data produced. In fact, important effects are sometimes apparent without any calculations. The efficiency stems from using settings of the independent variables that are completely uncorrelated with each other. In mathematical terms, the experimental designs are *orthogonal*. The consequence of the orthogonal design is that the main effect of each experimental factor, and also the interactions between factors, can be estimated independent of the other effects.

Case Study: Compaction of Fly Ash

There was a proposal to use pozzolanic fly ash from a large coal-fired electric generating plant to build impermeable liners for storage lagoons and landfills. Pozzolanic fly ash reacts with water and sets into a rock-like material. With proper compaction this material can be made very impermeable. A typical criterion is that the liner must have a permeability of no more than 10^{-7} cm/sec. This is easily achieved using small quantities of fly ash in the laboratory, but in the field there are difficulties because the rapid pozzolanic chemical reaction can start to set the fly ash mixture before it is properly compacted. If this happens, the permeability will probably exceed the target of 10^{-7} cm/sec.

As a first step it was decided to study the importance of water content (%), compaction effort (psi), and reaction time (min) before compaction. These three factors were each investigated at two levels. This is a *two-level, three-factor experimental design.* Three factors at two levels gives a total of eight experimental conditions. The eight conditions are given in Table 27.1, where W denotes water content (4% or 10%), C denotes compaction effort (60 psi or 260 psi), and T denotes reaction time (5 or 20 min). Also given are the measured densities, in lb/ft^3. The permeability of each test specimen was also measured. The data are not presented, but permeability was inversely proportional to density. The eight test specimens were made at the same time and the eight permeability tests started simultaneously (Edil et al., 1987).

The results of the experiment are presented as a cube plot in Figure 27.1. Each corner of the cube represents one experimental condition. The plus (+) and minus (−) signs indicate the levels of the factors. The top of the cube represents the four tests at high compression, whereas the bottom represents the four tests at low pressure. The front of the cube shows the four tests at low reaction time, while the back shows long reaction time.

It is apparent without any calculations that each of the three factors has some effect on density. Of the investigated conditions, the best is run 4 with high water content, high compaction effort, and short

TABLE 27.1

Experimental Conditions and Responses for Eight Fly Ash Specimens

Run	Factor W (%)	C (psi)	T (min)	Density (lb/ft^3)
1	4	60	5	107.9
2	10	60	5	120.8
3	4	260	5	118.6
4	10	260	5	126.5
5	4	60	20	99.8
6	10	60	20	117.5
7	4	260	20	107.6
8	10	260	20	118.9

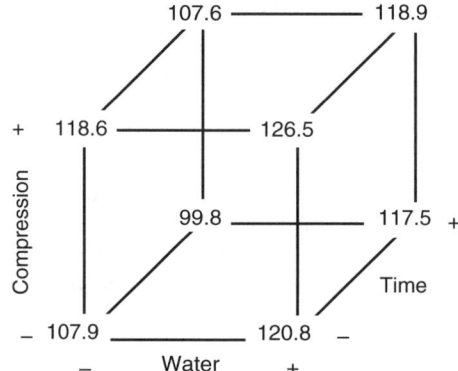

FIGURE 27.1 Cube plot showing the measured densities for the eight experimental conditions of the 2^3 factorial design.

reaction time. Densities are higher at the top of the cube than at the bottom, showing that higher pressure increases density. Density is lower at the back of the cube than at the front, showing that long reaction time reduces density. Higher water content increases density. The difference between the response at high and low levels is called a *main effect*. They can be quantified and tested for statistical significance.

It is possible that density is affected by how the factors act in combination. For example, the effect of water content at 20-min reaction time may not be the same as at 5 min. If it is not, there is said to be a *two-factor interaction* between water content and reaction time. Water content and compaction might interact, as might compaction and time.

Method: A Full 2^k Factorial Design

The k independent variables whose possible influence on a response variable is to be assessed are referred to as factors. An experiment with k factors, each set at two levels, is called a *two-level factorial design*. A *full factorial design* involves making runs at 2^k different experimental conditions which represent all combinations of the k factors at high and low levels. This is also called a *saturated design*. The high and low levels are conveniently denoted by + and −, or by +1 and −1. The factors can be continuous (pressure, temperature, concentration, etc.) or discrete (additive present, source of raw material, stirring used, etc.) The response variable (dependent variable) is y.

There are two-level designs that use less than 2^k runs to investigate k factors. These *fractional factorial designs* are discussed in Chapter 28. An experiment in which each factor is set at three levels would be a three-level factorial design (Box and Draper, 1987; Davies, 1960). Only two-level designs will be considered here.

Factorial Experimental Designs

TABLE 27.2

Design Matrices for 2^3 and 2^4 Full Factorial Designs

Run Number	Factor 1	Factor 2	Factor 3	Run Number	Factor 1	Factor 2	Factor 3	Factor 4
1	−	−	−	1	−	−	−	−
2	+	−	−	2	+	−	−	−
3	−	+	−	3	−	+	−	−
4	+	+	−	4	+	+	−	−
5	−	−	+	5	−	−	+	−
6	+	−	+	6	+	−	+	−
7	−	+	+	7	−	+	+	−
8	+	+	+	8	+	+	+	−
				9	−	−	−	+
				10	+	−	−	+
				11	−	+	−	+
				12	+	+	−	+
				13	−	−	+	+
				14	+	−	+	+
				15	−	+	+	+
				16	+	+	+	+

Experimental Design

The *design matrix* lists the setting of each factor in a standard order. Table 27.2 contains the design matrix for a full factorial design with $k = 3$ factors at two levels and a $k = 4$ factor design. The three-factor design uses $2^3 = 8$ experimental runs to investigate three factors. The 2^4 design uses 16 runs to investigate four factors. Note the efficiency: only 8 runs to investigate three factors, or 16 runs to investigate four factors.

The design matrix provides the information needed to set up each experimental test condition. Run number 5 in the 2^3 design, for example, is to be conducted with factor 1 at its low (−) setting, factor 2 at its low (−) setting, and factor 3 at its high (+) setting. If all the runs cannot be done simultaneously, they should carried out in *randomized* order to avoid the possibility that unknown or uncontrolled changes in experimental conditions might bias the factor effect. For example, a gradual increase in response over time might wrongly be attributed to factor 3 if runs were carried out in the standard order sequence. The lower responses would occur in the early runs where 3 is at the low setting, while the higher responses would tend to coincide with the + settings of factor 3.

Data Analysis

The statistical analysis consists of estimating the effects of the factors and assessing their significance. For a 2^3 experiment we can use the cube plots in Figure 27.2 to illustrate the nature of the estimates of the three main effects.

The main effect of a factor measures the average change in the response caused by changing that factor from its low to its high setting. This experimental design gives four separate estimates of each effect. Table 27.2 shows that the only difference between runs 1 and 2 is the level of factor 1. Therefore, the difference in the response measured in these two runs is an estimate of the effect of factor 1. Likewise, the effect of factor 1 is estimated by comparing runs 3 and 4, runs 5 and 6, and runs 7 and 8. These four estimates of the effect are averaged to estimate the main effect of factor 1.

This can also be shown graphically. The main effect of factor 1, shown in panel a of Figure 27.2, is the average of the responses measured where factor 1 is at its high (+) setting minus the average of the low (−) setting responses. Graphically, the average of the four corners with small dots are subtracted from the average of the four corners with large dots. Similarly, the main effects of factor 2 (panel b) and factor 3 (panel c) are the differences between the average at the high settings and the low settings for factors 2 and 3. Note that the effects are the changes in the response resulting from changing a factor from the low to the high level. It is not, as we are accustomed to seeing in regression models, the change associated with a one-unit change in the level of the factor.

TABLE 27.3

Model Matrix for a 2^3 Full Factorial Design

Run	X_0	X_1	X_2	X_3	X_{12}	X_{13}	X_{23}	X_{123}	y
1	+1	−1	−1	−1	+1	+1	+1	−1	y_1
2	+1	+1	−1	−1	−1	−1	+1	+1	y_2
3	+1	−1	+1	−1	−1	+1	−1	+1	y_3
4	+1	+1	+1	−1	+1	−1	−1	−1	y_4
5	+1	−1	−1	+1	+1	−1	−1	+1	y_5
6	+1	+1	−1	+1	−1	+1	−1	−1	y_6
7	+1	−1	+1	+1	−1	−1	+1	−1	y_7
8	+1	+1	+1	+1	+1	+1	+1	+1	y_8

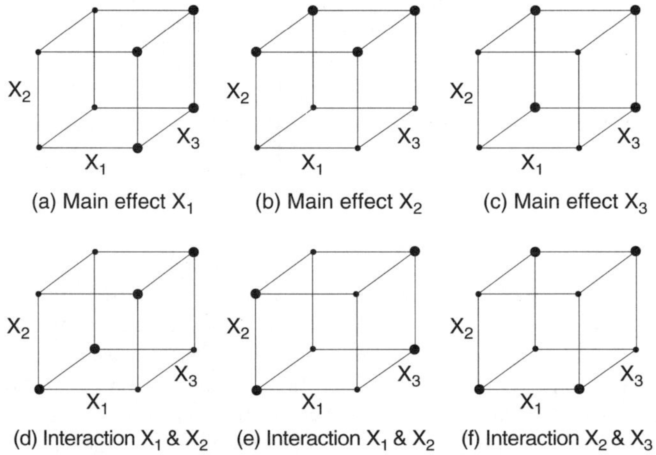

(a) Main effect X_1 (b) Main effect X_2 (c) Main effect X_3

(d) Interaction X_1 & X_2 (e) Interaction X_1 & X_2 (f) Interaction X_2 & X_3

FIGURE 27.2 Cube plots showing the main effects and two-factor interactions of a 2^3 factorial experimental design. The main effects and interactions are estimated by subtracting the average of the four values indicated with small dots from the average of the four values indicated by large dots.

The interactions measure the *non-additivity* of the effects of two or more factors. A significant *two-factor interaction* indicates antagonism or synergism between two factors; their combined effect is not the sum of their separate contributions. The interaction between factors 1 and 2 (panel d) is the average difference between the effect of factor 1 at the high setting of factor 2 and the effect of factor 1 at the low setting of factor 2. Equivalently, it is the effect of factor 2 at the high setting of factor 1 minus the effect of factor 2 at the low setting of factor 1. This interpretation holds for the two-factor interactions between factors 1 and 3 (panel e) and factors 2 and 3 (panel f). This is equivalent to subtracting the average of the four corners with small dots from the average of the four corners with large dots.

There is also a three-factor interaction. Ordinarily, this is expected to be small compared to the two factor interactions and the main effects. This is not diagrammed in Figure 27.2.

The effects are estimated using the *model matrix*, shown in Table 27.3. The structure of the matrix is determined by the model being fitted to the data. The model to be considered here is linear and it consists of the average plus three main effects (one for each factor) plus three two-factor interactions and a three-factor interaction. The model matrix gives the signs that are used to calculate the effects.

This model matrix consists of a column vector for the average, plus one column for each main effect, one column for each interaction effect, and a column vector of the response values. The number of columns is equal to the number of experimental runs because eight runs allow eight parameters to be estimated. The elements of the column vectors (X_i) can always be coded to be +1 or −1, and the signs are determined from the design matrix, Table 27.3. X_0 is always a vector of +1. X_1 has the signs associated with factor 1 in the design matrix, X_2 those associated with factor 2, and X_3 those of factor 3, etc. for higher-order full factorial designs. These vectors are used to estimate the main effects.

Factorial Experimental Designs

Interactions are represented in the model matrix by cross-products. The elements in X_{12} are the products of X_1 and X_2 (for example, $(-1)(-1) = 1$, $(1)(-1) = -1$, $(-1)(1) = -1$, $(1)(1) = 1$, etc.). Similarly, X_{13} is X_1 times X_3. X_{23} is X_2 times X_3. Likewise, X_{123} is found by multiplying the elements of X_1, X_2, and X_3 (or the equivalent, X_{12} times X_3, or X_{13} times X_2). The order of the X vectors in the model matrix is not important, but the order shown (a column of +1's, the factors, the two-factor interactions, followed by higher-order interactions) is a standard and convenient form.

From the eight response measurements $y_1, y_2, \ldots, y_8$, we can form eight statistically independent quantities by multiplying the y vector by each of the X vectors. The reason these eight quantities are statistically independent derives from the fact that the X vectors are orthogonal.[1] The independence of the estimated effects is a consequence of the orthogonal arrangement of the experimental design.

This multiplication is done by applying the signs of the X vector to the responses in the y vector and then adding the signed y's. For example, y multiplied by X_0 gives the sum of the responses: $X_0 \cdot y = y_1 + y_2 + \cdots + y_8$. Dividing the quantity $X_0 \cdot y$ by 8 gives the average response of the whole experiment. Multiplying the y vector by an X_i vector yields the sum of the four differences between the four y's at the +1 levels and the four y's at the −1 levels. The effect is estimated by the average of the four differences; that is, the effect of factor X_i is $X_i \cdot y/4$.

The eight effects and interactions that can be calculated from a full eight-run factorial design are:

Average $\quad X_0 \cdot y = \dfrac{y_1 + y_2 + y_3 + y_4 + y_5 + y_6 + y_7 + y_8}{8}$

Main effect of factor 1 $\quad X_1 \cdot y = \dfrac{-y_1 + y_2 - y_3 + y_4 - y_5 + y_6 - y_7 + y_8}{4}$

$\qquad = \dfrac{y_2 + y_4 + y_6 + y_8}{4} - \dfrac{y_1 + y_3 + y_5 + y_7}{4}$

Main effect of factor 2 $\quad X_2 \cdot y = \dfrac{y_3 + y_4 + y_7 + y_8}{4} - \dfrac{y_1 + y_2 + y_5 + y_6}{4}$

Main effect of factor 3 $\quad X_3 \cdot y = \dfrac{y_5 + y_6 + y_7 + y_8}{4} - \dfrac{y_1 + y_2 + y_3 + y_4}{4}$

Interaction of factors 1 and 2 $\quad X_{12} \cdot y = \dfrac{y_1 + y_4 + y_5 + y_8}{4} - \dfrac{y_2 + y_3 + y_6 + y_7}{4}$

Interaction factors 1 and 3 $\quad X_{13} \cdot y = \dfrac{y_1 + y_3 + y_6 + y_8}{4} - \dfrac{y_2 + y_4 + y_5 + y_7}{4}$

Interaction of factors 2 and 3 $\quad X_{23} \cdot y = \dfrac{y_1 + y_2 + y_7 + y_8}{4} - \dfrac{y_3 + y_4 + y_5 + y_6}{4}$

Interaction of factors 1, 2, and 3 $\quad X_{123} \cdot y = \dfrac{y_2 + y_3 + y_5 + y_8}{4} - \dfrac{y_1 + y_4 + y_6 + y_7}{4}$

If the variance of the individual measurements is σ^2, the variance of the mean is:

$$\text{Var}(\bar{y}) = \left(\frac{1}{8}\right)^2 [\text{Var}(y_1) + \text{Var}(y_2) + \cdots + \text{Var}(y_8)] = \left(\frac{1}{8}\right)^2 8\sigma^2 = \frac{\sigma^2}{8}$$

The variance of each main effect and interaction is:

$$\text{Var}(\text{effect}) = \left(\frac{1}{4}\right)^2 [\text{Var}(y_1) + \text{Var}(y_2) + \cdots + \text{Var}(y_8)] = \left(\frac{1}{4}\right)^2 8\sigma^2 = \frac{\sigma^2}{2}$$

---

[1] Orthogonal means that the product of any two-column vectors is zero. For example, $X_3 \cdot X_{123} = (-1)(-1) + \cdots + (+1)(+1) = 1 - 1 - 1 + 1 + 1 - 1 - 1 + 1 = 0$.

The experimental design just described does not produce an estimate of σ^2 because there is no replication at any experimental condition. In this case the significance of effects and interactions is determined from a normal plot of the effects (Box et al., 1978). This plot is illustrated later.

Case Study Solution

The responses at each setting and the calculation of the main effects are shown on the cube plots in Figure 27.3. As in Figure 27.1, each corner of the cube is the density measured at one of the eight experimental conditions.

The average density is ($X_0 \cdot y$):

$$\frac{107.9 + 120.8 + 118.6 + 126.5 + 99.8 + 117.5 + 107.6 + 118.9}{8} = 114.7$$

The estimates of the three main effects, the three two-factor interactions, and the one three-factor interaction are:

Main effect of water ($X_1 \cdot y$)

$$\frac{120.8 + 126.5 + 117.5 + 118.9}{4} - \frac{107.9 + 118.6 + 99.8 + 107.6}{4} = 12.45$$

Main effect of compaction ($X_2 \cdot y$)

$$\frac{118.6 + 126.5 + 107.6 + 118.9}{4} - \frac{107.9 + 120.8 + 99.8 + 117.5}{4} = 6.40$$

Main effect of time ($X_3 \cdot y$)

$$\frac{99.8 + 117.5 + 107.6 + 118.9}{4} - \frac{107.9 + 120.8 + 118.6 + 126.5}{4} = -7.50$$

Two-factor interaction of water × compaction ($X_{12} \cdot y$)

$$\frac{107.9 + 126.5 + 99.8 + 118.9}{4} - \frac{120.8 + 118.6 + 117.5 + 107.6}{4} = -2.85$$

Two-factor interaction of water × time ($X_{13} \cdot y$)

$$\frac{107.9 + 118.6 + 117.5 + 118.9}{4} - \frac{120.8 + 126.5 + 99.8 + 107.6}{4} = -2.05$$

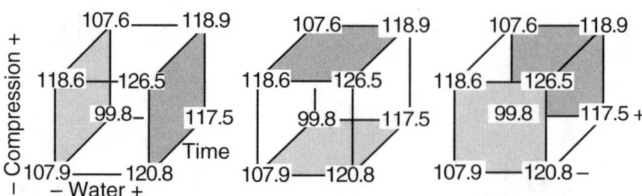

FIGURE 27.3 Cube plots of the 2^3 factorial experimental design. The values at the corners of the cube are the measured densities at the eight experimental conditions. The shaded faces indicate how the main effects are computed by subtracting the average of the four values at the low setting (− sign; light shading) from the average of the four values at the high setting (+ sign; dark shading).

Factorial Experimental Designs

Two-factor interaction of compaction × time ($X_{23} \cdot y$)

$$\frac{107.9 + 120.8 + 107.6 + 118.9}{4} - \frac{118.6 + 126.5 + 99.8 + 117.5}{4} = -1.80$$

Three-factor interaction of water × compaction × time ($X_{123} \cdot y$)

$$\frac{120.8 + 118.6 + 99.8 + 118.9}{4} - \frac{107.9 + 126.5 + 117.5 + 107.6}{4} = -0.35$$

Before interpreting these effects, we want to know whether they are large enough not to have arisen from random error. If we had an estimate of the variance of measurement error, the variance of each effect could be estimated and confidence intervals could be used to make this assessment. In this experiment there are no replicated measurements, so it is not possible to compute an estimate of the variance. Lacking a variance estimate, another approach is used to judge the significance of the effects.

If the effects are random (i.e., arising from random measurement errors), they might be expected to be normally distributed, just as other random variables are expected to be normally distributed. Random effects will plot as a straight line on normal probability paper. The *normal plot* is constructed by ordering the effects (excluding the average), computing the probability plotting points as shown in Chapter 5, and making a plot on normal probability paper. Because probability paper is not always handy, and many computer graphics programs do not make probability plots, it is handy to plot the effects against the *normal order scores* (or *rankits*). Table 27.4 shows both the probability plotting positions and the normal order scores for the effects.

Figure 27.4 is a plot of the estimated effects estimated against the normal order scores. Random effects will fall along a straight line on this plot. These are not *statistically significant*. We consider them to have values of zero. Nonrandom effects will fall off the line; these effects will be the largest (in absolute value). The nonrandom effects are considered to be statistically significant.

In this case a straight line covers the two- and three-factor interactions on the normal plot. None of the interactions are significant. The significant effects are the main effects of water content, compaction effort, and reaction time. Notice that it is possible to draw a straight line that covers the main effects and leaves the interactions off the line. Such an interpretation — significant interactions and insignificant main effects — is not physically plausible. Furthermore, effects of near-zero magnitude cannot be significant when effects with larger absolute values are not.

TABLE 27.4

Effects, Plotting Positions, and Normal Order Scores for Figure 27.4

Order number i	1	2	3	4	5	6	7
Identity of effect	3	12	23	123	13	2	1
Effect	−7.5	−2.85	−1.80	−0.35	2.05	6.40	12.45
$P = 100(i - 0.5)/7$	0.07	0.21	0.36	0.50	0.64	0.79	0.93
Normal order scores	−1.352	−0.757	−0.353	0	0.353	0.757	1.352

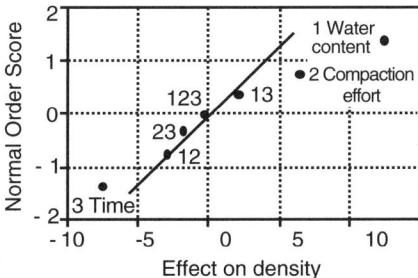

FIGURE 27.4 Normal probability plot of the estimated main effects and interactions.

The final interpretation of the results is:

1. The average density over the eight experimental design conditions is 114.7.
2. Increasing water content from 4 to 10% increases the density by an average of 12.45 lb/ft^3.
3. Increasing compaction effort from 60 to 260 psi increases density by an average of 6.40 lb/ft^3.
4. Increasing reaction time from 5 to 20 min decreases density by an average of 7.50 lb/ft^3.
5. These main effects are additive because the interactions are zero. Therefore, increasing both water content and compaction effort from their low to high values increases density by 12.45 + 6.40 = 18.85 lb/ft^3.

Comments

Two-level factorial experiments are a way of investigating a large number of variables with a minimum number of experiments. In general, a k variable two-level factorial experiment will require 2^k experimental runs. A 2^2 experiment evaluates two variables in four runs, a 2^3 experiment evaluates three variables in eight runs, a 2^4 design evaluates four variables in sixteen runs, etc. The designs are said to be *full* or *saturated*. From this small number of runs it is possible to estimate the average level of the response, k main effects, all two-factor interactions, and all higher-order interactions. Furthermore, these main effects and interactions are estimated independently of each other. Each main effect independently estimates the change associated with one experimental factor, and only one.

Why do so few experimental runs provide so much information? The strength and beauty of this design arise from its economy and balance. Each data point does triple duty (at least) in estimating main effects. Each observation is used in the computation of each factor main effect and each interaction. Main effects are averaged over more than one setting of the companion variables. This is the result of varying all experimental factors simultaneously. One-factor-at-a-time (OFAT) designs have none of this efficiency or power. An OFAT design in eight runs would provide only estimates of the main effects (no interactions) and the estimates of the main effects would be inferior to those of the two-level factorial design.

The statistical significance of the estimated effects can be evaluated by making the normal plot. If the effects represent only random variation, they will plot as a straight line. If a factor has caused an effect to be larger than expected due to random error alone, the effect will not fall on a straight line. Effects of this kind are interpreted as being significant. Another way to evaluate significance is to compute a confidence interval, or a reference distribution. This is shown in Chapter 28.

Factorial designs should be the backbone of an experimenter's design strategy. Chapter 28 shows how four factors can be evaluated with only eight runs. Experimental designs of this kind are called fractional factorials. Chapter 29 extends this idea. In Chapter 30 we show how the effects are estimated by linear algebra or regression, which is more convenient in larger designs and in experiments where the independent variables have not been set exactly according to the orthogonal design. Chapter 43 explains how factorial designs can be used sequentially to explore a process and optimize its performance.

References

Box, G. E. P., W. G. Hunter, and J. S. Hunter (1978). *Statistics for Experimenters: An Introduction to Design, Data Analysis, and Model Building,* New York, Wiley Interscience.

Box, G. E. P. and N. R. Draper (1987). *Empirical Model Building and Response Surfaces,* New York, John Wiley.

Davies, O. L. (1960). *Design and Analysis of Industrial Experiments,* New York, Hafner Co.

Edil, T. B., P. M. Berthouex, and K. Vesperman (1987). "Fly Ash as a Potential Waste Liner," *Proc. Conf. Geotechnical Practice in Waste Disposal,* Geotech. Spec. Pub. No. 13, ASCE, pp. 447–461.

Tiao, George et al., Eds. (2000). *Box on Quality and Discovery with Design, Control, and Robustness,* New York, John Wiley & Sons.

Exercises

27.1 Recycled Water Irrigation. Evaluate an irrigation system that uses recycled water to grow cucumbers and eggplant. Some field test data are given in the table below. Irrigation water was applied in two ways: sprinkle and drip. Evaluate the yield, yield per gallon, and biomass production

Vegetable	Irrigation Type	Irrigation Source	Yield (lb/ft²)	Yield (lb/gal)	Biomass (lb/plant)
Cucumber	Sprinkle	Tap water	6.6	0.15	5.5
		Recycled water	6.6	0.15	5.7
	Drip	Tap water	4.9	0.25	4.5
		Recycled water	4.8	0.25	4.0
Eggplant	Sprinkle	Tap water	2.9	0.07	3.0
		Recycled water	3.2	0.07	3.5
	Drip	Tap water	1.6	0.08	1.9
		Recycled water	2.3	0.12	2.3

27.2 Water Pipe Corrosion. Students at Tufts University collected the following data to investigate the concentration of iron in drinking water as a means of inferring water pipe corrosion. (a) Estimate the main effects and interactions of the age of building, type of building, and location. (b) Make the normal plot to judge the significance of the estimated effects. (c) Based on duplicate observations at each condition, the estimate of σ is 0.03. Use this value to calculate the variance of the average and the main and interaction effects. Use $\text{Var}(\bar{y}) = S_p^2/N$ and $\text{Var}(\text{Effect}) = 4S_p^2/N$, where N = total number of measurements (in this case $N = 16$) to evaluate the results. Compare your conclusions regarding significance with those made using the normal plot.

Age	Type	Location	Iron (mg/L)	
Old	Academic	Medford	0.23	0.28
New	Academic	Medford	0.36	0.29
Old	Residential	Medford	0.03	0.06
New	Residential	Medford	0.05	0.02
Old	Academic	Somerville	0.08	0.05
New	Academic	Somerville	0.03	0.08
Old	Residential	Somerville	0.04	0.07
New	Residential	Somerville	0.02	0.06

27.3 Bacterial Tests. Analysts A and B each made bacterial tests on samples of sewage effluent and water from a clean stream. The bacterial cultures were grown on two media: M1 and M2. The experimental design is given below. Each test condition was run in triplicate. The y values are logarithms of the measured bacterial populations. The s_i^2 are the variances of the three replicates at each test condition. (a) Calculate the main and interaction effects using the averages at each test condition. (b) Draw the normal plot to interpret the results. (c) Average the eight variances to estimate σ^2 for the experiment. Use $\text{Var}(\bar{y}) = \frac{\sigma^2}{24}$ and $\text{Var}(\text{effect}) = \frac{\sigma^2}{6}$ to evaluate the results. [Note that Var(effect) applies to main effects *and* interactions. These variance equations account for the replication in the design.]

Source	Analyst	Medium	y (3 Replicates)			$\bar{y}_i$	s_i^2
Effluent	A	M1	3.54	3.79	3.40	3.58	0.0390
Stream	A	M1	1.85	1.76	1.72	1.78	0.0044
Effluent	B	M1	3.81	3.82	3.79	3.81	0.0002
Stream	B	M1	1.72	1.75	1.55	1.67	0.0116
Effluent	A	M2	3.63	3.67	3.71	3.67	0.0016
Stream	A	M2	1.60	1.74	1.72	1.69	0.0057
Effluent	B	M2	3.86	3.86	4.08	3.93	0.0161
Stream	B	M2	2.05	1.51	1.70	1.75	0.0750

27.4 Reaeration. The data below are from an experiment that attempted to relate the rate of dissolution of an organic chemical to the reaeration rate (y) in a laboratory model stream channel. The three experimental factors are stream velocity (V, in m/sec), stream depth (D, in cm), and channel roughness (R). Calculate the main effects and interactions and interpret the results.

Run	V	D	R	y (Triplicates)			Average
1	0.25	10	Smooth	107	117	117	113.7
2	0.5	10	Smooth	190	178	179	182.3
3	0.25	15	Smooth	119	116	133	122.7
4	0.5	15	Smooth	188	191	195	191.3
5	0.25	10	Coarse	119	132	126	125.7
6	0.5	10	Coarse	187	173	166	175.3
7	0.25	15	Coarse	140	133	132	135.0
8	0.5	15	Coarse	164	145	144	151.0

27.5 Metal Inhibition. The results of a two-level, four-factor experiment to study the effect of zinc (Zn), cobalt (Co), and antimony (Sb) on the oxygen uptake rate of activated sludge are given below. Calcium (Ca) was added to some test solutions. The (−) condition is absence of Ca, Zn, Co, or Sb. The (+) condition is 10 mg/L Zn, 1 mg/L Co, 1 mg/L Sb, or 300 mg/L Ca (as $CaCO_3$). The control condition (zero Ca, Zn, Co, and Sb) was duplicated. The measured response is cumulative oxygen uptake (mg/L) in 20-hr reaction time. Interpret the data in terms of the main and interaction effects of the four factors.

Run	Zn	Co	Sb	Ca	Uptake (mg/L)
1	−1	−1	−1	−1	761
2	+1	−1	−1	−1	532
3	−1	+1	−1	−1	759
4	+1	+1	−1	−1	380
5	−1	−1	+1	−1	708
6	+1	−1	+1	−1	348
7	−1	+1	+1	−1	547
8	+1	+1	+1	−1	305
9	−1	−1	−1	+1	857
10	+1	−1	−1	+1	902
11	−1	+1	−1	+1	640
12	+1	+1	−1	+1	636
13	−1	−1	+1	+1	822
14	+1	−1	+1	+1	798
15	−1	+1	+1	+1	511
16	+1	+1	+1	+1	527
1 (rep)	−1	−1	−1	−1	600

Source: Hartz, K. E., *J. WPFC*, 57, 942–947.

27.6 Plant Lead Uptake. Anaerobically digested sewage sludge and commercial fertilizer were applied to garden plots (10 ft × 10 ft) on which were grown turnips or Swiss chard. Each treatment was done in triplicate. After harvesting, the turnip roots or Swiss chard leaves were washed, dried, and analyzed for total lead. Determine the main and interaction effects of the sludge and fertilizer on lead uptake by these plants.

Exp.	Sludge	Fertilizer	Turnip Root	Swiss Chard Leaf
1	None	None	0.46, 0.57, 0.43	2.5, 2.7, 3.0
2	110 gal/plot	None	0.56, 0.53, 0.66	2.0, 1.9, 1.4
3	None	2.87 lb/plot	0.29, 0.39, 0.30	3.1, 2.5, 2.2
4	110 gal/plot	2.87 lb/plot	0.31, 0.32, 0.40	2.5, 1.6, 1.8

Source: Auclair, M. S. (1976). M.S. thesis, Civil Engr. Dept., Tufts University.

28

Fractional Factorial Experimental Designs

KEY WORDS *alias structure, confounding, defining relation, dissolved oxygen, factorial design, fractional factorial design, half-fraction, interaction, main effect, reference distribution, replication, ruggedness testing,* t *distribution, variance.*

Two-level factorial experimental designs are very efficient but the number of runs grows exponentially as the number of factors increases.

3 factors at 2 levels	$2^3 = 8$ runs
4 factors at 2 levels	$2^4 = 16$ runs
5 factors at 2 levels	$2^5 = 32$ runs
6 factors at 2 levels	$2^6 = 64$ runs
7 factors at 2 levels	$2^7 = 128$ runs
8 factors at 2 levels	$2^8 = 256$ runs

Usually your budget cannot support 128 or 256 runs. Even if it could, you would not want to commit your entire budget to one very large experiment. As a rule-of-thumb, you should not commit more than 25% of the budget to preliminary experiments for the following reasons. Some of the factors may be inactive and you will want to drop them in future experiments; you may want to use different factor settings in follow-up experiments; a two-level design will identify interactions, but not quadratic effects, so you may want to augment the design and do more testing; you may need to repeat some experiments; and/or you may need to replicate the entire design to improve the precision of the estimates. These are reasons why *fractional factorial designs* are attractive. They provide flexibility by reducing the amount of work needed to conduct preliminary experiments that will screen for important variables and guide you toward more interesting experimental settings.

Fractional means that we do a fraction or a part of the full factorial design. We could do a half-fraction, a quarter-fraction, or an eighth-fraction. A half-fraction is to do half of the full factorial design, or $(1/2)2^4 = (1/2)16 = 8$ runs to investigate four factors; $(1/2)(2^5) = (1/2)32 = 16$ runs to investigate five factors; and so on. Examples of quarter-fractions are $(1/4)2^5 = (1/4)32 = 8$, or $(1/4)2^7 = (1/4)128 = 32$ runs. An example eighth-fraction is $(1/8)2^8 = (1/8)256 = 32$ runs. These five examples lead to designs that could investigate 4 variables in 8 runs, 5 factors in 16 runs or 8 runs, 7 factors in 32 runs, or 8 factors in 32 runs.

Of course, some information must be sacrificed in order to investigate 8 factors in 32 runs, instead of the full 256 runs, but you will be surprised how little is lost. The lost information is about interactions, if you select the right 32 runs out of the possible $2^8 = 256$. How to do this is explained fully in Box et al. (1978) and Box and Hunter (1961a, 1961b).

Case Study: Sampling High Dissolved Oxygen Concentrations

Ruggedness testing is a means of determining which of many steps in an analytical procedure must be carefully controlled and which can be treated with less care. Each aspect or step of the technique needs checking. These problems usually involve a large number of variables and an efficient experimental approach is needed. Fractional factorial designs provide such an approach.

It was necessary to measure the oxygen concentration in the influent to a pilot plant reactor. The influent was under 20 psig pressure and was aerated with pure oxygen. The dissolved oxygen (DO) concentration was expected to be about 40 mg/L. Sampling methods that are satisfactory at low DO levels (e.g., below saturation) will not work in this situation. Also, conventional methods for measuring dissolved oxygen are not designed to measure DO above about 20 mg/L. The sampling method that was developed involved withdrawing the highly oxygenated stream into a volume of deoxygenated water, thereby diluting the DO so it could be measured using conventional methods. The estimated *in situ* DO of the influent was the measured DO multiplied by the dilution factor.

There was a possibility that small bubbles would form and oxygen would be lost as the pressure dropped from 20 psig in the reactor to atmospheric pressure in the dilution bottle. It was essential to mix the pressurized solution with the dilution water in a way that would eliminate, or at least minimize, this loss. One possible technique would be to try to capture the oxygen before bubbles formed or escaped by introducing the sample at a high rate into a stirred bottle containing a large amount of dilution water. On the other hand, the technique would be more convenient if stirring could be eliminated, if a low sample flow rate could be used, and if only a small amount of dilution water was needed. Perhaps one or all of these simplifications could be made. An experiment was needed that would indicate which of these variables were important in a particular context. The outcome of this experiment should indicate how the sampling technique could be simplified without loss of accuracy.

Four variables in the sampling procedure seemed critical: (1) stirring rate S, (2) dilution ratio D, (3) specimen input location L, and (4) sample flow rate F. A two-level, four-variable fractional factorial design (2^{4-1}) was used to evaluate the importance of the four variables. This design required measurements at eight combinations of the independent variables. The high and low settings of the independent variables are shown in Table 28.1. The experiment was conducted according to the design matrix in Table 28.2, where the factors (variables) S, D, L, and F are identified as 1, 2, 3, and 4, respectively. The run order was randomized, and each test condition was run in duplicate. The average and difference between duplicates for each run are shown in Table 28.2.

TABLE 28.1

Experimental Settings for the Independent Variables

Setting	Stirring S	Dilution Ratio D	Sample Input Location L	Sample Flow Rate F
Low level (−)	Off	2:1	Surface	2.6 mL/sec
High level (+)	On	4:1	Bottom	8.2 mL/sec

TABLE 28.2

Experimental Design and Measured Dissolved Oxygen Concentrations

Run	S (1)	D (2)	L (3)	F (4)	Duplicates (mg/L) y_{1i}	y_{2i}	Avg. DO (mg/L) $\bar{y}_i$	Difference (mg/L) d_i
1	−	−	−	−	38.9	41.5	40.20	−2.6
2	+	−	−	+	45.7	45.4	45.55	0.3
3	−	+	−	+	47.8	48.8	48.30	−1.0
4	+	+	−	−	45.8	43.8	44.80	2.0
5	−	−	+	+	45.2	47.6	46.40	−2.4
6	+	−	+	−	46.9	48.3	47.60	−1.4
7	−	+	+	−	41.0	45.8	43.40	−4.8
8	+	+	+	+	53.5	52.4	52.95	1.1

Note: Defining relation: **I = 1234**.

Method: Fractional Factorial Designs

A fractional factorial design is an experimental layout where a full factorial design is augmented with one or more factors (independent variables) to be analyzed without increasing the number of experimental runs. These designs are labeled 2^{k-p}, where k is the number of factors that could be evaluated in a full factorial design of size 2^k and p is the number of additional factors to be included. When a fourth factor is to be incorporated in a 2^3 design of eight runs, the resulting design is a 2^{4-1} fractional factorial, which also has $2^3 = 8$ runs. The full 2^4 factorial would have 16 runs. The 2^{4-1} has only eight runs. It is a half-fraction of the full four-factor design. Likewise, a 2^{5-2} experimental design has eight runs; it is a quarter-fraction of the full five-factor design.

To design a half-fraction of the full four-factor design, we must determine which half of the $2^4 = 16$ experiments is to be done. To preserve the balance of the design, there must be four experiments at the high setting of X_4 and four experiments at the low setting. Note that any combination of four high and four low that we choose for factor 4 will correspond exactly to one of the column combinations for interactions among factors X_1, X_2, and X_3 already used in the matrix of the 2^3 factorial design (Table 28.2). Which combination should we select? Standard procedure is to choose the three-factor interaction $X_1X_2X_3$ for setting the levels of X_4. Having the levels of X_4 the same as the levels of $X_1X_2X_3$ means that the separate effects of X_4 and $X_1X_2X_3$ cannot be estimated. We can only estimate their combined effect. Their individual effects are confounded. *Confounded* means confused with, or tangled up with, in a way that we cannot separate without doing more experiments.

The design matrix for a 2^{4-1} design is shown in Table 28.3. The signs of the factor 4 column vector of levels are determined by the product of column vectors for the column 1, 2, and 3 factors. (Also, it is the same as the three-factor interaction column in the full 2^3 design.) For example, the signs for run 4 (row 4) are (+) (+) (−) (−), where the last (−) comes from the product (+) (+) (−) = (−).

The model matrix is given in Table 28.4. The eight experimental runs allow estimation of eight effects, which are computed as the product of a column vector X_i and the y vector just as was explained for the full factorial experiment discussed in Chapter 27. The other effects also are computed as for the full factorial experiment but they have a different interpretation, which will be explained now.

To evaluate four factors with only eight runs, we give up the ability to estimate independent main effects. Notice in the design matrix that column vector **1** is identical to the product of column vectors **2**, **3**, and **4**. The effect that is computed as $y \cdot X_1$ is not an independent estimate of the main effect of factor 1. It is the main effect of X_1 *plus* the three-way interaction of factors 2, 3, and 4. We say that the main effect of X_1 is confounded with the three-factor interaction of X_2, X_3, and X_4. Furthermore, each main effect is confounded with a three-factor interaction, as follows:

$$1 + 234 \qquad 2 + 134 \qquad 3 + 124 \qquad 4 + 123$$

The *defining relation* of the design allows us to determine all the confounding relationships in the fractional design. In this 2^{4-1} design, the defining relation is **I = 1234**. **I** indicates a vector of +1's.

TABLE 28.3

Design Matrix for a 2^{4-1} Fractional Factorial Design

Run	Factor (Independent Variable)			
	1	2	3	4
1	−	−	−	−
2	+	−	−	+
3	−	+	−	+
4	+	+	−	−
5	−	−	+	+
6	+	−	+	−
7	−	+	+	−
8	+	+	+	+

TABLE 28.4

Model Matrix for the 2^{4-1} Fractional Factorial Design

Run	Avg.	1 S	2 D	3 L	4 = 123 F (SDL)	12 = 34 SD (LF)	13 = 24 SL (DF)	23 = 14 DL (SF)
1	+	−	−	−	−	+	+	+
2	+	+	−	−	+	−	−	+
3	+	−	+	−	+	−	+	−
4	+	+	+	−	−	+	−	−
5	+	−	−	+	+	+	−	−
6	+	+	−	+	−	−	+	−
7	+	−	+	+	−	−	−	+
8	+	+	+	+	+	+	+	+

Note: Defining relation: **I** = **1234** (or **I** = SDLF).

Therefore, **I = 1234** means that multiplying the column vectors for factors 1, 2, 3, and 4, which consists of +1's and −1's, gives a vector that consists of +1 values. It also means that multiplying the column vectors of factors 2, 3, and 4 gives the column vector for factor 1. This means that the effect calculated using the column of +1 and −1 values for factor 1 is the same as the value that is calculated using the column vector of the $X_2 X_3 X_4$ interaction. Thus, the main effect of factor 1 is confounded with the three-factor interaction of factors 2, 3, and 4. Also, multiplying the column vectors of factors 1, 3, and 4 gives the column vector for factor 2, etc.

Having the main effects confounded with three-factor interactions is part of the price we pay we to investigate four factors in eight runs. Another price, which can be seen in the defining relation **I = 1234**, is that the two-factor interactions are confounded with each other:

$$12 + 34 \qquad 13 + 24 \qquad 23 + 14$$

The two-way interaction of factors 1 and 2 is confounded with the two-way interaction of factors 3 and 4, etc.

The consequence of this intentional confounding is that the estimated main effects are biased unless the three-factor interactions are negligible. Fortunately, three-way interactions are often small and can be ignored. There is no safe basis for ignoring any of the two-factor interactions, so the effects calculated as two-factor interactions must be interpreted with caution.

Understanding how confounding is identified by the defining relation reveals how the fractional design was created. Any fractional design will involve some confounding. The experimental designer wants to make this as painless as possible. The best we can do is to hope that the three-factor interactions are unimportant and arrange for the main effects to be confounded with three-factor interactions. Intentionally confounding factor 4 with the three-factor interaction of factors 1, 2, and 3 accomplishes that. By convention, we write the design matrix in the usual form for the first three factors. The fourth column becomes the product of the first three columns. Then we multiply pairs of columns to get the columns for the two-factor interactions, as shown in Table 28.4.

Case Study Solution

The average response at each experimental setting is shown in Figure 28.1. The small boxes identify the four tests that were conducted at the high flow rate (X_4); the low flow rate tests are the four unboxed values. Calculation of the effects was explained in Chapter 27 and are not repeated here. The estimated effects are given in Table 28.5.

This experiment has replication at each experimental condition so we can estimate the variance of the measurement error and of the estimated effects. The differences between duplicates (d_i) can be used to

Fractional Factorial Experimental Designs

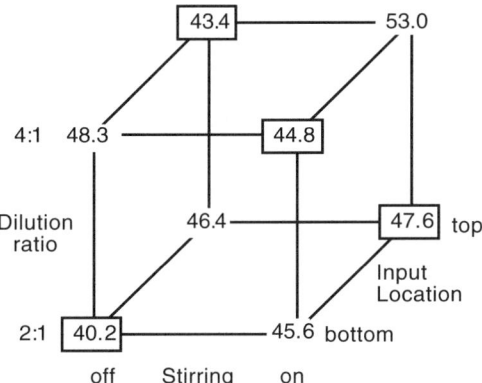

FIGURE 28.1 A 2^{4-1} fractional factorial design and the average of duplicated measurements at each of the eight design settings.

TABLE 28.5

Estimated Effects and Their Standard Errors

Effect	Contributing Factors and Interactions	Estimated Effect	Estimated Standard Error
Average + 1234	Average(I) + SDLF	46.2	0.41
1 + 234	S + DLF	3.2	0.82
2 + 134	D + SLF	2.4	0.82
3 + 124	L + SDF	2.9	0.82
4 + 123	F + SDL	4.3	0.82
12 + 34	SD + LF	−0.1	0.82
13 + 24	SL + DF	2.2	0.82
23 + 14	DL + SF	−1.2	0.82

estimate the variance of the average response for each run. For a single pair of duplicate observations (y_{1i} and y_{2i}), the sample variance is:

$$s_i^2 = \frac{1}{2}d_i^2$$

where $d_i = y_{1i} - y_{2i}$ is the difference between the two observations. The average of the duplicate observations is:

$$\bar{y}_i = \frac{y_{1i} - y_{2i}}{2}$$

and the variance of the average of the duplicates is:

$$s_{\bar{y}}^2 = \frac{s_i^2}{2} = \frac{d_i^2}{4}$$

The individual estimates for n pairs of duplicate observations can be combined to get a pooled estimate of the variance of the average:

$$s_{\bar{y}}^2 = \frac{1}{n}\sum_{i=1}^{n}\frac{d_i^2}{4} = \frac{1}{4n}\sum_{i=1}^{n}d_i^2$$

For this experiment, $n = 8$ gives:

$$s_{\bar{y}}^2 = \frac{1}{4(8)}((-2.6)^2 + 0.3^2 + \cdots + 1.1^2) = 1.332$$

and

$$s_{\bar{y}} = 1.15$$

The main and interaction effects are estimated using the model matrix given in Table 28.4. The average is:

$$\bar{y} = \frac{1}{8}\sum_{i=1}^{8} \bar{y}_i$$

and the estimate of each effect is:

$$\text{Effect}(j) = \frac{1}{4}\sum_{i=1}^{8} X_{ij}\bar{y}_i$$

where X_{ij} is the ith element of the vector in column j.

The variance of the average $\bar{y}$ is:

$$\text{Var}(\bar{y}) = (1/8)^2 8\sigma_{\bar{y}}^2 = \frac{\sigma_{\bar{y}}^2}{8}$$

and the variance of the main and interaction effects is:

$$\text{Var}(\text{Effect}) = (1/4)^2 8\sigma_{\bar{y}}^2 = \frac{\sigma_{\bar{y}}^2}{2}$$

Substituting $s_{\bar{y}}^2$ for $\sigma_{\bar{y}}^2$ gives the standard errors of the average and the estimated effects:

$$\text{SE}(\bar{y}) = \sqrt{\frac{s_{\bar{y}}^2}{8}} = \sqrt{\frac{1.332}{8}} = 0.41$$

and

$$\text{SE}(\text{Effect}_j) = \sqrt{\frac{s_{\bar{y}}^2}{2}} = \sqrt{\frac{1.332}{2}} = 0.82$$

The estimated effects and their standard errors are given in Table 28.5.

The 95% confidence interval for the true value of the effects is bounded by:

$$\text{Effect}_j \pm t_{\nu=8,\alpha/2=0.025}\text{SE}(\text{Effect}_j)$$

$$\text{Effect}_j \pm 2.306(0.82) = \text{Effect}_j \pm 1.9$$

from which we can state, with 95% confidence, that effects larger than 1.9, or smaller than −1.9, represent real effects. All four main effects and the two-factor interaction SL + DF are significant.

Alternately, the estimated effects can be viewed in relation to their relevant reference distribution shown in Figure 28.2. This distribution was constructed by scaling a t distribution with $\nu = 8$ degrees of freedom according to the estimated standard error of the effects, which means, in this case, using a scaling factor of 0.83. The calculations are shown in Table 28.6.

The main effects of all four variables are far out on the tails of the reference distribution, indicating that they are statistically significant. The bounds of the confidence interval (±1.9) could be plotted on this reference distribution, but this is not necessary because the results are clear. Stirring (S), on average,

Fractional Factorial Experimental Designs

TABLE 28.6

Constructing the Reference Distribution for Scale Factor = 0.82

t distribution ($v = 8$)		Scaled Reference Distribution[a]	
Value of t	t Ordinate	$t \times 0.82$	Ordinate/0.82
0	0.387	0.00	0.472
0.25	0.373	0.21	0.455
0.5	0.337	0.41	0.411
0.75	0.285	0.62	0.348
1.0	0.228	0.82	0.278
1.25	0.173	1.03	0.211
1.50	0.127	1.23	0.155
1.75	0.090	1.44	0.110
2.00	0.062	1.64	0.076
2.25	0.043	1.85	0.052
2.50	0.029	2.05	0.035
2.75	0.019	2.26	0.023
3.0	0.013	2.46	0.016

[a] Scaling both the abscissa and the ordinate makes the area under the reference distribution equal to 1.00.

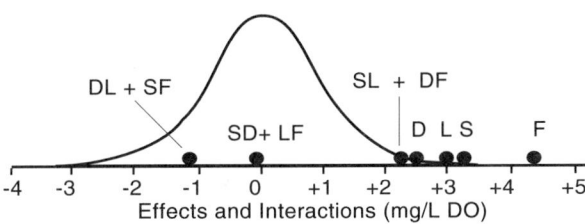

FIGURE 28.2 Reference distribution that would describe the effects and interactions if they were all random. Effects or interactions falling on the tails of the reference distribution are judged to be real.

elevates the response by 3.2 mg/L. Changing the dilution rate (D) from 2:1 to 4:1 causes an increase of 2.4 mg/L. Setting the sample input location (L) at the bottom yields a response 2.9 mg/L higher than a surface input location. And, increasing the sample flow rate (F) from 2.6 to 8.2 mL/sec causes an increase of about 4.3 mg/L.

Assuming that the three-factor interactions are negligible, the effects of the four main factors S, D, L, and F can be interpreted as being independent estimates (that is, free of confounding with any interactions). This assumption is reasonable because significant three-factor interactions rarely exist. By this we mean that it is likely that three interacting factors will have a tendency to offset each other and produce a combined effect that is comparable to experimental error. Thus, when the assumption of negligible three-factor interaction is valid, we achieve the main effects from eight runs instead of 16 runs in the full 2^4 factorial.

The two-factor interactions are confounded pairwise. The effect we have called DL is not the pure interaction of factors D and L. It is the interaction of D and L plus the interaction of S and F. This same problem exists for all three of the two-factor interaction effects. This is the price of running a 2^{4-1} fractional factorial experiment in eight runs instead of the full 2^4 design, which would estimate all effects without confounding.

Whenever a two-factor interaction appears significant, interpretation of the main effects must be reserved until the interactions have been examined. The reason for this is that a significant two-factor interaction means that the main effect of one interacting factor varies as a function of the level of the other factor. For example, factor X_1 might have a large positive effect at a low value of X_2, but a large negative effect at a high value of X_2. The estimated main effect of X_1 could be near zero (because of

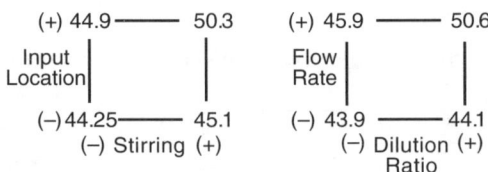

FIGURE 28.3 Two-factor interactions of S and L, and of F and D.

averaging the positive and negative effects) and looking at it alone might lead us to wrongly conclude that the factor is inactive.

In this case study, the SL + DF interaction effect appears large enough to be real. Interpreting the SL interaction is done by viewing the experiment as one conducted in the interacting factors only. This is shown in Figure 28.3, where the responses averaged over all runs with the same signs for L and S. The same examination must be made of the DF interaction, which is also shown in Figure 28.3.

The sample flow rate is not very important at the low dilution ratio and dilution ratio is not very important at the low sample flow rate. But when the sample flow rate is high and the dilution ratio is large, the response increases dramatically. A similar interpretation can be made for the SL interaction. Stirring in conjunction with injecting the sample at the bottom gives higher oxygen measurements while the other three combinations of L and S show much the same DO levels.

In short, the significant two-factor interaction means that the effect of any factor depends on the level of some other factor. It may be, however, that only one of the interactions SL or DF is real and the other is an artifact caused by the confounding of the two interactions. Additional experiments could be done to untangle this indefinite situation. One option would be to run the other half of the 2^{4-1} design.

For the case study problem, a resolution of the two-factor interaction is not needed because all four factors S, D, L, and F do influence the oxygen measurements and it is clear that all four should be set at their + levels. That is, the best of the measurement techniques investigated would be to inject the sample at the bottom of the bottle, stir, use the 4:1 dilution ratio, and use the high sample flow rate (8.2 mL/sec). Using even higher dilution ratios, faster stirring, and a higher sample flow rate might yield even better results.

Comments

This example has shown how four variables can be evaluated with just eight runs. The eight runs were a half-fraction of the sixteen runs that would be used to evaluate four factors in a full factorial design. This design provided estimates of the main effects of the four factors just as would have been obtained from the full 2^4 design, if we are willing to make the rather mild assumption that three-factor interactions are negligible.

There was a price paid for forcing the extra factor into the eight-run design. Only a total of eight effects and interactions can be estimated from the eight runs. These include the average and the four main effects. The estimated two-factor interactions are confounded pairwise with each other and their interpretation is not as clear as it would have been from the full factorial design.

Often we are interested primarily in the main effects, at least in the early iterations of the learning cycle. If the main effects are significant and some of the two-factor interactions hold interest, additional runs could be performed to untangle the interpretation. The most logical follow-up experiment would be to run the other half of the 2^{4-1} design. Table 28.7 shows how the full 2^4 design was divided into two half-fractions, one of which was used in the case study experiment, and the other which could be used as a follow-up. The two half-fractions combined are a saturated 2^4 design from which all main effects and all interactions can be estimated independently (i.e., without confounding). A further advantage is that the two half-fractions are *blocks* and any nuisance factors that enter the experiment between the execution of the half-fractions will not distort the final results. (This is not obvious. For details, see Box et al., 1978.)

Fractional Factorial Experimental Designs

TABLE 28.7

Two Half-Fractions of the Full 2^4 Factorial Design

	Factor			
	1	2	3	4
I - Run 1	−	−	−	−
II	+	−	−	−
II	−	+	−	−
I - Run 4	+	+	−	−
II	−	−	+	−
I - Run 6	+	−	+	−
I - Run 7	−	+	+	−
II	+	+	+	−
II	−	−	−	+
I - Run 2	+	−	−	+
I - Run 3	−	+	−	+
II	+	+	−	+
I - Run 5	−	−	+	+
II	+	−	+	+
II	−	+	+	+
I - Run 8	+	+	+	+

Note: The eight settings opposite I - Run 1, I - Run 2, etc. were used in the 2^{4-1} fractional design of the case study. The other eight combinations, marked II, are a second 2^{4-1} fractional design that could be run as a follow-up.

Fractional factorial designs are often used in an iterative experimental strategy. Chapter 29 illustrates a 2^{5-1} design for evaluating five factors in 16 runs. Box et al. (1978) give two examples of 2^{7-4} designs for evaluating eight variables in $2^3 = 8$ runs. They also show other fractional factorial designs, their confounding pattern, and give a detailed explanation of how the confounding pattern is discovered.

References

Box, G. E. P. and J. S. Hunter (1961a). "The 2^{k-p} Fractional Factorial Designs. Part I," *Technometrics*, 3(3), 311–351.

Box, G. E. P. and J. S. Hunter (1961b). "The 2^{k-p} Fractional Factorial Designs. Part II," *Technometrics*, 3(4), 449–458.

Box, G. E. P., W. G. Hunter, and J. S. Hunter (1978). *Statistics for Experimenters: An Introduction to Design, Data Analysis, and Model Building,* New York, Wiley Interscience.

Tiao, George et al., Eds. (2000). *Box on Quality and Discovery with Design, Control, and Robustness,* New York, John Wiley & Sons.

Exercises

28.1 Membrane Bioreactor. The ZeeWeed internal-membrane reactor was tested using a 2^{4-1} fractional factorial design to evaluate four factors in eight runs, as shown in the table below. The response is the permeate flow rate. Evaluate the main effects and interactions.

Backpulse Duration (min)	Backpulse Frequency (min)	Membrane Airflow (m³/hr)	MLSS (mg/L)	Permeate Flow Rate (L/min)
−1	−1	−1	−1	83
+1	−1	−1	+1	72
−1	+1	−1	+1	64
+1	+1	−1	−1	72

−1	−1	+1	+1	64
+1	−1	+1	−1	87
−1	+1	+1	−1	81
+1	+1	+1	+1	68

Note: Defining relation: **I = 1234**.

Source: Cantor, J. et al. (1999). *Ind. Wastewater,* Nov./Dec., pp. 18–22.

28.2 Corrosion. A study of the effect of construction metal corrosion under simulated acid fog/rain conditions used factorial experimental designs with pH, salinity, temperature, and exposure time as variables. The 2^{4-1} experiment was performed on galvanized steel and the response was corrosion rate. Each run was replicated $n = 17$ times. Estimate the main effects of the four factors and the two-factor interactions. Note that the design identity is **I = 1234**, so the two-factor interactions are confounded with other two-factor interactions.

Run	pH	Salinity (%NaCl)	Temp. (°C)	Time (hr)	Avg. Corrosion Rate (μm/yr)	St. Dev. (μm/yr)
1	2.5	1.0	35	48	501	65
2	3.5	1.0	35	120	330	60
3	2.5	6.0	35	120	561	67
4	3.5	6.0	35	48	666	95
5	2.5	1.0	45	120	218	85
6	3.5	1.0	45	48	247	57
7	2.5	6.0	45	48	710	102
8	3.5	6.0	45	120	438	51

Source: Fang, H. H. P. et al. (1990). *Water, Air, and Soil Poll.,* 53, 315–325.

28.3 Fly Ash Mixture. The table below describes a 2^{5-1} experiment in 16 runs to investigate five factors: (1) type of fly ash, (2) percentage fly ash in the mixture, (3) test specimens put through a wet/dry curing cycle, (4) test specimens put through a freeze/thaw curing cycle, and (5) percentage of bentonite in the mixture. High permeability is desirable. What combination of factors promotes this? Quantify the effect of the five factors. Evaluate and explain any interactions between factors.

Run	Type of Fly Ash	Fly Ash (%)	Wet/Dry Cycle	Freeze/Thaw Cycle	Bentonite (%)	Permeability
1	A	50	N	N	10	1000
2	B	50	N	N	0	160
3	A	100	N	N	0	1450
4	B	100	N	N	10	77
5	A	50	Y	N	0	1400
6	B	50	Y	N	10	550
7	A	100	Y	N	10	320
8	B	100	Y	N	0	22
9	A	50	N	Y	0	1400
10	B	50	N	Y	10	390
11	A	100	N	Y	10	580
12	B	100	N	Y	0	8
13	A	50	Y	Y	10	2800
14	B	50	Y	Y	0	160
15	A	100	Y	Y	0	710
16	B	100	Y	Y	10	19

28.4 Metal Inhibition. Divide the 2^4 factorial experiment in Exercise 27.5 into two half-fractions. Calculate the main effects and interactions of each half-fraction and compare them.

28.5 Water Pipe Corrosion. Environmental engineering students at Tufts University performed a 2^{4-1} fractional factorial experiment to investigate the concentration of iron in drinking water in various campus buildings as a means of inferring water pipe corrosion. The four experimental factors were age of building, building type, location, and day of week. Estimate the main and interaction effects and their 95% confidence intervals. The design identity is $\mathbf{I = 1234}$, and the results are:

Age	Type	Location	Day	Iron (mg/L)
Old	Academic	Medford	Wednesday	0.26, 0.21
New	Academic	Medford	Monday	0.37, 0.32
Old	Residential	Medford	Monday	0.01, 0.05
New	Residential	Medford	Wednesday	0.03, 0.07
Old	Academic	Somerville	Monday	0.11, 0.05
New	Academic	Somerville	Wednesday	0.06, 0.03
Old	Residential	Somerville	Wednesday	0.03, 0.05
New	Residential	Somerville	Monday	0.07, 0.02

29

Screening of Important Variables

KEY WORDS *confounding, constant variance, defining relation, fly ash, interactions, main effect, normal plot, permeability, factorial design, fractional factorial design, log transformation, replication, resolution, variance.*

Often, several independent variables are potentially important in determining the performance of a process, or the properties of a material. The goal is to efficiently screen these variables to discover which, if any, alter performance. Fractional factorial experiments are efficient for this purpose. The designs and case study presented here are an extension of the factorial experiment designs discussed in Chapters 27 and 28.

Case Study: Using Fly Ash to Make an Impermeable Barrier

Fly ash from certain kinds of coal is pozzolanic, which means that it will set into a rock-like material when mixed with proper amounts of water. Preliminary tests showed that pozzolanic fly ash can be mixed with sand or soil to form a material that has a permeability of 10^{-7} cm/sec or lower. Such mixtures can be used to line storage lagoons and landfills if the permeability is not reduced by being frozen and thawed, or wetted and dried. The effect of these conditions on the permeability of fly ash and sand/fly ash mixtures needed to be tested. Two types of fly ash were being considered for this use. The addition of a clay (bentonite) that has been frequently used to build impermeable barriers was also tested.

A two-level experiment was planned to evaluate the five factors at the levels listed below. The goal was to formulate a durable mixture that had a low permeability.

1. Type of fly ash: A or B
2. Percentage of fly ash in the mixture: 100% ash, or 50% ash and 50% sand
3. Bentonite addition (percent of total mixture weight): none or 10%
4. Wet/dry cycle: yes or no
5. Freeze/thaw cycle: yes or no

A full two-level, five-factor experiment would require testing $2^5 = 32$ different conditions. Each permeability test would take one to three weeks to complete, so doing 32 runs was not attractive. Reducing the number of variables to be investigated was not acceptable. How could five factors be investigated without doing 32 runs? A fractional factorial design provided the solution. Table 29.1 shows the experimental settings for a 16-run 2^{5-1} fractional factorial experimental design.

Method: Designs for Screening Important Variables

A full factorial experiment using two levels to investigate k factors requires 2^k experimental runs. The data produced are sufficient to independently estimate 2^k parameters, in this case the average (k main effects) and $2^k - k - 1$ interactions. The number of main effects and interactions for a few full designs are tabulated in Table 29.2.

TABLE 29.1

2^{5-1} Fractional Factorial Design and the Measured Permeability of the Fly Ash Mixtures

Run No.	Type of Fly Ash (X_1)	% Fly Ash (X_2)	Wet/Dry Cycle (X_3)	Freeze/Thaw Cycle (X_4)	Bentonite Addition (X_5)	Permeability (cm/sec $\times 10^{10}$) (y)
1	A	50	No	No	10%	1025
2	B	50	No	No	None	190
3	A	100	No	No	None	1490
4	B	100	No	No	10%	105
5	A	50	Yes	No	None	1430
6	B	50	Yes	No	10%	580
7	A	100	Yes	No	10%	350
8	B	100	Yes	No	None	55
9	A	50	No	Yes	None	1420
10	B	50	No	Yes	10%	410
11	A	100	No	Yes	10%	610
12	B	100	No	Yes	None	40
13	A	50	Yes	Yes	10%	2830
14	B	50	Yes	Yes	None	1195
15	A	100	Yes	Yes	None	740
16	B	100	Yes	Yes	10%	45

TABLE 29.2

Number of Main Effects and Interactions That Can Be Estimated from a Full 2^k Factorial Design

				Interactions					
k	Number of Runs	Avg.	Main Effects	2-Factor	3-Factor	4-Factor	5-Factor	6-Factor	7-Factor
3	8	1	3	3	1				
4	16	1	4	6	4	1			
5	32	1	5	10	10	5	1		
7	128	1	7	21	35	35	21	7	1

There are good reasons other than the amount of work involved to not run large designs like 2^5, 2^6, etc. First, three-factor and higher-level interactions are almost never significant so we have no interest in getting data just to estimate them. Second, most two-factor interactions are not significant either, and some of the main effects will not be significant. Fractional factorial designs provide an efficient strategy to reduce the work when relatively few effects are realistically expected to be important.

Suppose that out of 32 main effects and interactions that could be estimated, we expect that 5 or 6 might be important. If we intelligently select the right subset of experimental runs, we can estimate these few effects. The problem, then, is how to select the subset of experiments.

We could do half the full design, which gives what is called a half-fraction. If there are 5 factors, the full design requires 32 runs, but the half-fraction requires only 16 runs. Sixteen effects can be estimated with this design. Halving the design again would give $2^{5-2} = 8$ runs.

Doing a half-fraction design means giving up independent estimates of the higher-order interactions. At some level of fractioning, we also give up the independent estimates of the two-factor interactions. (See Chapter 28 for more details.) If our primary interest is in knowing the main effects, this price is more than acceptable. It is a terrific bargain. A screening experiment is designed to identify the most important variables so we are satisfied to know only the main effects. If we later want to learn about the interactions, we could run the missing half-fraction of the full design.

We now show how this works for the 2^5 design of the case study problem. The full design is shown in the left-hand part of Table 29.3. All 32 combinations of the five factors set at two levels are included. The right-hand section of Table 29.3 shows one of the two equivalent half-fraction 2^{5-1} designs that can be selected from the full design. The runs selected from the full design are marked with asterisks (∗) in the left-most

TABLE 29.3

Comparison of the Full 2^5 Factorial Design and the Fractional 2^{5-1} Design

	Full 2^5 Factorial Design						Fractional 2^{5-1} Design				
	Factor						Factor				
Run	1	2	3	4	5	Run	1	2	3	4	5
1	−1	−1	−1	−1	−1	17	−1	−1	−1	−1	1
*2	1	−1	−1	−1	−1	2	1	−1	−1	−1	−1
*3	−1	1	−1	−1	−1	3	−1	1	−1	−1	−1
4	1	1	−1	−1	−1	20	1	1	−1	−1	1
*5	−1	−1	1	−1	−1	5	−1	−1	1	−1	−1
6	1	−1	1	−1	−1	22	1	−1	1	−1	1
7	−1	1	1	−1	−1	23	−1	1	1	−1	1
*8	1	1	1	−1	−1	8	1	1	1	−1	−1
*9	−1	−1	−1	1	−1	9	−1	−1	−1	1	−1
10	1	−1	−1	1	−1	26	1	−1	−1	1	1
11	−1	1	−1	1	−1	27	−1	1	−1	1	1
*12	1	1	−1	1	−1	12	1	1	−1	1	−1
13	−1	−1	1	1	−1	29	−1	−1	1	1	1
*14	1	−1	1	1	−1	14	1	−1	1	1	−1
*15	−1	1	1	1	−1	15	−1	1	1	1	−1
16	1	1	1	1	−1	32	1	1	1	1	1
*17	−1	−1	−1	−1	1						
18	1	−1	−1	−1	1						
19	−1	1	−1	−1	1						
*20	1	1	−1	−1	1						
21	−1	−1	1	−1	1						
*22	1	−1	1	−1	1						
*23	−1	1	1	−1	1						
24	1	1	1	−1	1						
25	−1	−1	−1	1	1						
*26	1	−1	−1	1	1						
*27	−1	1	−1	1	1						
28	1	1	−1	1	1						
*29	−1	−1	1	1	1						
30	1	−1	1	1	1						
31	−1	1	1	1	1						
*32	1	1	1	1	1						

Note: The runs selected from the full design are marked with an asterisk (*), and are identified by run number in the fractional design.

column and they are identified by run number in the fractional design. (An equivalent fractional design consists of the 16 runs that are not marked with asterisks.) Notice that in each column of the full design, half the values are +1 and half are −1. This balance must be preserved when we select the half-fraction.

The 16 runs of the fractional design constructed as follows. A full 2^4 design was written for the four variables **1, 2, 3,** and **4**. The column of signs for the **1234** interaction was written and these were used to define the levels of variable **5**. Thus, we made **5 = 1234**.

The consequence of making **5 = 1234** is that the quantity we calculate as the main effect of factor 5 is the main effect plus the four-factor interaction of factors **1, 2, 3,** and **4**. The main effect of **5** is *confounded* with the **1234** interaction. Or we can say that **5** and **1234** are *aliases* of each other. There are other interactions that are confounded and we need to understand what they are.

The confounding pattern between the columns in the model matrix is determined by the *defining relation* **I = 12345**. That means that multiplying columns **1234** gives column **5**; that is, **5 = 1234**. It is also true that **1 = 2345, 2 = 1345, 3 = 1245,** and **4 = 1235**. Each of the main effects is confounded with a four-factor interaction. These high order effects are expected to be negligible, in which case the design produces independent estimates of the main effects.

The two-factor effects are also confounded in the design. Multiplying columns **123** gives a column identical to **45**. As a consequence, the quantity that estimates the **45** interaction also includes the three-factor interaction of **123** (the **123** and **45** interactions are confounded with each other). Also, **12 = 345, 13 = 245, 14 = 235, 15 = 234, 23 = 145, 24 = 135, 25 = 134, 34 = 125**, and **35 = 124**. Each two-factor interaction is confounded with a three-factor interaction. If a two-factor interaction appeared to be significant, its interpretation would have to take into account the possibility that the effect may be due in part to the third-order interaction. Fortunately, the third-order interactions are usually small and can be neglected.

Using the 2^{5-1} design instead of the full 2^5 saves us 16 runs, but at the cost of having the main effects confounded with four-factor interactions and having the two-factor interactions confounded with three-factor interactions. If the objective is mainly to learn about the main effects, and if the four-factor interactions are small, the design is highly efficient for identifying the most important factors. Furthermore, because each estimated main effect is the average of eight virtually independent comparisons, the precision of the estimates can be excellent.

Case Study Solution

The measured permeabilities are plotted in Figure 29.1. Because the permeability vary over several orders of magnitude, the data are best displayed on a logarithmic plot. Fly ash A (solid circles) clearly has higher permeabilities than fly ash B (open circles).

The main effects of each variable were of primary interest. Two-factor interactions were of minor interest. Three-factor and higher-order interactions were expected to be negligible. A half-fraction, or 2^{5-1} fractional factorial design, consisting of 16 runs was used. There are 16 data points, so it is possible to estimate 16 parameters. The "parameters" in this case are the mean, five main effects, and 10 two-factor interactions. Table 29.4 gives the model matrix in terms of the coded variables. The products X_1X_2, X_1X_3, etc. indicate two-factor interactions. Also listed are the permeability (y) and $\ln(y)$.

The computation of the average and the effects was explained in Chapters 27 and 28. Duplicate tests that are not reported here indicated that the variance tended to be proportional to the permeability. Because of this, a log transformation was used to stabilize the variance. The average permeability is 5.92 on the log scale. The estimated main effects (confounded with four-factor interactions) and two-factor interactions (confounded with three-factor interactions) are given in Table 29.5. These are also on the log scale.

In the absence of an estimate of σ^2, the significance of these effects was judged by making a normal plot, as shown in Figure 29.2. If the effects arise only from random measurement error, they will fall along

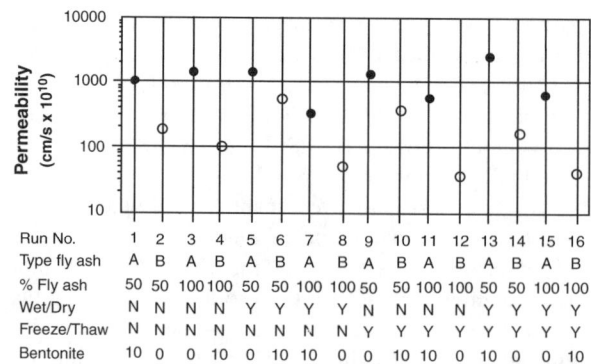

FIGURE 29.1 Logarithmic plot of the permeability data from the 2^{5-1} factorial experiment. (Solid circles are fly ash A; open circles are fly ash B.)

TABLE 29.4
Model Matrix for the 2^{5-1} Fractional Factorial Experiment on Permeability of Fly Ash Mixtures with the Measured and Log-Transformed Permeabilities

X_0 Avg.	X_1 1	X_2 2	X_3 3	X_4 4	X_5 5	X_1X_2 12	X_1X_3 13	X_1X_4 14	X_1X_5 15	X_2X_3 23	X_2X_4 24	X_2X_5 25	X_3X_4 34	X_3X_5 35	X_4X_5 45	y	$\ln(y)$
1	-1	-1	-1	-1	1	1	1	1	-1	1	1	-1	1	-1	-1	1025	6.932
1	1	-1	-1	-1	-1	-1	-1	-1	-1	1	1	1	1	1	1	190	5.247
1	-1	1	-1	-1	-1	-1	1	1	1	-1	-1	-1	1	1	1	1490	7.307
1	1	1	-1	-1	1	1	-1	-1	1	-1	-1	1	1	-1	-1	105	4.654
1	-1	-1	1	-1	-1	1	-1	1	1	-1	1	1	-1	-1	1	1430	7.265
1	1	-1	1	-1	1	-1	1	-1	1	-1	1	-1	-1	1	-1	580	6.363
1	-1	1	1	-1	1	-1	-1	1	-1	1	-1	1	-1	1	-1	350	5.858
1	1	1	1	-1	-1	1	1	-1	-1	1	-1	-1	-1	-1	1	55	4.007
1	-1	-1	-1	1	-1	1	1	-1	1	1	-1	1	-1	1	-1	1420	7.258
1	1	-1	-1	1	1	-1	-1	1	1	1	-1	-1	-1	-1	1	410	6.016
1	-1	1	-1	1	1	-1	1	-1	-1	-1	1	1	-1	-1	1	610	6.413
1	1	1	-1	1	-1	1	-1	1	-1	-1	1	-1	-1	1	-1	40	3.689
1	-1	-1	1	1	1	1	-1	-1	-1	-1	-1	-1	1	1	1	2830	7.948
1	1	-1	1	1	-1	-1	1	1	-1	-1	-1	1	1	-1	-1	195	5.273
1	-1	1	1	1	-1	-1	-1	-1	1	1	1	-1	1	-1	-1	740	6.607
1	1	1	1	1	1	1	1	1	1	1	1	1	1	1	1	45	3.807

Note: The defining relation is **I = 12345**.

TABLE 29.5

Main Effects and Two-Factor Interactions for the Fly Ash Permeability Study

Effect of Factor	Confounded Factors	Estimated Effect (natural log scale)
Average		5.92
Main effects (+four-factor interactions)		
Factor 1	1 + 2345	−2.07
Factor 2	2 + 1345	−1.24
Factor 3	3 + 1245	−0.05
Factor 4	4 + 1235	−0.08
Factor 5	5 + 1234	0.17
Two-factor interactions (+three-factor interactions)		
Factors 1 and 2	12 + 345	−0.44
Factors 1 and 3	13 + 245	0.01
Factors 1 and 4	14 + 235	−0.29
Factors 1 and 5	15 + 234	0.49
Factors 2 and 3	23 + 145	−0.40
Factors 2 and 4	24 + 135	−0.25
Factors 2 and 5	25 + 134	−0.39
Factors 3 and 4	34 + 125	0.11
Factors 3 and 5	35 + 124	0.04
Factors 4 and 5	45 + 123	0.17

Note: Data were transformed by taking natural logarithms.

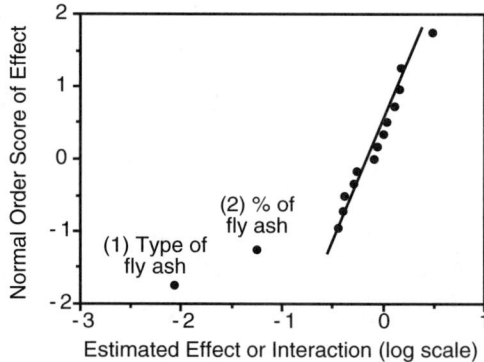

FIGURE 29.2 Normal plot of the effects computed from the log-transformed permeability values.

a straight line. The significant (i.e., nonrandom) effects fall off the line. These are large enough to be considered a real result of changing the factor, and not the result of random measurement error.

Two factors are significant: factor 1 = type of fly ash and factor 2 = percentage of fly ash in the mixture. The permeability of fly ash A is higher than fly ash B, as can be clearly seen in Figure 29.1. Using 100% fly ash gives a lower permeability than the 50% mixture. Freeze/thaw cycle, wet/dry cycle, and the additional of bentonite did not affect the permeability of the compacted specimens. None of the two-factor interactions are significant.

The effects were calculated on log-transformed permeability values. The reason given was that unreported replicate tests indicated nonconstant variance. Variance was never calculated in the case study example, so you may wonder why this matters. If the analysis is done on the original permeability values, the normal plot cannot be interpreted as we have done. The effects that should appear random, and thus fall on a straight line, define a curve and one cannot easily identify the nonrandom effects.

Comments

Fractional factorial experiments offer an efficient way of evaluating a large number of variables with a reasonable number of experimental runs. In this example we have evaluated the importance of five variables in 16 runs. This was a half-fraction of a full $2^5 = 32$-run factorial experiment. Recalling the adage that there is "no free lunch," we wonder what was given up in order to study five variables with just 16 runs. In the case study, the main effect of each factor was confounded with a four-factor interaction, and each two-factor interaction was confounded with a three-factor interaction. If the higher-order interactions are small, which is expected, the design produces excellent estimates of the main effects and identifies the most important factors. In other words, we did have a free lunch.

In the case study, all interactions and three main effects were insignificant. This means that the experiment can be interpreted by collapsing the design onto the two significant factors. Figure 29.3 shows permeability in terms of factors 1 and 2, which now appear to have been replicated four times. You can confirm that the main effects calculated from this view of the experiment are exactly as obtained from the previous analysis.

This gain in apparent replication is common in screening experiments. It is one reason they are so efficient, despite the confounding that the inexperienced designer fears will weaken the experiment. To appreciate this, suppose that three factors had been significant in the case study. Now the collapsed design is equivalent to a 2^3 design that is replicated twice at each condition. Or suppose that we had been even more ambitious with the fractional design and had investigated the five factors in just eight runs with a 2^{5-2} design. If only two factors proved to be significant, the collapsed design would be equivalent to a 2^2 experiment replicated twice at each condition.

Finding an insignificant factor in a fractional factorial experiment always has the effect of creating apparent replication. Screening experiments are designed with the expectation that some factors will be inactive. Therefore, confounding usually produces a bonus instead of a penalty. This is not the case in an experiment where all factors are known to be important, that is, in an experiment where the objective is to model the effect of changing prescreened variables.

Table 29.6 summarizes some other fractional factorial designs. It shows five designs that use only eight runs. In eight runs we can evaluate three or four factors and get independent estimates of the main effects. If we try to evaluate five, six, or seven factors in just eight runs, the main effects will be confounded with second-order interactions. This can often be an efficient design for a screening experiment. The table also shows seven designs that use sixteen runs. These can handle four or five factors without confounding the main effects. Lack of confounding with three-factor interactions (or higher) is indicated by "OK" in the last two columns of the table, while "Confounded" indicates that the mentioned effect is confounded with at least one second-order interaction.

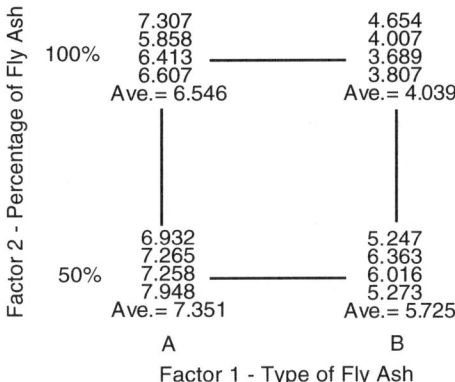

FIGURE 29.3 The experimental results shown in terms of the two significant main effects. Because three main effects are not significant, the fractional design is equivalent to a 2^2 design replicated four times.

TABLE 29.6

Some Fractional Factorial Designs

No. of Factors	Design	No. of Runs	Resolution	Main Effects	Second-Order Interactions
3	2^3	8		OK	OK
4	2^{4-1}	8	IV	OK	Confounded
5	2^{5-2}	8	III	Confounded	Confounded
6	2^{6-3}	8	III	Confounded	Confounded
7	2^{7-4}	8	III	Confounded	Confounded
4	2^4	16		OK	OK
5	2^{5-1}	16	V	OK	OK
6	2^{6-2}	16	IV	OK	Confounded
7	2^{7-3}	16	IV	OK	Confounded
8	2^{8-4}	16	IV	OK	Confounded
9	2^{9-5}	16	III	Confounded	Confounded
10	2^{10-6}	16	III	Confounded	Confounded
9	2^{9-4}	32	IV	OK	Confounded
10	2^{10-5}	32	IV	OK	Confounded
11	2^{11-6}	32	IV	OK	Confounded
11	2^{11-7}	16	III	Confounded	Confounded

Note: Resolution V: All main effects and two-factor interactions can be estimated. No two-factor interactions are confounded with other two-factor interactions or main effects. No main effects are confounded with each other.

Resolution IV: No main effects are confounded with two-factor interactions or other main effects. Two-factor interactions may be confounded with other two-factor interactions.

Resolution III: No main effects are confounded with other main effects. Main effects may be confounded with two-factor interactions.

The *resolution* of the design, indicated by roman numerals, describes the extent of confounding among the main effects and the two-factor interactions. The higher the resolution, the better the main effects and two-factor interactions can be estimated. The definitions are:

Design Resolution	Main Effects are Confounded with:	Two-Factor Interactions are Confounded with:
III	Two-factor interactions	Main effects
IV	Three-factor interactions	Other two-factor interactions
V	Four-factor interactions	Three-factor interactions
VI	Five-factor interactions	Four-factor interactions

The 2^{5-1} design used in the fly ash case study is a resolution V design. The 2^{4-1} design used in Chapter 28 is a resolution IV design.

This chapter presented only the main ideas about confounding in fractional experimental designs. A great deal more is worth knowing and Box et al. (1978) is worth careful study. "A Practical Aid for Experimenters" (Bisgaard, 1987) summarizes more than 40 factorial and fractional designs and gives the effects that can be estimated and their confounding arrangement.

References

Bisgaard, S. (1987). *A Practical Aid for Experimenters,* Madison, WI, Center for Quality and Productivity Improvement, University of Wisconsin–Madison.

Box, G. E. P. and J. S. Hunter (1961). "The 2^{k-p} Fractional Factorial Designs. Part I," *Technometrics,* 3(3), 311–351.

Box, G. E. P. and J. S. Hunter (1961). "The 2^{k-p} Fractional Factorial Designs. Part II," *Technometrics,* 3(4), 449–458.

Box, G. E. P., W. G. Hunter, and J. S. Hunter (1978). *Statistics for Experimenters: An Introduction to Design, Data Analysis, and Model Building,* New York, Wiley Interscience.

Draper, N. R. and H. Smith, (1998). *Applied Regression Analysis,* 3rd ed., New York, John Wiley.

Exercises

29.1 Fly Ash Case Study. Compute the main and two-factor effects for the fly ash permeability data without making the log transformation. Make the normal plot and discuss the difficulty of identifying significant and random effects.

29.2 Adsorption of Cu. The removal by adsorption of copper from desulfurization scrubber wastewater by treatment with fly ash and lime was studied using the 2^{7-4} fractional factorial design given below. The factors are pH, metal/adsorbent ratio (M/A), fly ash/lime ratio (FA/L), fly ash washed or unwashed (W/U), stirring speed (SS), fly ash origin (FAO), and salinity (S). There are three responses: adsorption rate (mg/L-hr), equilibrium concentration (mg/L), and equilibrium capacity (mg/g). (a) Calculate the average and the main effects for each response. (b) Write the model matrix and calculate the interaction effects. To do this, you must first discover the confounding pattern. The first three factors are in standard order.

	Factors							Responses		
Run	pH 1	M/A 2	FA/L 3	W/U 4	SS 5	FAO 6	S 7	Rate (mg/L-hr)	Eq. Conc. (mg/L)	Eq. Cap. (mg/g)
1	3	0.005	1	U	400	S	5	−2.8	458.1	0.6
2	5	0.005	1	W	200	L	5	−13.9	38.3	4.9
3	3	0.01	1	U	200	L	0	−5.8	361.8	3.2
4	5	0.01	1	W	400	S	0	−15.4	148.1	7.4
5	3	0.005	4	W	200	S	0	−2.7	468.3	0.7
6	5	0.005	4	U	400	L	0	−16.1	119.1	4.1
7	3	0.01	4	W	400	L	5	−5.1	403.4	2.3
8	5	0.01	4	U	200	S	5	−16.7	143.9	7.8

Source: Ricou et al. (1999). *Water Sci. & Tech.,* 39, 239–247.

29.3 Ruggedness Testing. The 2^{7-4} design below was given in Chapter 9 for ruggedness testing a laboratory method. The seven factors are studied in eight runs. This is a one-sixteenth fraction.

(a) How many different one-sixteenth designs can be constructed? Is there any reason to prefer this design to the other possibilities? (b) Show that column 7 is the product of columns 1, 2, and 3. (c) Columns 4, 5, and 6 are each products of two columns. Which two?

	Factor							Observed
Run	1	2	3	4	5	6	7	Response
1	−	−	−	+	+	+	−	y_1
2	+	−	−	−	−	+	+	y_2
3	−	+	−	−	+	−	+	y_3
4	+	+	−	+	−	−	−	y_4
5	−	−	+	+	−	−	+	y_5
6	+	−	+	−	+	−	−	y_6
7	−	+	+	−	−	+	−	y_7
8	+	+	+	+	+	+	+	y_8

29.4 Fly Ash. Verify that calculating the main effects of factors 1 and 2 in the case study from the collapsed (apparently replicated) design in Figure 29.3 gives the same results reported in Table 29.5.

29.5 Oxygen Transfer. Reaeration experiments in a laboratory flume were studied using a screening 2^{7-4} design and its "fold-over." A fold-over design simply does a second set of eight experiments with all the signs switched (+ to −, and − to +). Main effects are estimated free of two-factor interactions, and two-factor interactions are confounded in groups of three. The factors studied were depth (3 and 6 in.), velocity (1 and 1.5 fps), roughness (smooth and rough), temperature (20 and 25°C), surface area (4.6 and 9.2 ft^2), straightening grid (not used, used), and division vane (area, flow). The response is oxygen transfer coefficient, k_2, (day^{-1}). The experimental design and results are given below. Evaluate the main and interaction effects of these factors. Which are most important in affecting the oxygen transfer coefficient?

Run	Depth	Vel.	Roughness	Temp.	Area	Grid	Vane	k_2
1	−	−	−	−	+	+	+	15.0
2	+	−	−	+	+	−	−	10.2
3	−	+	−	+	−	−	+	28.9
4	+	+	−	−	−	+	−	12.1
5	−	−	+	+	−	+	−	24.4
6	+	−	+	−	−	−	+	9.6
7	−	+	+	−	+	−	−	28.9
8	+	+	+	+	+	+	+	11.4
9	+	+	+	+	−	−	−	18.0
10	−	+	+	−	−	+	+	34.1
11	+	−	+	−	+	+	−	8.3
12	−	−	+	+	+	−	+	21.6
13	+	+	−	−	+	−	+	13.4
14	−	+	−	+	+	+	−	23.7
15	+	−	−	+	−	+	+	10.1
16	−	−	−	−	−	−	−	18.1

Note: The design identity for the 2^{7-4} design is **I = 1234 = 235 = 126 = 137**.
The design identity for the fold-over design is **I = 1234 = −235 = −126 = −137**.

30

Analyzing Factorial Experiments by Regression

KEY WORDS augmented design, center points, confidence interval, coded variables, cube plots, design matrix, effects, factorial design, interaction, intercept, least squares, linear model, log transformation, main effect, matrix, matrix of independent variables, inverse, nitrate, PMA, preservative, quadratic model, regression, regression coefficients, replication, standard deviation, standard error, transformation, transpose, star points, variance, variance-covariance matrix, vector.

Many persons who are not acquainted with factorial experimental designs know linear regression. They may wonder about using regression to analyze factorial or fractional factorial experiments. It is possible and sometimes it is necessary.

If the experiment is a balanced two-level factorial, we have a free choice between calculating the effects as shown in the preceding chapters and using regression. Calculating effects is intuitive and easy. Regression is also easy when the data come from a balanced factorial design. The calculations, if done using matrix algebra, are almost identical to the calculation of effects. The similarity and difference will be explained.

Common experimental problems, such as missing data and failure to precisely set the levels of independent variables, will cause a factorial design to be unbalanced or messy (Milliken and Johnson, 1992). In these situations, the simple algorithm for calculating the effects is not exactly correct and regression analysis is advised.

Case Study: Two Methods for Measuring Nitrate

A large number of nitrate measurements were needed on a wastewater treatment project. Method A was the standard method for measuring nitrate concentration in wastewater. The newer Method B was more desirable (faster, cheaper, safer, etc.) than Method A, but it could replace Method A only if shown to give equivalent results over the applicable range of concentrations and conditions. The evaluation of phenylmercuric acetate (PMA) as a preservative was also a primary objective of the experiment.

A large number of trials with each method was done at the conditions that were routinely being monitored. A representative selection of these trials is shown in Table 30.1 and in the cube plots of Figure 30.1. Panel (a) shows the original duplicate observations and panel (b) shows the average of the log-transformed observations on which the analysis is actually done. The experiment is a fully replicated 2^3 factorial design. The three factors were nitrate level, use of PMA preservative, and analytical method.

The high and low nitrate levels were included in the experimental design so that the interaction of concentration with method and PMA preservative could be evaluated. It could happen that PMA affects one method but not the other, or that the PMA has an effect at high but not at low concentrations. The low level of nitrate concentration (1–3 mg/L NO_3-N) was obtained by taking influent samples from a conventional activated sludge treatment process. The high level (20–30 mg/L NO_3-N) was available in samples from the effluent of a nitrifying activated sludge process.

TABLE 30.1

Results for Comparative Tests of Methods A and B

Nitrate Level	Method	PMA	NO$_3$ (mg/L) y_1	y_2	Average NO$_3$ $\bar{y}$ (mg/L)
Low	A	None	2.9	2.8	2.85
High	A	None	26.0	27.0	26.50
Low	B	None	3.1	2.8	2.95
High	B	None	30.0	32.0	31.00
Low	A	Yes	2.9	3.0	2.95
High	A	Yes	28.0	27.0	27.50
Low	B	Yes	3.3	3.1	3.20
High	B	Yes	30.4	31.1	30.75

TABLE 30.2

Design Matrix Expressed in Terms of the Coded Variables

Nitrate Level X_1	Method X_2	PMA Level X_3	ln(NO$_3$) x_1	x_2	Average $\bar{x}$	Variance s^2
−1	−1	−1	1.0647	1.0296	1.0472	0.0006157
1	−1	−1	3.2581	3.2958	3.2770	0.0007122
−1	1	−1	1.1314	1.0296	1.0805	0.0051799
1	1	−1	3.4012	3.4657	3.4335	0.0020826
−1	−1	1	1.0647	1.0986	1.0817	0.0005747
1	−1	1	3.3322	3.2958	3.3140	0.0006613
−1	1	1	1.1939	1.1314	1.1627	0.0019544
1	1	1	3.4144	3.4372	3.4258	0.0002591

Note: Values shown for the logarithms of the duplicate observations, $x = \ln(y)$, and their average and variance.

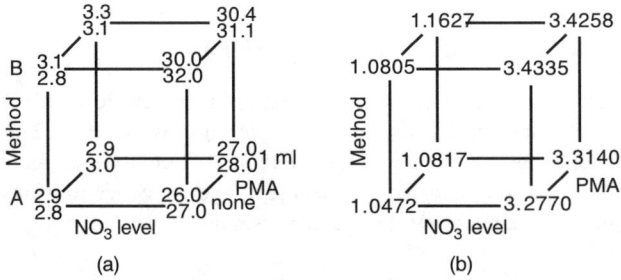

FIGURE 30.1 Cube plots for the nitrate measurement data. (a) Duplicate observations at the eight combinations of settings for a 2^3 factorial design to compare Methods A and B at two levels of nitrate and two levels of PMA preservative. (b) Average of the duplicate log-transformed at the eight experimental conditions.

Factorial designs can be conveniently represented by coding the high and low nitrate levels of each variable as −1 and +1 instead of using the actual values. The design matrix of Table 30.1, expressed in terms of the coded variables and in standard order, is shown in Table 30.2. The natural logarithms of the duplicate observations are listed. Also given are the averages and the variance of each duplicate pair of log-transformed concentrations. The log-transformation is needed to achieve constant variance over the tenfold range in nitrate concentration.

Method A seems to gives lower values than Method B. PMA does not seem to show any effect. We do not want to accept these impressions without careful analysis.

Method

Examples in Chapters 27 and 28 have explained that one main effect or interaction effect can be estimated for each experimental run. A 2^3 factorial has eight runs and thus eight effects can be estimated. Making two replicates at each condition gives a total of 16 observations but does not increase the number of effects that can be estimated. The replication, however, gives an internal estimate of the experimental error and increases the precision of the estimated effects

The experimental design provides information to estimate eight parameters, which previously were called main effects and interactions. In the context of regression, they are coefficients or parameters of the regression model. The mathematical model of the 2^3 factorial design is:

$$\eta = \beta_0 + \beta_1 x_1 + \beta_2 x_2 + \beta_3 x_3 + \beta_{12} x_1 x_2 + \beta_{13} x_1 x_3 + \beta_{23} x_2 x_3 + \beta_{123} x_1 x_2 x_3$$

where the x_1, x_2, and x_3 are the levels of the three experimental variables and the β's are *regression coefficients* that indicate the magnitude of the effects of each of the variables and the interactions of the variables. These coefficients will be estimated using the method of least squares, considering the model in the form:

$$y = b_0 + b_1 x_1 + b_2 x_2 + b_3 x_3 + b_{12} x_1 x_2 + b_{13} x_1 x_3 + b_{23} x_2 x_3 + b_{123} x_1 x_2 x_3 + e$$

where e is the residual error. If the model is adequate to fit the data, the e's are random experimental error and they can be used to estimate the standard error of the effects. If some observations are replicated, we can make an independent estimate of the experimental error variance.

We will develop the least squares estimation procedure using matrix algebra. The matrix algebra is general for all linear regression problems (Draper and Smith, 1998). What is special for the balanced two-level factorial designs is the ease with which the matrix operations can be done (i.e., almost by inspection). Readers who are not familiar with matrix operations will still find the calculations in the solution section easy to follow.

The model written in matrix form is:

$$y = \mathbf{X}\beta + e$$

where $\mathbf{X}$ is the *matrix of independent variables*, β is a vector of the coefficients, and $\mathbf{y}$ is the vector of observed values.

The least squares estimates of the coefficients are:

$$\mathbf{b} = (\mathbf{X}'\mathbf{X})^{-1}\mathbf{X}'\mathbf{y}$$

The variance of the coefficients is:

$$\text{Var}(\mathbf{b}) = (\mathbf{X}'\mathbf{X})^{-1}\sigma^2$$

Ideally, replicate measurements are made to estimate σ^2.

$\mathbf{X}$ is formed by augmenting the design matrix. The first column of +1's is associated with the coefficient β_0, which is the grand mean when coded variables are used. Additional columns are added based on the form of the mathematical model. For the model shown above, three columns are added for the two-factor interactions. For example, column 5 represents $x_1 x_2$ and is the product of the columns for x_1 and x_2. Column 8 represents the three-factor interaction.

The *matrix of independent variables* is:

$$\mathbf{X} = \begin{vmatrix} 1 & -1 & -1 & -1 & +1 & 1 & 1 & -1 \\ 1 & 1 & -1 & -1 & -1 & -1 & 1 & 1 \\ 1 & -1 & 1 & -1 & -1 & 1 & -1 & 1 \\ 1 & 1 & 1 & -1 & +1 & -1 & -1 & -1 \\ 1 & -1 & -1 & 1 & 1 & -1 & -1 & 1 \\ 1 & 1 & -1 & 1 & -1 & 1 & -1 & -1 \\ 1 & -1 & 1 & 1 & -1 & -1 & 1 & -1 \\ 1 & 1 & 1 & 1 & 1 & 1 & 1 & 1 \end{vmatrix}$$

Notice that this matrix is the same as the model matrix for the 2^3 factorial shown in Table 27.3.

To calculate **b** and Var(**b**) we need the *transpose* of **X**, denoted as **X**′. The transpose is created by making the first column of **X** the first row of **X**′; the second column of **X** becomes the second row of **X**′, etc. This is shown below. We also need the product of **X** and **X**′, denoted as **X**′**X**, the and the inverse of this, which is $(\mathbf{X}'\mathbf{X})^{-1}$.

The *transpose* of the **X** matrix is:

$$\mathbf{X}' = \begin{bmatrix} 1 & 1 & 1 & 1 & 1 & 1 & 1 & 1 \\ -1 & 1 & -1 & 1 & -1 & 1 & -1 & 1 \\ -1 & -1 & 1 & 1 & -1 & -1 & 1 & 1 \\ -1 & -1 & -1 & -1 & 1 & 1 & 1 & 1 \\ 1 & -1 & -1 & 1 & 1 & -1 & -1 & 1 \\ 1 & -1 & 1 & -1 & -1 & 1 & -1 & 1 \\ 1 & 1 & -1 & -1 & -1 & -1 & 1 & 1 \\ -1 & 1 & 1 & -1 & 1 & -1 & -1 & 1 \end{bmatrix}$$

The **X**′**X** matrix is:

$$\mathbf{X}'\mathbf{X} = \begin{bmatrix} 8 & 0 & 0 & 0 & 0 & 0 & 0 & 0 \\ 0 & 8 & 0 & 0 & 0 & 0 & 0 & 0 \\ 0 & 0 & 8 & 0 & 0 & 0 & 0 & 0 \\ 0 & 0 & 0 & 8 & 0 & 0 & 0 & 0 \\ 0 & 0 & 0 & 0 & 8 & 0 & 0 & 0 \\ 0 & 0 & 0 & 0 & 0 & 8 & 0 & 0 \\ 0 & 0 & 0 & 0 & 0 & 0 & 8 & 0 \\ 0 & 0 & 0 & 0 & 0 & 0 & 0 & 8 \end{bmatrix}$$

The inverse of the **X**′**X** matrix is called the *variance-covariance matrix*. It is:

$$(\mathbf{X}'\mathbf{X})^{-1} = \begin{bmatrix} 1/8 & 0 & 0 & 0 & 0 & 0 & 0 & 0 \\ 0 & 1/8 & 0 & 0 & 0 & 0 & 0 & 0 \\ 0 & 0 & 1/8 & 0 & 0 & 0 & 0 & 0 \\ 0 & 0 & 0 & 1/8 & 0 & 0 & 0 & 0 \\ 0 & 0 & 0 & 0 & 1/8 & 0 & 0 & 0 \\ 0 & 0 & 0 & 0 & 0 & 1/8 & 0 & 0 \\ 0 & 0 & 0 & 0 & 0 & 0 & 1/8 & 0 \\ 0 & 0 & 0 & 0 & 0 & 0 & 0 & 1/8 \end{bmatrix}$$

These matrices are easy to create and manipulate for a factorial experimental design. **X** is an *orthogonal matrix*, that is, the inner product of any two columns of vectors is zero. Because **X** is an orthogonal matrix, **X'X** is a *diagonal matrix*, that is, all elements are zero except diagonal elements. If **X** has n columns and m rows, **X'** has m columns and n rows. The product **X'X** will be a *square matrix* with n rows and n columns. If **X'X** is a diagonal matrix, its inverse $(\mathbf{X'X})^{-1}$ is just the reciprocal of the elements of **X'X**.

Case Study Solution

The variability of the nitrate measurements is larger at the higher concentrations. This is because the logarithmic scale of the instrument makes it possible to read to 0.1 mg/L at the low concentration but only to 1 mg/L at the high level. The result is that the measurement errors are proportional to the measured concentrations. The appropriate transformation to stabilize the variance in this case is to use the natural logarithm of the measured values. Each value was transformed by taking its natural logarithm and then the logs of the replicates were averaged.

Parameter Estimation

Using the matrix algebra defined above, the coefficients b are calculated as:

$$\mathbf{b} = \begin{vmatrix} 1/8 & 0 & 0 & 0 & 0 & 0 & 0 & 0 \\ 0 & 1/8 & 0 & 0 & 0 & 0 & 0 & 0 \\ 0 & 0 & 1/8 & 0 & 0 & 0 & 0 & 0 \\ 0 & 0 & 0 & 1/8 & 0 & 0 & 0 & 0 \\ 0 & 0 & 0 & 0 & 1/8 & 0 & 0 & 0 \\ 0 & 0 & 0 & 0 & 0 & 1/8 & 0 & 0 \\ 0 & 0 & 0 & 0 & 0 & 0 & 1/8 & 0 \\ 0 & 0 & 0 & 0 & 0 & 0 & 0 & 1/8 \end{vmatrix} \begin{bmatrix} 1 & 1 & 1 & 1 & 1 & 1 & 1 & 1 \\ -1 & 1 & -1 & 1 & -1 & 1 & -1 & 1 \\ -1 & -1 & 1 & 1 & -1 & -1 & 1 & 1 \\ -1 & -1 & -1 & -1 & 1 & 1 & 1 & 1 \\ 1 & -1 & -1 & 1 & 1 & -1 & -1 & 1 \\ 1 & -1 & 1 & -1 & -1 & 1 & -1 & 1 \\ 1 & 1 & -1 & -1 & -1 & -1 & 1 & 1 \\ -1 & 1 & 1 & -1 & 1 & -1 & -1 & 1 \end{bmatrix} \begin{bmatrix} 1.0472 \\ 3.2770 \\ 1.0805 \\ 3.4335 \\ 1.0817 \\ 3.3140 \\ 1.1627 \\ 3.4258 \end{bmatrix}$$

which gives:

$b_0 = 1/8(1.0472 + 3.2770 + \cdots + 1.1627 + 3.4258) = 2.2278$

$b_1 = 1/8(-1.0472 + 3.2770 - 1.0805 + 3.4335 - 1.0817 + 3.3140 - 1.1627 + 3.4258) = 1.1348$

$b_2 = 1/8(-1.0472 - 3.2770 + 1.0805 + 3.4335 - 1.0817 - 3.3140 + 1.1627 + 3.4258) = 0.0478$

and so on.

The estimated coefficients are:

$$b_0 = 2.2278 \qquad b_1 = 1.1348 \qquad b_2 = 0.0478 \qquad b_3 = 0.0183$$
$$b_{12} = 0.0192 \qquad b_{13} = -0.0109 \qquad b_{23} = 0.0004 \qquad b_{123} = -0.0115$$

The subscripts indicate which factor or interaction the coefficient multiplies in the model. Because we are working with coded variables, b_0 is the average of the observed values. Intrepreting b_0 as the intercept where all x's are zero is mathematically correct, but it is physical nonsense. Two of the factors are discrete variables. There is no method between A and B. Using half the amount of PMA preservative (i.e., $x_2 = 0$) would either be effective or ineffective; it cannot be half-effective.

This arithmetic is reminiscent of that used to estimate main effects and interactions. One difference is that in estimating the effects, division is by 4 instead of by 8. This is because there were four differences used to estimate each effect. The effects indicate how much the response is changed by moving from the low level to the high level (i.e., from -1 to $+1$). The regression model coefficients indicated how much the response changes by moving one unit (i.e., from -1 to 0 or from 0 to $+1$). The regression coefficients are exactly half as large as the effects estimated using the standard analysis of two-level factorial designs.

Precision of the Estimated Parameters

The variance of the coefficients is:

$$\text{Var}(\mathbf{b}) = \sigma^2/16$$

The denominator is 16 is because there are $n = 16$ observations. In this replicated experiment, σ^2 is estimated by s^2, which is calculated from the logarithms of the duplicate observations (Table 30.2). If there were no replication, the variance would be $\text{Var}(\mathbf{b}) = \sigma^2/8$ for a 2^3 experimental design, and σ^2 would be estimated from data external to the design.

The variances of the duplicate pairs are shown in the table below. These can be averaged to estimate the variance for each method.

Method	s_i^2 ($\times 10^3$) of Duplicate Pairs				$s^2_{\text{Method}} = \Sigma s^2/4$ ($\times 10^3$)
A	0.6157	0.7122	0.5747	0.6613	$s_A^2 = 0.641$
B	5.1799	2.0826	1.9544	0.2591	$s_B^2 = 2.369$

The variances of A and B can be pooled (averaged) to estimate the variance of the entire experiment if they are assumed to come from populations having the same variance. The data suggest that the variance of Method A may be smaller than that of Method B, so this should be checked.

The hypothesis that the population variances are equal can be tested using the F statistic. The upper 5% level of the F statistic for (4, 4) degrees of freedom is $F_{4,4} = 6.39$. A ratio of two variances as large as this is expected to occur by chance one in twenty times. The ratio of the two variances in this problem is $F_{\text{exp}} = s_B^2/s_A^2 = 2.369/0.641 = 3.596$, which is less than $F_{4,4} = 6.596$. The conclusion is that a ratio of 3.596 is not exceptional. It is accepted that the variances for Methods A and B are estimating the same population variance and they are pooled to give:

$$s^2 = [4(0.000641) + 4(0.002369)]/8 = 0.001505$$

The variance of each coefficient is:

$$\text{Var}(b) = 0.001505/16 = 0.0000941$$

and the standard error of the true value of each coefficient is:

$$\text{SE}(b) = \sqrt{\text{Var}(b)} = \sqrt{0.0000941} = 0.0097$$

The half-width of the 95% confidence interval for each coefficient is:

$$\text{SE}(b) \times t_{8, 0.025} = 0.0097(2.306) = 0.0224$$

Judging the magnitude of each estimated coefficient against the width of the confidence interval, we conclude:

$b_0 = 2.2278 \pm 0.0224$ Average — significant
$b_1 = 1.1348 \pm 0.0224$ Nitrate level — significant
$b_2 = 0.0478 \pm 0.0224$ Method — significant
$b_3 = 0.0183 \pm 0.0224$ PMA — not significant

Coefficients b_0 and b_1 were expected to be significant. There is nothing additional to note about them.

The interactions are not significant:

$b_{12} = 0.0192 \pm 0.0224$ $b_{13} = -0.0109 \pm 0.0224$
$b_{23} = 0.0004 \pm 0.0224$ $b_{123} = -0.0115 \pm 0.0224$

Method A gives results that are from 0.025 to 0.060 mg/L lower than Method B (on the log-transformed scale). This is indicated by the coefficient $b_2 = 0.0478 \pm 0.0224$. The difference between A and B on the log scale is a percentage on the original measurement scale.[1] Method A gives results that are 2.5 to 6% lower than Method B.

If a 5% difference between methods is not important in the context of a particular investigation, and if Method B offers substantial advantages in terms of cost, speed, convenience, simplicity, etc., one might decide to adopt Method B although it is not truly equivalent to Method A. This highlights the difference between "statistically significant" and "practically important." The statistical problem was important to learn whether A and B were different and, if so, by how much and in which direction. The practical problem was to decide whether a real difference could be tolerated in the application at hand.

Using PMA as a preservative caused no measurable effect or interference. This is indicated by the confidence interval [−0.004, 0.041] for b_3, which includes zero. This does not mean that wastewater specimens could be held without preservation. It was already known that preservation was needed, but it was not known how PMA would affect Method B. This important result meant that the analyst could do nitrate measurements twice a week instead of daily and holding wastewater over the weekend was possible. This led to economies of scale in processing.

This chapter began by saying that Method A, the widely accepted method, was considered to give accurate measurements. It is often assumed that widely used methods are accurate, but that is not necessarily true. For many analyses, no method is known *a priori* to be correct. In this case, finding that Methods A and B are equivalent would not prove that either or both give correct results. Likewise, finding them different would not mean necessarily that one is correct. Both might be wrong.

At the time of this study, all nitrate measurement methods were considered tentative (i.e., not yet proven accurate). Therefore, Method A actually was not known to be correct. A 5% difference between Methods A and B was of no practical importance in the application of interest. Method B was adopted because it was sufficiently accurate and it was simpler, faster, and cheaper.

Comments

The arithmetic of fitting a regression model to a factorial design and estimating effects in the standard way is virtually identical. The main effect indicates the change in response that results from moving from the low level to the high level (i.e., from −1 to +1). The coefficients in the regression model indicate the change in response associated with moving only one unit (e.g., from 0 to +1). Therefore, the regression coefficients are exactly half as large as the effects.

Obviously, the decision of analyzing the data by regression or by calculating effects is largely a matter of convenience or personal preference. Calculating the effects is more intuitive and, for many persons, easier, but it is not really different or better.

There are several common situations where linear regression must be used to analyze data from a factorial experiment. The factorial design may not have been executed precisely as planned. Perhaps one run has failed so there is a missing data point. Or perhaps not all runs were replicated, or the number of replicates is different for different runs. This makes the experiment unbalanced, and matrix multiplication and inversion cannot be done by inspection as in the case of a balanced two-level design.

---

[1] Suppose that Method B measures 3.0 mg/L, which is 1.0986 on the log scale, and Method A measures 0.0477 less on the log scale, so it would give 1.0986 − 0.0477 = 1.0509. Transform this to the original metric by taking the antilog; exp(1.0509) = 2.86 mg/L. The difference 3.0 − 2.86 = 0.14, expressed as a percentage is 100(0.14/3.0) = 4.77%. This is the same as the effect of method (0.0477) on the log-scale that was computed in the analysis.

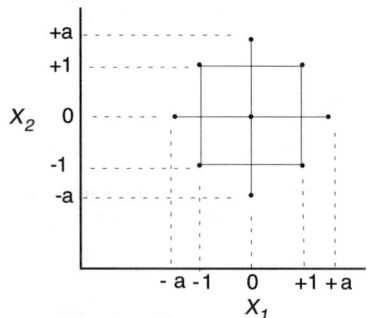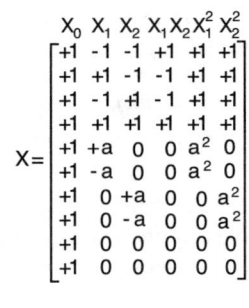

FIGURE 30.2 Experimental design and matrix of independent variables for a composite design with star points and center points. This design allows a quadratic model to be fitted by regression.

Another common situation results from our inability to set the independent variables at the levels called for by the design. As an example of this, suppose that a design specifies four runs at pH 6 and four at pH 7.7, but the actual pH values at the low-level runs were 5.9, 6.0, 6.1, 6.0, and similar variation existed at the high-level runs. These give a design matrix that is not orthogonal; it is fuzzy. The data can be analyzed by regression.

Another situation, which is discussed further in Chapter 43, is when the two-level design is augmented by adding "star points" and center points. Figure 30.2 shows an augmented design in two factors and the matrix of independent variables. This design allows us to fit a quadratic model of the form:

$$y = b_0 + b_1 x_1 + b_2 x_2 + b_{11} x_1^2 + b_{22} x_2^2 + b_{12} x_1 x_2$$

The matrix of independent variables is shown in Figure 30.2. This design is not orthogonal, but almost, because the covariance is very small.

The center points are at (0, 0). The star points are a distance a from the center, where $a > 1$. Without the center points there would be an information hole in the center of the experimental region. Replicate center points are used to improve the balance of information obtained over the experimental region, and also to provide an estimate of the experimental error.

How do we pick a? It cannot be too big because this model is intended only to describe a limited region. If $a = 1.414$, then all the corner and star points fall on a circle of diameter 1.414 and the design is balanced and rotatable. Another common augmented design is to use $a = 2$.

References

Box, G. E. P., W. G. Hunter, and J. S. Hunter (1978). *Statistics for Experimenters: An Introduction to Design, Data Analysis, and Model Building,* New York, Wiley Interscience.

Draper, N. R. and H. Smith, (1998). *Applied Regression Analysis,* 3rd ed., New York, John Wiley.

Milliken, G. A. and D. E. Johnson (1992). *Analysis of Messy Data, Vol. I: Designed Experiments,* New York, Van Nostrand Reinhold.

Exercises

30.1 Nitrate Measurement. A 2^3 factorial experiment with four replicates at a center point was run to compare two methods for measuring nitrate and the use of a preservative. Tests were done on two types of wastewater. Use the log-transformed data and evaluate the main and interaction effects.

X_1	X_2	X_3	y	$\ln(y)$
−1	1	−1	1.88	0.631
−1	−1	−1	2.1	0.742
−1	1	1	6.1	1.808
−1	−1	1	6.4	1.856
1	−1	−1	16	2.773
1	1	−1	17	2.833
1	−1	1	19	2.944
1	1	1	19.5	2.970
0	0	0	10.1	2.313
0	0	0	10.1	2.313
0	0	0	10.5	2.351
0	0	0	10.9	2.389

Note: X_1 is the type of wastewater:

$\quad\quad\quad$ −1 = influent $\quad$ +1 = effluent

X_2 is preservative: $\quad$ −1 = none $\quad$ +1 = added

X_3 is method: $\quad$ −1 = Method A $\quad$ +1 = Method B

30.2 Fly Ash Density. The 16 runs in the table below are from a study on the effect of water content (W), compaction effort (C), and time of curing (T) on the density of a material made from pozzolanic fly ash and sand. Two runs were botched so the effects and interactions must be calculated by regression. Do this and report your analysis.

Run	Factor W	Factor C	Factor T	Density (lb/ft^3)
1	−	−	−	107.3
2	+	−	−	Missing
3	−	+	−	115.9
4	+	+	−	121.4
5	−	−	+	101.8
6	+	−	+	115.6
7	−	+	+	109.3
8	+	+	+	121.1
9	−	−	−	Missing
10	+	−	−	120.8
11	−	+	−	118.6
12	+	+	−	126.5
13	−	−	+	99.8
14	+	−	+	117.5
15	−	+	+	107.6
16	+	+	+	118.9

30.3 Metal Inhibition. Solve Exercise 27.5 using regression.

31

Correlation

KEY WORDS *BOD, COD, correlation, correlation coefficient, covariance, nonparametric correlation, Pearson product-moment correlation coefficient, R^2, regression, serial correlation, Spearman rank correlation coefficient, taste, chlorine.*

Two variables have been measured and a plot of the data suggests that there is a *linear* relationship between them. A statistic that quantifies the strength of the linear relationship between the two variables is the *correlation coefficient*.

Care must be taken lest correlation is confused with causation. Correlation may, but does not necessarily, indicate causation. Observing that y increases when x increases does not mean that a change in x causes the increase in y. Both x and y may change as a result of change in a third variable, z.

Covariance and Correlation

A measure of the linear dependence between two variables x and y is the covariance between x and y. The sample covariance of x and y is:

$$\text{Cov}(x, y) = \frac{\sum (x_i - \eta_x)(y_i - \eta_y)}{N}$$

where η_x and η_y are the population means of the variables x and y, and N is the size of the population. If x and y are independent, $\text{Cov}(x, y)$ would be zero. Note that the converse is not true. Finding $\text{Cov}(x, y) = 0$ does not mean they are independent. (They might be related by a quadratic or exponential function.)

The covariance is dependent on the scales chosen. Suppose that x and y are distances measured in inches. If x is converted from inches to feet, the covariance would be divided by 12. If both x and y are converted to feet, the covariance would be divided by $12^2 = 144$. This makes it impossible in practice to know whether a value of covariance is large, which would indicate a strong linear relation between two variables, or small, which would indicate a weak association.

A *scaleless covariance*, called the *correlation coefficient* $\rho(x, y)$, or simply ρ, is obtained by dividing the covariance by the two population standard deviations σ_x and σ_y, respectively. The possible values of ρ range from -1 to $+1$. If x is independent of y, ρ would be zero. Values approaching -1 or $+1$ indicate a strong correspondence of x with y. A positive correlation ($0 < \rho \leq 1$) indicates the large values of x are associated with large values of y. In contrast, a negative correlation ($-1 \leq \rho < 0$) indicates that large values of x are associated with small values of y.

The true values of the population means and standard deviations are estimated from the available data by computing the means $\bar{x}$ and $\bar{y}$. The *sample correlation coefficient* between x and y is:

$$r = \frac{\sum (x_i - \bar{x})(y_i - \bar{y})}{\sqrt{\sum (x_i - \bar{x})^2 \sum (y_i - \bar{y})^2}}$$

This is the Pearson product-moment correlation coefficient, usually just called the *correlation coefficient*. The range of r is from -1 to $+1$.

Case Study: Correlation of BOD and COD Measurements

Figure 31.1 shows $n = 90$ pairs of effluent BOD_5 and COD concentrations (mg/L) from Table 31.1, and the same data after a log transformation. We know that these two measures of wastewater strength are related. The purpose of calculating a correlation coefficient is to quantify the strength of the relationship.

We find $r = 0.59$, indicating a moderate positive correlation, which is consistent with the impression gained from the graphical display. It makes no difference whether COD or BOD is plotted on the x-axis; the sample correlation coefficient is still $r = 0.59$. The log-transformed data transformation have symmetry about the median, but they also appear variable, and perhaps curvilinear, and the correlation coefficient is reduced ($r = 0.53$).

It is tempting to use ordinary regression to fit a straight line for predicting BOD from COD, as shown in Figure 31.2. The model would be BOD = 2.5 + 1.6 COD, with $R^2 = 0.35$. Fitting COD = 2.74 + 0.22 BOD also gives $R^2 = 0.35$. Notice that R^2 is the same in both cases and that it happens to be the squares of the correlation coefficient between the two variables ($r^2 = 0.59^2 = 0.35$). In effect, regression has revealed the same information about the strength of the association although R^2 and r are different statistics with different interpretations. This correspondence between r and R^2 is true only for straight-line relations.

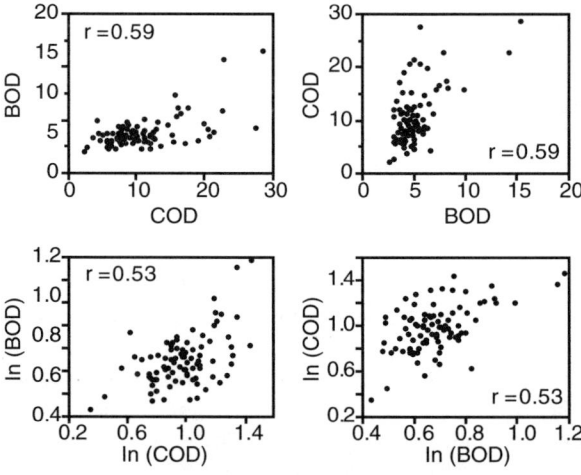

FIGURE 31.1 Scatterplot for 90 pairs of effluent five-day BOD vs. COD measurements, and ln(BOD) vs. ln(COD).

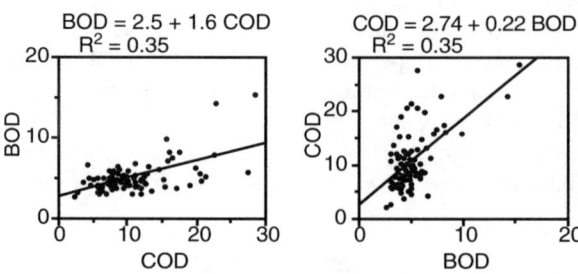

FIGURE 31.2 Two possible regressions on the COD and BOD_5 data. Both are invalid because the x and y variables have substantial measurement error.

Correlation

TABLE 31.1

Ninety Paired Measurements of Effluent Five-Day BOD and Effluent COD Concentrations

COD	BOD	COD	BOD	COD	BOD	COD	BOD	COD	BOD
9.1	4.5	6.0	3.6	7.6	4.4	11.2	3.8	16.5	7.5
5.7	3.3	4.5	5.0	8.1	5.9	10.1	5.9	13.6	3.4
15.8	7.2	4.7	4.1	7.3	4.9	17.5	8.2	12.0	3.1
7.6	4.0	4.3	6.7	8.5	4.9	16.0	8.3	11.6	3.9
6.5	5.1	9.7	5.0	8.6	5.5	11.2	6.9	12.5	5.1
5.9	3.0	5.8	5.0	7.8	3.5	9.6	5.1	12.0	4.6
10.9	5.0	6.3	3.8	7.2	4.3	6.4	3.4	20.7	4.6
9.9	4.3	8.8	6.1	8.5	3.8	10.3	4.1	28.6	15.3
8.3	4.7	5.7	4.1	7.0	3.1	11.2	4.4	2.2	2.7
8.1	4.2	6.3	4.2	22.8	14.2	7.9	4.9	14.6	6.0
12.4	4.6	9.7	4.3	5.0	4.8	13.1	6.4	15.2	4.8
12.1	4.8	15.4	4.0	3.7	4.4	8.7	6.3	12.8	5.6
10.2	4.7	12.0	3.7	6.2	3.9	22.7	7.9	19.8	6.3
12.6	4.4	7.9	5.4	7.1	4.5	9.2	5.2	9.5	5.4
10.1	4.1	6.4	4.2	5.9	3.8	5.7	4.0	27.5	5.7
9.4	5.2	5.7	3.9	7.5	5.9	17.2	3.7	20.5	5.6
8.1	4.9	8.0	5.7	10.0	5.2	10.7	3.1	19.1	4.1
15.7	9.8	11.1	5.4	2.8	3.1	9.5	3.7	21.3	5.1

Note: Concentrations are expressed as mg/L.

The regression is not strictly valid because both BOD and COD are subject to considerable measurement error. The regression correctly indicates the strength of a linear relation between BOD and COD, but any statements about probabilities on confidence intervals and prediction would be wrong.

Spearman Rank-Order Correlation

Sometimes, data can be expressed only as ranks. There is no numerical scale to express one's degree of disgust to odor. Taste, appearance, and satisfaction cannot be measured numerically. Still, there are situations when we must interpret nonnumeric information available about odor, taste, appearance, or satisfaction. The challenge is to relate these intangible and incommensurate factors to other factors that can be measured, such as amount of chlorine added to drinking water for disinfection, or the amount of a masking agent used for odor control, or degree of waste treatment in a pulp mill.

The *Spearman rank correlation* method is a *nonparametric* method that can be used when one or both of the variables to be correlated are expressed in terms of rank order rather than in quantitative units (Miller and Miller, 1984; Siegel and Castallan, 1988). If one of the variables is numeric, it will be converted to ranks. The ranks are simply "A is better than B, B is better than D, etc." There is no attempt to say that A is twice as good as B. The ranks therefore are not scores, as if one were asked to rate the taste of water on a scale of 1 to 10.

Suppose that we have rankings on n samples of wastewater for odor $[x_1, x_2,\ldots, x_n]$ and color $[y_1, y_2,\ldots, y_n]$. If odor and color are perfectly correlated, the ranks would agree perfectly with $x_i = y_i$ for all i. The difference between each pair of x,y rankings will be zero: $d_i = x_i - y_i = 0$. If, on the other hand, sample 8 has rank $x_i = 10$ and rank $y_i = 14$, the difference in ranks is $d_8 = x_8 - y_8 = 10 - 14 = -4$. Therefore, it seems logical to use the differences in rankings as a measure of disparity between the two variables.

The magnitude of the discrepancies is an index of disparity, but we cannot simply sum the difference because the positives would cancel out the negatives. This problem is eliminated if d_i^2 is used instead of d_i.

If we had two series of values for x and y and did not know they were ranks, we would calculate $r = \frac{\sum x_i y_i}{\sqrt{\sum x_i^2 \sum y_i^2}}$, where x_i is replaced by $x_i - \bar{x}$ and y_i by $y_i - \bar{y}$. The sums are over the n observed values.

Knowing that the data are rankings, we can simplify this using $d_i^2 = (x_i - y_i)^2$, which gives $x_i y_i = \frac{1}{2}(x_i^2 + y_i^2 - d_i^2)$ and:

$$r_S = \frac{\sum(x_i^2 + y_i^2 - d_i^2)}{2\sqrt{\sum x_i^2 \sum y_i^2}} = \frac{\sum x_i^2 + \sum y_i^2 - \sum d_i^2}{2\sqrt{\sum x_i^2 \sum y_i^2}}$$

The above equation can be used even when there are tied ranks. If there are no ties, then $\sum x_i^2 = \sum y_i^2 = n(n^2-1)/12$ and:

$$r_S = 1 - \frac{6\sum d_i^2}{n(n^2-1)}$$

The subscript S indicates the Spearman rank-order correlation coefficient. Like the Pearson product-moment correlation coefficient, r_S can vary between -1 and $+1$.

Case Study: Taste and Odor

Drinking water is treated with seven concentrations of a chemical to improve taste and reduce odor. The taste and odor resulting from the seven treatments could not be measured quantitatively, but consumers could express their opinions by ranking them. The consumer ranking produced the following data, where rank 1 is the most acceptable and rank 7 is the least acceptable.

Water Sample	A	B	C	D	E	F	G
Taste and odor ranking	1	2	3	4	5	6	7
Chemical added (mg/L)	0.9	2.8	1.7	2.9	3.5	3.3	4.7

The chemical concentrations are converted into rank values by assigning the lowest (0.9 mg/L) rank 1 and the highest (4.7 mg/L) rank 7. The table below shows the ranks and the calculated differences. A perfect correlation would have identical ranks for the taste and the chemical added, and all differences would be zero. Here we see that the differences are small, which means the correlation is strong.

Water Sample	A	B	C	D	E	F	G
Taste ranking	1	2	3	4	5	6	7
Chemical added	1	3	2	4	6	5	7
Difference, d_i	0	−1	1	0	−1	1	0

The Spearman rank correlation coefficient is:

$$r_s = 1 - \frac{6\sum(-1)^2 + 1^2 + 1^2 + (-1)^2}{7(7^2-1)} = 1 - \frac{24}{336} = 0.93$$

From Table 31.2, when $n = 7$, r_s must exceed 0.786 if the null hypothesis of "no correlation" is to be rejected at 95% confidence level. Here we conclude there is a correlation and that the water is better when less chemical is added.

Comments

Correlation coefficients are a familiar way of characterizing the association between two variables. Correlation is valid when both variables have random measurement errors. There is no need to think of one variable as x and the other as y, or of one as predictor and the other predicted. The two variables stand equal and this helps remind us that correlation and causation are not equivalent concepts.

TABLE 31.2

The Spearman Rank Correlation Coefficient Critical Values for 95% Confidence

n	One-Tailed Test	Two-Tailed Test	n	One-Tailed Test	Two-Tailed Test
5	0.900	1.000	13	0.483	0.560
6	0.829	0.886	14	0.464	0.538
7	0.714	0.786	15	0.446	0.521
8	0.643	0.738	16	0.429	0.503
9	0.600	0.700	17	0.414	0.488
10	0.564	0.649	18	0.401	0.472
11	0.536	0.618	19	0.391	0.460
12	0.504	0.587	20	0.380	0.447

Familiarity sometimes leads to misuse so we remind ourselves that:

1. The correlation coefficient is a valid indicator of association between variables only when that association is linear. If two variables are functionally related according to $y = a + bx + cx^2$, the computed value of the correlation coefficient is not likely to approach ±1 even if the experimental errors are vanishingly small. A scatterplot of the data will reveal whether a low value of r results from large random scatter in the data, or from a nonlinear relationship between the variables.
2. Correlation, no matter how strong, does not prove causation. Evidence of causation comes from knowledge of the underlying mechanistic behavior of the system. These mechanisms are best discovered by doing experiments that have a sound statistical design, and not from doing correlation (or regression) on data from unplanned experiments.

Ordinary linear regression is similar to correlation in that there are two variables involved and the relation between them is to be investigated. In regression, the two variables of interest are assigned particular roles. One (x) is treated as the independent (predictor) variable and the other (y) is the dependent (predicted) variable. Regression analysis assumes that only y is affected by measurement error, while x is considered to be controlled or measured without error. Regression of x on y is not strictly valid when there are errors in both variables (although it is often done). The results are useful when the errors in x are small relative to the errors in y. As a rule-of-thumb, "small" means $s_x < 1/3 s_y$. When the errors in x are large relative to those in y, statements about probabilities of confidence intervals on regression coefficients will be wrong. There are special regression methods to deal with the errors-in-variables problem (Mandel, 1964; Fuller, 1987; Helsel and Hirsch, 1992).

References

Chatfield, C. (1983). *Statistics for Technology,* 3rd ed., London, Chapman & Hall.
Folks, J. L. (1981). *Ideas of Statistics,* New York, John Wiley.
Fuller, W. A. (1987). *Measurement Error Models,* New York, Wiley.
Helsel, D. R. and R. M. Hirsch (1992). *Studies in Environmental Science 49: Statistical Models in Water Resources,* Amsterdam, Elsevier.
Mandel, J. (1964). *The Statistical Analysis of Experimental Data,* New York, Interscience Publishers.
Miller, J. C. and J. N. Miller (1984). *Statistics for Analytical Chemistry,* Chichester, England, Ellis Horwood Ltd.
Siegel, S. and N. J. Castallan (1988). *Nonparametric Statistics for the Behavioral Sciences,* 2nd ed., New York, McGraw-Hill.

Exercises

31.1 BOD/COD Correlation. The table gives $n = 24$ paired measurements of effluent BOD_5 and COD. Interpret the data using graphs and correlation.

COD (mg/L)	4.5	4.7	4.2	9.7	5.8	6.3	8.8	5.7	6.3	9.7	15.4	12.0
BOD (mg/L)	5.0	4.1	6.7	5.0	5.0	3.8	6.1	4.1	4.2	4.3	4.0	3.7
COD (mg/L)	8.0	11.1	7.6	8.1	7.3	8.5	8.6	7.8	7.2	7.9	6.4	5.7
BOD (mg/L)	5.7	5.4	4.4	5.9	4.9	4.9	5.5	3.5	4.3	5.4	4.2	3.9

32.2 Heavy Metals. The data below are 21 observations on influent and effluent lead (Pb), nickel (Ni), and zinc (Zn) at a wastewater treatment plant. Examine the data for correlations.

Inf. Pb	Eff. Pb	Inf. Ni	Eff. Ni	Inf. Zn	Eff. Zn
18	3	33	25	194	96
3	1	47	41	291	81
4	1	26	8	234	63
24	21	33	27	225	65
35	34	23	10	160	31
31	2	28	16	223	41
32	4	36	19	206	40
14	6	41	43	135	47
40	6	47	18	329	72
27	9	42	16	221	72
8	6	13	14	235	68
14	7	21	3	241	54
7	20	13	13	207	41
19	9	24	15	464	67
17	10	24	27	393	49
19	4	24	25	238	53
24	7	49	13	181	54
28	5	42	17	389	54
25	4	48	25	267	91
23	8	69	21	215	83
30	6	32	63	239	61

31.3 Influent Loadings. The data below are monthly average influent loadings (lb/day) for the Madison, WI, wastewater treatment plant in the years 1999 and 2000. Evaluate the correlation between BOD and total suspended solids (TSS).

1999	BOD	TSS	2000	BOD	TSS
Jan	68341	70506	Jan	74237	77018
Feb	74079	72140	Feb	79884	83716
Mar	70185	67380	Mar	75395	77861
Apr	76514	78533	Apr	74362	76132
May	71019	68696	May	74906	81796
Jun	70342	73006	Jun	71035	84288
Jul	69160	73271	Jul	76591	82738
Aug	72799	73684	Aug	78417	85008
Sep	69912	71629	Sep	76859	74226
Oct	71734	66930	Oct	78826	83275
Nov	73614	70222	Nov	73718	73783
Dec	75573	76709	Dec	73825	78242

31.4 Rounding. Express the data in Exercise 31.3 as thousands, rounded to one decimal place, and recalculate the correlation; that is, the Jan. 1999 BOD becomes 68.3.

31.5 Coliforms. Total coliform (TC), fecal coliform (FC), and chlorine residual (Cl_2 Res.) were measured in a wastewater effluent. Plot the data and evaluate the relationships among the three variables.

Cl_2 Res. (mg/L)	ln(TC)	ln(FC)	Cl_2 Res. (mg/L)	ln(TC)	ln(FC)	Cl_2 Res. (mg/L)	ln(TC)	ln(FC)
2.40	4.93	1.61	1.80	5.48	1.61	1.90	4.38	1.61
1.90	2.71	1.61	2.90	1.61	1.61	2.60	1.61	1.61
1.00	7.94	1.61	2.80	1.61	1.61	3.30	1.61	1.61
0.07	16.71	12.61	2.90	1.61	1.61	2.00	3.00	1.61
0.03	16.52	14.08	3.90	1.61	1.61	2.70	3.00	1.61
0.14	10.93	5.83	2.30	2.71	1.61	2.70	1.61	1.61
3.00	4.61	1.61	0.40	8.70	1.61	2.80	1.61	1.61
5.00	3.69	1.61	3.70	1.61	1.61	1.70	2.30	1.61
5.00	3.69	1.61	0.90	2.30	1.61	0.90	5.30	2.30
2.30	6.65	1.61	0.90	5.27	1.61	0.50	8.29	1.61
3.10	4.61	4.32	3.00	2.71	1.61	3.10	1.61	1.61
1.20	6.15	1.61	1.00	4.17	1.61	0.03	16.52	13.82
1.80	2.30	1.61	1.80	3.40	1.61	2.90	5.30	1.61
0.03	16.91	14.04	3.30	1.61	1.61	2.20	1.61	1.61
2.50	5.30	1.61	3.90	5.25	1.61	0.60	7.17	2.30
2.80	4.09	1.61	2.30	1.61	1.61	1.40	5.70	2.30
3.20	4.01	1.61	3.00	4.09	1.61	2.80	4.50	1.61
1.60	3.00	1.61	1.70	3.00	1.61	1.50	5.83	2.30
2.30	2.30	1.61	2.80	3.40	1.61	1.30	5.99	1.61
2.50	2.30	1.61	3.10	1.61	3.00	2.40	7.48	1.61

31.6 AA Lab. A university laboratory contains seven atomic absorption spectrophotometers (A–G). Research students rate the instruments in this order of preference: B, G, A, D, C, F, E. The research supervisors rate the instruments G, D, B, E, A, C, F. Are the opinions of the students and supervisors correlated?

31.7 Pump Maintenance. Two expert treatment plant operators (judges 1 and 2) were asked to rank eight pumps in terms of ease of maintenance. Their rankings are given below. Find the coefficient of rank correlation to assess how well the judges agree in their evaluations.

Judge 1	5	2	8	1	4	6	3	7
Judge 2	4	5	7	3	2	8	1	6

32

Serial Correlation

KEY WORDS ACF, autocorrelation, autocorrelation coefficient, BOD, confidence interval, correlation, correlation coefficient, covariance, independence, lag, sample size, sampling frequency, serial correlation, serial dependence, variance.

When data are collected sequentially, there is a tendency for observations taken close together (in time or space) to be more alike than those taken farther apart. Stream temperatures, for example, may show great variation over a year, while temperatures one hour apart are nearly the same. Some automated monitoring equipment make measurements so frequently that adjacent values are practically identical. This tendency for neighboring observations to be related is *serial correlation* or *autocorrelation*. One measure of the serial dependence is the *autocorrelation coefficient*, which is similar to the Pearson correlation coefficient discussed in Chapter 31. Chapter 51 will deal with autocorrelation in the context of time series modeling.

Case Study: Serial Dependence of BOD Data

A total of 120 biochemical oxygen demand (BOD) measurements were made at two-hour intervals to study treatment plant dynamics. The data are listed in Table 32.1 and plotted in Figure 32.1. As one would expect, measurements taken 24 h apart (12 sampling intervals) are similar. The task is to examine this daily cycle and the assess the strength of the correlation between BOD values separated by one, up to at least twelve, sampling intervals.

Correlation and Autocorrelation Coefficients

Correlation between two variables x and y is estimated by the sample correlation coefficient:

$$r = \frac{\sum (x_i - \bar{x})(y_i - \bar{y})}{\sqrt{\sum (x_i - \bar{x})^2 \sum (y_i - \bar{y})^2}}$$

where $\bar{x}$ and $\bar{y}$ are the sample means. The *correlation coefficient* (r) is a dimensionless number that can range from -1 to $+1$.

Serial correlation, or autocorrelation, is the correlation of a variable with itself. If sufficient data are available, serial dependence can be evaluated by plotting each observation y_t against the immediately preceding one, y_{t-1}. (Plotting y_t vs. y_{t+1} is equivalent to plotting y_t vs. y_{t-1}.) Similar plots can be made for observations two units apart (y_t vs. y_{t-2}), three units apart, etc.

If measurements were made daily, a plot of y_t vs. y_{t-7} might indicate serial dependence in the form of a weekly cycle. If y represented monthly averages, y_t vs. y_{t-12} might reveal an annual cycle. The distance between the observations that are examined for correlation is called the *lag*. The convention is to measure lag as the number of intervals between observations and not as real time elapsed. Of course, knowing the time between observations allows us to convert between real time and lag time.

TABLE 32.1

120 BOD Observations Made at 2-h Intervals

Day	Sampling Interval											
	1	2	3	4	5	6	7	8	9	10	11	12
1	200	122	153	176	129	168	165	119	113	110	113	98
2	180	122	156	185	163	177	194	149	119	135	113	129
3	160	105	127	162	132	184	169	160	115	105	102	114
4	112	148	217	193	208	196	114	138	118	126	112	117
5	180	160	151	88	118	129	124	115	132	190	198	112
6	132	99	117	164	141	186	137	134	120	144	114	101
7	140	120	182	198	171	170	155	165	131	126	104	86
8	114	83	107	162	140	159	143	129	117	114	123	102
9	144	143	140	179	174	164	188	107	140	132	107	119
10	156	116	179	189	204	171	141	123	117	98	98	108

Note: Time runs left to right.

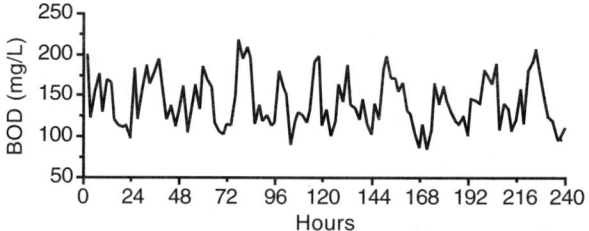

FIGURE 32.1 A record of influent BOD data sampled at 2-h intervals.

The correlation coefficients of the lagged observations are called autocorrelation coefficients, denoted as ρ_k. These are estimated by the *lag k sample autocorrelation coefficient* as:

$$r_k = \frac{\Sigma(y_t - \bar{y})(y_{t-k} - \bar{y})}{\Sigma(y_t - \bar{y})^2}$$

Usually the autocorrelation coefficients are calculated for $k = 1$ up to perhaps $n/4$, where n is the length of the time series. A series of $n \geq 50$ is needed to get reliable estimates. This set of coefficients (r_k) is called the *autocorrelation function (ACF)*. It is common to graph r_k as a function of lag k. Notice that the correlation of y_t with y_t is $r_0 = 1$. In general, $-1 < r_k < +1$.

If the data vary about a fixed level, the r_k die away to small values after a few lags. The approximate 95% confidence interval for r_k is $\pm 1.96/\sqrt{n}$. The confidence interval will be ± 0.28 for $n = 50$, or less for longer series. Any r_k smaller than this is attributed to random variation and is disregarded.

If the r_k do not die away, the time series has a persistent trend (upward or downward), or the series slowly drifts up and down. These kinds of time series are fairly common. The shape of the autocorrelation function is used to identify the form of the time series model that describes the data. This will be considered in Chapter 51.

Case Study Solution

Figure 32.2 shows plots of BOD at time t, denoted as BOD_t, against the BOD at 1, 3, 6, and 12 sampling intervals earlier. The sampling interval is 2 h so the time intervals between these observations are 2, 6, 12, and 24 h.

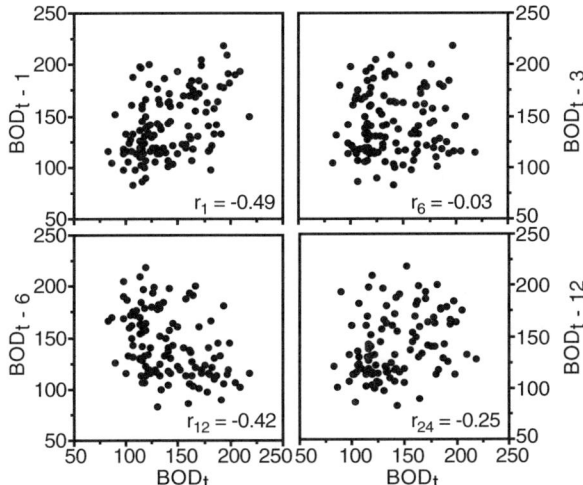

FIGURE 32.2 Plots of BOD at time t, denoted as BOD_t, against the BOD at lags of 1, 3, 6, and 12 sampling intervals, denoted as BOD_{t-1}, BOD_{t-3}, BOD_{t-6}, and BOD_{t-12}. The observations are 2 h apart, so the time intervals between these observations are 2, 6, 12, and 24 h apart, respectively.

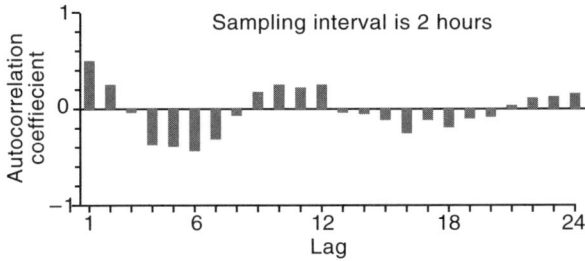

FIGURE 32.3 The autocorrelation coefficients for lags $k = 1 - 24$ h. Each observation is 2 h apart so the lag 12 autocorrelation indicates a 24-h cycle.

The sample autocorrelation coefficients are given on each plot. There is a strong correlation at lag 1 (2 h). This is clear in the plot of BOD_t vs BOD_{t-1}, and also by the large autocorrelation coefficient ($r_1 = 0.49$). The graph and the autocorrelation coefficient ($r_3 = -0.03$) show no relation between observations at lag 3 (6 h apart). At lag 6 (12 h), the autocorrelation is strong and negative ($r_6 = -0.42$). The negative correlation indicates that observations taken 12 h apart tend to be opposite in magnitude, one being high and one being low. Samples taken 24 h apart are positively correlated ($r_{12} = 0.25$). The positive correlation shows that when one observation is high, the observation 24 h ahead (or 24 h behind) is also high. Conversely, if the observation is low, the observation 24 h distant is also low.

Figure 32.3 shows the *autocorrelation function* for observations that are from lag 1 to lag 24 (2 to 48 h apart). The approximate 95% confidence interval is $\pm 1.96 \sqrt{120} = \pm 0.18$. The correlations for the first 12 lags show a definite diurnal pattern. The correlations for lags 13 to 24 repeat the pattern of the first 12, but less strongly because the observations are farther apart. Lag 13 is the correlation of observations 26 h apart. It should be similar to the lag 1 correlation of samples 2 h apart, but less strong because of the greater time interval between the samples. The lag 24 and lag 12 correlations are similar, but the lag 24 correlation is weaker. This system behavior makes physical sense because many factors (e.g., weather, daily work patterns) change from day to day, thus gradually reducing the strength of the system memory.

Implications for Sampling Frequency

The sample mean of autocorrelated data ($\bar{y}$) is unaffected by autocorrelation. It is still an unbiased estimator of the true mean. This is not true of the variance of y or the sample mean $\bar{y}$, as calculated by:

$$s_y^2 = \frac{\sum(y_t - \bar{y})^2}{n-1} \quad \text{and} \quad s_{\bar{y}}^2 = s_y^2/n$$

With autocorrelation, s_y^2 is the purely random variation *plus* a component due to drift about the mean (or perhaps a cyclic pattern).

The estimate of the variance of $\bar{y}$ that accounts for autocorrelation is:

$$s_{\bar{y}}^2 = \frac{s_y^2}{n} + \frac{2 s_y^2}{n^2} \sum_{k=1}^{n-1} (n-1) r_k$$

If the observations are independent, then all r_k are zero and this becomes $s_{\bar{y}}^2 = s_y^2/n$, the usual expression for the variance of the sample mean. If the r_k are positive (>0), which is common for environmental data, the variance is inflated. This means that n correlated observations will not give as much information as n independent observations (Gilbert, 1987).

Assuming the data vary about a fixed mean level, the number of observations required to estimate $\bar{y}$ with maximum error E and $(1 - \alpha)100\%$ confidence is approximately:

$$n = \left(\frac{z_{\alpha/2}\sigma}{E}\right)^2 \left(1 + 2 \sum_{k=1}^{n-1} r_k\right)$$

The lag at which r_k becomes negligible identifies the time between samples at which observations become independent. If we sample at that interval, or at a greater interval, the sample size needed to estimate the mean is reduced to $n = (z_{\alpha/2}\sigma/E)^2$.

If there is a regular cycle, sample at half the period of the cycle. For a 24-h cycle, sample every 12 h. If you sample more often, select multiples of the period (e.g., 6 h, 3 h).

Comments

Undetected serial correlation, which is a distinct possibility in small samples ($n < 50$), can be very upsetting to statistical conclusions, especially to conclusions based on t-tests and F-tests. This is why randomization is so important in designed experiments. The t-test is based on an assumption that the observations are normally distributed, random, and independent. Lack of independence (serial correlation) will bias the estimate of the variance and invalidate the t-test. A sample of $n = 20$ autocorrelated observations may contain no more information than ten independent observations. Thus, using $n = 20$ makes the test appear to be more sensitive than it is. With moderate autocorrelation and moderate sample sizes, what you think is a 95% confidence interval may be in fact a 75% confidence interval. Box et al. (1978) present a convincing example. Montgomery and Loftis (1987) show how much autocorrelation can distort the error rate.

Linear regression also assumes that the residuals are independent. If serial correlation exists, but we are unaware and proceed as though it is absent, all statements about probabilities (hypothesis tests, confidence intervals, etc.) may be wrong. This is illustrated in Chapter 41. Chapter 54 on intervention analysis discusses this problem in the context of assessing the shift in the level of a time series related to an intentional intervention in the system.

References

Box, G. E. P., W. G. Hunter, and J. S. Hunter (1978). *Statistics for Experimenters: An Introduction to Design, Data Analysis, and Model Building,* New York, Wiley Interscience.

Box, G. E. P., G. M. Jenkins, and G. C. Reinsel (1994). *Time Series Analysis, Forecasting and Control,* 3rd ed., Englewood Cliffs, NJ, Prentice-Hall.

Cryer, J. D. (1986). *Time Series Analysis,* Boston, MA, Duxbury Press.

Gilbert, R. O. (1987). *Statistical Methods for Environmental Pollution Monitoring,* New York, Van Nostrand Reinhold.

Montgomery, R. H. and J. C. Loftis, Jr. (1987). "Applicability of the *t*-Test for Detecting Trends in Water Quality Variables," *Water Res. Bull.,* 23, 653–662.

Exercises

32.1 Arsenic in Sludge. Below are annual average arsenic concentrations in municipal sewage sludge, measured in units of milligrams (mg) As per kilogram (kg) dry solids. Time runs from left to right, starting with 1979 (9.4 mg/kg) and ending with 2000 (4.8 mg/kg). Calculate the lag 1 autocorrelation coefficient and prepare a scatterplot to explain what this coefficient means.

```
9.4  9.7  4.9  8.0  7.8  8.0  6.4  5.9  3.7  9.9  4.2
7.0  4.8  3.7  4.3  4.8  4.6  4.5  8.2  6.5  5.8  4.8
```

32.2 Diurnal Variation. The 70 BOD values given below were measured at 2-h intervals (time runs from left to right). (a) Calculate and plot the autocorrelation function. (b) Calculate the approximate 95% confidence interval for the autocorrelation coefficients. (c) If you were to redo this study, what sampling interval would you use?

```
189  118  157  183  138  177  171  119  118  128  132  135  166  113  171  194  166
179  177  163  117  126  118  122  169  116  123  163  144  184  174  169  118  122
112  121  121  162  189  184  194  174  128  166  139  136  139  129  188  181  181
143  132  148  147  136  140  166  197  130  141  112  126  160  154  192  153  150
133  150
```

32.3 Effluent TSS. Determine the autocorrelation structure of the effluent total suspended solids (TSS) data in Exercise 3.4.

33

The Method of Least Squares

KEY WORDS confidence interval, critical sum of squares, dependent variable, empirical model, experimental error, independent variable, joint confidence region, least squares, linear model, linear least squares, mechanistic model, nonlinear model, nonlinear least squares, normal equation, parameter estimation, precision, regression, regressor, residual, residual sum of squares.

One of the most common problems in statistics is to fit an equation to some data. The problem might be as simple as fitting a straight-line calibration curve where the independent variable is the known concentration of a standard solution and the dependent variable is the observed response of an instrument. Or it might be to fit an unsteady-state nonlinear model, for example, to describe the addition of oxygen to wastewater with a particular kind of aeration device where the independent variables are water depth, air flow rate, mixing intensity, and temperature.

The equation may be an *empirical model* (simply descriptive) or *mechanistic model* (based on fundamental science). A *response variable* or *dependent variable* (y) has been measured at several settings of one or more *independent variables* (x), also called *input variables, regressors,* or *predictor variables*. *Regression* is the process of fitting an equation to the data. Sometimes, regression is called *curve fitting* or *parameter estimation*.

The purpose of this chapter is to explain that certain basic ideas apply to fitting both linear and nonlinear models. Nonlinear regression is neither conceptually different nor more difficult than linear regression. Later chapters will provide specific examples of linear and nonlinear regression. Many books have been written on regression analysis and introductory statistics textbooks explain the method. Because this information is widely known and readily available, some equations are given in this chapter without much explanation or derivation. The reader who wants more details should refer to books listed at the end of the chapter.

Linear and Nonlinear Models

The fitted model may be a simple function with one independent variable, or it may have many independent variables with higher-order and nonlinear terms, as in the examples given below.

Linear models $\quad \eta = \beta_0 + \beta_1 x + \beta_2 x^2 \quad\quad \eta = \beta_0 + \beta_1 x_1 + \beta_2 x_2 + \beta_2 x_1 x_2$

Nonlinear models $\quad \eta = \dfrac{\theta_1}{1 - \exp(-\theta_2 x)} \quad\quad \eta = \exp(-\theta x_1)(1 - x_2)^{\theta_2}$

To maintain the distinction between linear and nonlinear we use a different symbol to denote the parameters. In the general linear model, $\eta = f(x, \beta)$, x is a vector of independent variables and β are parameters that will be estimated by regression analysis. The estimated values of the parameters $\beta_1, \beta_2, \ldots$ will be denoted by $b_1, b_2, \ldots$. Likewise, a general nonlinear model is $\eta = f(x, \theta)$ where θ is a vector of parameters, the estimates of which are denoted by $k_1, k_2, \ldots$.

The terms *linear* and *nonlinear* refer to the parameters in the model and not to the independent variables. Once the experiment or survey has been completed, the numerical values of the dependent and independent variables are known. It is the parameters, the β's and θ's, that are unknown and must be computed. The model $y = \beta x^2$ is nonlinear in x; but once the known value of x^2 is provided, we have an equation that is linear in the parameter β. This is a linear model and it can be fitted by linear regression. In contrast, the model $y = x^\theta$ is nonlinear in θ, and θ must be estimated by nonlinear regression (or we must transform the model to make it linear).

It is usually assumed that a well-conducted experiment produces values of x_i that are essentially without error, while the observations of y_i are affected by random error. Under this assumption, the y_i observed for the ith experimental run is the sum of the true underlying value of the response (η_i) and a residual error (e_i):

$$y_i = \eta_i + e_i \quad i = 1, 2, \ldots, n$$

Suppose that we know, or tentatively propose, the linear model $\eta = \beta_0 + \beta_1 x$. The observed responses to which the model will be fitted are:

$$y_i = \beta_0 + \beta_1 x_i + e_i$$

which has residuals:

$$e_i = y_i - \beta_0 - \beta_1 x_i$$

Similarly, if one proposed the nonlinear model $\eta = \theta_1 \exp(-\theta_2 x)$, the observed response is:

$$y_i = \theta_1 \exp(-\theta_2 x_i) + e_i$$

with residuals:

$$e_i = y_i - \theta_1 \exp(-\theta_2 x_i)$$

The relation of the residuals to the data and the fitted model is shown in Figure 33.1. The lines represent the model functions evaluated at particular numerical values of the parameters. The residual ($e_i = y_i - \eta_i$) is the vertical distance from the observation to the value on the line that is calculated from the model. The residuals can be positive or negative.

The position of the line obviously will depend upon the particular values that are used for β_0 and β_1 in the linear model and for θ_1 and θ_2 in the nonlinear model. The regression problem is to select the values for these parameters that best fit the available observations. "Best" is measured in terms of making the residuals small according to a least squares criterion that will be explained in a moment.

If the model is correct, the residual $e_i = y_i - \eta_i$ will be nothing more than *random measurement error*. If the model is incorrect, e_i will reflect lack-of-fit due to all terms that are needed but missing from the model specification. This means that, after we have fitted a model, the residuals contain diagnostic information.

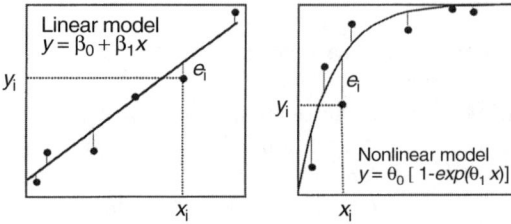

FIGURE 33.1 Definition of residual error for a linear model and a nonlinear model.

Residuals that are normally and independently distributed with constant variance over the range of values studied are persuasive evidence that the proposed model adequately fits the data. If the residuals show some pattern, the pattern will suggest how the model should be modified to improve the fit. One way to check the adequacy of the model is to check the properties of the residuals of the fitted model by plotting them against the predicted values and against the independent variables.

The Method of Least Squares

The best estimates of the model parameters are those that minimize the sum of the squared residuals:

$$S = \sum_{i=1}^{n}(e_i)^2 = \sum_{i=1}^{n}(y_i - \eta_i)^2$$

The minimum sum of squares is called the *residual sum of squares* (S_R). This approach to estimating the parameters is known as the *method of least squares*. The method applies equally to linear and nonlinear models. The difference between linear and nonlinear regression lies in how the least squares parameter estimates are calculated. The essential difference is shown by example.

Each term in the summation is the difference between the observed y_i and the η computed from the model at the corresponding values of the independent variables x_i. If the residuals (e_i) are normally and independently distributed with constant variance, the parameter estimates are unbiased and have minimum variance.

For models that are linear in the parameters, there is a simple algebraic solution for the least squares parameter estimates. Suppose that we wish to estimate β in the model $\eta = \beta x$. The sum of squares function is:

$$S(\beta) = \sum(y_i - \beta x_i)^2 = \sum(y_i^2 - 2\beta x_i y_i + \beta^2 x_i^2)$$

The parameter value that minimizes S is the *least squares estimate* of the true value of β. This estimate is denoted by b. We can solve the sum of squares function for this estimate (b) by setting the derivative with respect to β equal to zero and solving for b:

$$\frac{dS(\beta)}{d\beta} = 0 = 2\sum(bx_i^2 - x_i y_i)$$

This equation is called the *normal equation*. Note that this equation is linear with respect to b. The algebraic solution is:

$$b = \frac{\sum x_i y_i}{\sum x_i^2}$$

Because x_i and y_i are known once the experiment is complete, this equation provides a generalized method for direct and exact calculation of the least squares parameter estimate. (Warning: This is not the equation for estimating the slope in a two-parameter model.)

If the linear model has two (or more) parameters to be estimated, there will be two (or more) normal equations. Each normal equation will be linear with respect to the parameters to be estimated and therefore an algebraic solution is possible. As the number of parameters increases, an algebraic solution is still possible, but it is tedious and the linear regression calculations are done using linear algebra (i.e., matrix operations). The matrix formulation was given in Chapter 30.

Unlike linear models, no unique algebraic solution of the normal equations exists for nonlinear models. For example, if $\eta = \exp(-\theta x)$, the method of least squares requires that we find the value of θ that minimizes S:

$$S(\theta) = \sum(y_i - \exp(-\theta x_i))^2 = \sum[y_i^2 - 2y_i \exp(-\theta x_i) + (\exp(-\theta x_i))^2]$$

TABLE 33.1
Example Data and the Sum of Squares Calculations for a One-Parameter Linear Model and a One-Parameter Nonlinear Model

	Linear Model: $\eta = \beta x$				Nonlinear Model: $\eta_i = \exp(-\theta x_i)$			
x_i	$y_{obs,i}$	$y_{calc,i}$	e_i	$(e_i)^2$	x_i	$y_{obs,i}$	$y_{calc,i}$	e_i	$(e_i)^2$
Trial value: $b = 0.115$					Trial value: $k = 0.32$				
2	0.150	0.230	−0.080	0.0064	2	0.620	0.527	0.093	0.0086
4	0.461	0.460	0.001	0.0000	4	0.510	0.278	0.232	0.0538
6	0.559	0.690	−0.131	0.0172	6	0.260	0.147	0.113	0.0129
10	1.045	1.150	−0.105	0.0110	10	0.180	0.041	0.139	0.0194
14	1.364	1.610	−0.246	0.0605	14	0.025	0.011	0.014	0.0002
19	1.919	2.185	−0.266	0.0708	19	0.041	0.002	0.039	0.0015
		Sum of squares = 0.1659					Sum of squares = 0.0963		
Trial value: $b = 0.1$ (optimal)					Trial value: $k = 0.2$ (optimal)				
2	0.150	0.200	−0.050	0.0025	2	0.620	0.670	−0.050	0.0025
4	0.461	0.400	0.061	0.0037	4	0.510	0.449	0.061	0.0037
6	0.559	0.600	−0.041	0.0017	6	0.260	0.301	−0.041	0.0017
10	1.045	1.000	0.045	0.0020	10	0.180	0.135	0.045	0.0020
14	1.364	1.400	−0.036	0.0013	14	0.025	0.061	−0.036	0.0013
19	1.919	1.900	0.019	0.0004	19	0.041	0.022	0.019	0.0003
		Minimum sum of squares = 0.0116					Minimum sum of squares = 0.0115		

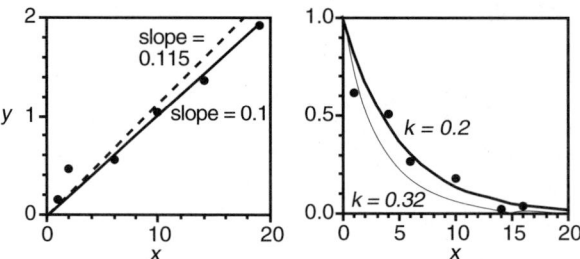

FIGURE 33.2 Plots of data to be fitted to linear (left) and nonlinear (right) models and the curves generated from the initial parameter estimates of $b = 0.115$ and $k = 0.32$ and the minimum least squares values ($b = 0.1$ and $k = 0.2$).

The least squares estimate of θ still satisfies $\partial S/\partial \theta = 0$, but the resulting derivative does not have an algebraic solution. The value of θ that minimizes S is found by iterative numerical search.

Examples

The similarities and differences of linear and nonlinear regression will be shown with side-by-side examples using the data in Table 33.1. Assume there are theoretical reasons why a linear model ($\eta_i = \beta x_i$) fitted to the data in Figure 33.2 should go through the origin, and an exponential decay model ($\eta_i = \exp(-\theta x_i)$) should have $y = 1$ at $t = 0$. The models and their sum of squares functions are:

$$y_i = \beta x_i + e_i \qquad \min S(\beta) = (y_i - \beta x_i)^2$$
$$y_i = \exp(-\theta x_i) + e_i \qquad \min S(\theta) = \sum (y_i - \exp(-\theta x_i))^2$$

For the linear model, the sum of squares function expanded in terms of the observed data and the parameter β is:

$$S(\beta) = (0.15 - 2\beta)^2 + (0.461 - 4\beta)^2 + (0.559 - 6\beta)^2$$
$$+ (1.045 - 10\beta)^2 + (1.361 - 14\beta)^2 + (1.919 - 19\beta)^2$$

The Method of Least Squares

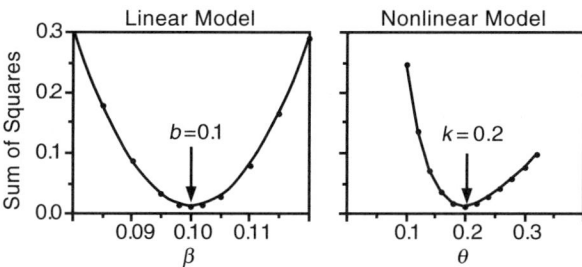

FIGURE 33.3 The values of the sum of squares plotted as a function of the trial parameter values. The least squares estimates are $b = 0.1$ and $k = 0.2$. The sum of squares function is symmetric (parabolic) for the linear model (left) and asymmetric for the nonlinear model (right).

For the nonlinear model it is:

$$S(\theta) = (0.62 - e^{-2\theta})^2 + (0.51 - e^{-4\theta})^2 + (0.26 - e^{6\theta})^2$$
$$+ (0.18 - e^{-10\theta})^2 + (0.025 - e^{-14\theta})^2 + (0.041 - e^{-19\theta})^2$$

An algebraic solution exists for the linear model, but to show the essential similarity between linear and nonlinear parameter estimation, the least squares parameter estimates of both models will be determined by a straightforward numerical search of the sum of squares functions. We simply plot $S(\beta)$ over a range of values of β, and do the same for $S(\theta)$ over a range of θ.

Two iterations of this calculation are shown in Table 33.1. The top part of the table shows the trial calculations for initial parameter estimates of $b = 0.115$ and $k = 0.32$. One clue that these are poor estimates is that the residuals are not random; too many of the linear model regression residuals are negative and all the nonlinear model residuals are positive. The bottom part of the table is for $b = 0.1$ and $k = 0.2$, the parameter values that give the minimum sum of squares.

Figure 33.3 shows the smooth sum of squares curves obtained by following this approach. The minimum sum of squares — the minimum point on the curve — is called the *residual sum of squares* and the corresponding parameter values are called the *least squares estimates*. The least squares estimate of β is $b = 0.1$. The least squares estimate of θ is $k = 0.2$. The fitted models are $\hat{y} = 0.1x$ and $\hat{y} = \exp(-0.2x)$. $\hat{y}$ is the predicted value of the model using the least squares parameter estimate.

The sum of squares function of a linear model is always symmetric. For a univariate model it will be a parabola. The curve in Figure 33.3a is a parabola. The sum of squares function for nonlinear models is not symmetric, as can be seen in Figure 33.3b.

When a model has two parameters, the sum of squares function can be drawn as a surface in three dimensions, or as a contour map in two dimensions. For a two-parameter linear model, the surface will be a parabaloid and the contour map of S will be concentric ellipses. For nonlinear models, the sum of squares surface is not defined by any regular geometric function and it may have very interesting contours.

The Precision of Estimates of a Linear Model

Calculating the "best" values of the parameters is only part of the job. The precision of the parameter estimates needs to be understood. Figure 33.3 is the basis for showing the confidence interval of the example one-parameter models.

For the *one-parameter* linear model *through the origin*, the variance of b is:

$$\text{Var}(b) = \frac{\sigma^2}{\sum x_i^2}$$

The summation is over all squares of the settings of the independent variable x. σ^2 is the *experimental error variance*. (Warning: This equation does not give the variance for the slope of a two-parameter linear model.)

Ideally, σ^2 would be estimated from independent replicate experiments at some settings of the x variable. There are no replicate measurements in our example, so another approach is used. The residual sum of squares can be used to estimate σ^2 if one is willing to assume that the model is correct. In this case, the residuals are random errors and the average of these residuals squared is an estimate of the error variance σ^2. Thus, σ^2 may be estimated by dividing the residual sum of squares (S_R) by its degrees of freedom ($\nu = n - p$), where n is the number of observations and p is the number of estimated parameters.

In this example, $S_R = 0.0116$, $p = 1$ parameter, $n = 6$, $\nu = 6 - 1 = 5$ degrees of freedom, and the estimate of the experimental error variance is:

$$s^2 = \frac{S_R}{n-p} = \frac{0.0116}{5} = 0.00232$$

The estimated variance of b is:

$$\text{Var}(b) = \frac{s^2}{\sum x_i^2} = \frac{0.00232}{713} = 0.0000033$$

and the standard error of b is:

$$\text{SE}(b) = \sqrt{\text{Var}(b)} = \sqrt{0.0000032} = 0.0018$$

The $(1-\alpha)100\%$ confidence limits for the true value β are:

$$b \pm t_{\nu,\alpha/2}\text{SE}(b)$$

For $\alpha = 0.05$, $\nu = 5$, we find $t_{5,0.025} = 2.571$, and the 95% confidence limits are $0.1 \pm 2.571(0.0018) = 0.1 \pm 0.0046$.

Figure 33.4a expands the scale of Figure 33.3a to show more clearly the confidence interval computed from the t statistic. The sum of squares function and the confidence interval computed using the t statistic are both symmetric about the minimum of the curve. The upper and lower bounds of the confidence interval define two intersections with the sum of squares curve. The sum of squares at these two points is identical because of the symmetry that always exists for a linear model. This level of the sum of squares function is the *critical sum of squares, S_c*. All values of β that give $S < S_c$ fall within the 95% confidence interval.

Here we used the easily calculated confidence interval to define the critical sum of squares. Usually the procedure is reversed, with the critical sum of squares being used to determine the boundary of the confidence region for two or more parameters. Chapters 34 and 35 explain how this is done. The F statistic is used instead of the t statistic.

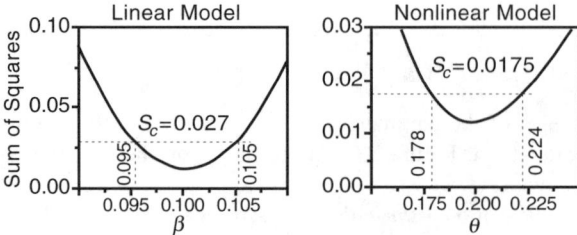

FIGURE 33.4 Sum of squares functions from Figure 33.3 replotted on a larger scale to show the confidence intervals of β for the linear model (left) and θ for the nonlinear model (right).

The Precision of Estimates of a Nonlinear Model

The sum of squares function for the nonlinear model (Figure 33.3) is not symmetrical about the least squares parameter estimate. As a result, the confidence interval for the parameter θ is not symmetric. This is shown in Figure 33.4, where the confidence interval is $0.20 - 0.022$ to $0.20 + 0.024$, or [0.178, 0.224].

The asymmetry near the minimum is very modest in this example, and a symmetric *linear approximation* of the confidence interval would not be misleading. This usually is not the case when two or more parameters are estimated. Nevertheless, many computer programs do report confidence intervals for nonlinear models that are based on symmetric linear approximations. These intervals are useful as long as one understands what they are.

This asymmetry is one difference between the linear and nonlinear parameter estimation problems. The essential similarity, however, is that we can still define a critical sum of squares and it will still be true that all parameter values giving $S \leq S_c$ fall within the confidence interval. Chapter 35 explains how the critical sum of squares is determined from the minimum sum of squares and an estimate of the experimental error variance.

Comments

The method of least squares is used in the analysis of data from planned experiments and in the analysis of data from unplanned happenings. For the least squares parameter estimates to be unbiased, the residual errors ($e = y - \eta$) must be random and independent with constant variance. It is the tacit assumption that these requirements are satisfied for unplanned data that produce a great deal of trouble (Box, 1966). Whether the data are planned or unplanned, the residual (e) includes the effect of latent variables (lurking variables) which we know nothing about.

There are many conceptual similarities between linear least squares regression and nonlinear regression. In both, the parameters are estimated by minimizing the sum of squares function, which was illustrated in this chapter using one-parameter models. The basic concepts extend to models with more parameters.

For linear models, just as there is an exact solution for the parameter estimates, there is an exact solution for the $100(1 - \alpha)\%$ confidence interval. In the case of linear models, the linear algebra used to compute the parameter estimates is so efficient that the work effort is not noticeably different to estimate one or ten parameters.

For nonlinear models, the sum of squares surface can have some interesting shapes, but the precision of the estimated parameters is still evaluated by attempting to visualize the sum of squares surface, preferably by making contour maps and tracing approximate joint confidence regions on this surface.

Evaluating the precision of parameter estimates in multiparameter models is discussed in Chapters 34 and 35. If there are two or more parameters, the sum of squares function defines a surface. A joint confidence region for the parameters can be constructed by tracing along this surface at the critical sum of squares level. If the model is linear, the joint confidence regions are still based on parabolic geometry. For two parameters, a contour map of the joint confidence region will be described by ellipses. In higher dimensions, it is described by ellipsoids.

References

Box, G. E. P. (1966). "The Use and Abuse of Regression," *Technometrics,* 8, 625–629.
Chatterjee, S. and B. Price (1977). *Regression Analysis by Example,* New York, John Wiley.
Draper, N. R. and H. Smith, (1998). *Applied Regression Analysis,* 3rd ed., New York, John Wiley.
Meyers, R. H. (1986). *Classical and Modern Regression with Applications,* Boston, MA, Duxbury Press.

Mosteller, F. and J. W. Tukey (1977). *Data Analysis and Regression: A Second Course in Statistics,* Reading, MA, Addison-Wesley Publishing Co.
Neter, J., W. Wasserman, and M. H. Kutner (1983). *Applied Regression Models,* Homewood, IL, Richard D. Irwin Co.
Rawlings, J. O. (1988). *Applied Regression Analysis: A Research Tool,* Pacific Grove, CA, Wadsworth and Brooks/Cole.

Exercises

33.1 Model Structure. Are the following models linear or nonlinear in the parameters?
   (a) $\eta = \beta_0 + \beta_1 x^2$
   (b) $\eta = \beta_0 + \beta_1 2^x$
   (c) $\eta = \beta_0 + \beta_1 x + \beta_2 x^2 + \beta_3 x^3 + \dfrac{\beta_4}{x - 60}$
   (d) $\eta = \dfrac{\beta_0}{x + \beta_1 x}$
   (e) $\eta = \beta_0 (1 + \beta_1 x_1)(1 + \beta_2 x_2)$
   (f) $\eta = \beta_0 + \beta_1 x_1 + \beta_2 x_2 + \beta_3 x_3 + \beta_{12} x_1 x_2 + \beta_{13} x_1 x_3 + \beta_{23} x_2 x_3 + \beta_{123} x_1 x_2 x_3$
   (g) $\eta = \beta_0 [1 - \exp(-\beta_1 x)]$
   (h) $\eta = \beta_0 [1 - \beta_1 \exp(-x)]$
   (i) $\ln(\eta) = \beta_0 + \beta_1 x$
   (j) $\dfrac{1}{\eta} = \beta_0 + \dfrac{\beta_1}{x}$

33.2 Fitting Models. Using the data below, determine the least squares estimates of β and θ by plotting the sum of squares for these models: $\eta_1 = \beta x^2$ and $\eta_2 = 1 - \exp(-\theta x)$.

x	y_1	y_2
2	2.8	0.44
4	6.2	0.71
6	10.4	0.81
8	17.7	0.93

33.3 Normal Equations. Derive the two normal equations to obtain the least squares estimates of the parameters in $y = \beta_0 + \beta_1 x$. Solve the simultaneous equations to get expressions for b_0 and b_1, which estimate the parameters β_0 and β_1.

34

Precision of Parameter Estimates in Linear Models

KEY WORDS *confidence interval, critical sum of squares, joint confidence region, least squares, linear regression, mean residual sum of squares, nonlinear regression, parameter correlation, parameter estimation, precision, prediction interval, residual sum of squares, straight line.*

Calculating the best values of the parameters is only half the job of fitting and evaluating a model. The precision of these estimates must be known and understood. The precision of estimated parameters in a linear or nonlinear model is indicated by the size of their *joint confidence region*. Joint indicates that all the parameters in the model are considered simultaneously.

The Concept of a Joint Confidence Region

When we fit a model, such as $\eta = \beta_0 + \beta_1 x$ or $\eta = \theta_1[1 - \exp(-\theta_2 x)]$, the regression procedure delivers a set of parameter values. If a different sample of data were collected using the same settings of x, different y values would result and different parameter values would be estimated. If this were repeated with many data sets, many pairs of parameter estimates would be produced. If these pairs of parameter estimates were plotted as x and y on Cartesian coordinates, they would cluster about some central point that would be very near the true parameter values. Most of the pairs would be near this central value, but some could fall a considerable distance away. This happens because of random variation in the y measurements.

The data (if they are useful for model building) will restrict the plausible parameter values to lie within a certain region. The intercept and slope of a straight line, for example, must be within certain limits or the line will not pass through the data, let alone fit it reasonably well. Furthermore, if the slope is decreased somewhat in an effort to better fit the data, inevitably the intercept will increase slightly to preserve a good fit of the line. Thus, low values of slope paired with high values of intercept are plausible, but high slopes paired with high intercepts are not. This relationship between the parameter values is called *parameter correlation*. It may be strong or weak, depending primarily on the settings of the x variables at which experimental trials are run.

Figure 34.1 shows some joint confidence regions that might be observed for a two-parameter model. Panels (a) and (b) show typical elliptical confidence regions of linear models; (c) and (d) are for nonlinear models that may have confidence regions of irregular shape. A small joint confidence region indicates precise parameter estimates. The orientation and shape of the confidence region are also important. It may show that one parameter is estimated precisely while another is only known roughly, as in (b) where β_2 is estimated more precisely than β_1. In general, the size of the confidence region decreases as the number of observations increases, but it also depends on the actual choice of levels at which measurements are made. This is especially important for nonlinear models. The elongated region in (d) could result from placing the experimental runs in locations that are not informative.

The *critical sum of squares* value that bounds the $(1 - \alpha)100\%$ joint confidence region is:

$$S_c = S_R + S_R\left(\frac{p}{n-p} F_{p,n-p,\alpha}\right) = S_R\left(1 + \frac{p}{n-p} F_{p,n-p,\alpha}\right)$$

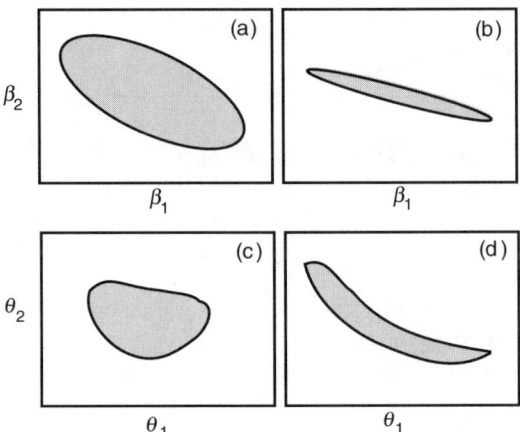

FIGURE 34.1 Examples of joint confidence regions for two parameter models. The elliptical regions (a) and (b) are typical of linear models. The irregular shapes of (c) and (d) might be observed for nonlinear models.

where p is the number of parameters estimated, n is the number of observations, and $F_{p,n-p,\alpha}$ is the upper α percent value of the F distribution with p and $n - p$ degrees of freedom, and S_R is the residual sum of squares. Here $S_R/(n - p)$ is used to estimate σ^2. If there were replicate observations, an independent estimate of σ^2 could be calculated.

This defines an exact $(1 - \alpha)100\%$ confidence region for a linear model; it is only approximate for nonlinear models. This is discussed in Chapter 35.

Theory: A Linear Model

Standard statistics texts all give a thorough explanation of linear regression, including a discussion of how the precision of the estimated parameters is determined. We review these ideas in the context of a straight-line model $y = \beta_0 + \beta_1 x + e$. Assuming the errors ($e$) are normally distributed with mean zero and constant variance, the best parameter estimates are obtained by the method of least squares. The parameters β_0 and β_1 are estimated by b_0 and b_1:

$$b_0 = \bar{y} - b_1 \bar{x}$$

$$b_1 = \frac{\sum(x_i - \bar{x})(y_i - \bar{y})}{\sum(x_i - \bar{x})^2}$$

The true response (η) estimated from a measured value of x_0 is $\hat{y} = b_0 - b_1 x_0$.

The statistics b_0, b_1, and $\hat{y}$ are normally distributed random variables with means equal to β_0, β_1, and η, respectively, and variances:

$$\text{Var}(b_0) = \left(\frac{1}{n} + \frac{\bar{x}^2}{\sum(x_i - \bar{x})^2}\right)\sigma^2$$

$$\text{Var}(b_1) = \left(\frac{1}{\sum(x_i - \bar{x})^2}\right)\sigma^2$$

$$\text{Var}(\hat{y}_0) = \left(\frac{1}{n} + \frac{(x_0 - \bar{x})^2}{\sum(x_i - \bar{x})^2}\right)\sigma^2$$

The value of σ^2 is typically unknown and must be estimated from the data; replicate measurements will provide an estimate. If there is no replication, σ^2 is estimated by the mean residual sum of squares (s^2) which has $v = n - 2$ degrees of freedom (two degrees of freedom are lost by estimating the two parameters β_0 and β_1):

$$s^2 = \frac{\Sigma(y_i - \hat{y})^2}{n-2} = \frac{S_R}{n-2}$$

The $(1 - \alpha)100\%$ confidence intervals for β_0 and β_1 are given by:

$$b_0 \pm t_{v,\alpha/2}s\sqrt{\frac{1}{n} + \frac{\bar{x}^2}{\Sigma(x_i - \bar{x})^2}}$$

$$b_1 \pm t_{v,\alpha/2}s\sqrt{\frac{1}{\Sigma(x_i - \bar{x})^2}}$$

These interval estimates suggest that the joint confidence region is rectangular, but this is not so. The joint confidence region is elliptical. The exact solution for the $(1 - \alpha)100\%$ joint confidence region for β_0 and β_1 is enclosed by the ellipse given by:

$$n(b_0 - \beta_0)^2 + 2\left(\sum x_i\right)(b_0 - \beta_0)(b_1 - \beta_1) + \left(\sum x_i^2\right)(b_1 - \beta_1)^2 = 2s^2 F_{2,n-2,\alpha}$$

where $F_{2,n-2,\alpha}$ is the tabulated value of the F statistic with 2 and $n - 2$ degrees of freedom.

The *confidence interval* for the mean response (η_0) at a particular value x_0 is:

$$(b_0 + b_1 x_0) \pm t_{v,\alpha/2}s\sqrt{\frac{1}{n} + \frac{(x_0 - \bar{x})^2}{\Sigma(x_i - \bar{x})^2}}$$

The *prediction interval* for the future single observation ($\hat{y}_f = b_0 + b_1 x_f$) to be recorded at a setting x_f is:

$$(b_0 + b_1 x_f) \pm t_{v,\alpha/2}s\sqrt{1 + \frac{1}{n} + \frac{(x_f - \bar{x})^2}{\Sigma(x_i - \bar{x})^2}}$$

Note that this prediction interval is larger than the confidence interval for the mean response (η_0) because the prediction error includes the error in estimating the mean response plus measurement error in y. This introduces the additional "1" under the square root sign.

Case Study: A Linear Model

Data from calibration of an HPLC instrument and the fitted model are shown in Table 34.1 and in Figure 34.2. The results of fitting the model $y = \beta_0 + \beta_1 x + e$ are shown in Table 34.2. The fitted equation:

$$\hat{y} = b_0 + b_1 x = 0.566 + 139.759x$$

TABLE 34.1

HPLC Calibration Data (in run order from left to right)

Dye Conc.	0.18	0.35	0.055	0.022	0.29	0.15	0.044	0.028
HPLC Peak Area	26.666	50.651	9.628	4.634	40.206	21.369	5.948	4.245
Dye Conc.	0.044	0.073	0.13	0.088	0.26	0.16	0.10	
HPLC Peak Area	4.786	11.321	18.456	12.865	35.186	24.245	14.175	

Source: Bailey, C. J., E. A. Cox, and J. A. Springer (1978). *J. Assoc. Off. Anal. Chem.*, 61, 1404–1414.

TABLE 34.2

Results of the Linear Regression Analysis

Variable	Coefficient	Standard Error	t	P (2-tail)
Constant	0.566	0.473	1.196	0.252
x	139.759	2.889	48.38	0.000

Analysis of Variance

Source	Sum of Squares	Degrees of Freedom	Mean Square	F-Ratio	P
Regression	2794.309	1	2794.309	2340	0.000000
Residual	15.523	13	1.194		

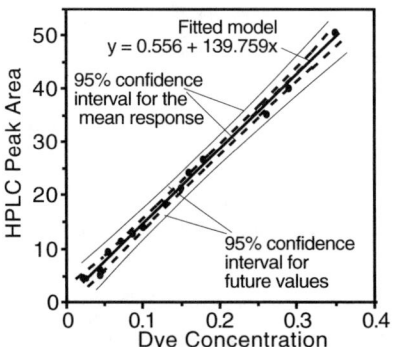

FIGURE 34.2 Fitted calibration line with 95% confidence bounds for the mean and future values.

is shown with the data in Figure 34.2. Also shown are the 95% confidence bounds for the mean and future values.

An estimate of the variance of the measured values is needed to make any statements about the precision of the estimated parameters, or to compute confidence intervals for the line. Because there is no true replication in this experiment, *the mean residual sum of squares* is used as an estimate of the variance σ^2. The mean residual sum of squares is the residual sum of squares divided by the degrees of freedom ($s^2 = \frac{15.523}{13} = 1.194$), which is estimated with $\nu = 15 - 2 = 13$ degrees of freedom. Using this value, the estimated variances of the parameters are:

$$\text{Var}(b_0) = 0.2237 \quad \text{and} \quad \text{Var}(b_1) = 8.346$$

The appropriate value of the t statistic for estimation of the 95% confidence intervals of the parameters is $t_{v=13, \alpha/2=0.025} = 2.16$. The individual confidence intervals estimates are:

$$\beta_0 = 0.566 \pm 1.023 \quad \text{or} \quad -0.457 < \beta_0 < 1.589$$

$$\beta_1 = 139.759 \pm 6.242 \quad \text{or} \quad 133.52 < \beta_1 < 146.00$$

The joint confidence interval for the parameter estimates is given by the shaded area in Figure 34.2. Notice that it is elliptical and not rectangular, as suggested by the individual interval estimates. It is bounded by the contour with sum of squares value:

$$S_c = 15.523\left(1 + \frac{2}{13}(3.81)\right) = 24.62$$

The equation of this ellipse, based on $n = 15$, $b_0 = 0.566$, $b_1 = 139.759$, $s^2 = 1.194$, $F_{2,13,0.05} = 3.8056$, $\sum x_i = 1.974$, $\sum x_i^2 = 0.40284$, is:

$$15(0.566 - \beta_0)^2 + 2(1.974)(0.566 - \beta_0)(139.759 - \beta_1) + (0.40284)(139.759 - \beta_0)^2 = 2(1.194)(3.8056)$$

This simplifies to:

$$15\beta_0^2 - 568.75\beta_0 + 3.95\beta_0\beta_1 - 281.75\beta_1 + 0.403\beta_0^2 + 8176.52 = 0$$

The *confidence interval* for the mean response η_0 at a single chosen value of $x_0 = 0.2$ is:

$$0.566 + 139.759(0.2) \pm 2.16(1.093)\sqrt{\frac{1}{15} + \frac{(0.2 - 0.1316)^2}{0.1431}} = 28.518 \pm 0.744$$

The interval 27.774 to 29.262 can be said with 95% confidence to contain η when $x_0 = 0.2$.
The *prediction interval* for a future single observation recorded at a chosen value (i.e., $x_f = 0.2$) is:

$$0.566 + 139.759(0.2) \pm 2.16(1.093)\sqrt{1 + \frac{1}{15} + \frac{(0.2 - 0.1316)^2}{0.1431}} = 28.518 \pm 2.475$$

It can be stated with 95% confidence that the interval 26.043 to 30.993 will contain the future single observation recorded at $x_f = 0.2$.

Comments

Exact joint confidence regions can be developed for linear models but they are not produced automatically by most statistical software. The usual output is interval estimates as shown in Figure 34.3. These do help interpret the precision of the estimated parameters as long as we remember the ellipse is probably tilted.

Chapters 35 to 40 have more to say about regression and linear models.

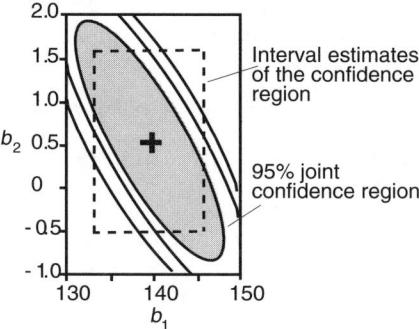

FIGURE 34.3 Contour map of the mean sum of squares surface. The rectangle is bounded by the marginal confidence limits of the parameters considered individually. The shaded area is the 95% joint confidence region for the two parameters and is enclosed by the contour $S_c = 15.523[1 + (2/13)(3.81)] = 24.62$.

References

Bailey, C. J., E. A. Cox, and J. A. Springer (1978). "High Pressure Liquid Chromatographic Determination of the Immediate/Side Reaction Products in FD&C Red No. 2 and FD&C Yellow No. 5: Statistical Analysis of Instrument Response," *J. Assoc. Off. Anal. Chem.*, 61, 1404–1414.

Draper, N. R. and H. Smith (1998). *Applied Regression Analysis*, 3rd ed., New York, John Wiley.

Exercises

34.1 Nonpoint Pollution. The percentage of water collected by a water and sediment sampler was measured over a range of flows. The data are below. (a) Estimate the parameters in a linear model to fit the data. (b) Calculate the variance and 95% confidence interval of each parameter. (c) Find a 95% confidence interval for the mean response at flow = 32 gpm. (d) Find a 95% prediction interval for a measured value of percentage of water collected at 32 gpm.

Percentage	2.65	3.12	3.05	2.86	2.72	2.70	3.04	2.83	2.84	2.49	2.60	3.19	2.54
Flow (gpm)	52.1	19.2	4.8	4.9	35.2	44.4	13.2	25.8	17.6	47.4	35.7	13.9	41.4

Source: Dressing, S. et al. (1987). *J. Envir. Qual.*, 16, 59–64.

34.2 Calibration. Fit the linear (straight line) calibration curve for the following data and evaluate the precision of the estimate slope and intercept. Assume constant variance over the range of the standard concentrations. Plot the 95% joint confidence region for the parameters.

Standard Conc.	0.00	0.01	0.100	0.200	0.500
Absorbance	0.000	0.004	0.041	0.082	0.196

34.3 Reaeration Coefficient. The reaeration coefficient (k_2) depends on water temperature. The model is $k_2(T) = \theta_1 \theta_2^{T-20}$, where T is temperature and θ_1 and θ_2 are parameters. Taking logarithms of both sides gives a linear model: $\ln[k_2(T)] = \ln[\theta_1] + (T - 20) \ln \theta_2$. Estimate θ_1 and θ_2. Plot the 95% joint confidence region. Find 95% prediction intervals for a measured value of k_2 at temperatures of 8.5 and 22°C.

Temp. (°C)	5.27	5.19	5.19	9.95	9.95	9.99	15.06	15.06	15.04
k_2	0.5109	0.4973	0.4972	0.5544	0.5496	0.5424	0.6257	0.6082	0.6304
Temp. (°C)	20.36	20.08	20.1	25.06	25.06	24.85	29.87	29.88	29.66
k_2	0.6974	0.7096	0.7143	0.7876	0.7796	0.8064	0.8918	0.8830	0.8989

Source: Tennessee Valley Authority (1962). *Prediction of Stream Reaeration Rates,* Chattanooga, TN.

34.4 Diesel Fuel Partitioning. The data below describe organic chemicals that are found in diesel fuel and that are soluble in water. Fit a linear model that relates partition coefficient (K) and the aqueous solubility (S) of these chemicals. It is most convenient to work with the logarithms of K and S.

Compound	log(K)	log(S)
Naphthalene	3.67	−3.05
1-Methyl-naphthalene	4.47	−3.72
2-Methyl-naphthalene	4.31	−3.62
Acenaphthene	4.35	−3.98
Fluorene	4.45	−4.03
Phenanthrene	4.60	−4.50
Anthracene	5.15	−4.49
Fluoranthene	5.32	−5.19

Source: Lee, L. S. et al. (1992). *Envir. Sci. Technol.,* 26, 2104–2110.

34.5 Background Lead. Use the following data to estimate the background concentration of lead in the wastewater effluent to which the indicated spike amounts of lead were added. What is the confidence interval for the background concentration?

Pb Added (μg/L)	Five Replicate Measurements of Pb (μg/L)				
0	1.8	1.2	1.3	1.4	1.7
1.25	1.7	1.9	1.7	2.7	2.0
2.5	3.3	2.4	2.7	3.2	3.3
5.0	5.6	5.6	5.6	5.4	6.2
10.0	11.9	10.3	9.3	12.0	9.8

35

Precision of Parameter Estimates in Nonlinear Models

KEY WORDS *biokinetics, BOD, confidence interval, confidence region, joint confidence region, critical sum of squares, Monod model, nonlinear least squares, nonlinear regression, parameter correlation, parameter estimates, precision, residuals sum of squares.*

The precision of parameter estimates in nonlinear models is defined by a boundary on the sum of squares surface. For linear models this boundary traces symmetric shapes (parabolas, ellipses, or ellipsoids). For nonlinear models the shapes are not symmetric and they are not defined by any simple geometric equation.

The *critical sum of squares* value:

$$S_c = S_R + S_R\left(\frac{p}{n-p}F_{p,n-p,\alpha}\right) = S_R\left(1 + \frac{p}{n-p}F_{p,n-p,\alpha}\right)$$

bounds an *exact* $(1 - \alpha)100\%$ *joint confidence region* for a linear model. In this, p is the number of parameters estimated, n is the number of observations, $F_{p,n-p,\alpha}$ is the upper α percent value of the F distribution with p and $n - p$ degrees of freedom, and S_R is the residual sum of squares. We are using $s^2 = S_R/(n - p)$ as an estimate of σ^2.

Joint confidence regions for nonlinear models can be defined by this expression but the confidence level will not be exactly $1 - \alpha$. In general, the exact confidence level is not known, so the defined region is called an *approximate* $1 - \alpha$ *confidence region* (Draper and Smith, 1998). This is because $s^2 = S_R/(n - p)$ is no longer an unbiased estimate of σ^2.

Case Study: A Bacterial Growth Model

Some data obtained by operating a continuous flow biological reactor at steady-state conditions are:

x (mg/L COD)	28	55	83	110	138
y (1/h)	0.053	0.060	0.112	0.105	0.099

The Monod model has been proposed to fit the data:

$$y_i = \frac{\theta_1 x_i}{\theta_2 + x_i} + e_i$$

where y_i = growth rate (h^{-1}) obtained at substrate concentration x, θ_1 = maximum growth rate (h^{-1}), and θ_2 = saturation constant (in units of the substrate concentration).

The parameters θ_1 and θ_2 were estimated by minimizing the sum of squares function:

$$\text{minimize } S = \sum\left(y_i - \frac{\theta_1 x}{\theta_2 + x}\right)^2$$

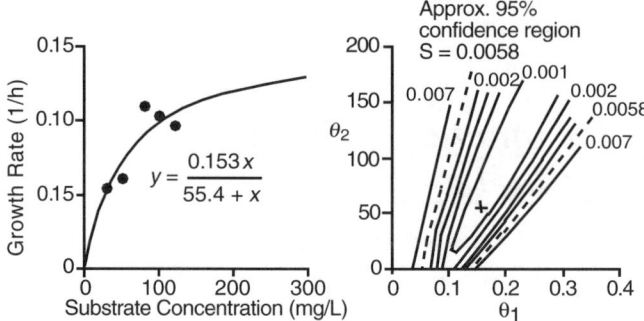

FIGURE 35.1 Monod model fitted to the example data (left) and the contour map of the sum of squares surface and the approximate 95% joint confidence region for the Monod model (right).

to obtain the fitted model:

$$\hat{y} = \frac{0.153x}{55.4 + x}$$

This is plotted over the data in the left-hand panel of Figure 35.1.

The right-hand panel is a contour map of the sum of squares surface that shows the approximate 95% joint confidence region. The contours were mapped from sum of squares values calculated over a grid of paired values for θ_1 and θ_2. For the case study data, $S_R = 0.00079$. For $n = 5$ and $p = 2$, $F_{2,3,0.05} = 9.55$, the critical sum of squares value that bounds the approximate 95% joint confidence region is:

$$S_c = 0.00079\left(1 + \frac{2}{5-2}(9.55)\right) = 0.00582$$

This is a joint confidence region because it considers the parameters as pairs. If we collected a very large number of data sets with $n = 5$ observations at the locations used in the case study, 95% of the pairs of estimated parameter values would be expected to fall within the joint confidence region.

The size of the region indicates how precisely the parameters have been estimated. This confidence region is extremely large. It does not close even when θ_2 is extended to 500. This indicates that the parameter values are poorly estimated.

The Size and Shape of the Confidence Region

Suppose that we do not like the confidence region because it is large, unbounded, or has a funny shape. In short, suppose that the parameter estimates are not sufficiently precise to have adequate predictive value. What can be done?

The size and shape of the confidence region depends on (1) the measurement precision, (2) the number of observations made, and (3) the location of the observations along the scale of the independent variable. Great improvements in measurement precision are not likely to be possible, assuming measurement methods have been practiced and perfected before running the experiment. The number of observations can be relatively less important than the location of the observations. In the case study example of the Monod model, doubling the number of observations by making duplicate tests at the five selected settings

Do More Observations Improve Precision?

Figure 35.2 shows the Monod model fitted to the original data set augmented with two observations at high substrate concentrations, specifically at ($x = 225$ mg/L, $y = 0.122$ h^{-1}) and ($x = 375$ mg/L and $y = 0.125$ h^{-1}). It also shows the resulting approximate 95% joint confidence region. The least squares parameter estimates are $\hat{\theta}_1 = 0.146$ h^{-1} and $\hat{\theta}_2 = 49.1$ mg/L, with a residual sum of squares $S_R = 0.000817$. These values compare closely with $\hat{\theta}_1 = 0.153$, $\hat{\theta}_2 = 55.4$, and $S_R = 0.00079$ obtained with the original five observations.

The differences in the estimated parameter values are small (e.g., 0.146 vs. 0.153 for $\hat{\theta}_1$) and one might wonder whether the two additional observations did lead to more precise parameter estimates. The substantial gain is apparent when the joint confidence region is examined. Figure 35.2 (right-hand panel) shows that both the size and the shape of the confidence region are improved. Whereas the original joint confidence region was so large that the parameter estimates were virtually unbounded, the joint confidence region is now reasonably small. In particular, θ_1 is estimated with good precision and we are highly confident the true value lies between about 0.1 and 0.2. θ_2 is between about 30 and 150 mg/L. The improvement is due mainly to having the new observations at large concentrations because they contribute directly to estimating θ_1. The additional benefit is that any improvement in the precision of θ_1 simultaneously improves the estimate of θ_2.

One reason the confidence region is smaller is because $n = 7$. This gives a value of the F statistic of $F = 5.79$, compared with $F = 9.55$ for $n = 5$. The sum of squares value that bounds the confidence region for $n = 7$ is $S_c = 3.316 S_R$ compared with $S_c = 7.367 S_R$ for $n = 5$. This would explain a reduction of about half in the size of the joint confidence region. The observed reduction is many times more than this.

To see that the improvement in precision is not due entirely to the smaller value of the F statistic, compare Figure 35.1 with Figure 35.3. Figure 35.3 was obtained from $n = 5$ observations (four of the original data points at $x = 28, 55, 83$, and 138 plus the one new observation at $x = 375$). Indeed, the confidence region is larger, being bounded by $S_c = 0.0057$ instead of $S_c = 0.0027$, but it is much smaller than the region in Figure 35.1, which was also for $n = 5$ but without an observation at a high substrate level. Clearly, it is possible to get a small and well-conditioned confidence region without doing a massive number of experiments. Also clear is the corollary: a large number of observations does not guarantee a small confidence region.

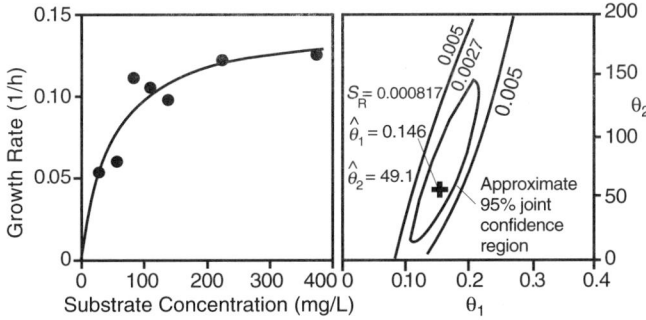

FIGURE 35.2 Monod model fitted to the original five points plus two more at higher substrate concentrations (left) and the resulting joint confidence region of the parameters (right).

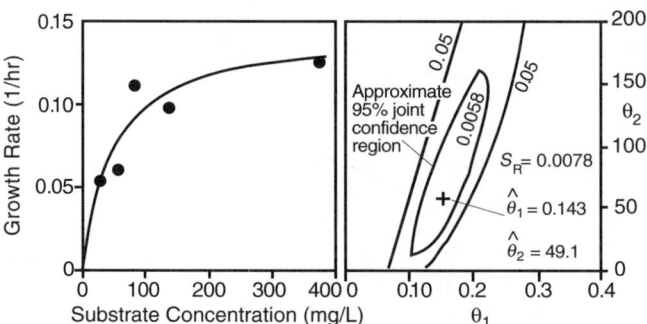

FIGURE 35.3 Monod model fitted to four of the original five points and one additional point at a higher substrate concentration (left) and the resulting joint confidence region of the parameters (right).

The Problem of Parameter Correlation

Parameter correlation means that the estimate of one parameter is related to the estimate of another. Parameter correlation is what causes elongated joint confidence regions. Here we look at the importance of experimental design in reducing parameter correlation.

The location of observations is crucial and making a large number of observations at the wrong locations does not overcome the weakness of a bad experimental design. A great many articles on the effect of temperature, pH, metal concentration, etc. on reactions contain parameters estimated from weak designs (Berthouex and Szewczyk, 1984). The efficiency of aeration equipment was sometimes estimated using experimental designs that could not yield precise parameter estimates (Boyle et al., 1974).

Asymptotic functions, which are common in environmental modeling, present a particular problem in parameter estimation that we will illustrate with the classical first-order model for long-term BOD:

$$y_i = \theta_1[1 - \exp(-\theta_2 t_i)] + e_i$$

where y_i is the BOD measured at time t_i. The ultimate BOD (θ_1) is approached asymptotically as time goes on. θ_2 is the first-order reaction rate coefficient. The reaction is slow and t is measured in days. Each observation of y comes from incubating a test specimen for time t and, as a result, the y values and their errors (e) are independent. We further assume that the errors are normally distributed and have constant variance.

Many published papers show estimates of θ_1 and θ_2 derived from measurements at just a few early times, that is, at days 1 through 5. The experimenters may have reasoned something like this: "I am interested mainly in an estimate of the rate constant, and I want to get the data in the shortest time possible. If θ_2 is known, I can compute θ_1. Because the reaction rate is high on days 1 through 5, data over this range will give a good estimate of θ_2 and the experiment can be finished within a few days."

This plausible argument is wrong. The experiment gives poor estimates of both θ_1 and θ_2. The rate constant can only be precisely estimated if the asymptotic level is well estimated, and this requires having measurements at 15 to 20 days no matter how many measurements are made at early times. The problem is that data at early times only allow estimation of the initial slope of the curve, which is:

$$\left.\frac{dy}{dt}\right|_{t=0} = \theta_2\theta_1\exp(-\theta_2 t)|_{t=0} = \theta_2\theta_1 = \text{constant}$$

Thus, any values of θ_1 and θ_2 that have a product nearly equal to the slope of the curve over the first few days will reasonably fit the data collected on the first few days. The shape of the joint confidence region will be similar to Figure 34.1d. The hyperbolic shape reflects the parameter correlation and shows that neither parameter is well estimated. Well-designed experiments will yield precise, uncorrelated parameter estimates and the joint confidence region will tend toward being elliptical.

Comments

Computing estimates of parameters values is easy. Hand-held calculators can do it for a straight line, and inexpensive commercial software for personal computers can do it for large linear and nonlinear models. Unfortunately, most of these computing resources do not compute or plot the joint confidence region for the parameters.

For linear models, exact confidence regions can be calculated. If the model is nonlinear, we construct an approximate joint confidence region. Although the confidence level (i.e., 95%) associated with the region is approximate, the size and shape are correct. The effort required to determine these confidence regions is rewarded with useful information.

The highly correlated parameter estimates that result from weak experimental designs can lead to serious disputes. Suppose that two laboratories made independent experiments to estimate the parameters in a model and, furthermore, that both used a poor design. One might report $\theta_1 = 750$ and $\theta_2 = 0.3$ while the other reports $\theta_1 = 13,000$ and $\theta_2 = 0.1$. Each might suspect the other of doing careless work, or of making a mistake in calculating the parameter values. If the experimental design produced an elongated confidence region with a high degree of correlation, both sets of parameter values could be within the confidence region. In a sense, both could be considered correct. And both could be useless. If the joint confidence region were presented along with the estimated parameter values, it would be clear that the disagreement arises from flawed experimental designs and not from flawed analytical procedures.

The examples presented here have, for graphical convenience, been based on two-parameter models. We can obtain a picture of the joint confidence region when there are three parameters by making contour maps of slices through the volume that is bounded by the critical sum of squares value.

An interesting variation on nonlinear least squares is fitting kinetic models that cannot be solved explicitly for the measured response. For example, the change in substrate concentration for Michaelis-Menten biokinetics in a batch reactor is $K_m \ln\left(\frac{S_0}{S}\right)(S_0 - S) = v_m t$. There is no analytic solution for S, so the predicted value of S must be calculated iteratively and then compared with the observed value. Goudar and Devlin (2001) discuss this problem.

References

Berthouex, P. M. and J. E. Szewczyk (1984). "Discussion of Influence of Toxic Metals on the Repression of Carbonaceous Oxygen Demand," *Water Res.*, 18, 385–386.

Boyle, W. C., P. M. Berthouex, and T. C. Rooney (1974). "Pitfalls in Parameter Estimation for Oxygen Transfer Data," *J. Envir. Engr. Div., ASCE*, 100, 391–408.

Draper, N. R. and H. Smith (1998). *Applied Regression Analysis*, 3rd ed., New York, John Wiley.

Goudar, C. T. and J. F. Devlin (2001), "Nonlinear Estimation of Microbial and Enzyme Kinetic Parameters from Progress Curve Data," *Water Envir. Res.*, 73, 260–265.

Exercises

35.1 Soybean Oil. The data below are from a long-term BOD test on soybean oil, where y = g BOD per g of soybean oil, and t = time in days. There are four replicates at each time. (a) Fit the first-order BOD model $y = \theta_1[1 - \exp(-\theta_2 t)]$ to estimate the ultimate BOD θ_1 and the rate coefficient, θ_2. (b) Plot the data and the fitted model. (c) Plot the residuals as a function of the predicted values and as a function of time. (d) Estimate σ^2 from $s^2 = S_R/(n - 2)$ and from the replicate observations. Compare the two values obtained. (e) Map the sum of squares surface and indicate the approximate 95% joint confidence region.

1 day	2 days	3 days	5 days	7 days	12 days	20 days
0.40	0.99	0.95	1.53	1.25	2.12	2.42
0.55	0.95	1.00	1.77	1.35	2.21	2.28
0.61	0.98	1.05	1.75	1.90	2.34	1.96
0.66	0.95	1.20	1.95	1.95	1.95	1.92

35.2 BOD Experimental Design I. (a) Fit the BOD model $y = \theta_1[1 - \exp(-\theta_2 t)]$ to the data for days 1, 2, 3, and 5 in Exercise 35.1. (b) Plot the data and the fitted model. (c) Plot the approximate 95% joint confidence region. (d) Select three pairs of parameter values from within the joint confidence region and plot the curves obtained by using them in the model.

35.3 BOD Experimental Design II. Fit the BOD model to the first two rows of data (two replicates) for days 3, 5, 7, 12, and 20, plot the joint confidence region, and compare the result with the confidence region obtained in Exercise 35.1 or 35.2. Discuss how changing the experimental design has affected the size and shape of the confidence region.

35.4 Monod Model. Fit the Monod model $y_i = \theta_1 x_i/(\theta_2 + x_i) + e_i$ to the following data and plot the joint confidence region of the parameters.

Substrate	5.1	20.3	37	74.5	96.5	112	266	386
Growth Rate	0.059	0.177	0.302	0.485	0.546	0.610	0.792	0.852

35.5 Haldane Model. Fit the Monod model $y_i = \theta_1 x_i/(\theta_2 + x_i) + e_i$ and the Haldane model $y_i = \dfrac{\theta_1 x_i}{x_i + \theta_2 + x_i^2/\theta_3} + e_i$ to the following data. y is the growth rate; x is substrate (phenol) concentration. Which model provides the better fit? Support your conclusions with plots of the data, the residuals, and the joint confidence regions.

x	10	25	30	100	450	550	760	870	950
y	0.04	0.085	0.15	0.18	0.15	0.12	0.09	0.11	0.09

35.6 Model Building. You have been asked to identify a biological growth model that fits the following data. Fit some candidate models and discuss the difficulties you encounter.

Substrate Conc.	28	55	83	110	138
Growth Rate	0.053	0.060	0.112	0.105	0.090

35.7 Reactions in Series. Reactant A decomposes to form Product B, which in turn decomposes to form Product C. That is A → B → C. Both reactions are first order with respect to the reactant. The model for the concentration of B is:

$$\eta_B = \dfrac{\theta_1}{\theta_2 - \theta_1} C_A[\exp(-\theta_1 t) - \exp(-\theta_2 t)]$$

where θ_1 is the reaction rate coefficient for the transformation of A to B and θ_2 is the reaction rate coefficient for the transformation of B to C. Three experiments have been done, each with initial conditions $C_A = 1$, $C_B = 0$, and $C_C = 0$ at $t = 0$.

	Concentration of B		
Time (min)	Exp 1	Exp 2	Exp 3
0.65	0.010	—	0.010
1.25	0.022	—	0.022
2.5	0.08	—	0.08
5.0	0.15	—	0.15

10.0	0.22	0.22	0.22
20.0	0.51	0.51	0.51
40.0	—	0.48	0.48
50.0	—	0.29	0.29
60.0	—	0.20	0.20
70.0	—	0.12	0.12

For each experiment, use nonlinear least squares to (a) estimate the two parameters in the model, (b) plot the predicted and observed concentrations of B vs. time, (c) plot the residuals vs. time and vs. the predicted concentration of B, and (d) plot the approximate 95% joint confidence region for the estimated parameters. Discuss the difference between the result obtained for the three experiments.

35.8 Modeling. Discuss two environmental engineering systems that are described by the model in Exercise 35.7. The systems may be natural or constructed. Optional: Find some data for one of the systems and analyze it.

35.9 Chick's Law. The table below gives some data on fraction (F) of coliform bacteria surviving for a given contact time (T, min.) at a given chlorine dosage (C, mg/L). The classical model for bacterial die-off is Chick's law: $\ln(F) = -kCT$, so this might be a starting model. An alternate model is $\ln(F) = -bTC^a$. Which model is best?

	Chlorine Dosage (mg/L)			
Contact Time (min)	1 mg/L	2 mg/L	4 mg/L	8 mg/L
0	1.000	1.000	1.000	1.000
15	0.690	0.280	0.400	0.200
30	0.400	0.110	0.090	0.0390
45	0.250	0.090	0.032	0.0075
60	0.150	0.0320	0.008	0.0012
75	0.080	0.0080	0.0016	0.0003
90	0.050	0.0045	0.0009	0.00005

Source: Tikhe, M. L. (1976). *J. Envir. Engr. Div.,* ASCE, 102, 1019–1028.

36

Calibration

KEY WORDS *calibration, confidence band, confidence interval, inverse prediction, least squares, measurement error, precision, prediction, straight line, regression, Working-Hotelling.*

Instrumental methods of chemical analysis (and some "wet" analyses) are based on being able to construct a calibration curve that will translate the instrument's highly precise physical signal (light absorption, peak height, voltage, etc.) into an estimate of the true concentration of the specimen being tested. The usual procedure to construct a calibration curve is for the analyst to take a series of measurements on prepared specimens (at least three and usually more) for which the concentration is known. It is important that these calibration standards cover the entire range of concentrations required in subsequent analyses. Concentrations of test specimens are to be determined by interpolation and not by extrapolation.

The calibration standards are measured under the same conditions as those that subsequently will be used to test the unknown specimens. Also, a blank specimen should be included in the calibration curve. The blank contains no deliberately added analyte, but it does contain the same solvents, reagents, etc. as the other calibration standards and test specimens, and it is subjected to the same sequence of preparatory steps. The instrumental signal given by the blank sample often will not be zero. It is, however, subject to error and it should be treated in the same manner as any of the other points on the calibration plot. It is wrong to subtract the blank value from the other standard values before plotting the calibration curve.

Let us assume that the best calibration line is to be determined and that the measured concentrations will be accompanied by a statement of their precision. This raises several statistical questions (Miller and Miller, 1984):

1. Does the plot of the calibration curve appear to be well described by a straight line? If it appears to be curved, what is the mathematical form of the curve?
2. Bearing in mind that each point on the calibration curve is subject to error, what is the best straight line (or curved line) through these points?
3. If the calibration curve is actually linear, what are the estimated errors and confidence limits for the slope and intercept of the line?
4. When the calibration curve is used for the analysis of a test specimen, what are the error and confidence limits for the determined concentration?

The calibration curve is always plotted with the instrument response on the vertical (y) axis and the standard concentrations on the horizontal (x) axis. This is because the standard statistical procedures used to fit a calibration line to the data and to interpret the precision of estimated concentrations assume that the y values contain experimental errors but the x values are error-free. Another assumption implicit in the usual analysis of the calibration data is that the magnitude of the errors in the y values is independent of the analyte concentrations. Common sense and experience indicate that measurement error is often proportional to analyte concentration. This can be checked once data are available and the model has been fitted. If the assumption is violated, fit the calibration curve using weighted least squares as explained in Chapter 37.

TABLE 36.1

Calibration Data for HPLC Measurement of Dye

Dye Conc.	0.18	0.35	0.055	0.022	0.29	0.15	0.044	0.028
HPLC Peak Area	26.666	50.651	9.628	4.634	40.206	21.369	5.948	4.245
Dye Conc.	0.044	0.073	0.13	0.088	0.26	0.16	0.10	
HPLC Peak Area	4.786	11.321	18.456	12.865	35.186	24.245	14.175	

Note: In run order reading from left to right.

Source: Bailey, C. J., E. A. Cox, and J. A. Springer (1978). *J. Assoc. Off. Anal. Chem.*, 61, 1404–1414; Hunter, J. S. (1981). *J. Assoc. Off. Anal. Chem.*, 64(3), 574–583.

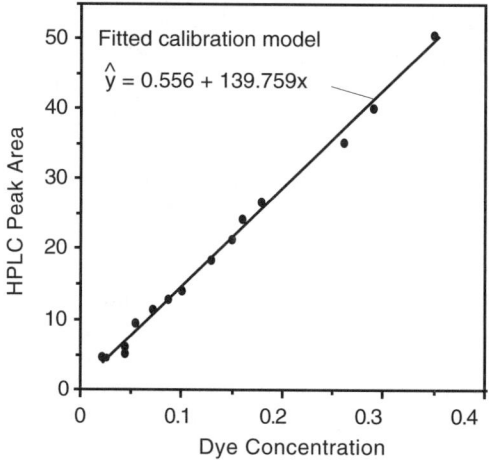

FIGURE 36.1 Plot of the calibration data and the fitted line.

Case Study: HPLC Calibration

A chemist will use the straight-line calibration in Figure 36.1 to predict dye concentration from peak areas measured on a high-pressure liquid chromatograph (HPLC). The calibration data are given in Table 36.1; fitting the calibration line was discussed in Chapter 34. This chapter shows how to obtain a confidence band for the calibration line that gives a confidence interval for the predicted dye concentration.

Theory: A Straight-Line Calibration Curve

The calibration curve will relate the concentration of the standard solution (x) and the instrument response (η). Assume that the functional relationship between these two variables can be well approximated by a straight line of the form $\eta = \beta_0 + \beta_1 \xi$, where β_0 is the intercept and β_1 is the slope of the calibration line. In practice, the true values of η and ξ are not known. Instead, we have the observations x and y, where $x = \xi + e_x$ and $y = \eta + e_y$. Here e_x is a random measurement error associated with the attempt to realize the true concentration (ξ) and e_y is another random measurement error associated with the response η. Assuming this error structure, the model is:

$$y = \beta_0 + \beta_1(x - e_x) + e_y = \beta_0 + \beta_1 x + (e_y - \beta_1 e_x)$$

In the usual straight-line model, it is assumed that the error in x is zero ($e_x = 0$) or, if that is not literally true, that the error in x is much smaller than the error in y (i.e., $e_x \ll e_y$). In terms of the experiment, this means that the settings of the x values are controlled and the experiment can be repeated at any

Calibration

desired x value. In most calibration problems, it is accepted that the random variability of the measurement system can be attributed solely to the y values. Thus, the model becomes the familiar:

$$y = \beta_0 + \beta_1 x + e_y$$

Usually the e_y are assumed to be independent and normally distributed with mean zero and constant variance σ^2.

It is now clear that fitting the best straight-line calibration curve is a problem of linear regression. The slope and intercept of the straight line are estimated and their precision is evaluated using the procedures described in Chapter 34. An excellent summary of single component calibration is given by Danzer and Currie (1998).

Using the Calibration Curve

In practice, the objective is to estimate concentrations (x values) from measured peak areas (y values). The calibration curve is a means to this end. This is the *inverse prediction* problem; that is, the inverse of the common use of a fitted model to predict y from a known value of x.

An interesting problem arises in using the calibration curve in this inverse fashion. Of course it is a simple matter to use the fitted equation ($y = b_0 + b_1 x$) to compute x for any given value of y. Specifically, the estimate of x for any future observation y is $\hat{x} = (y - b_0)/b_1$. Because the calibration curve is to be used over and over again, what is required for a future value of y, in addition to the predicted value $\hat{x}$, is an interval estimate that has the property that at least $100P\%$ of the intervals will contain the true concentration value ξ with $100(1 - \alpha)\%$ confidence. For example, for $P = 0.9$ and $\alpha = 0.05$, we can assert with 95% confidence that 90% of the computed interval estimates of concentration will contain the true value. This problem was resolved by Leiberman et al. (1967). Hunter (1981) gives details.

The error in $\hat{x}$ clearly will depend in some way on the error in measuring y and on how closely b_0 and b_1 estimate the true slope and intercept of the calibration line. The uncertainties associated with estimating β_0 and β_1 imply that the regression line is not unique. Another set of standards and measurements would yield a different line. We should consider, instead of a regression line, that there is a regression band (Sharaf et al., 1986), which can be represented by a *confidence band* for the calibration line. The confidence band for the entire calibration line (i.e., valid for any value of x) can be used to translate the confidence interval for the true response η, based on a future observation of y, into a confidence interval for the abcissa value.

The Working-Hotelling confidence band for the entire line is computed from:

$$b_0 + b_1 x \pm \sqrt{2F_{2,v,\alpha} s^2 \left(\frac{1}{n} + \frac{(x - \bar{x})^2}{\sum(x_i - \bar{x})^2}\right)}$$

in which

$$s^2 = \frac{\sum(y_i - \hat{y})^2}{n - 2}$$

Note that this confidence band is not the same as the confidence interval for y_f predicted from a future value of x_f, a case that was described in Chapter 34. This confidence region can be narrowed somewhat by limiting the range of the abscissa variable (Wynn and Bloomfield, 1971; Hunter, 1981).

Leiberman et al. (1967) proposed a practical approach to using the Working-Hotelling confidence region for the calibration line to obtain a confidence interval for the predicted concentration. A $100P\%$ confidence interval for the true response (η) based upon a future observation y is computed using:

$$y \pm z_p \sqrt{\frac{vs^2}{\chi^2_{v,\alpha/2}}}$$

where y is the measured instrument response and z_p is the standard normal deviate for percentage P. The quantity under the square root is an estimate of the upper $(1 - \alpha)100\%$ confidence limit on the standard deviation of the instrument response (y) where s^2 is the variance of replicate measurements of y.

$\chi^2_{\nu,\alpha/2}$ is the lower $\alpha/2$ percentile point of the χ^2 distribution, where $\nu = n - 2$ degrees of freedom associated with the variance estimate (s^2).

This interval for y and the Working–Hotelling confidence band for the calibration line are used to estimate a confidence interval about $\hat{x}$ for the true concentration ξ. This typically is done graphically. Hunter (1981) provides a full derivation and explanation of the procedure.

Case Study: Solution

The fitted model calibration line is:

$$y = b_0 + b_1 x = 0.556 + 139.759 x$$

The estimate of the variance is $s^2 = 1.194$, which has $\nu = 15 - 2 = 13$ degrees of freedom. The estimated variances of the parameters are $\text{Var}(b_0) = 0.2237$ and $\text{Var}(b_1) = 8.346$. The appropriate value of the t statistic for estimation of the 95% confidence intervals of the parameters is $t_{13,0.025} = 2.16$, giving 95% confidence intervals as follows:

$$\beta_0 = 0.566 \pm 1.023 \quad \text{or} \quad -0.457 < \beta_0 < 1.589$$
$$\beta_1 = 139.759 \pm 6.242 \quad \text{or} \quad 133.52 < \beta_1 < 146.00$$

Using the Calibration Curve to Predict Concentrations

Suppose that the chemist measured a peak area of 22.0. The predicted concentration is read from the calibration line or computed from the calibration equation:

$$\hat{x} = (y - b_0)/b_1 = (22.0 - 0.566)/139.76 = 0.153.$$

To establish confidence limits for the true concentration, first determine the confidence limits for the value $y = 22.0$. Then, using the Working-Hotelling confidence band for the calibration line, translate the confidence limits for y into confidence limits for x. Figure 36.2 shows the fitted calibration line with Working-Hotelling 95% confidence band. Only part of the range of dye concentration is shown in order to expand the scale.

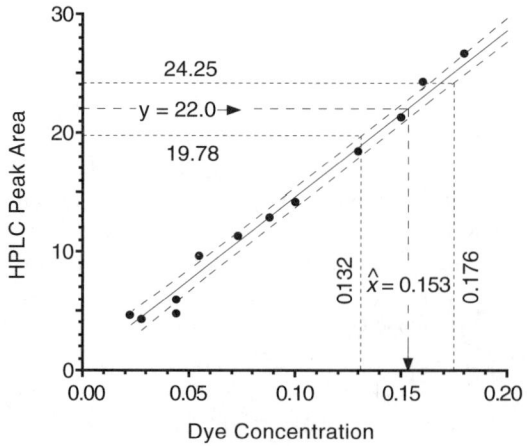

FIGURE 36.2 The lower-concentration portion of the fitted calibration curve with its Working-Hotelling 95% band showing an example of predicting x from $y = 22.0$.

Calibration

The specific steps to construct the Working-Hotelling confidence band and confidence interval are as follows. The measured peak area of 22.0 yields an estimated concentration of 0.153. The confidence limits for the y value that will contain the true value of the peak area (ordinate) 90% of the time are calculated using $z_{p=0.9} = 1.28$. To have a confidence level of $100(1 - \alpha)\% = 95\%$, specify $\alpha/2 = 0.025$, which gives $\chi^2_{13,0.025} = 5.01$. The corresponding interval for the peak area is:

$$y \pm 1.28 \sqrt{\frac{13(1.194)}{5.01}} = y \pm 2.253$$

For the observed peak area ($y = 22.0$), the relevant interval estimate of the ordinate variable is 22 ± 2.253. Or, we can state with 95% confidence that the bounds:

$$19.75 \le \eta \le 24.25$$

contain the true value of peak area 90% of the time.

The bounds for the true concentration, read from the Working-Hotelling confidence limits in Figure 36.2, are:

$$0.132 \le \xi \le 0.176$$

These bounds contain the true value of the abscissa variable 90% of the time, a statement we can make with 95% confidence.

Comments

Suppose that the calibration curve is constructed and the confidence interval for predicted concentration values is found to be undesirably large. Aside from simply using more care to make the measurements, the confidence band of the calibration curve can be narrowed by increasing the number of calibration points on the straight line (this includes replication of points). Instead of trying to shrink the confidence band of the calibration line, we could shrink the confidence interval of the measured y that will be translated into a prediction of x.

To explain the benefits of replication as simply as possible, we will use an approximate method (instead of Leiberman's method) to evaluate the error in the predicted concentrations (Miller and Miller, 1984). Approximate confidence limits on ξ can be computed using $x_0 \pm ts_{x_0}$, where t has $n - 2$ degrees of freedom, and:

$$s_{x_0} = \frac{s}{b_1} \sqrt{\frac{1}{m} + \frac{1}{n} + \frac{(y_0 - \bar{y})^2}{b_1^2 \sum(x_i - \bar{x})^2}}$$

All terms have been defined earlier except m, which is the number of replicates of y that are averaged to get y_0.

Two approaches to shrinking this confidence interval can be assessed by considering the terms under the square root sign in the equation above. If only one measurement of y is made, $m = 1$ and the first term will be large compared with the $1/n$ of the second term. Making m replicate measurements of y reduces the first term to $1/m$. Just a few replicates give a rapid gain. For example, suppose that we have $b_1 = 2.0$, $\sum(x_i - \bar{x})^2 = 100$, $\bar{y} = 20$, $s = 0.4$, and $n = 5$. The confidence interval on x is directly proportional to the standard deviation:

$$s_{x_0} = \frac{0.4}{2} \sqrt{\frac{1}{m} + \frac{1}{5} + \frac{(y_0 - 20)^2}{(2)^2(100)}}$$

Suppose we are interested in predicting concentration from a measured value of $y = 30$. For $m = 1$, $s_{x_0} = 0.24$. If we had $m = 4$ measurements that average $y_o = 30$, s_{x_0} would be reduced to 0.17. This is a considerable gain in precision.

The effect of increasing the number of calibration points (n) is more complex because changing n also changes the value of the t and the χ^2 statistics that are used to compute the confidence intervals. The motivation to reduce n is to limit the amount of work involved with preparing standards at many concentrations. It is clear, however, that large values of n will decrease the confidence interval by making the term $1/n$ small, while simultaneously decreasing the corresponding t statistic. In most calibrations, six or so calibration points will be adequate and if extra precision is needed, it can be gained most efficiently by making repeated measurements at some or all calibration levels.

The calibration model may not be a straight line, and the variance of the signal may increase as the values of x and y increase. This is the case study in Chapter 37 which deals with weighted regression.

There are other interesting problems associated with calibrations. In some analytical applications (e.g., photometric titrations), it is necessary to locate the intersection of two regression lines. In the standard addition method, the concentration is estimated by extrapolating a straight line down to the abscissa. References at the end of the chapter provide guidance on these problems.

A special case is where there are errors in both x and y. Errors in x can be ignored if $\sigma_x^2 \ll \sigma_y^2$. Carroll and Spiegelman (1986) examine this criteria in some detail. The effect of errors in x is to pull the regression line down so the estimated slope is less than would be estimated if errors in x were taken into account. In going from y to x, this could badly overestimate x. This emphasizes the importance of using accurate standards in preparing calibration curves.

Values of the correlation coefficient (r) and the coefficient of determination (R^2) are often cited as evidence that the calibration relation is strong and useful, or that the calibration is in fact a straight line. An R^2 value near 1.00 does not prove these points and in the context of calibration curves it has little meaning of any kind. Values of $R^2 = 0.99+$ are to be expected in calibration curves. If the relation between standard and instrumental response is not clean and strong, there is simply no useful measurement method. Second, the value of R^2 value can be increased without increasing the precision of the measurements or of the predictions. This is done simply by expanding the range of concentrations covered by the standards. Third, R^2 can be large (>0.98) although the curve deviates slightly from the linear. The coefficient of determination (R^2) is discussed in Chapter 39.

References

Bailey, C. J., E. A. Cox, and J. A. Springer (1978). "High Pressure Liquid Chromatographic Determination of the Immediate/Side Reaction Products in FD&C Red No. 2 and FD&C Yellow No. 5: Statistical Analysis of Instrument Response," *J. Assoc. Off. Anal. Chem.*, 61, 1404–1414.

Carroll, R. J. and C. H. Spiegelman (1986). "The Effect of Ignoring Small Measurement Errors in Precision Instrument Calibration," *J. Quality Tech.*, 18(3), 170–173.

Danzer, K. and L. A. Currie (1998). "Guidelines for Calibration in Analytical Chemistry," *Pure Appl. Chem.*, 70, 993–1014.

Hunt, D. T. E. and A. L. Wilson (1986). *The Chemical Analysis of Water*, 2nd ed., London, The Royal Society of Chemistry.

Hunter, J. S. (1981). "Calibration and the Straight Line: Current Statistical Practice," *J. Assoc. Off. Anal. Chem.*, 64(3), 574–583.

Leiberman, G. J., R. G. Miller, Jr., and M. A. Hamilton (1967). "Unlimited Simultaneous Discrimination Intervals in Regression," *Biometrika*, 54, 133–145.

Mandel, J. (1964). *The Statistical Analysis of Experimental Data*, New York, Interscience Publishers.

Miller, J. C. and J. N. Miller (1984). *Statistics for Analytical Chemistry*, Chichester, England, Ellis Horwood Ltd.

Sharaf, M. A., D. L. Illman, and B. Kowalski (1986). *Chemometrics*, New York, John Wiley.

Wynn, H. P. and P. Bloomfield (1971). "Simultaneous Confidence Bands in Regression Analysis," *J. Royal Stat. Soc. B*, 33, 202–217.

Exercises

36.1 Cyanide Measurement. Calibration of a photometric method for measuring cyanide yielded the following data. Does a straight line provide an adequate calibration curve?

Conc.	0	10	20	30	40	50	60	70	80
Abs.	0.000	0.049	0.099	0.153	0.203	0.258	0.310	0.356	0.406
Conc.	90	100	110	120	130	140	150	160	170
Abs.	0.460	0.504	0.561	0.609	0.671	0.708	0.761	0.803	0.863
Conc.	180	190	200	210	220	230	240	250	
Abs.	0.904	0.956	0.997	1.053	1.102	1.158	1.186	1.245	

36.2 Sulfate Calibration. The table gives calibration data for HPLC measurement of sulfate. (a) Fit the calibration curve and plot the Working-Hotelling confidence band for the calibration line. (b) Determine the 95% prediction interval for a peak height at a sulfate concentration of 2.5 mg/L. (c) An unknown sample gives a peak height of 1500. Estimate the sulfate concentration and its 95% confidence interval.

Sulfate (mg/L)	0.2	0.4	0.7	1	2	3.5	5	10
Peak Height	237	438	787	1076	2193	3821	5519	11192

36.3 UV Absorbance. The data below are for calibration of an instrument that measures absorbance of light at 245 nm. (a) Fit the calibration curve and plot the Working-Hotelling confidence band for the calibration line. (b) Determine the 95% prediction interval for absorbance at a COD concentration of 275 mg/L. (c) An unknown sample gives an absorbance of 1.10. Estimate the COD concentration and its 95% confidence interval.

COD	Abs.	COD	Abs.	COD	Abs.	COD	Abs.
60	0.30	140	0.70	375	1.80	525	2.50
70	0.30	195	0.95	375	1.65	550	2.20
90	0.35	250	1.30	380	1.70	550	2.30
100	0.45	250	1.50	450	1.75	550	2.50
100	0.70	300	1.60	480	1.50	550	2.70
120	0.50	300	1.65	500	2.30	600	2.35
130	0.48	350	1.70	500	2.45	650	2.45
130	0.70	350	1.80	525	2.35	675	2.70

Source: Briggs and Gatter (1992). *ISA Trans.*, 31, 111–123.

36.4 TSS Measurement. The data below are for calibration of an instrument to measure total suspended solids (TSS) in wastewater. The data were obtained by reading the instrument and simultaneously grabbing a wastewater sample for a later analysis. Derive the calibration curve for instrument signal as a function of TSS. Discuss how this method of calibrating an instrument differs from usual practice.

Signal	TSS	Signal	TSS	Signal	TSS	Signal	TSS
30	94	60	58	80	40	100	34
40	80	60	62	80	46	120	19
40	86	60	68	90	38	120	22
50	70	70	52	90	41	120	23
50	76	70	56	100	28	130	12

Source: Briggs and Gatter (1992). *ISA Trans.*, 31, 111–123.

37

Weighted Least Squares

KEY WORDS *calibration, confidence interval, constant variance, ion chromatography, least squares, nitrate, regression, simulation, variance, weighted least squares, weights, weighted regression.*

Weighted least squares (*weighted regression*) is used to fit a model $y = f(\beta, x)$ when the variance of the response y is not constant. This method is often needed to fit calibration curves where x and y cover wide ranges of concentration and response signal. As the range of values gets larger, it becomes more likely that there will be some proportionality between the level of the response and the variance of the response.

Figure 37.1 shows four possible situations: (a) a straight line with random measurement errors of equal variance at all concentrations, (b) a straight line with random measurement errors that are proportional to concentration, (c) a curved function with random measurement errors of constant variance, and (d) a curved function with random measurement errors that are proportional to concentration.

For curves a and c the magnitude of the error in y is the same over the full range of x. We can think of these curves as having random measurement error in the range of $\pm \Delta$ mg/L added or subtracted from the true response, where Δ does not change as the concentration changes. In contrast, for curves b and d the magnitude of the error increases as x and y increase. The error for these curves tends to be a percentage of the response. We can think of these curves as having random measurement error in the range of $\pm \Delta\%$ of the true response level added or subtracted from the true response. Then the measurement error in units of milliliters per liter (mg/L) does change over the range of concentration values.

Calibration curve a is the most familiar (see Chapter 36) but there is no theoretical reason for assuming that a straight line will be the correct calibration model. There are factors inherent in some measurement processes that make curve b much more likely than curve a (e.g., ICP instruments). Also, there are some instruments for which a straight-line calibration is not expected (e.g., HPLC instruments). Therefore, the analyst must know how to recognize and how to treat calibration data for these four patterns.

The analyst must decide whether the data have constant variance and if not, what weights should be assigned to each y value. In addition, the analyst is also trying to decide the form of the fitted function (straight line or some curved function). These decisions will have a profound effect on confidence intervals and prediction intervals. The confidence intervals of parameter estimates will be wrong if weighted regression has not been used when there is nonconstant variance. Also, prediction intervals for the model line and specimen concentrations will be wrong unless proper weighting and model form have been used.

Case Study: Ion Chromatograph Calibration

The nitrate calibration data in Table 37.1 were made on a Dionex-120 ion chromatograph. There are triplicate measurements at 13 concentration levels covering a range from 0 to 40 mg/L nitrate.

Suppose we assume that the calibration model is a straight line and fit it using ordinary least squares to obtain:

$$\hat{y} = -3246 + 8795x$$

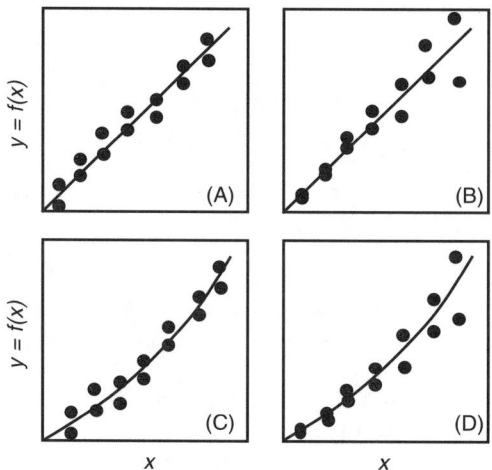

FIGURE 37.1 Four possible calibration curves: (a) linear with random measurement errors of equal magnitude at all standard concentrations, (b) linear with random measurement errors that are larger at high standard concentrations, (c) curvilinear with random measurement errors of equal magnitude at all standard concentrations, and (d) curvilinear with random measurement errors that are larger at high standard concentrations.

TABLE 37.1

Ion Chromatograph Data for Nitrate

Nitrate (mg/L)	Peak 1	Peak 2	Peak 3	Average of Replicates	Standard Deviation	Variance
0.05	390	354	371	372	18	324
0.15	1030	1028	1051	1036	12	162
0.275	1912	1909	1914	1912	2	6
0.4	2784	2779	2678	2747	60	3577
0.8	5616	5637	5612	5622	13	180
1.4	9733	9821	9786	9780	44	1963
2.0	14213	14238	14236	14229	14	193
4.0	29138	29504	29473	29372	203	41190
7.0	53065	53701	53326	53364	320	102207
10.0	75967	78499	78422	77629	1440	2073996
20.0	166985	169620	169379	168661	1457	2122090
30.0	257119	262286	262765	260723	3131	9800774
40.0	351149	353419	355161	353243	2012	4047268

Source: Greg Zelinka, Madison Metropolitan Sewerage District, Madison, WI.

with $R^2 = 0.999$. Figure 37.2 shows the data and the fitted line. The graph shows no obvious discrepancies. The large R^2 is often taken as evidence that the straight line is the correct calibration curve. This is wrong. A large R^2 proves nothing about whether the model adequately fits the data. This straight-line model does not fit the data.

Figure 37.3 shows two problems with this calibration. The enlarged view of lower concentrations (left panel) shows that the straight line does not fit the low concentrations. A plot of the residuals against the predicted peak value shows that the residuals are not random. The straight line underestimates at the low and very high values and overestimates in the mid-range. Furthermore, the vertical spread of the three residuals at each concentration increases as the peak height increases.

Either (1) the straight line must be replaced with a curve, (2) weighted least squares must be used to give the high concentrations less influence, or (3) weighted least squares must be used to fit a curved function.

Weighted Least Squares 329

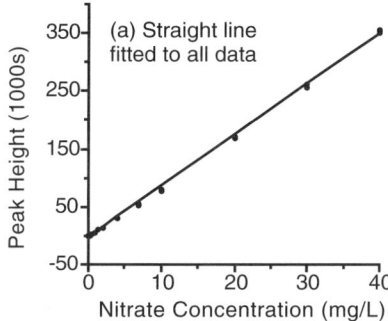

FIGURE 37.2 Plot of the nitrate calibration data and a straight line fitted by ordinary (unweighted) least squares.

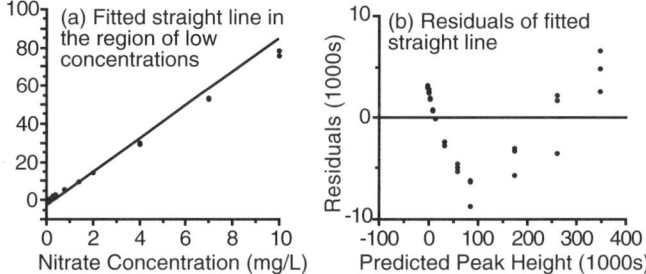

FIGURE 37.3 (a) Expanded view of the straight-line calibration shows lack of fit at low concentrations. (b) Residuals of the straight-line model show lack of fit and suggest that a quadratic or cubic calibration model should be tried. The greater spread of triplicates at higher values of peak height also suggests that weighting would be appropriate.

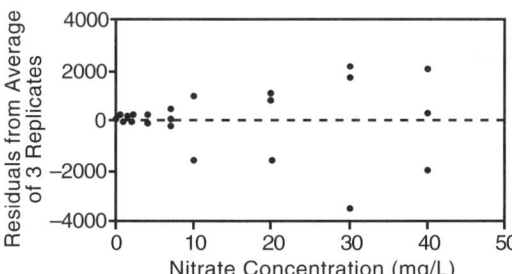

FIGURE 37.4 Residuals from the average at each of the 13 concentration levels used to construct the calibration curve show that the variance increases in proportion to nitrate concentrations.

Diagnosing the need for weighted least squares is easy in this case because there are triplicate measurements. The variation within replicates is evident in the tabulated data, but it is hidden in Figure 37.2. Figure 37.3 suggests the nonconstant variance, but Figure 37.4 shows it better by flattening the curve to show the residuals with respect to the average at each of the 13 standard concentration levels. The residuals are larger when the analyte concentration is large, which means that the variance is not constant at all concentration levels.

There is a further problem with the straight-line analysis given above. A check on the confidence interval of the intercept would support keeping the negative value. This confidence interval is wrong because the residuals of the fitted model are not random and they *do not* have constant variance. The violation of the constant variance condition of regression distorts all statements about confidence intervals, prediction intervals, and probabilities associated with these quantities. Thus, weighting is important even if it does not make a notable difference in the position of the line.

Theory: Weighted Least Squares

The following is a general statement of the least squares criterion. It is used for all models, linear and otherwise, and for constant or nonconstant variance. If the values of the response are $y_1, y_2, \ldots y_n$ and if the variances of these observations are $\sigma_1^2, \sigma_2^2, \ldots, \sigma_n^2$, then the parameter estimates that individually and uniquely have the smallest variance will be obtained by minimizing the weighted sum of squares:

$$\text{minimize } S = \sum w_i (y_i - \eta)^2$$

where η is the response calculated from the proposed model, y_i is the observation at a specified value of x_i, and w_i is the weight assigned to observation y_i. The w_i will be proportional to $1/\sigma_i^2$. If the variance is constant ($\sigma_1^2 = \sigma_2^2 = \cdots = \sigma_n^2$), all $w_i = 1$, and each observation has an equal opportunity to determine the calibration curve. If the variance is not constant, the least accurate measurements are assigned a small weight and the most accurate measurements are assigned large weights. This prevents the least accurate measurements from dominating the outcome of the regression.

The least squares parameter estimates for a general linear model $\eta = \beta_0 + \beta_1 x + \beta_2 x^2 + \cdots + \beta_n x^n$ are obtained from:

$$\text{minimize } S = \sum w_i (y_i - \eta)^2 = \sum w_i [y_i - (\beta_0 + \beta_1 x_i + \cdots + \beta_n x_i^n)]^2$$

The analytical solution for a straight-line model applied to calibration is given in Gibbons (1994), Otto (1999), and Zorn et al. (1997, 1999).

Determining the Appropriate Weights

If the variance is not constant, the magnitude of the weights will depend somehow on the magnitude of the variance. We present two ways in which the weights might be assigned.

Method 1

The weights are inversely proportional to the variance of each observation ($w_i = 1/\sigma_i^2$) where s_i^2 is used as an estimate of σ_i^2. Obviously this method can only be used when there are replicate measurements to calculate the s_i^2. The weights may increase smoothly from low to high levels of the analyte, as shown in the left-hand panel of Figure 37.5, or they might be as shown in the other two panels. Using weights that are inversely proportional to the variance will deal with any of these cases.

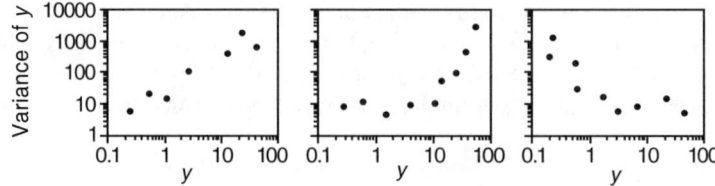

FIGURE 37.5 Possible patterns of variance and concentration. Using weights that are inversely proportional to the variance ($w_i = 1/s_i^2$) will deal with any of these cases.

Method 2

The weights are a smooth function of the dependent variable, such as:

$$\sigma_i^2 = a y_i^b$$

and

$$w_i = \frac{1}{\sigma_i^2} = \frac{1}{a y_i^b}$$

In the instrument calibration problem, the relation between x and y is linear, or very nearly so, which makes it possible to express equivalent weights in terms of x instead of y. That is:

$$\sigma_i^2 = a' x_i^{b'}$$

and

$$w_i = \frac{1}{\sigma_i^2} = \frac{1}{a' x_i^{b'}}$$

The value of a (or a') is irrelevant because it appears in each term of the sum of squares function and can be factored out. Therefore, the weights will be:

$$w_i \propto \frac{1}{\sigma_i^2} \propto \frac{1}{y_i^b}$$

or

$$w_i \propto \frac{1}{\sigma_i^2} \propto \frac{1}{x_i^{b'}}$$

The value of b is the slope of a plot of $\log(\sigma^2)$ against $\log(y)$, or b' is the slope of a plot of $\log(\sigma^2)$ against $\log(x)$. For some instruments (e.g., ICP), we expect $b = 1$ (or $b' = 1$) and the weights are proportional to the inverse of the concentration. For others (e.g., HPLC), we expect $b = 2$.

The absolute values of the weights are not important. The numerical values of the weights can be scaled in any way that is convenient, as long as they convey the relative precision of the different measurements. The absolute values of weights $w_i = 1/x_i^2$ will be numerically much different than $w_i = 1/s_i^2$, but in both cases the weights generally decrease with concentration and have the same relative values.

Case Study: Solution

Table 37.2 shows the variances and the weights calculated by Methods 1 and 2. The relative weights in the two right-hand columns have been scaled so the least precisely measured concentration level has $w = 1$. The calculated calibration curve will be the same whether we use the weights or the relative weights. For this nitrate calibration data, the measurements at low concentrations are, roughly speaking, 10^5 times more precise (in terms of variance) than those at high concentrations.

The calibration curve obtained using Minitab to do weighted least squares with weights $w_i = 1/s_i^2$ is:

$$\hat{y} = 4.402 + 6908.57x + 105.501x^2 - 1.4428x^3$$

The diagnostic output is given in Table 37.3. The constant (intercept) has a small t value and can be made zero. The quadratic and the cubic terms are justified in the equation. This can be seen by examining

TABLE 37.2

Variance, Weights, and Relative Weights for the Nitrate Calibration Data

Nitrate (mg/L) y	Variance s^2	Weights		Relative Weights	
		$w = 1/s^2$	$w = 1/x^2$	$w = 1/s^2$	$w = 1/x^2$
0.05	324	0.0030832	400.0	30218	640000
0.15	162	0.0061602	44.4	60374	71111
0.28	6	0.1578947	13.2	1547491	21157
0.40	3577	0.0002796	6.2	2740	10000
0.80	180	0.0055453	1.6	54348	2500
1.40	1963	0.0005094	0.51	4993	816
2.00	193	0.0051813	0.25	50781	400
4.00	41190	0.0000243	0.062	238	100
7.00	102207	0.0000098	0.020	96	33
10.00	2073996	0.0000005	0.010	45	16
20.00	2122090	0.0000005	0.0025	5	4
30.00	9800774	0.0000001	0.0011	1	2
40.00	4047268	0.0000002	0.0006	2	1

TABLE 37.3

Diagnostic Statistics for the Cubic Calibration Curve Fitted Using Weights $w_i = 1/s_i^2$

Predictor Variable	Parameter Estimate	Std. Dev. of Estimate	t-ratio	p
Constant	4.402	3.718	1.18	0.244
x	6908.57	11.98	576.83	0.000
x^2	105.501	4.686	22.51	0.000
x^3	−1.4428	0.1184	−12.18	0.000

the t statistic, the value of p, or by computing the confidence interval, for each parameter. Roughly speaking, a t value less than 2.5 means that the parameter is not significant. p is the probability that the parameter is not significant. Small t corresponds to large p; $t = 2.5$ corresponds to $p = 0.05$, or 95% confidence in the statement about significance. The half-width of approximate confidence interval is two times the standard deviation of the estimate.

Setting the constant equal to zero and refitting the model gives:

$$\hat{y} = 6920.8x + 101.68x^2 - 1.35x^3$$

Figure 37.6 shows the weighted residuals for the cubic model plotted as a function of the predicted (fitted) peak value. A logarithmic horizontal axis was used to better display the residuals at the low concentrations. (There was no log transformation involved in fitting the data.) The magnitude of the residuals is the same over the range of the predicted variable. This is the condition that weighting was supposed to create. Therefore, the weighted least squares is the correct approach. (It is left as an exercise for the reader to do the unweighted regression and see that the residuals increase as y increases.)

Figure 37.7 shows log-log plots of $\sigma_i^2 = a y_i^b$ for the nitrate calibration data. The estimated slope is $a = 2.0$, which corresponds to weights of $w_i = 1/y_i^2$. Because the relation between x and y is nearly linear, an excellent approximation would be $w_i = 1/x_i^2$. Obviously, the weights $w_i = 1/y_i^2$ and $w_i = 1/x_i^2$ will be numerically different, and the estimated parameter values will be slightly different as well. The weighting and the results, nevertheless, are valid. Weighting with respect to x is convenient because it can be done when there are no replicate measurements with which to calculate variances of the y's. Also, the weights with respect to x will increase in a smooth pattern, whereas the *calculated* variances of the y's do not.

Weighted Least Squares

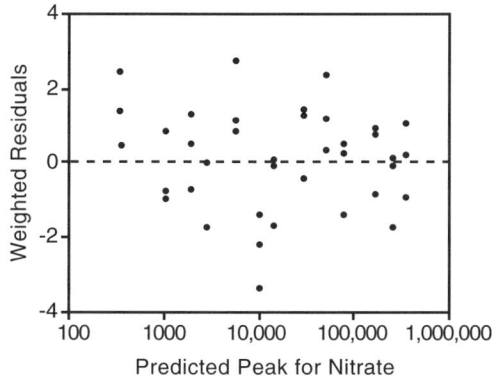

FIGURE 37.6 Plots of the weighted residuals $(y_i - \hat{y})/s_i^2$ for the cubic equation fitted to the nitrate calibration data using weighted regression. A logarithmic horizontal axis is used to better display the residuals at the low concentrations.

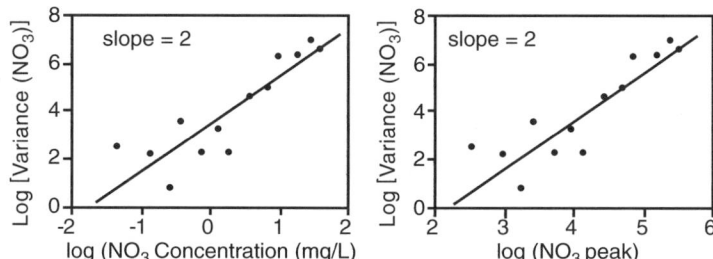

FIGURE 37.7 Plot of the sample variance as a function of concentration (left) and peak value (right).

What To Do If You Have No Replicates?

Having replicates allows the analyst to check for nonconstant variance and to quantify it. It is common (but not recommended) for analysts to have no replication. Not having replicates does not eliminate the nonconstant variance. It merely hides the problem and misleads the analyst.

Prior experience or special knowledge of an instrument may lead one to do weighted regression even if there are no replicates. If the precision of measurements at high concentrations is likely to be much poorer than the precision at low concentrations, the regression should be weighted to reflect this.

Without replication it is impossible to calculate the variances that are needed to use Method 1 ($w_i = 1/s_i^2$). Method 2 is still available, however, and it should be used. Using either $w = 1/x$ or $w = 1/x^2$ will be better than using $w = 1$ for all levels. Do not avoid weighted regression just because you are unsure of which is best. Either weighting will be better than none.

Constant Variance: How Variable is the Variance?

Constant variance means that the underlying true variance (σ^2) does not change. It does not mean that the sample variance (s_i^2) will be exactly the same at all levels of x and y. In fact, the sample variances may be quite different even when the true variance is constant. This is because the statistic s_i^2 is not a very precise estimator of σ^2 when the sample size is small. This will be shown by simulation.

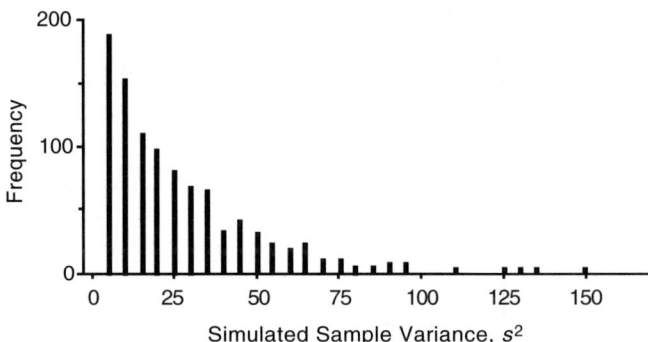

FIGURE 37.8 Distribution of 1000 simulated sample variances, each calculated using three observations drawn at random from a normal distribution with $\sigma^2 = 25$. The average of the 1000 simulated values is 25.3, with 30 variances above 100 and 190 variances of five or less.

Figure 37.8 is the histogram of 1000 sample variances, each calculated using three observations drawn from a normal distribution with $\sigma^2 = 25$. The average of the simulated sample variances was 25.3, with 30 values above 100 and 190 values of five or less. This is the range of variation in s_i^2 for sample size $n = 3$.

A formal comparison of the equality of two sample variances uses the F statistic. Comparing two samples variances, each estimated with three degrees of freedom, would use the upper 5% value of $F_{3,3} = 9.28$. If the ratio of the larger to the smaller of two variances is less than this F value, the two variances would be considered equal. For $F_{3,3} = 9.28$, this would include variances from $25/9.28 = 2.7$ to $25(9.28) = 232$.

This shows that the variance of repeat observations in a calibration experiment will be quite variable due to random experimental error. If triplicate observations in a calibration experiment did have *true constant variance* $\sigma^2 = 25$, replicates at one concentration level could have $s^2 = 3$, and at another level (not necessarily a higher concentration) the variance could be $s^2 = 200$. Therefore, our interest is not in "unchanging" variance, but rather in the pattern of change over the range of x or y. If change from one level of y to another is random, the variances are probably just reflecting random sampling error. If the variance increases in proportion to one of the variables, weighted least squares should be used.

Making the slopes in Figure 37.7 integer values was justified by saying that the variance is estimated with low precision when there are only three replicates. Box (personal communication) has shown that the percent error in the variance is % error = $100/\sqrt{2\nu}$, where ν is the degrees of freedom. From this, about 200 observations of y would be needed to estimate the variance with an error of 5%.

Comments

Nonconstant variance may occur in a variety of situations. It is common in calibration data because they cover a wide range of concentration, and also because certain measurement errors tend to be multiplicative instead of additive.

Using unweighted least squares when there is nonconstant variance will distort all calculated t statistics, confidence intervals, and prediction intervals. It will lead to wrong decisions about the form of the calibration model and which parameters should be included in the model, and give biased estimates of analyte concentrations.

The appropriate weights can be determined from the data if replicate measurements have been made at some settings of x. These should be true replicates and not merely multiple measurements on the same standard solution.

If there is no replication, one may falsely assume that the variance is constant when it is not. If you suspect nonconstant variance, based on prior experience or knowledge about an instrument, apply reasonable weights. Any reasonable weighting is likely to be better than none.

Weighted Least Squares

One reason analysts often make many measurements at low concentrations is to use the calibration data to calculate the limit of detection for the measurement process. If this is to be done, proper weighting is critical (Zorn et al., 1997 and 1999).

References

Currie, L. A. (1984). "Chemometrics and Analytical Chemistry," in *Chemometrics: Mathematics and Statistics in Chemistry,* NATO ASI Series C, 138, 115–146.
Danzer, K. and L. A. Currie (1998). "Guidelines for Calibration in Analytical Chemistry," *Pure Appl. Chem.,* 70, 993–1014.
Gibbons, R. D. (1994). *Statistical Methods for Groundwater Monitoring,* New York, John Wiley.
Draper, N. R. and H. Smith, (1998). *Applied Regression Analysis,* 3rd ed., New York, John Wiley.
Otto, M. (1999). *Chemometrics,* Weinheim, Germany, Wiley-VCH.
Zorn, M. E., R. D. Gibbons, and W. C. Sonzogni (1999). "Evaluation of Approximate Methods for Calculating the Limit of Detection and Limit of Quantitation," *Envir. Sci. & Tech.,* 33(13), 2291–2295.
Zorn, M. E., R. D. Gibbons, and W. C. Sonzogni (1997). "Weighted Least Squares Approach to Calculating Limits of Detection and Quantification by Modeling Variability as a Function of Concentration," *Anal. Chem.,* 69(15), 3069–3075.

Exercises

37.1 ICP Calibration. Fit the ICP calibration data for iron (Fe) below using weights that are inversely proportional to the square of the peak intensity (I).

Standard Fe Conc. (mg/L)	0	50	100	200
Peak Intensity (I)	0.029	109.752	217.758	415.347

37.2 Nitrate Calibration I. For the case study nitrate data (Table 37.1), plot the residuals obtained by fitting a cubic calibration curve using unweighted regession.

37.3 Nitrate Calibration II. For the case study nitrate data (Table 37.1), compare the results of fitting the calibration curve using weights $1/x^2$ with those obtained using $1/s^2$ and $1/y^2$.

37.4 Chloride Calibration. The following table gives triplicate calibration peaks for HPLC measurement of chloride. Determine appropriate weights and fit the calibration curve. Plot the residuals to check the adequacy of the calibration model.

Chloride (mg/L)	Peak 1	Peak 2	Peak 3
0.2	1112	895	1109
0.5	1892	1806	1796
0.7	3242	3162	3191
1.0	4519	4583	4483
2.0	9168	9159	9146
3.5	15,915	16,042	15,935
5.0	23,485	23,335	23,293
10.0	49,166	50,135	49,439
17.5	92,682	93,288	92,407
25.0	137,021	140,137	139,938
50.0	318,984	321,468	319,527
75.0	505,542	509,773	511,877
100.0	700,231	696,155	699,516

Source: Greg Zelinka, Madison Metropolitan Sewerage District.

37.5 BOD Parameter Estimation. The data below are duplicate measurements of the BOD of fresh bovine manure. Use weighted nonlinear least squares to estimate the parameters in the model $\eta = \theta_1(1 - \exp(-\theta_2 t))$.

Day	1	3	5	7	10	15
BOD (mg/L)	11,320	20,730	28,000	32,000	35,200	33,000
	11,720	22,320	29,600	33,600	32,000	36,600

Source: Marske, D. M. and L. B. Polkowski (1972). *J. WPCF,* 44, 1987–1992.

38

Empirical Model Building by Linear Regression

KEY WORDS all possible regressions, analysis of variance, coefficient of determination, confidence interval, diagnostic checking, empirical models, F test, least squares, linear regression, overfitting, parsimonious model, polynomial, regression sum of squares, residual plot, residual sum of squares, sedimentation, solids removal, standard error, t statistic, total sum of squares.

Empirical models are widely used in engineering. Sometimes the model is a straight line; sometimes a mathematical French curve — a smooth interpolating function — is needed. Regression provides the means for selecting the complexity of the French curve that can be supported by the available data.

Regression begins with the specification of a model to be fitted. One goal is to find a *parsimonious model* — an adequate model with the fewest possible terms. Sometimes the proposed model turns out to be too simple and we need to augment it with additional terms. The much more common case, however, is to start with more terms than are needed or justified. This is called *overfitting*. Overfitting is harmful because the prediction error of the model is proportional to the number of parameters in the model.

A fitted model is always checked for inadequacies. The statistical output of regression programs is somewhat helpful in doing this, but a more satisfying and useful approach is to make diagnostic plots of the residuals. As a minimum, the residuals should be plotted against the predicted values of the fitted model. Plots of residuals against the independent variables are also useful. This chapter illustrates how this diagnosis is used to decide whether terms should be added or dropped to improve a model. If a tentative model is modified, it is refitted and rechecked. The model builder thus works iteratively toward the simplest adequate model.

A Model of Sedimentation

Sedimentation removes solid particles from a liquid by allowing them to settle under quiescent conditions. An ideal sedimentation process can be created in the laboratory in the form of a batch column. The column is filled with the suspension (turbid river water, industrial wastewater, or sewage) and samples are taken over time from sampling ports located at several depths along the column. The measure of sedimentation efficiency will be solids concentrations (or fraction of solids removed), which will be measured as a function of time and depth.

The data come from a quiescent batch settling test. At the beginning of the test, the concentration is uniform over the depth of the test settling column. The mass of solids in the column initially is $M = C_0 Z A$, where C_0 is the initial concentration (g/m^3), Z is the water depth in the settling column (m), and A is the cross-sectional area of the column (m^2). This is shown in the left-hand panel of Figure 38.1.

After settling has progressed for time t, the concentration near the bottom of the column has increased relative to the concentration at the top to give a solids concentration profile that is a function of depth at any time t. The mass of solids remaining above depth z is $M = A \int C(z, t) dz$. The total mass of solids in the column is still $M = C_0 Z A$. This is shown in the right-hand panel of Figure 38.1.

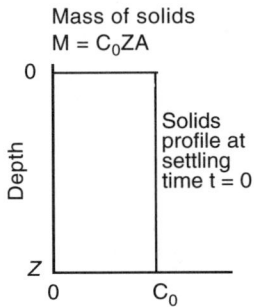 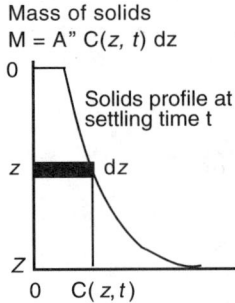

FIGURE 38.1 Solids concentration as a function of depth at time. The initial condition ($t = 0$) is shown on the left. The condition at time t is shown on the right.

The fraction of solids removed in a settling tank at any depth z, that has a detention time t, is estimated as:

$$R(z, t) = \frac{AZC_0 - A\int_0^z C(z, t)dz}{AZC_0} = 1 - \frac{1}{ZC_0}\int_0^z C(z, t)dz$$

This integral could be calculated graphically (Camp, 1946) or an approximating polynomial can be derived for the concentration curve and the fraction of solids removed (R) can be calculated algebraically.

Suppose, for example, that:

$$C(z, t) = 167 - 2.74t + 11.9z - 0.08zt + 0.014t^2$$

is a satisfactory empirical model and we want to use this model to predict the removal that will be achieved with 60-min detention time, for a depth of 8 ft and an initial concentration of 500 mg/L. The solids concentration profile as a function of depth at $t = 60$ min is:

$$C(z, t) = 167 - 2.74(60) + 11.9z - 0.08z(60) + 0.014(60)^2 = 53.0 + 7.1z$$

This is integrated over depth ($Z = 8$ ft) to give the fraction of solids that are expected to be removed:

$$R(z = 8, t = 60) = 1 - \frac{1}{8(500)} \int_{z=0}^{z=8} (53.0 + 7.1z)dz$$

$$= 1 - \frac{1}{8(500)} (53(8) + 3.55(8^2)) = 0.84$$

The model building problem is to determine the form of the polynomial function and to estimate the coefficients of the terms in the function.

Method: Linear Regression

Suppose the correct model for the process is $\eta = f(\beta, x)$ and the observations are $y_i = f(\beta, x) + e_i$, where the e_i are random errors. There may be several parameters (β) and several independent variables (x). According to the least squares criterion, the best estimates of the β's minimize the sum of the squared residuals:

$$\text{minimize } S(\beta) = \sum (y_i - \eta_i)^2$$

where the summation is over all observations.

Empirical Model Building by Linear Regression

The minimum sum of squares is called the *residual sum of squares, RSS*. The *residual mean square* (*RMS*) is the residual sum of squares divided by its degrees of freedom. $RMS = RSS/(n - p)$, where $n =$ number of observations and $p =$ number of parameters estimated.

Case Study: Solution

A column settling test was done on a suspension with initial concentration of 560 mg/L. Samples were taken at depths of 2, 4, and 6 ft (measured from the water surface) at times 20, 40, 60, and 120 min; the data are in Table 38.1. The simplest possible model is:

$$C(z, t) = \beta_0 + \beta_1 t$$

The most complicated model that might be needed is a full quadratic function of time and depth:

$$C(z, t) = \beta_0 + \beta_1 t + \beta_2 t^2 + \beta_3 z + \beta_4 z^2 + \beta_5 zt$$

We can start the model building process with either of these and add or drop terms as needed.

Fitting the simplest possible model involving time and depth gives:

$$\hat{y} = 132.3 + 7.12z - 0.97t$$

which has $R^2 = 0.844$ and residual mean square = 355.82. R^2, the *coefficient of determination*, is the percentage of the total variation in the data that is accounted for by fitting the model (Chapter 39).

Figure 38.2a shows the diagnostic residual plots for the model. The residuals plotted against the predicted values are not random. This suggests an inadequacy in the model, but it does not tell us how

TABLE 38.1

Data from a Laboratory Settling Column Test

Depth (ft)	Suspended Solids Concentration at Time t (min)			
	20	40	60	120
2	135	90	75	48
4	170	110	90	53
6	180	126	96	60

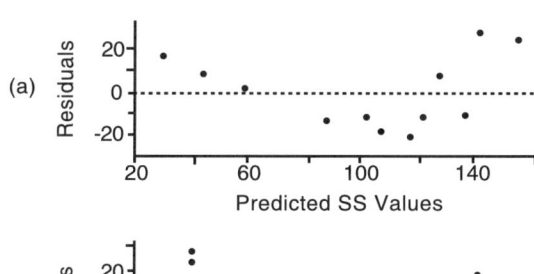

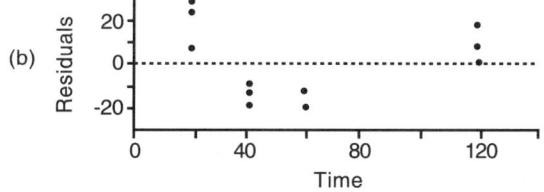

FIGURE 38.2 (a) Residuals plotted against the predicted suspended solids concentrations are not random. (b) Residuals plotted against settling time suggest that a quadratic term is needed in the model.

TABLE 38.2
Analysis of Variance for the Six-Parameter Settling Linear Model

Due to	df	SS	MS = SS/df
Regression (Reg SS)	5	20255.5	4051.1
Residuals (RSS)	6	308.8	51.5
Total (Total SS)	11	20564.2	

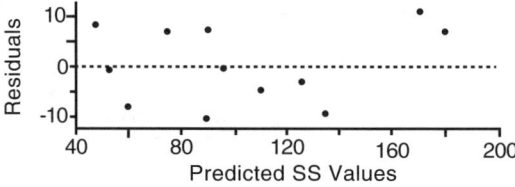

FIGURE 38.3 Plot of residuals against the predicted values of the regression model $\hat{y} = 185.97 + 7.125t + 0.014t^2 - 3.057z$.

the model might be improved. The pattern of the residuals plotted against time (Figure 38.2b) suggests that adding a t^2 term may be helpful. This was done to obtain:

$$\hat{y} = 186.0 + 7.12z - 3.06t + 0.0143t^2$$

which has $R^2 = 0.97$ and residual mean square = 81.5. A diagnostic plot of the residuals (Figure 38.3) reveals no inadequacies. Similar plots of residuals against the independent variables also support the model. This model is adequate to describe the data.

The most complicated model, which has six parameters, is:

$$\hat{y} = 152 + 20.9z - 2.74t - 1.13z^2 - 0.0143t^2 - 0.080zt$$

The model contains quadratic terms for time and depth and the interaction of depth and time (zt). The *analysis of variance* for this model is given in Table 38.2. This information is produced by computer programs that do linear regression. For now we do not need to know how to calculate this, but we should understand how it is interpreted.

Across the top, SS is sum of squares and df = degrees of freedom associated with a sum of squares quantity. MS is mean square, where MS = SS/df. The sum of squares due to regression is the regression sum of squares (RegSS): RegSS = 20,255.5. The sum of squares due to residuals is the *residual sum of squares* (RSS); RSS = 308.8. The *total sum of squares*, or Total SS, is:

$$\text{Total SS} = \text{RegSS} + \text{RSS}$$

Also:

$$\text{Total SS} = \sum (y_i - \bar{y})^2$$

The residual sum of squares (RSS) is the minimum sum of squares that results from estimating the parameters by least squares. It is the variation that is not explained by fitting the model. If the model is correct, the RSS is the variation in the data due to random measurement error. For this model, RSS = 308.8. The residual mean square is the RSS divided by the degrees of freedom of the residual sum of squares. For RSS, the degrees of freedom is $df = n - p$, where n is the number of observations and p is the number of parameters in the fitted model. Thus, RMS = RSS/$(n - p)$. The residual sum of squares (RSS = 308.8) and the

TABLE 38.3

Summary of All Possible Regressions for the Settling Test Model

Model	b_0	Coefficient of the Term b_1z	b_2t	b_3z^2	b_4t^2	b_5tz	R^2	RegSS	Decrease in RegSS
A	152	20.9	−2.74	−1.13	0.014	−0.08	0.985	20256	
(t ratio)		(2.3)	(8.3)	(1.0)	(7.0)	(2.4)			
[SE]		[9.1]	[0.33]	[1.1]	[0.002]	[0.03]			
B	167	11.9	−2.74		0.014	−0.08	0.982	20202	54
C	171	16.1	−3.06	−1.13	0.014		0.971	19966	289
D	186	7.1	−3.06		0.143		0.968	19912	343
E	98	20.9	−0.65	−1.13		−0.08	0.864	17705	2550
F	113	11.9	−0.65			−0.08	0.858	17651	2605
G	117	16.1	−0.97	−1.13			0.849	17416	2840
H	132	7.1	−0.97				0.844	17362	2894

Note: () indicates t ratios of the estimated parameters. [] indicates standard errors of the estimated parameters.

residual mean square (RMS = 308.8/6 = 51.5) are the key statistics in comparing this model with simpler models.

The regression sum of squares (RegSS) shows how much of the total variation (i.e., how much of the Total SS) has been explained by the fitted equation. For this model, RegSS = 20,255.5.

The coefficient of determination, commonly denoted as R^2, is the regression sum of squares expressed as a fraction of the total sum of squares. For the complete six-parameter model (Model A in Table 38.3), $R^2 = (20256/20564) = 0.985$, so it can be said that this model accounts for 98.5% of the total variation in the data.

It is natural to be fascinated by high R^2 values and this tempts us to think that the goal of model building is to make R^2 as high as possible. Obviously, this can be done by putting more high-order terms into a model, but it should be equally obvious that this does not necessarily improve the predictions that will be made using the model. Increasing R^2 is the wrong goal. Instead of worrying about R^2 values, we should seek the simplest adequate model.

Selecting the "Best" Regression Model

The "best" model is the one that adequately describes the data with the fewest parameters. Table 38.3 summarizes parameter estimates, the coefficient of determination R^2, and the regression sum of squares for all eight possible linear models. The total sum of squares, of course, is the same in all eight cases because it depends on the data and not on the form of the model. Standard errors [SE] and t ratios (in parentheses) are given for the complete model, Model A.

One approach is to examine the t ratio for each parameter. Roughly speaking, if a parameter's t ratio is less than 2.5, the true value of the parameter could be zero and that term could be dropped from the equation.

Another approach is to examine the confidence intervals of the estimated parameters. If this interval includes zero, the variable associated with the parameter can be dropped from the model. For example, in Model A, the coefficient of z^2 is $b_3 = -1.13$ with standard error = 1.1 and 95% confidence interval [−3.88 to +1.62]. This confidence interval includes zero, indicating that the true value of b_3 is very likely to be zero, and therefore the term z^2 can be tentatively dropped from the model. Fitting the simplified model (without z^2) gives Model B in Table 38.3.

The standard error [SE] is the number in brackets. The half-width of the 95% confidence interval is a multiple of the *standard error* of the estimated value. The multiplier is a t statistic that depends on the selected level of confidence and the degrees of freedom. This multiplier is not the same value as the t ratio given in Table 38.3. Roughly speaking, if the degrees of freedom are large ($n - p \geq 20$), the half-width of the confidence interval is about 2SE for a 95% confidence interval. If the degrees of freedom are small ($n - p < 10$), the multiplier will be in the range of 2.3SE to 3.0SE.

After modifying a model by adding, or in this case dropping, a term, an additional test should be made to compare the regression sum of squares of the two models. Details of this test are given in texts on regression analysis (Draper and Smith, 1998) and in Chapter 40. Here, the test is illustrated by example.

The regression sum of squares for the complete model (Model A) is 20,256. Dropping the z^2 term to get Model B reduced the regression sum of squares by only 54. We need to consider that a reduction of 54 in the regression sum of squares may not be a statistically significant difference.

The reduction in the regression sum of squares due to dropping z^2 can be thought of as a variance associated with the z^2 term. If this variance is small compared to the variance of the pure experimental error, then the term z^2 contributes no real information and it should be dropped from the model. In contrast, if the variance associated with the z^2 term is large relative to the pure error variance, the term should remain in the model.

There were no repeated measurements in this experiment, so an independent estimate of the variance due to pure error variance cannot be computed. The best that can be done under the circumstances is to use the residual mean square of the complete model as an estimate of the pure error variance. The residual mean square for the complete model (Model A) is 51.5. This is compared with the difference in regression sum of squares of the two models; the difference in regression sum of squares between Models A and B is 54. The ratio of the variance due to z^2 and the pure error variance is $F = 54/51.5 = 1.05$. This value is compared against the upper 5% point of the F distribution (1, 6 degrees of freedom). The degrees of freedom are 1 for the numerator (1 degree of freedom for the one parameter that was dropped from the model) and 6 for the denominator (the mean residual sum of squares). From Table C in the appendix, $F_{1,6} = 5.99$. Because $1.05 < 5.99$, we conclude that removing the z^2 term does not result in a significant reduction in the regression sum of squares. Therefore, the z^2 term is not needed in the model.

The test used above is valid to compare any two of the models that have one less parameter than Model A. To compare Models A and E, notice that omitting t^2 decreases the regression sum of squares by $20256 - 17705 = 2551$. The F statistic is $2551/51.5 = 49.5$. Because $49.5 \gg 5.99$ (the upper 95% point of the F distribution with 1 and 6 degrees of freedom), this change is significant and t^2 needs to be included in the model.

The test is modified slightly to compare Models A and D because Model D has two less terms than Model A. The decrease of 343 in the regression sum of squares results from dropping to terms (z^2 and zt). The F statistic is now computed using 343/2 in the numerator and 51.5 in the denominator: $F = (343/2)/51.5 = 3.33$. The upper 95% point of the appropriate reference distribution is $F = 5.14$, which has 2 degrees of freedom for the numerator and 6 degrees of freedom for the denominator. Because F for the model is less than the reference F ($F = 3.33 < 5.14$), the terms z^2 and zt are not needed.

Model D is as good as Model A. Model D is the simplest adequate model:

$$\text{Model D} \quad \hat{y} = 186 + 7.12t - 3.06z + 0.143t^2$$

This is the same model that was obtained by starting with the simplest possible model and adding terms to make up for inadequacies.

Comments

The model building process uses regression to estimate the parameters, followed by diagnosis to decide whether the model should be modified by adding or dropping terms. The goal is not to maximize R^2, because this puts unneeded high-order terms into the polynomial model. The best model should have the fewest possible parameters because this will minimize the prediction error of the model.

One approach to finding the simplest adequate model is to start with a simple tentative model and use diagnostic checks, such as residuals plots, for guidance. The alternate approach is to start by overfitting the data with a highly parameterized model and to then find appropriate simplifications. Each time a

Empirical Model Building by Linear Regression

term is added or deleted from the model, a check is made on whether the difference in the regression sum of squares of the two models is large enough to justify modification of the model.

References

Berthouex, P. M. and D. K. Stevens (1982). "Computer Analysis of Settling Data," *J. Envr. Engr. Div.*, ASCE, 108, 1065–1069.
Camp, T. R. (1946). "Sedimentation and Design of Settling Tanks," *Trans. Am. Soc. Civil Engr.*, 3, 895–936.
Draper, N. R. and H. Smith (1998). *Applied Regression Analysis,* 3rd ed., New York, John Wiley.

Exercises

38.1 Settling Test. Find a polynomial model that describes the following data. The initial suspended solids concentration was 560 mg/L. There are duplicate measurements at each time and depth.

Depth (ft)	Susp. Solids Conc. at Time t (min)			
	20	40	60	120
2	135	90	75	48
	140	100	66	40
4	170	110	90	53
	165	117	88	46
6	180	126	96	60
	187	121	90	63

38.2 Solid Waste Fuel Value. Exercise 3.5 includes a table that relates solid waste composition to the fuel value. The fuel value was calculated from the Dulong model, which uses elemental composition instead of the percentages of paper, food, metal, and plastic. Develop a model to relate the percentages of paper, food, metals, glass, and plastic to the Dulong estimates of fuel value. One proposed model is E(Btu/lb) = 23 Food + 82.8 Paper + 160 Plastic. Compare your model to this.

38.3 Final Clarifier. An activated sludge final clarifier was operated at various levels of overflow rate (OFR) to evaluate the effect of overflow rate (OFR), feed rate, hydraulic detention time, and feed slurry concentration on effluent total suspended solids (TSS) and underflow solids concentration. The temperature was always in the range of 18.5 to 21°C. Runs 11–12, 13–14, and 15–16 are duplicates, so the pure experimental error can be estimated. (a) Construct a polynomial model to predict effluent TSS. (b) Construct a polynomial model to predict underflow solids concentration. (c) Are underflow solids and effluent TSS related?

Run	OFR (m/d)	Feed Rate (m/d)	Detention Time (h)	Feed Slurry (kg/m^3)	Underflow (kg/m^3)	Effluent TSS (mg/L)
1	11.1	30.0	2.4	6.32	11.36	3.5
2	11.1	30.0	1.2	6.05	10.04	4.4
3	11.1	23.3	2.4	7.05	13.44	3.9
4	11.1	23.3	1.2	6.72	13.06	4.8
5	16.7	30.0	2.4	5.58	12.88	3.8
6	16.7	30.0	1.2	5.59	13.11	5.2
7	16.7	23.3	2.4	6.20	19.04	4.0
8	16.7	23.3	1.2	6.35	21.39	4.5
9	13.3	33.3	1.8	5.67	9.63	5.4

10	13.3	20.0	1.8	7.43	20.55	3.0
11	13.3	26.7	3.0	6.06	12.20	3.7
12	13.3	26.7	3.0	6.14	12.56	3.6
13	13.3	26.7	0.6	6.36	11.94	6.9
14	13.3	26.7	0.6	5.40	10.57	6.9
15	13.3	26.7	1.8	6.18	11.80	5.0
16	13.3	26.7	1.8	6.26	12.12	4.0

Source: Adapted from Deitz J. D. and T. M. Keinath, *J. WPCF*, 56, 344–350. (Original values have been rounded.)

38.4 Final Clarification. The influence of three factors on clarification of activated sludge effluent was investigated in 36 runs. Three runs failed because of overloading. The factors were solids retention time = SRT, hydraulic retention time = HRT, and overflow rate (OFR). Interpret the data.

Run	SRT (d)	HRT (h)	OFR (m/d)	Eff TSS (mg/L)	Run	SRT (d)	HRT (h)	OFR (m/d)	Eff TSS (mg/L)
1	8	12	32.6	48	19	5	12	16.4	18
2	8	12	32.6	60	20	5	12	8.2	15
3	8	12	24.5	55	21	5	8	40.8	47
4	8	12	16.4	36	22	5	8	24.5	41
5	8	12	8.2	45	23	5	8	8.2	57
6	8	8	57.0	64	24	5	4	40.8	39
7	8	8	40.8	55	25	5	4	24.5	41
8	8	8	24.5	30	26	5	4	8.2	43
9	8	8	8.2	16	27	2	12	16.4	19
10	8	8	8.2	45	28	2	12	16.4	36
11	8	4	57.0	37	29	2	12	8.2	23
12	8	4	40.8	21	30	2	8	40.8	26
13	8	4	40.8	14	31	2	8	40.8	15
14	8	4	24.5	4	32	2	8	24.5	17
15	8	4	8.2	11	33	2	8	8.2	14
16	5	12	32.6	20	34	2	4	40.8	39
17	5	12	24.5	28	35	2	4	24.5	43
18	5	12	24.5	12	36	2	4	8.2	48

Source: Cashion B. S. and T. M. Keinath, *J. WPCF*, 55, 1331–1338.

39

The Coefficient of Determination, R^2

KEY WORDS *coefficient of determination, coefficient of multiple correlation, confidence interval, F ratio, hapenstance data, lack of fit, linear regression, nested model, null model, prediction interval, pure error, R^2, repeats, replication, regression, regression sum of squares, residual sum of squares, spurious correlation.*

Regression analysis is so easy to do that one of the best-known statistics is the coefficient of determination, R^2. Anderson-Sprecher (1994) calls it "...a measure many statistician's love to hate."

Every scientist knows that R^2 is the *coefficient of determination* and R^2 is that proportion of the total variability in the dependent variable that is explained by the regression equation. This is so seductively simple that we often assume that a high R^2 signifies a useful regression equation and that a low R^2 signifies the opposite. We may even assume further that high R^2 indicates that the observed relation between independent and dependent variables is true and can be used to predict new conditions.

Life is not this simple. Some examples will help us understand what R^2 really reveals about how well the model fits the data and what important information can be overlooked if too much reliance is placed on the interpretation of R^2.

What Does "Explained" Mean?

Caution is recommended in interpreting the phrase "R^2 explains the variation in the dependent variable." R^2 is the proportion of variation in a variable Y that can be accounted for by fitting Y to a particular model instead of viewing the variable in isolation. R^2 does not explain anything in the sense that "Aha! Now we know why the response indicated by y behaves the way we have observed in this set of data." If the data are from a well-designed controlled experiment, with proper replication and randomization, it is reasonable to infer that an significant association of the variation in y with variation in the level of x is a causal effect of x. If the data had been observational, what Box (1966) calls *happenstance data*, there is a high risk of a causal interpretation being wrong. With observational data there can be many reasons for associations among variables, only one of which is causality.

A value of R^2 is not just a rescaled measure of variation. It is a comparison between two models. One of the models is usually referred to as *the model*. The other model — *the null model* — is usually never mentioned. The null model ($y = \beta_0$) provides the reference for comparison. This model describes a horizontal line at the level of the mean of the y values, which is the simplest possible model that could be fitted to any set of data.

- The model ($y = \beta_0 + \beta_1 x + \beta_2 x + \cdots + e_i$) has residual sum of squares $\sum (y_i - \hat{y})^2 = \text{RSS}_{\text{model}}$.
- The null model ($y = \beta_0 + e_i$) has residual sum of squares $\sum (y_i - \bar{y})^2 = \text{RSS}_{\text{null model}}$.

The comparison of the residual sums of squares (RSS) defines:

$$R^2 = 1 - \frac{\text{RSS}_{\text{model}}}{\text{RSS}_{\text{null model}}}$$

This shows that R^2 is a model comparison and that large R^2 measures only how much the model improves the null model. It does not indicate how good the model is in any absolute sense. Consequently, the common belief that a large R^2 demonstrates model adequacy is sometimes wrong.

The definition of R^2 also shows that comparisons are made only between *nested models*. The concept of proportionate reduction in variation is untrustworthy unless one model is a special case of the other. This means that R^2 cannot be used to compare models with an intercept with models that have no intercept: $y = \beta_0$ is not a reduction of the model $y = \beta_1 x$. It is a reduction of $y = \beta_0 + \beta_1 x$ and $y = \beta_0 + \beta_1 x + \beta_2 x^2$.

A High R^2 Does Not Assure a Valid Relation

Figure 39.1 shows a regression with $R^2 = 0.746$, which is statistically significant at almost the 1% level of confidence (a 1% chance of concluding significance when there is no true relation). This might be impressive until one knows the source of the data. X is the first six digits of pi, and Y is the first six Fibonocci numbers. There is no true relation between x and y. The linear regression equation has no predictive value (the seventh digit of pi does not predict the seventh Fibonocci number).

Anscombe (1973) published a famous and fascinating example of how R^2 and other statistics that are routinely computed in regression analysis can fail to reveal the important features of the data. Table 39.1

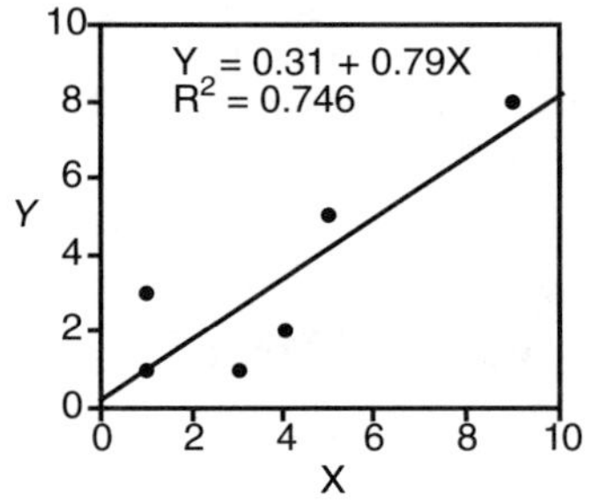

FIGURE 39.1 An example of nonsense in regression. X is the first six digits of pi and Y is the first six Fibonocci numbers. R^2 is high although there is no actual relation between x and y.

TABLE 39.1

Anscombe's Four Data Sets

A		B		C		D	
x	y	x	y	x	y	x	y
10.0	8.04	10.0	9.14	10.0	7.46	8.0	6.58
8.0	6.95	8.0	8.14	8.0	6.77	8.0	5.76
13.0	7.58	13.0	8.74	13.0	12.74	8.0	7.71
9.0	8.81	9.0	8.77	9.0	7.11	8.0	8.84
11.0	8.33	11.0	9.26	11.0	7.81	8.0	8.47
14.0	9.96	14.0	8.10	14.0	8.84	8.0	7.04
6.0	7.24	6.0	6.13	6.0	6.08	8.0	5.25
4.0	4.26	4.0	3.10	4.0	5.39	19.0	12.50
12.0	10.84	12.0	9.13	12.0	8.15	8.0	5.56
7.0	4.82	7.0	7.26	7.0	6.42	8.0	7.91
5.0	5.68	5.0	4.74	5.0	5.73	8.0	6.89

Note: Each data set has $n = 11$, mean of $\bar{x} = 9.0$, mean of $\bar{y} = 7.5$, equation of the regression line $y = 3.0 + 0.5x$, standard error of estimate of the slope $= 0.118$ (t statistic $= 4.24$, regression sum of squares (corrected for mean) $= 110.0$, residual sum of squares $= 13.75$, correlation coefficient $r = 0.82$ and $R^2 = 0.67$).

Source: Anscombe, F. J. (1973). *Am. Stat.*, 27, 17–21.

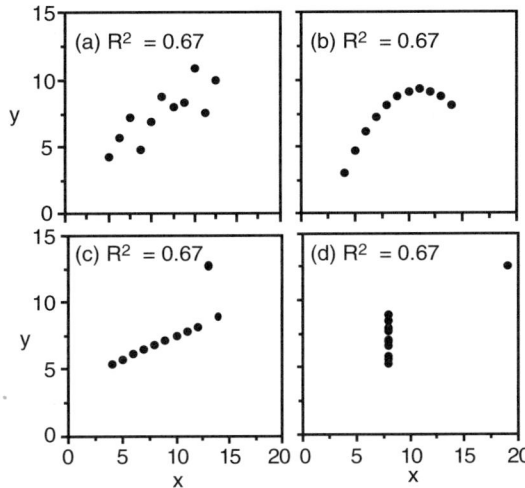

FIGURE 39.2 Plot of Anscombe's four data sets which all have $R^2 = 0.67$ and identical results from simple linear regression analysis (data from Anscombe 1973).

gives Anscombe's four data sets. Each data set has $n = 11$, $\bar{x} = 9.0$, $\bar{y} = 7.5$, fitted regression line $\hat{y} = 3 + 0.5x$, standard error of estimate of the slope = 0.118 (t statistic = 4.24), regression sum of squares (corrected for mean) = 110.0, residual sum of squares = 13.75, correlation coefficient = 0.82, and $R^2 = 0.67$. All four data sets appear to be described equally well by exactly the same linear model, at least until the data are plotted (or until the residuals are examined). Figure 39.2 shows how vividly they differ. The example is a persuasive argument for always plotting the data.

A Low R^2 Does Not Mean the Model is Useless

Hahn (1973) explains that the chances are one in ten of getting R^2 as high as 0.9756 in fitting a simple linear regression equation to the relation between an independent variable x and a normally distributed variable y based on only three observations, even if x and y are totally unrelated. On the other hand, with 100 observations, a value of $R^2 = 0.07$ is sufficient to establish statistical significance at the 1% level.

Table 39.2 lists the values of R^2 required to establish statistical significance for a simple linear regression equation. Table 39.2 applies only for the straight-line model $y = \beta_0 + \beta_1 x + e$; for multi-variable regression models, statistical significance must be determined by other means. This tabulation gives values at the 10, 5, and 1% significance levels. These correspond, respectively, to the situations where one is ready to take one chance in 10, one chance in 20, and one chance in 100 of incorrectly concluding there is evidence of a statistically significant linear regression when, in fact, x and y are unrelated.

A Significant R^2 Doesn't Mean the Model is Useful

Practical significance and statistical significance are not equivalent. Statistical significance and importance are not equivalent. A regression based on a modest and unimportant true relationship may be established as statistically significant if a sufficiently large number of observations are available. On the other hand, with a small sample it may be difficult to obtain statistical evidence of a strong relation.

It generally is good news if we find R^2 large and also statistically significant, but it does not assure a useful equation, especially if the equation is to be used for prediction. One reason is that the coefficient of determination is not expressed on the same scale as the dependent variable. A particular equation

TABLE 39.2

Values of R^2 Required to Establish Statistical Significance of a Simple Linear Regression Equation for Various Sample Sizes

Sample Size	Statistical Significance Level		
n	10%	5%	1%
3	0.98	0.99	0.99
4	0.81	0.90	0.98
5	0.65	0.77	0.92
6	0.53	0.66	0.84
8	0.39	0.50	0.70
10	0.30	0.40	0.59
12	0.25	0.33	0.50
15	0.19	0.26	0.41
20	0.14	0.20	0.31
25	0.11	0.16	0.26
30	0.09	0.13	0.22
40	0.07	0.10	0.16
50	0.05	0.08	0.13
100	0.03	0.04	0.07

Source: Hahn, G. J. (1973). *Chemtech,* October, pp. 609–611.

may explain a large proportion of the variability in the dependent variable, and thus have a high R^2, yet unexplained variability may be too large for useful prediction. It is not possible to tell from the magnitude of R^2 how accurate the predictions will be.

The Magnitude of R^2 Depends on the Range of Variation in X

The value of R^2 decreases with a decrease in the range of variation of the independent variable, other things being equal, and assuming the correct model is being fitted to the data. Figure 39.3 (upper left-hand panel) shows a set of 50 data points that has $R^2 = 0.77$. Suppose, however, that the range of x that could be investigated is only from 14 to 16 (for example, because a process is carefully constrained within narrow operating limits) and the available data are those shown in the upper right-hand panel of Figure 39.3. The underlying relationship is the same, and the measurement error in each observation is the same, but R^2 is now only 0.12. This dramatic reduction in R^2 occurs mainly because the range of x is restricted and not because the number of observations is reduced. This is shown by the two lower panels. Fifteen points (the same number as found in the range of $x = 14$ to 16), located at $x = 10, 15$, and 20, give $R^2 = 0.88$. Just 10 points, at $x = 10$ and 20, gives an even larger value, $R^2 = 0.93$.

These examples show that a large value of R^2 might reflect the fact that data were collected over an unrealistically large range of the independent variable x. This can happen, especially when x is time. Conversely, a small value might be due to a limited range of x, such as when x is carefully controlled by a process operator. In this case, x is constrained to a narrow range because it is known to be highly important, yet this importance will not be revealed by doing regression on typical data from the process.

Linear calibration curves always have a very high R^2, usually 0.99 and above. One reason is that the x variable covers a wide range (see Chapter 36.)

The Coefficient of Determination, R^2

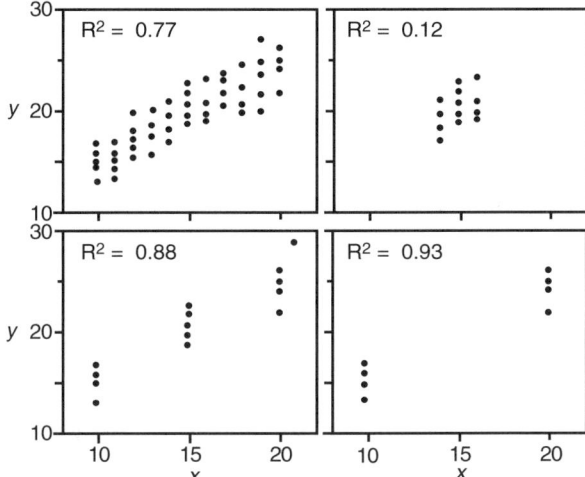

FIGURE 39.3 The full data set of 50 observations (upper-left panel) has $R^2 = 0.77$. The other three panels show how R^2 depends on the range of variation in the independent variable.

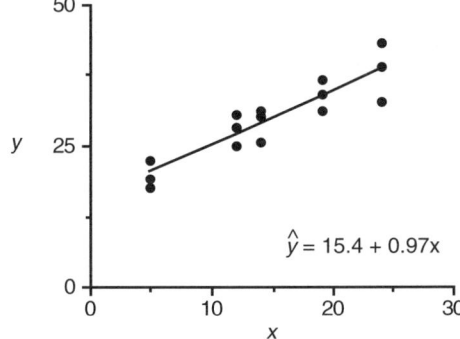

FIGURE 39.4 Linear regression with repeated observations. The regression sum of squares is 581.12. The residual sum of squares (RSS = 116.38) is divided into pure error sum of squares (SS_{PE} = 112.34) and lack-of-fit sum of squares (SS_{LOF} = 4.04). $R^2 = 0.833$, which explains 99% of the amount of residual error that can be explained.

The Effect of Repeated Runs on R^2

If regression is used to fit a model to n settings of x, it is possible for a model with n parameters to fit the data exactly, giving $R^2 = 1$. This kind of *overfitting* is not recommended but it is mathematically possible. On the other hand, if repeat measurements are made at some or all of the n settings of the independent variables, a perfect fit will not be possible. This assumes, of course, that the repeat measurements are not identical.

The data in Figure 39.4 are given in Table 39.3. The fitted model is $\hat{y} = 15.45 + 0.97x$. The relevant statistics are presented in Table 39.4. The fraction of the variation explained by the regression is $R^2 = 581.12/697.5 = 0.833$. The residual sum of squares (RSS) is divided into the *pure error sum of squares* (SS_{PE}), which is calculated from the repeated measurements, and the *lack-of-fit sum of squares* (SS_{LOF}). That is:

$$RSS = SS_{PE} + SS_{LOF}$$

TABLE 39.3

Linear Regression with Repeated Observations

x	y_1	y_2	y_3
5	17.5	22.4	19.2
12	30.4	28.4	25.1
14	30.1	25.8	31.1
19	36.6	31.3	34.0
24	38.9	43.2	32.7

TABLE 39.4

Analysis of Variance of the Regression with Repeat Observations Shown in Figure 39.4

Source	df	Sum of Sq.	Mean Sq.	F Ratio	Comments
Regression	1	581.12	581.12	64.91	
Residual	13	116.38	8.952		$= s^2$
Lack of fit (LOF)	3	4.04	1.35	0.12	$= s_L^2$
Pure error (PE)	10	112.34	11.23		$= s_e^2$
Total (Corrected)	14	697.50			

Suppose now that there had been only five observations (that is, no repeated measurements) and furthermore that the five values of y fell at the average of the repeated values in Figure 39.4. Now the fitted model would be exactly the same: $\hat{y} = 15.45 + 0.97x$ but the R^2 value would be 0.993. This is because the variance due to the repeats has been removed.

The maximum possible value for R^2 when there are repeat measurements is:

$$\max R^2 = \frac{\text{Total SS (corrected)} - \text{Pure error SS}}{\text{Total SS (corrected)}}$$

The pure error SS does not change when terms are added or removed from the model in an effort to improve the fit. For our example:

$$\max R^2 = \frac{697.5 - 112.3}{697.5} = 0.839$$

The actual $R^2 = 581.12/697.5 = 0.83$. Therefore, the regression has explained $100(0.833/0.839) = 99\%$ of the amount of variation that can be explained by the model.

A Note on Lack-Of-Fit

If repeat measurements are available, a lack-of-fit (LOF) test can be done. The lack-of-fit mean square $s_L^2 = SS_{LOF}/df_{LOF}$ is compared with the pure error mean square $s_e^2 = SS_{PE}/df_{PE}$. If the model gives an adequate fit, these two sums of squares should be of the same magnitude. This is checked by comparing the ratio s_L^2/s_e^2 against the F statistic with the appropriate degrees of freedom. Using the values in Table 39.4 gives $s_L^2/s_e^2 = 1.35/11.23 = 0.12$. The F statistic for a 95% confidence test with three degrees of freedom to measure lack of fit and ten degrees of freedom to measure the pure error is $F_{3,10} = 3.71$. Because $s_L^2/s_e^2 = 0.12$ is less than $F_{3,10} = 3.71$, there is no evidence of lack-of-fit. For this lack-of-fit test to be valid, true repeats are needed.

A Note on Description vs. Prediction

Is the regression useful? We have seen that a high R^2 does not guarantee that a regression has meaning. Likewise, a low R^2 may indicate a statistically significant relationship between two variables although the regression is not explaining much of the variation. Even less does *statistically significant* mean that the regression will predict future observations with much accuracy. "In order for the fitted equation to be regarded as a satisfactory predictor, the observed F ratio (regression mean square/residual mean square) should exceed not merely the selected percentage point of the F distribution, but several times the selected percentage point. How many times depends essentially on how great a ratio (prediction range/error of prediction) is specified" (Box and Wetz, 1973). Draper and Smith (1998) offer this rule-of-thumb: unless the observed F for overall regression exceeds the chosen test percentage point by at least a factor of four, and preferably more, the regression is unlikely to be of practical value for prediction purposes. The regression in Figure 39.4 has an F ratio of $581.12/8.952 = 64.91$ and would have some practical predictive value.

Other Ways to Examine a Model

If R^2 does not tell all that is needed about how well a model fits the data and how good the model may be for prediction, what else could be examined?

Graphics reveal information in data (Tufte 1983): always examine the data and the proposed model graphically. How sad if this advice was forgotten in a rush to compute some statistic like R^2.

A more useful single measure of the prediction capability of a model (including a *k*-variate regression model) is the standard error of the estimate. The standard error of the estimate is computed from the variance of the predicted value ($\hat{y}$) and it indicates the precision with which the model estimates the value of the dependent variable. This statistic is used to compute intervals that have the following meanings (Hahn, 1973).

- The *confidence interval for the dependent variable* is an interval that one expects, with a specified level of confidence, to contain the average value of the dependent variable at a set of specified values for the independent variables.
- A *prediction interval for the dependent variable* is an interval that one expects, with a specified probability, to contain a single future value of the dependent variable from the sampled population at a set of specified values of the independent variables.
- A *confidence interval around a parameter in a model* (i.e., a regression coefficient) is an interval that one expects, with a specified degree of confidence, to contain the true regression coefficient.

Confidence intervals for parameter estimates and prediction intervals for the dependent variable are discussed in Chapters 34 and 35. The exact method of obtaining these intervals is explained in Draper and Smith (1998). They are computed by most statistics software packages.

Comments

Widely used methods have the potential to be frequently misused. Linear regression, the most widely used statistical method, can be misused or misinterpreted if one relies too much on R^2 as a characterization of how well a model fits.

R^2 is a measure of the proportion of variation in *y* that is accounted for by fitting *y* to a particular linear model instead of describing the data by calculating the mean (a horizontal straight line). High R^2 does not prove that a model is correct or useful. A low R^2 may indicate a statistically significant relation between two variables although the regression has no practical predictive value. Replication dramatically improves the predictive error of a model, and it makes possible a formal lack-of-fit test, but it reduces the R^2 of the model.

Totally *spurious correlations*, often with high R^2 values, can arise when unrelated variables are combined. Two examples of particular interest to environmental engineers are presented by Sherwood (1974) and Rowe (1974). Both emphasize graphical analysis to stimulate and support any regression analysis. Rowe discusses the particular dangers that arise when sets of variables are combined to create new variables such as dimensional numbers (Froude number, etc.). Benson (1965) points out the same kinds of dangers in the context of hydraulics and hydrology.

References

Anderson-Sprecher, R. (1994). "Model Comparison and R^2," *Am. Stat.,* 48(2), 113–116.
Anscombe, F. J. (1973). "Graphs in Statistical Analysis," *Am. Stat.,* 27, 17–21.
Benson, M. A. (1965). "Spurious Correlation in Hydraulics and Hydrology," *J. Hydraulics Div.,* ASCE, 91, HY4, 35–45.
Box, G. E. P. (1966). "The Use and Abuse of Regression," *Technometrics,* 8, 625–629.
Box, G. E. P. and J. Wetz (1973). "Criteria for Judging Accuracy of Estimation by an Approximating Response Function," Madison, WI, University of Wisconsin Statistics Department, Tech. Rep. No. 9.
Draper, N. R. and H. Smith (1998). *Applied Regression Analysis,* 3rd ed., New York, John Wiley.
Hahn, G. J. (1973). "The Coefficient of Determination Exposed," *Chemtech,* October, pp. 609–611.
Rowe, P. N. (1974). "Correlating Data," *Chemtech,* January, pp. 9–14.
Sherwood, T. K. (1974). "The Treatment and Mistreatment of Data," *Chemtech,* December, pp. 736–738.
Tufte, E. R. (1983). *The Visual Display of Quantitative Information,* Cheshire, CT, Graphics Press.

Exercises

39.1 COD Calibration. The ten pairs of readings below were obtained to calibrate a UV spectrophotometer to measure chemical oxygen demand (COD) in wastewater.

COD (mg/L)	60	90	100	130	195	250	300	375	500	600
UV Absorbance	0.30	0.35	0.45	0.48	0.95	1.30	1.60	1.80	2.3	2.55

(a) Fit a linear model to the data and obtain the R^2 value. (b) Discuss the meaning of R^2 in the context of this calibration problem. (c) Exercise 36.3 contains a larger calibration data set for the same instrument. (d) Fit the model to the larger sample and compare the values of R^2. Will the calibration curve with the highest R^2 best predict the COD concentration? Explain why or why not.

39.2 Stream pH. The data below are $n = 200$ monthly pH readings on a stream that cover a period of almost 20 years. The data read from left to right. The fitted regression model is $\hat{y} = 7.1435 - 0.0003776t$; $R^2 = 0.042$. The confidence interval of the slope is [−0.00063, −0.000013]. Why is R^2 so low? Is the regression statistically significant? Is stream pH decreasing? What is the practical value of the model?

7.0	7.2	7.2	7.3	7.2	7.2	7.2	7.2	7.0	7.1	7.3	7.1	7.1	7.1	7.2	7.3	7.2	7.3	7.2	7.2
7.1	7.4	7.1	6.8	7.3	7.3	7.0	7.0	6.9	7.2	7.2	7.3	7.0	7.0	7.1	7.1	7.0	7.2	7.2	7.2
7.2	7.1	7.2	7.0	7.0	7.2	7.1	7.1	7.2	7.2	7.2	7.0	7.1	7.1	7.2	7.1	7.2	7.0	7.1	7.2
7.1	7.0	7.1	7.4	7.2	7.2	7.2	7.2	7.1	7.0	7.2	7.0	6.9	7.2	7.0	7.0	7.1	7.0	6.9	6.9
7.0	7.0	7.2	6.9	7.4	7.0	6.9	7.0	7.1	7.0	7.2	7.2	7.0	7.0	7.1	7.1	7.0	7.2	7.2	7.0
7.0	7.2	7.1	7.1	7.1	7.0	7.0	7.0	7.1	7.3	7.1	7.2	7.2	7.2	7.1	7.2	7.2	7.1	7.1	7.1
7.2	6.8	7.2	7.2	7.0	7.1	7.1	7.2	7.0	7.1	7.1	7.1	7.0	7.2	7.1	7.1	7.3	6.9	7.2	7.2
7.1	7.1	7.0	7.0	7.1	7.1	7.0	7.0	7.0	7.1	7.0	7.1	7.1	7.2	7.2	7.1	7.0	7.0	7.2	7.2
7.0	7.1	7.2	7.1	7.1	7.0	7.1	7.0	7.2	7.1	7.1	7.1	7.2	7.1	7.0	7.1	7.2	7.2	7.1	7.2
7.0	7.1	7.0	7.1	7.0	6.9	6.9	7.2	7.1	7.2	7.1	7.1	7.0	7.0	6.9	7.1	6.8	7.1	7.0	7.0

39.3 Replication. Fit a straight-line calibration model to y_1 and then fit the straight line to the three replicate measures of y. Suppose a colleague in another lab had the y_1 data only and you had all three replicates. Who will have the higher R^2 and who will have the best fitted calibration curve? Compare the values of R^2 obtained. Estimate the pure error variance. How much of the variation in y has been explained by the model?

x	y_1	y_2	y_3
2	0.0	1.7	2.0
5	4.0	2.0	4.5
8	5.1	4.1	5.8
12	8.1	8.9	8.4
15	9.2	8.3	8.8
18	11.3	9.5	10.9
20	11.7	10.7	10.4

39.4 Range of Data. Fit a straight-line calibration model to the first 10 observations in the Exercise 36.3 data set, that is for COD between 60 and 195 mg/L. Then fit the straight line to the full data set (COD from 60 to 675 mg/L). Interpret the change in R^2 for the two cases.

40

Regression Analysis with Categorical Variables

KEY WORDS   *acid rain, pH, categorical variable, F test, indicator variable, east squares, linear model, regression, dummy variable, qualitative variables, regression sum of squares, t-ratio, weak acidity.*

Qualitative variables can be used as explanatory variables in regression models. A typical case would be when several sets of data are similar except that each set was measured by a different chemist (or different instrument or laboratory), or each set comes from a different location, or each set was measured on a different day. The qualitative variables — chemist, location, or day — typically take on discrete values (i.e., chemist Smith or chemist Jones). For convenience, they are usually represented numerically by a combination of zeros and ones to signify an observation's membership in a category; hence the name *categorical variables*.

One task in the analysis of such data is to determine whether the same model structure and parameter values hold for each data set. One way to do this would be to fit the proposed model to each individual data set and then try to assess the similarities and differences in the goodness of fit. Another way would be to fit the proposed model to all the data as though they were one data set instead of several, assuming that each data set has the same pattern, and then to look for inadequacies in the fitted model.

Neither of these approaches is as attractive as using categorical variables to create a collective data set that can be fitted to a single model while retaining the distinction between the individual data sets. This technique allows the model structure and the model parameters to be evaluated using statistical methods like those discussed in the previous chapter.

Case Study: Acidification of a Stream During Storms

Cosby Creek, in the southern Appalachian Mountains, was monitored during three storms to study how pH and other measures of acidification were affected by the rainfall in that region. Samples were taken every 30 min and 19 characteristics of the stream water chemistry were measured (Meinert et al., 1982). Weak acidity (WA) and pH will be examined in this case study.

Figure 40.1 shows 17 observations for storm 1, 14 for storm 2, and 13 for storm 3, giving a total of 44 observations. If the data are analyzed without distinguishing between storms one might consider models of the form $pH = \beta_0 + \beta_1 WA + \beta_2 WA^2$ or $pH = \theta_3 + (\theta_1 - \theta_3)\exp(-\theta_2 WA)$. Each storm might be described by $pH = \beta_0 + \beta_1 WA$, but storm 3 does not have the same slope and intercept as storms 1 and 2, and storms 1 and 2 might be different as well. This can be checked by using categorical variables to estimate a different slope and intercept for each storm.

Method: Regression with Categorical Variables

Suppose that a model needs to include an effect due to the category (storm event, farm plot, treatment, truckload, operator, laboratory, etc.) from which the data came. This effect is included in the model in the form of categorical variables (also called *dummy* or *indicator variables*). In general $m - 1$ categorical variables are needed to specify m categories.

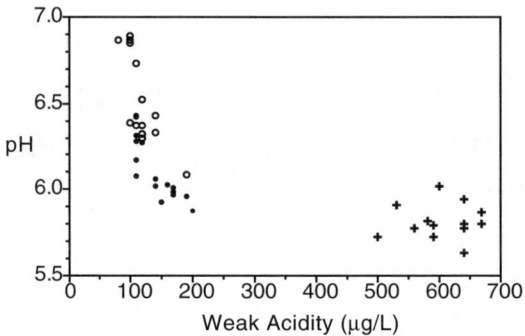

FIGURE 40.1 The relation of pH and weak acidity data of Cosby Creek after three storms.

Begin by considering data from a single category. The quantitative predictor variable is x_1 which can predict the independent variable y_1 using the linear model:

$$y_{1i} = \beta_0 + \beta_1 x_{1i} + e_i$$

where β_0 and β_1 are parameters to be estimated by least squares.

If there are data from two categories (e.g., data produced at two different laboratories), one approach would be to model the two sets of data separately as:

$$y_{1i} = \alpha_0 + \alpha_1 x_{1i} + e_i$$

and

$$y_{2i} = \beta_0 + \beta_1 x_{2i} + e_i$$

and then to compare the estimated intercepts (α_0 and β_0) and the estimated slopes (α_1 and β_1) using confidence intervals or t-tests.

A second, and often better, method is to simultaneously fit a single augmented model to all the data. To construct this model, define a categorical variable Z as follows:

$Z = 0$ if the data are in the first category
$Z = 1$ if the data are in the second category

The augmented model is:

$$y_i = \alpha_0 + \alpha_1 x_i + Z(\beta_0 + \beta_1 x_i) + e_i$$

With some rearrangement:

$$y_i = \alpha_0 + \beta_0 Z + \alpha_1 x_i + \beta_1 Z x_i + e_i$$

In this last form the regression is done as though there are three independent variables, x, Z, and Zx. The vectors of Z and Zx have to be created from the categorical variables defined above. The four parameters α_0, β_0, α_1, and β_1 are estimated by linear regression.

A model for each category can be obtained by substituting the defined values. For the first category, $Z = 0$ and:

$$y_i = \alpha_0 + \alpha_1 x_i + e_i$$

Regression Analysis with Categorical Variables

	Slopes Diffferent	Slopes Equal
Intercepts Different	$y_i=(\alpha_0+\beta_0)+(\alpha_1+\beta_1)x_i+e_i$	$y_i=(\alpha_0+\beta_0)+\alpha_1 x_i + e_i$
Intercepts Equal	$y_i=\alpha_0+(\alpha_1+\beta_1)x_i+e_i$	$y_i=\alpha_0+\alpha_1 x_i+e_i$

FIGURE 40.2 Four possible models to fit a straight line to data in two categories.

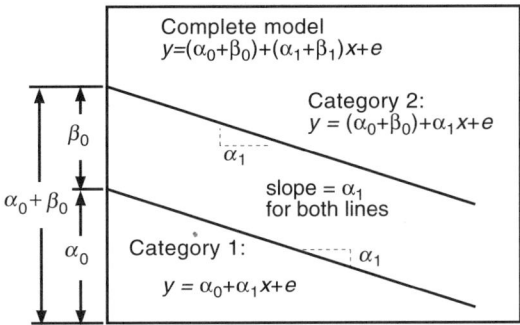

FIGURE 40.3 Model with two categories having different intercepts but equal slopes.

For the second category, $Z = 1$ and:

$$y_i = (\alpha_0 + \beta_0) + (\alpha_1 + \beta_1)x_i + e_i$$

The regression might estimate either β_0 or β_1 as zero, or both as zero. If $\beta_0 = 0$, the two lines have the same intercept. If $\beta_1 = 0$, the two lines have the same slope. If both β_1 and β_0 equal zero, a single straight line fits all the data. Figure 40.2 shows the four possible outcomes. Figure 40.3 shows the particular case where the slopes are equal and the intercepts are different.

If simplification seems indicated, a simplified version is fitted to the data. We show later how the full model and simplified model are compared to check whether the simplification is justified.

To deal with three categories, two categorical variables are defined:

Category 1:	$Z_1 = 1$	and	$Z_2 = 0$
Category 2:	$Z_1 = 0$	and	$Z_2 = 1$

This implies $Z_1 = 0$ and $Z_2 = 0$ for category 3.

The model is:

$$y_i = (\alpha_0 + \alpha_1 x_i) + Z_1(\beta_0 + \beta_1 x_i) + Z_2(\gamma_0 + \gamma_1 x_i) + e_i$$

The parameters with subscript 0 estimate the intercept and those with subscript 1 estimate the slopes. This can be rearranged to give:

$$y_i = \alpha_0 + \beta_0 Z_1 + \gamma_0 Z_2 + \alpha_1 x_i + \beta_1 Z_1 x_i + \gamma_1 Z_2 x_i + e_i$$

The six parameters are estimated by fitting the original independent variable x_i plus the four created variables Z_1, Z_2, $Z_1 x_i$, and $Z_2 x_i$.

Any of the parameters might be estimated as zero by the regression analysis. A couple of examples explain how the simpler models can be identified. In the simplest possible case, the regression would estimate $\beta_0 = 0$, $\gamma_0 = 0$, $\beta_1 = 0$, and $\gamma_1 = 0$ and the same slope (α_1) and intercept (α_0) would apply to all three categories. The fitted simplified model is $y_i = \alpha_0 + \alpha_1 x_i + e_i$.

If the intercepts are different for the three categories but the slopes are the same, the regression would estimate $\beta_1 = 0$ and $\gamma_1 = 0$ and the model becomes:

$$y_i = (\alpha_0 + \beta_0 Z_1 + \gamma_0 Z_2) + \alpha_1 x_i + e_i$$

For category 1: $y_i = (\alpha_0 + \beta_0 Z_1) + \alpha_1 x_i + e_i$
For category 2: $y_i = (\alpha_0 + \gamma_0 Z_2) + \alpha_1 x_i + e_i$
For category 3: $y_i = \alpha_0 + \alpha_1 x_i + e_i$

Case Study: Solution

The model under consideration allows a different slope and intercept for each storm. Two dummy variables are needed:

$Z_1 = 1$ for storm 1 and zero otherwise
$Z_2 = 1$ for storm 2 and zero otherwise

The model is:

$$\text{pH} = \alpha_0 + \alpha_1 WA + Z_1(\beta_0 + \beta_1 WA) + Z_2(\gamma_0 + \gamma_1 WA)$$

where the α's, β's, and γ's are estimated by regression. The model can be rewritten as:

$$\text{pH} = \alpha_0 + \beta_0 Z_1 + \gamma_0 Z_2 + \alpha_1 WA + \beta_1 Z_1 WA + \gamma_1 Z_2 WA$$

The dummy variables are incorporated into the model by creating the new variables $Z_1 WA$ and $Z_2 WA$. Table 40.1 shows how this is done.

Fitting the full six-parameter model gives:

Model A: pH = 5.77 − 0.00008WA + 0.998Z_1 + 1.65Z_2 − 0.005Z_1WA − 0.008Z_2WA
(*t*-ratios)   (0.11)   (2.14)   (3.51)   (3.63)   (4.90)

which is also shown as Model A in Table 40.2 (top row). The numerical coefficients are the least squares estimates of the parameters. The small numbers in parentheses beneath the coefficients are the *t*-ratios for the parameter values. Terms with $t < 2$ are candidates for elimination from the model because they are almost certainly not significant.

The term *WA* appears insignificant. Dropping this term and refitting the simplified model gives Model B, in which all coefficients are significant:

Model B: pH = 5.82 + 0.95Z_1 + 1.60Z_2 − 0.005Z_1WA − 0.008Z_2WA
(*t*-ratios)   (6.01)   (9.47)   (4.35)   (5.54)
[95% conf. interval]   [0.63 to 1.27] [1.26 to 1.94] [−0.007 to −0.002] [−0.01 to −0.005]

The regression sum of squares, listed in Table 40.2, is the same for Model A and for Model B (Reg SS = 4.278). Dropping the *WA* term caused no decrease in the regression sum of squares. Model B is equivalent to Model A.

Is any further simplification possible? Notice that the 95% confidence intervals overlap for the terms −0.005 Z_1WA and −0.008 Z_2WA. Therefore, the coefficients of these two terms might be the same. To check this, fit Model C, which has the same slope but different intercepts for storms 1 and 2. This is

TABLE 40.1

Weak Acidity (WA), pH, and Categorical Variables for Three Storms

Storm	WA	Z_1	Z_2	Z_1WA	Z_2WA	pH	Z_3	Z_3WA
1	190	1	0	190	0	5.96	1	190
1	110	1	0	110	0	6.08	1	110
1	150	1	0	150	0	5.93	1	150
1	170	1	0	170	0	5.99	1	170
1	170	1	0	170	0	6.01	1	170
1	170	1	0	170	0	5.97	1	170
1	200	1	0	200	0	5.88	1	200
1	140	1	0	140	0	6.06	1	140
1	140	1	0	140	0	6.06	1	140
1	160	1	0	160	0	6.03	1	160
1	140	1	0	140	0	6.02	1	140
1	110	1	0	110	0	6.17	1	110
1	110	1	0	110	0	6.31	1	110
1	120	1	0	120	0	6.27	1	120
1	110	1	0	110	0	6.42	1	110
1	110	1	0	110	0	6.28	1	110
1	110	1	0	110	0	6.43	1	110
2	140	0	1	0	140	6.33	1	140
2	140	0	1	0	140	6.43	1	140
2	120	0	1	0	120	6.37	1	120
2	190	0	1	0	190	6.09	1	190
2	120	0	1	0	120	6.32	1	120
2	110	0	1	0	110	6.37	1	110
2	110	0	1	0	110	6.73	1	110
2	100	0	1	0	100	6.89	1	100
2	100	0	1	0	100	6.87	1	100
2	120	0	1	0	120	6.30	1	120
2	120	0	1	0	120	6.52	1	120
2	100	0	1	0	100	6.39	1	100
2	80	0	1	0	80	6.87	1	80
2	100	0	1	0	100	6.85	1	100
3	580	0	0	0	0	5.82	0	0
3	640	0	0	0	0	5.94	0	0
3	500	0	0	0	0	5.73	0	0
3	530	0	0	0	0	5.91	0	0
3	670	0	0	0	0	5.87	0	0
3	670	0	0	0	0	5.80	0	0
3	640	0	0	0	0	5.80	0	0
3	640	0	0	0	0	5.78	0	0
3	560	0	0	0	0	5.78	0	0
3	590	0	0	0	0	5.73	0	0
3	640	0	0	0	0	5.63	0	0
3	590	0	0	0	0	5.79	0	0
3	600	0	0	0	0	6.02	0	0

Note: The two right-hand columns are used to fit the simplified model.

Source: Meinert, D. L., S. A. Miller, R. J. Ruane, and H. Olem (1982). "A Review of Water Quality Data in Acid Sensitive Watersheds in the Tennessee Valley," Rep. No. TVA.ONR/WR-82/10, TVA, Chattanooga, TN.

done by combining columns Z_1WA and Z_2WA to form the two columns on the right-hand side of Table 40.1. Call this new variable Z_3WA. $Z_3 = 1$ for storms 1 and 2, and 0 for storm 3.

The fitted model is:

Model C: $\quad$ pH = 5.82 + 1.11Z_1 + 1.38Z_2 − 0.0057Z_3WA

(*t*-ratios) $\qquad\qquad\qquad$ (8.43) $\quad$ (12.19) $\quad$ (6.68)

TABLE 40.2

Alternate Models for pH at Cosby Creek

Model		Reg SS	Res SS	R^2
A	pH = 5.77 − 0.00008WA + 0.998Z_1 + 1.65Z_2 − 0.005Z_1WA − 0.008Z_2WA	4.278	0.662	0.866
B	pH = 5.82 + 0.95Z_1 + 1.60Z_2 − 0.005Z_1WA − 0.008Z_2WA	4.278	0.662	0.866
C	pH = 5.82 + 1.11Z_1 + 1.38Z_2 − 0.0057Z_3WA	4.229	0.712	0.856

This simplification of the model can be checked in a more formal way by comparing regression sums of squares of the simplified model with the more complicated one. The regression sum of squares is a measure of how well the model fits the data. Dropping an important term will cause the regression sum of squares to decrease by a noteworthy amount, whereas dropping an unimportant term will change the regression sum of squares very little. An example shows how we decide whether a change is "noteworthy" (i.e., statistically significant).

If two models are equivalent, the difference of their regression sums of squares will be small, within an allowance for variation due to random experimental error. The variance due to experimental error can be estimated by the mean residual sum of squares of the full model (Model A).

The variance due to the deleted term is estimated by the difference between the regression sums of squares of Model A and Model C, with an adjustment for their respective degrees of freedom. The ratio of the variance due to the deleted term is compared with the variance due to experimental error by computing the F statistic, as follows:

$$F = \frac{(\text{Reg SS}_A - \text{Reg SS}_C)/(\text{Reg df}_A - \text{Reg df}_C)}{\text{Res SS}_A/\text{Res df}_A}$$

where
 Reg SS = regression sum of squares
 Reg df = degrees of freedom associated with the regression sum of squares
 Res SS = residual sum of squares
 Res df = degrees of freedom associated with the residual sum of squares

Model A has five degrees of freedom associated with the regression sum of squares (Reg df = 5), one for each of the six parameters in the model minus one for computing the mean. Model C has three degrees of freedom. Thus:

$$F = \frac{(4.278 - 4.229)/(5-3)}{0.66/38} = \frac{0.0245}{0.017} = 1.44$$

For a test of significance at the 95% confidence level, this value of F is compared with the upper 5% point of the F distribution with the appropriate degrees of freedom (5 − 3 = 2 in the numerator and 38 in the denominator): $F_{2,38,0.05} = 3.25$. The computed value ($F = 1.44$) is smaller than the critical value $F_{2,38,0.05} = 3.25$, which confirms that omitting WA from the model and forcing storms 1 and 2 to have the same slope has not significantly worsened the fit of the model. In short, Model C describes the data as well as Model A or Model B. Because it is simpler, it is preferred.

Models for the individual storms are derived by substituting the values of Z_1, Z_2, and Z_3 into Model C:

 Storm 1 $Z_1 = 1, Z_2 = 0, Z_3 = 1$ pH = 6.93 − 0.0057WA
 Storm 2 $Z_1 = 0, Z_2 = 1, Z_3 = 1$ pH = 7.20 − 0.0057WA
 Storm 3 $Z_1 = 0, Z_2 = 0, Z_3 = 0$ pH = 5.82

The model indicates a different intercept for each storm, a common slope for storms 1 and 2, and a slope of zero for storm 3, as shown by Figure 40.4. In storm 3, the variation in pH was random about a mean

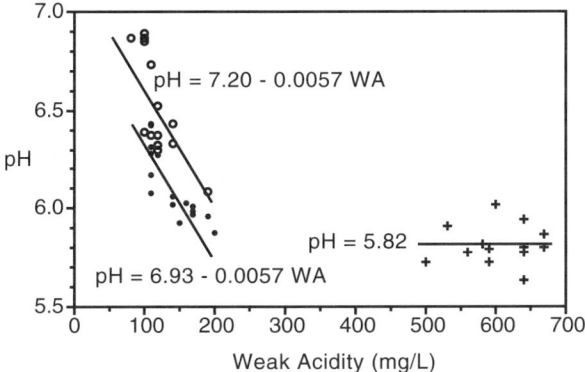

FIGURE 40.4 Stream acidification data fitted to Model C (Table 40.2). Storms 1 and 2 have the same slope.

of 5.82. For storms 1 and 2, increased *WA* was associated with a lowering of the pH. It is not difficult to imagine conditions that would lead to two different storms having the same slope but different intercepts. It is more difficult to understand how the same stream could respond so differently to storm 3, which had a range of *WA* that was much higher than either storm 1 or 2, a lower pH, and no change of pH over the observed range of *WA*. Perhaps high *WA* depresses the pH and also buffers the stream against extreme changes in pH. But why was the *WA* so much different during storm 3? The data alone, and the statistical analysis, do not answer this question. They do, however, serve the investigator by raising the question.

Comments

The variables considered in regression equations usually take numerical values over a continuous range, but occasionally it is advantageous to introduce a factor that has two or more discrete levels, or categories. For example, data may arise from three storms, or three operators. In such a case, we cannot set up a continuous measurement scale for the variable storm or operator. We must create categorical variables (dummy variables) that account for the possible different effects of separate storms or operators. The levels assigned to the categorical variables are unrelated to any physical level that might exist in the factors themselves.

Regression with categorical variables was used to model the disappearance of PCBs from soil (Berthouex and Gan, 1991; Gan and Berthouex, 1994). Draper and Smith (1998) provide several examples on creating efficient patterns for assigning categorical variables. Piegorsch and Bailer (1997) show examples for nonlinear models.

References

Berthouex, P. M. and D. R. Gan (1991). "Fate of PCBs in Soil Treated with Contaminated Municipal Sludge," *J. Envir. Engr. Div., ASCE,* 116(1), 1–18.

Daniel, C. and F. S. Wood (1980). *Fitting Equations to Data: Computer Analysis of Multifactor Data,* 2nd ed., New York, John Wiley.

Draper, N. R. and H. Smith, (1998). *Applied Regression Analysis,* 3rd ed., New York, John Wiley.

Gan, D. R. and P. M. Berthouex (1994). "Disappearance and Crop Uptake of PCBs from Sludge-Amended Farmland," *Water Envir. Res.,* 66, 54–69.

Meinert, D. L., S. A. Miller, R. J. Ruane, and H. Olem (1982). "A Review of Water Quality Data in Acid Sensitive Watersheds in the Tennessee Valley," Rep. No. TVA.ONR/WR-82/10, TVA, Chattanooga, TN.

Piegorsch, W. W. and A. J. Bailer (1997). *Statistics for Environmental Biology and Toxicology,* London, Chapman & Hall.

Exercises

40.1 **PCB Degradation in Soil.** PCB-contaminated sewage sludge was applied to test plots at three different loading rates (kg/ha) at the beginning of a 5-yr experimental program. Test plots of farmland where corn was grown were sampled to assess the rate of disappearance of PCB from soil. Duplicate plots were used for each treatment. Soil PCB concentration (mg/kg) was measured each year in the fall after the corn crop was picked and in the spring before planting. The data are below. Estimate the rate coefficients of disappearance (k) using the model $PCB_t = PCB_0 \exp(-kt)$. Are the rates the same for the four treatment conditions?

Time	Treatment 1		Treatment 2		Treatment 3	
0	1.14	0.61	2.66	2.50	0.44	0.44
5	0.63	0.81	2.69	2.96	0.25	0.31
12	0.43	0.54	1.14	1.51	0.18	0.22
17	0.35	0.51	1.00	0.48	0.15	0.12
24	0.35	0.34	0.93	1.16	0.11	0.09
29	0.32	0.30	0.73	0.96	0.08	0.10
36	0.23	0.20	0.47	0.46	0.07	0.06
41	0.20	0.16	0.57	0.36	0.03	0.04
48	0.12	0.09	0.40	0.22	0.03	0.03
53	0.11	0.08	0.32	0.31	0.02	0.03

40.2 **1,1,1-Trichloroethane Biodegradation.** Estimates of biodegradation rate (k_b) of 1,1,1-trichloroethane were made under three conditions of activated sludge treatment. The model is $y_i = bx_i + e_i$, where the slope b is the estimate of k_b. Two dummy variables are needed to represent the three treatment conditions, and these are arranged in the table below. Does the value of k_b depend on the activated sludge treatment condition?

x ($\times 10^{-6}$)	Z_1	Z_2	Z_1^*x	Z_2^*x	$y(\times 10^{-3})$
61.2	0	0	0	0	142.3
9.8	0	0	0	0	140.8
8.9	0	0	0	0	62.7
44.9	0	0	0	0	32.5
6.3	0	0	0	0	82.3
20.3	0	0	0	0	58.6
7.5	0	0	0	0	15.5
1.2	0	0	0	0	2.5
159.8	1	0	159.8	0	1527.3
44.4	1	0	44.4	0	697.5
57.4	1	0	57.4	0	429.9
25.9	1	0	25.9	0	215.2
37.9	1	0	37.9	0	331.6
55.0	1	0	55.0	0	185.7
151.7	1	0	151.7	0	1169.2
116.2	1	0	116.2	0	842.8
129.9	1	0	129.9	0	712.9
19.4	0	1	0	19.4	49.3
7.7	0	1	0	7.7	21.6
36.7	0	1	0	36.7	53.3
17.8	0	1	0	17.8	59.4
8.5	0	1	0	8.5	112.3

40.3 Diesel Fuel. Four diesel fuels were tested to estimate the partition coefficient K_{dw} of eight organic compounds as a function of their solubility in water (S). The compounds are (1) naphthalene, (2) 1-methyl-naphthalene, (3) 2-methyl-naphthalene, (4) acenaphthene, (5) fluorene, (6) phenanthrene, (7) anthracene, and (8) fluoranthene. The table is set up to do linear regression with dummy variables to differentiate between diesel fuels. Does the partitioning relation vary from one diesel fuel to another?

Compound	$y = \log(K_{dw})$	$x = \log(S)$	Z_1	Z_2	Z_3	$Z_1\log(S)$	$Z_2\log(S)$	$Z_3\log(S)$
Diesel fuel #1								
1	3.67	−3.05	0	0	0	0	0	0
2	4.47	−3.72	0	0	0	0	0	0
3	4.31	−3.62	0	0	0	0	0	0
4	4.35	−3.98	0	0	0	0	0	0
5	4.45	−4.03	0	0	0	0	0	0
6	4.6	−4.50	0	0	0	0	0	0
7	5.15	−4.49	0	0	0	0	0	0
8	5.32	−5.19	0	0	0	0	0	0
Diesel fuel #2								
1	3.62	−3.05	1	0	0	−3.05	0	0
2	4.29	−3.72	1	0	0	−3.72	0	0
3	4.21	−3.62	1	0	0	−3.62	0	0
4	4.46	−3.98	1	0	0	−3.98	0	0
5	4.41	−4.03	1	0	0	−4.03	0	0
6	4.61	−4.50	1	0	0	−4.50	0	0
7	5.38	−4.49	1	0	0	−4.49	0	0
8	4.64	−5.19	1	0	0	−5.19	0	0
Diesel fuel #3								
1	3.71	−3.05	0	1	0	0	−3.05	0
2	4.44	−3.72	0	1	0	0	−3.72	0
3	4.36	−3.62	0	1	0	0	−3.62	0
4	4.68	−3.98	0	1	0	0	−3.98	0
5	4.52	−4.03	0	1	0	0	−4.03	0
6	4.78	−4.50	0	1	0	0	−4.50	0
7	5.36	−4.49	0	1	0	0	−4.49	0
8	5.61	−5.19	0	1	0	0	−5.19	0
Diesel fuel #4								
1	3.71	−3.05	0	0	1	0	0	−3.05
2	4.49	−3.72	0	0	1	0	0	−3.72
3	4.33	−3.62	0	0	1	0	0	−3.62
4	4.62	−3.98	0	0	1	0	0	−3.98
5	4.55	−4.03	0	0	1	0	0	−4.03
6	4.78	−4.50	0	0	1	0	0	−4.50
7	5.20	−4.49	0	0	1	0	0	−4.49
8	5.60	−5.19	0	0	1	0	0	−5.19

Source: Lee, L. S. et al. (1992). *Envir. Sci. Tech.*, 26, 2104–2110.

40.4 Threshold Concentration. The data below can be described by a hockey-stick pattern. Below some threshold value (τ) the response is a constant plateau value ($\eta = \gamma_0$). Above the threshold, the response is linear $\eta = \gamma_0 + \beta_1(x - \tau)$. These can be combined into a continuous segmented model using a dummy variable z such that $z = 1$ when $x > \tau$ and $z = 0$ when $x \leq \tau$. The dummy variable formulation is $\eta = \gamma_0 + \beta_1(x - \tau)z$, where z is a dummy variable. This gives $\eta = \gamma_0$ for $x \leq \tau$ and $\eta = \gamma_0 + \beta_1(x - \tau) = \gamma_0 + \beta_1 x - \beta_1 \tau$ for $x \geq \tau$. Estimate the plateau value γ_0, the post-threshold slope β_1, and the unknown threshold dose τ.

x	2.5	22	60	90	105	144	178	210	233	256	300	400
y	16.6	15.3	16.9	16.1	17.1	16.9	18.6	19.3	25.8	28.4	35.5	45.3

40.5 Coagulation. Modify the hockey-stick model of Exercise 40.4 so it describes the intersection of two straight lines with nonzero slopes. Fit the model to the coagulation data (dissolved organic carbon, DOC) given below to estimate the slopes of the straight-line segments and the chemical dose (alum) at the intersection.

Alum Dose (mg/L)	DOC (mg/L)	Alum Dose (mg/L)	DOC (mg/L)
0	6.7	35	3.3
5	6.4	40	3.3
10	6.0	49	3.1
15	5.2	58	2.8
20	4.7	68	2.7
25	4.1	78	2.6
30	3.9	87	2.6

Source: White, M. W. et al. (1997). *J. AWWA,* 89(5).

41

The Effect of Autocorrelation on Regression

KEY WORDS *autocorrelation, autocorrelation coefficient, drift, Durbin-Watson statistic, randomization, regression, time series, trend analysis, serial correlation, variance (inflation).*

Many environmental data exist as sequences over time or space. The time sequence is obvious in some data series, such as daily measurements on river quality. A characteristic of such data can be that neighboring observations tend to be somewhat alike. This tendency is called *autocorrelation*. Autocorrelation can also arise in laboratory experiments, perhaps because of the sequence in which experimental runs are done or drift in instrument calibration. Randomization reduces the possibility of autocorrelated results. Data from unplanned or unrandomized experiments should be analyzed with an eye open to detect autocorrelation.

Most statistical methods, estimation of confidence intervals, ordinary least squares regression, etc. depend on the residual errors being independent, having constant variance, and being normally distributed. *Independent* means that the errors are not autocorrelated. The errors in statistical conclusions caused by violating the condition of independence can be more serious than those caused by not having normality.

Parameter estimates may or may not be seriously affected by autocorrelation, but unrecognized (or ignored) autocorrelation will bias estimates of variances and any statistics calculated from variances. Statements about probabilities, including confidence intervals, will be wrong.

This chapter explains why ignoring or overlooking autocorrelation can lead to serious errors and describes the Durbin-Watson test for detecting autocorrelation in the residuals of a fitted model. Checking for autocorrelation is relatively easy although it may go undetected even when present in small data sets. Making suitable provisions to incorporate existing autocorrelation into the data analysis can be difficult. Some useful references are given but the best approach may be to consult with a statistician.

Case Study: A Suspicious Laboratory Experiment

A laboratory experiment was done to demonstrate to students that increasing factor X by one unit should cause factor Y to increase by one-half a unit. Preliminary experiments indicated that the standard deviation of repeated measurements on Y was about 1 unit. To make measurement errors small relative to the signal, the experiment was designed to produce 20 to 25 units of y. The procedure was to set x and, after a short time, to collect a specimen on which y would be measured. The measurements on y were not started until all 11 specimens had been collected. The data, plotted in Figure 41.1, are:

$x =$	0	1	2	3	4	5	6	7	8	9	10
$y =$	21.0	21.8	21.3	22.1	22.5	20.6	19.6	20.9	21.7	22.8	23.6

Linear regression gave $\hat{y} = 21.04 + 0.12x$, with $R^2 = 0.12$. This was an unpleasant surprise. The 95% confidence interval of the slope was -0.12 to 0.31, which does not include the theoretical slope of 0.5 that the experiment was designed to reveal. Also, this interval includes zero so we cannot even be sure that x and y are related.

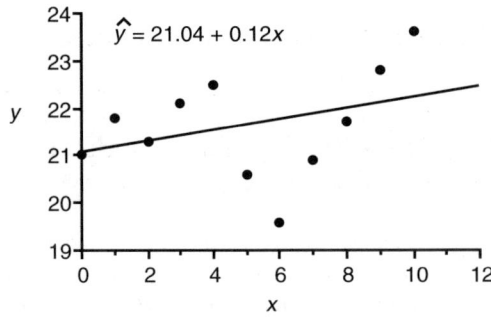

FIGURE 41.1 The original data from a suspicious laboratory experiment.

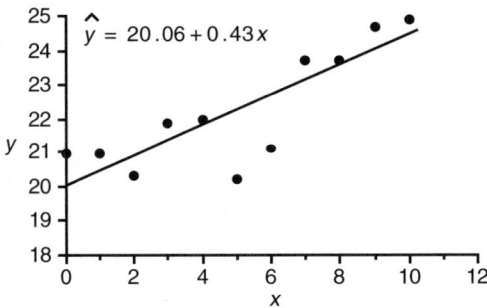

FIGURE 41.2 Data obtained from a repeated experiment with randomization to eliminate autocorrelation.

One might be tempted to blame the peculiar result entirely on the low value measured at $x = 6$, but the experimenters did not leap to conclusions. Discussion of the experimental procedure revealed that the tests were done starting with $x = 0$ first, then with $x = 1$, etc., up through $x = 10$. The measurements of y were also done in order of increasing concentration. It was also discovered that the injection port of the instrument used to measure y might not have been thoroughly cleaned between each run. The students knew about randomization, but time was short and they could complete the experiment faster by not randomizing. The penalty was autocorrelation and a wasted experiment.

They were asked to repeat the experiment, this time randomizing the order of the runs, the order of analyzing the specimens, and taking more care to clean the injection port. This time the data were as shown in Figure 41.2. The regression equation is $\hat{y} = 20.06 + 0.43x$, with $R^2 = 0.68$. The confidence interval of the slope is 0.21 to 0.65. This interval includes the expected slope of 0.5 and shows that x and y are related.

Can the dramatic difference in the outcome of the first and second experiments possibly be due to the presence of autocorrelation in the experimental data? It is both possible and likely, in view of the lack of randomization in the order of running the tests.

The Consequences of Autocorrelation on Regression

An important part of doing regression is obtaining a valid statement about the precision of the estimates. Unfortunately, autocorrelation acts to destroy our ability to make such statements. If the error terms are positively autocorrelated, the usual confidence intervals and tests using t and F distributions are no longer strictly applicable because the variance estimates are distorted (Neter et al., 1983).

Why Autocorrelation Distorts the Variance Estimates

Suppose that the system generating the data has the true underlying relation $\eta = \beta_0 + \beta_1 x$, where x could be any independent variable, including time as in a times series of data. We observe n values: $y_1 = \eta + e_1, \ldots, y_{i-2} = \eta + e_{i-2}, y_{i-1} = \eta + e_{i-1}, y_i = \eta + e_i, \ldots, y_n = \eta + e_n$. The usual assumption is that the residuals (e_i) are independent, meaning that the value of e_i is not related to e_{i-1}, e_{i-2}, etc. Let us examine what happens when this is not true.

Suppose that the residuals (e_i), instead of being random and independent, are correlated in a simple way that is described by $e_i = \rho \, e_{i-1} + a_i$, in which the errors ($a_i$) are independent and normally distributed with constant variance σ^2. The strength of the autocorrelation is indicated by the *autocorrelation coefficient* (ρ), which ranges from -1 to $+1$. If $\rho = 0$, the e_i are independent. If ρ is positive, successive values of e_i are similar to each other and:

$$e_i = \rho e_{i-1} + a_i$$
$$e_{i-1} = \rho e_{i-2} + a_{i-1}$$
$$e_{i-2} = \rho e_{i-3} + a_{i-2}$$

and so on. By recursive substitution we can show that:

$$e_i = \rho(\rho e_{i-2} + a_{i-1}) + a_i = \rho^2 e_{i-2} + \rho a_{i-1} + a_i$$

and

$$e_i = \rho^3 e_{i-3} + \rho^2 a_{i-2} + \rho a_{i-1} + a_i$$

This shows that the process is "remembering" past conditions to some extent, and the strength of this memory is reflected in the value of ρ.

Reversing the order of the terms and continuing the recursive substitution gives:

$$e_i = a_i + \rho a_{i-1} + \rho^2 a_{i-2} + \rho^3 a_{i-3} + \cdots \rho^n a_{i-n}$$

The expected values of $a_i, a_{i-1}, \ldots$ are zero and so is the expected value of e_i. The variance of e_i and the variance of a_i, however, are not the same. The variance of e_i is the sum of the variances of each term:

$$\sigma_e^2 = \text{Var}(a_i) + \rho^2 \text{Var}(a_{i-1}) + \rho^4 \text{Var}(a_{i-2}) + \cdots + \rho^{2n} \text{Var}(a_{i-n}) + \cdots$$

By definition, the a's are independent so $\sigma_a^2 = \text{Var}(a_i) = \text{Var}(a_{i-1}) = \ldots = \text{Var}(a_{i-n})$. Therefore, the variance of e_i is:

$$\sigma_e^2 = \sigma_a^2 (1 + \rho^2 + \rho^4 + \cdots + \rho^{2n} + \cdots)$$

For positive correlation ($\rho > 0$), the power series converges and:

$$(1 + \rho^2 + \rho^4 + \cdots + \rho^{2n} + \cdots) = \frac{1}{1-\rho^2}$$

Thus:

$$\sigma_e^2 = \sigma_a^2 \frac{1}{1-\rho^2}$$

This means that when we do not recognize and account for positive autocorrelation, the estimated variance σ_e^2 will be larger than the true variance of the random independent errors (σ_a^2) by the factor $1/(1 - \rho^2)$. This inflation can be impressive. If ρ is large (i.e., $\rho = 0.8$), $\sigma_e^2 = 2.8\sigma_a^2$.

An Example of Autocorrelated Errors

The laboratory data presented for the case study were created to illustrate the consequences of autocorrelation on regression. The true model of the experiment is $\eta = 20 + 0.5x$. The data structure is shown in Table 41.1. If there were no autocorrelation, the observed values would be as shown in Figure 41.2. These are the third column in Table 41.1, which is computed as $y_i + 20 + 0.5x_i + a_i$, where the a_i are independent values drawn randomly from a normal distribution with mean zero and variance of one (the a_i's actually selected have a variance of 1.00 and a mean of −0.28).

In the flawed experiment, hidden factors in the experiment were assumed to introduce autocorrelation. The data were computed assuming that the experiment generated errors having first-order autocorrelation with $\rho = 0.8$. The last three columns in Table 41.1 show how independent random errors are converted to correlated errors. The function producing the flawed data is:

$$y_i = \eta + e_i = 20 + 0.5x_i + 0.8e_{i-1} + a_i$$

If the data were produced by the above model, but we were unaware of the autocorrelation and fit the simpler model $\eta = \beta_0 + \beta_0 x$, the estimates of β_0 and β_1 will reflect this misspecification of the model. Perhaps more serious is the fact that t-tests and F-tests on the regression results will be wrong, so we may be misled as to the significance or precision of estimated values. Fitting the data produced from the autocorrelation model of the process gives $y_i = 21.0 + 0.12x_i$. The 95% confidence interval of the slope is [−0.12 to 0.35] and the t-ratio for the slope is 1.1. Both of these results indicate the slope is not significantly different from zero. Although the result is reported as statistically insignificant, it is wrong because the true slope is 0.5.

This is in contrast to what would have been obtained if the experiment had been conducted in a way that prevented autocorrelation from entering. The data for this case are listed in the "no autocorrelation" section of Table 41.1 and the results are shown in Table 41.2. The fitted model is $y_i = 20.06 + 0.43x_i$, the confidence interval of the slope is [0.21 to 0.65] and the t-ratio for the slope is 4.4. The slope is statistically significant and the true value of the slope ($\beta = 0.5$) falls within the confidence interval.

Table 41.2 summarizes the results of these two regression examples ($\rho = 0$ and $\rho = 0.8$). The Durbin-Watson statistic (explained in the next section) provided by the regression program indicates independence in the case where $\rho = 0$, and shows serial correlation in the other case.

TABLE 41.1

Data Created Using True Values of $y_i = 20 + 0.5x_i + a_i$ with $a_i = N(0,1)$

	No Autocorrelation			Autocorrelation, $\rho = 0.8$					
x	η	a_i	$y_i = \eta + a_i$	$0.8e_{i-1}$	+	a_i	=	e_i	$y_i = \eta + e_i$
0	20.0	1.0	21.0	0.00	+	1.0	=	1.0	21.0
1	20.5	0.5	21.0	0.80	+	0.5	=	1.3	21.8
2	21.0	−0.7	20.3	1.04	+	−0.7	=	0.3	21.3
3	21.5	0.3	21.8	0.27	+	0.3	=	0.6	22.1
4	22.0	0.0	22.0	0.46	+	0.0	=	0.5	22.5
5	22.5	−2.3	20.2	0.37	+	−2.3	=	−1.9	20.6
6	23.0	−1.9	21.1	−1.55	+	−1.9	=	−3.4	19.6
7	23.5	0.2	23.7	−2.76	+	0.2	=	−2.6	20.9
8	24.0	−0.3	23.7	−2.05	+	−0.3	=	−2.3	21.7
9	24.5	0.2	24.7	−1.88	+	0.2	=	−1.7	22.8
10	25.0	−0.1	24.9	−1.34	+	−0.1	=	−1.4	23.6

TABLE 41.2

Summary of the Regression "Experiments"

Estimated Statistics	No Autocorrelation $\rho = 0$	Autocorrelation $\rho = 0.8$
Intercept	20.1	21.0
Slope	0.43	0.12
Confidence interval of slope	[0.21, 0.65]	[−0.12, 0.35]
Standard error of slope	0.10	0.10
R^2	0.68	0.12
Mean square error	1.06	1.22
Durbin–Watson D	1.38	0.91

TABLE 41.3

Durbin-Watson Test Bounds for the 0.05 Level of Significance

| | $p = 2$ | | $p = 3$ | | $p = 4$ | |
n	d_L	d_U	d_L	d_U	d_L	d_U
15	1.08	1.36	0.95	1.54	0.82	1.75
20	1.20	1.41	1.10	1.54	1.00	1.68
25	1.29	1.45	1.21	1.55	1.12	1.66
30	1.35	1.49	1.28	1.57	1.21	1.65
50	1.50	1.59	1.46	1.63	1.42	1.67

Note: n = number of observations; p = number of parameters estimated in the model.

Source: Durbin, J. and G. S. Watson (1951). *Biometrika,* 38, 159–178.

A Statistic to Indicate Possible Autocorrelation

Detecting autocorrelation in a small sample is difficult; sometimes it is not possible. In view of this, it is better to design and conduct experiments to exclude autocorrelated errors. Randomization is our main weapon against autocorrelation in designed experiments. Still, because there is a possibility of autocorrelation in the errors, most computer programs that do regression also compute the *Durbin-Watson statistic*, which is based on an examination of the residual errors for autocorrelation. The Durbin-Watson test assumes a first-order model of autocorrelation. Higher-order autocorrelation structure is possible, but less likely than first-order, and verifying higher-order correlation would be more difficult. Even detecting the first-order effect is difficult when the number of observations is small and the Durbin-Watson statistic cannot always detect correlation when it exists.

The test examines whether the first-order autocorrelation parameter ρ is zero. In the case where $\rho = 0$, the errors are independent. The test statistic is:

$$D = \frac{\sum_{i=2}^{n}(e_i - e_{i-1})^2}{\sum_{i=1}^{n}(e_i)^2}$$

where the e_i are the residuals determined by fitting a model using least squares.

Durbin and Watson (1971) obtained approximate upper and lower bounds (d_L and d_U) on the statistic D. If $d_L \leq D \leq d_U$, the test is inconclusive. However, if $D > d_U$, conclude $\rho = 0$; and if $D < d_L$, conclude $\rho > 0$. A few Durbin-Watson test bounds for the 0.05 level of significance are given in Table 41.3. Note that this test is for positive ρ. If $\rho < 0$, a test for negative correlation is required; the test statistic to be used is $4 - D$, where D is calculated as before.

TABLE 41.4
Results of Trend Analysis of Data in Figure 41.3

Result	Time Series A	Time Series B
Generating model	$y_t = 10 + a_t$	$y_t = 10 + 0.8e_{t-1} + a_t$
Fitted model	$\hat{y}_t = 9.98 + 0.005t$	$\hat{y}_t = 9.71 + 0.033t$
Confidence interval of β_1	[−0.012 to 0.023]	[0.005 to 0.061]
Standard error of β_1	0.009	0.014
Conclusion regarding β_1	$\beta_1 = 0$	$\beta_1 > 0$
Durbin-Watson statistic	2.17	0.44

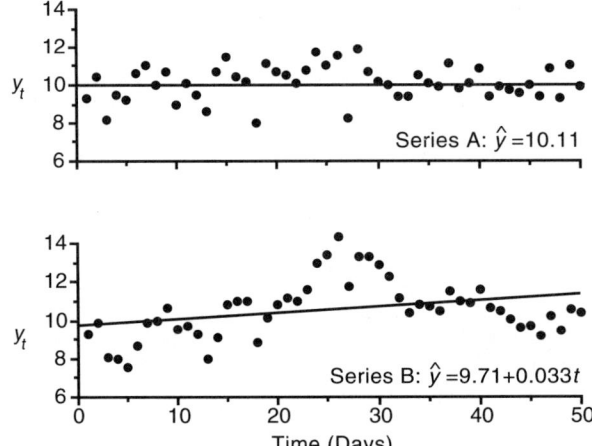

FIGURE 41.3 Time series of simulated environmental data. Series A is random, normally distributed values with $\eta = 10$ and $\sigma = 1$. Series B was constructed using the random variates of Series A to construct serially correlated values with $\rho = 0.8$, to which a constant value of 10 was added.

Autocorrelation and Trend Analysis

Sometimes we are tempted to take an existing record of environmental data (pH, temperature, etc.) and analyze it for a trend by doing linear regression to estimate a slope. A slope statistically different from zero is taken as evidence that some long-term change has been occurring. Resist the temptation, because such data are almost always serially correlated. *Serial correlation* is autocorrelation between data that constitute a time series. An example, similar to the regression example, helps make the point.

Figure 41.3 shows two time series of simulated environmental data. There are 50 values in each series. The model used to construct Series A was $y_t = 10 + a_t$, where a_t is a random, independent variable with N(0,1). The model used to construct Series B was $y_t = 10 + 0.8e_{t-1} + a_t$. The a_i are the same as in Series A, but the e_i variates are serially correlated with $\rho = 0.8$.

For both data sets, the true underlying trend is zero (the models contain no term for slope). If trend is examined by fitting a model of the form $\eta = \beta_0 + \beta_1 t$, where t is time, the results are in Table 41.4.

For Series A in Figure 41.3, the fitted model is $\hat{y} = 9.98 + 0.005t$, but the confidence interval for the slope includes zero and we simplify the model to $\hat{y} = 10.11$, the average of the observed values.

For Series B in Figure 41.3, the fitted model is $\hat{y} = 9.71 + 0.033t$. The confidence interval of the slope does not include zero and the nonexistent upward trend seems verified. This is caused by the serial correlation. The serial correlation causes the time series to *drift* and over a short period of time this drift looks like an upward trend. There is no reason to expect that this upward drift will continue. A series generated with a different set of a_i's could have had a downward trend. The Durbin-Watson statistic did give the correct warning about serial correlation.

Comments

We have seen that autocorrelation can cause serious problems in regression. The Durbin-Watson statistic might indicate when there is cause to worry about autocorrelation. It will not always detect autocorrelation, and it is especially likely to fail when the data set is small. Even when autocorrelation is revealed as a problem, it is too late to eliminate it from the data and one faces the task of deciding how to model it.

The pitfalls inherent with autocorrelated errors provide a strong incentive to plan experiments to include proper randomization whenever possible. If an experiment is intended to define a relationship between x and y, the experiments should not be conducted by gradually increasing (or decreasing) the x's. Randomize over the settings of x to eliminate autocorrelation due to time effects in the experiments.

Chapter 51 discusses how to deal with serial correlation.

References

Box, G. E. P., W. G. Hunter, and J. S. Hunter (1978). *Statistics for Experimenters: An Introduction to Design, Data Analysis, and Model Building,* New York, Wiley Interscience.

Durbin, J. and G. S. Watson (1951). "Testing for Serial Correlation in Least Squares Regression, II," *Biometrika,* 38, 159–178.

Durbin, J. and G. S. Watson (1971). "Testing for Serial Correlation in Least Squares Regression, III," *Biometrika,* 58, 1–19.

Neter, J., W. Wasserman, and M. H. Kutner (1983). *Applied Regression Models,* Homewood, IL, Richard D. Irwin Co.

Exercises

41.1 Blood Lead. The data below relate the lead level measured in the umbilical cord blood of infants born in a Boston hospital in 1980 and 1981 to the total amount of leaded gasoline sold in Massachusetts in the same months. Do you think autocorrelation might be a problem in this data set? Do you think the blood levels are related directly to the gasoline sales in the month of birth, or to gasoline sales in the previous several months? How would this influence your model building strategy?

Month	Year	Leaded Gasoline Sold	Pb in Umbilical Cord Blood (μg/dL)
3	1980	141	6.4
4	1980	166	6.1
5	1980	161	5.7
6	1980	170	6.9
7	1980	148	7.0
8	1980	136	7.2
9	1980	169	6.6
10	1980	109	5.7
11	1980	117	5.7
12	1980	87	5.3
1	1981	105	4.9
2	1981	73	5.4
3	1981	82	4.5
4	1981	75	6.0

41.2 pH Trend. Below are $n = 40$ observations of pH in a mountain stream that may be affected by acidic precipitation. The observations are weekly averages made 3 months apart, giving a record that covers 10 years. Discuss the problems inherent in analyzing these data to assess whether there is a trend toward lower pH due to acid rain.

6.8	6.8	6.9	6.5	6.7	6.8	6.8	6.7	6.9	6.8	6.7	6.8	6.9	6.7	6.9	6.8	6.7	6.9	6.7	7.0
6.6	7.1	6.6	6.8	7.0	6.7	6.7	6.9	6.9	6.9	6.7	6.6	6.4	6.4	7.0	7.0	6.9	7.0	6.8	6.9

41.3 Simulation. Simulate several time series of $n = 50$ using $y_t = 10 + 0.8e_{t-1} + a_t$ for different series of a_t, where $a_t = N(0,1)$. Fit the series using linear regression and discuss your results.

41.4 Laboratory Experiment. Describe a laboratory experiment, perhaps one that you have done, in which autocorrelation could be present. Explain how randomization would protect against the conclusions being affected by the correlation.

42

The Iterative Approach to Experimentation

KEY WORDS biokinetics, chemostat, dilution rate, experimental design, factorial designs, iterative design, model building, Monod model, parameter estimation, sequential design.

The dilemma of model building is that what needs to be known in order to design good experiments is exactly what the experiments are supposed to discover. We could be easily frustrated by this if we imagined that success depended on one grand experiment. Life, science, and statistics do not work this way. Knowledge is gained in small steps. We begin with a modest experiment that produces information we can use to design the second experiment, which leads to a third, etc. Between each step there is need for reflection, study, and creative thinking. Experimental design, then, is a philosophy as much as a technique.

The iterative (or sequential) philosophy of experimental investigation diagrammed in Figure 42.1 applies to mechanistic model building and to empirical exploration of operating conditions (Chapter 43). The iterative approach is illustrated for an experiment in which each observation requires a considerable investment.

Case Study: Bacterial Growth

The material balance equations for substrate (S) and bacterial solids (X) in a completely mixed reactor operated without recycle are:

$$V\frac{dX}{dt} = 0 - QX + \left(\frac{\theta_1 S}{\theta_2 + S}\right)XV \qquad \text{Material balance on bacterial solids}$$

$$V\frac{dS}{dt} = QS_0 - QX - \frac{1}{\theta_3}\left(\frac{\theta_1 S}{\theta_2 + S}\right)XV \qquad \text{Material balance on substrate}$$

where Q = liquid flow rate, V = reactor volume, $D = Q/V$ = dilution rate, S_0 = influent substrate concentration, X = bacterial solids concentration in the reactor and in the effluent, and S = substrate concentration in the reactor and in the effluent. The parameters of the Monod model for bacterial growth are the maximum growth rate (θ_1); the half-saturation constant (θ_2), and the yield coefficient (θ_3). This assumes there are no bacterial solids in the influent.

After dividing by V, the equations are written more conveniently as:

$$\frac{dX}{dt} = \frac{\theta_1 SX}{\theta_2 + S} - DX \quad \text{and} \quad \frac{dS}{dt} = DS_0 - DS - \left(\frac{\theta_1 X}{\theta_3}\right)\left(\frac{S}{\theta_2 + S}\right)$$

The steady-state solutions ($dX/dt = 0$ and $dS/dt = 0$) of the equations are:

$$S = \frac{\theta_2 D}{\theta_1 - D} \quad \text{and} \quad X = \theta_3(S_0 - S) = \theta_3\left(S_0 - \frac{\theta_2 D}{\theta_1 - D}\right)$$

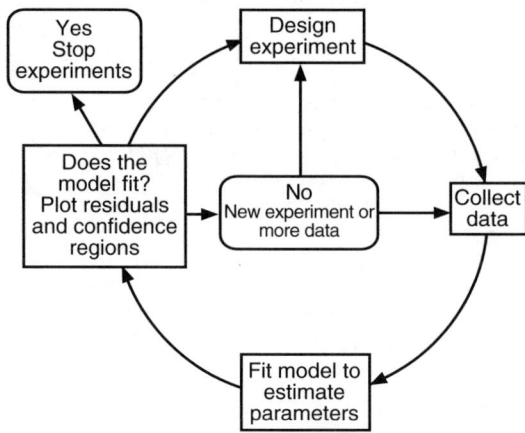

FIGURE 42.1 The iterative cycle of experimentation. (From Box, G. E. P. and W. G. Hunter (1965). *Technometrics,* 7, 23.)

If the dilution rate is sufficiently large, the organisms will be washed out of the reactor faster than they can grow. If all the organisms are washed out, the effluent concentration will equal the influent concentration, $S = S_0$. The lowest dilution rate at which washout occurs is called the critical dilution rate (D_c) which is derived by substituting $S = S_0$ into the substrate model above:

$$D_c = \frac{\theta_1 S_0}{\theta_2 + S_0}$$

When $S_0 \gg \theta_2$, which is often the case, $D_c \approx \theta_1$.

Experiments will be performed at several dilution rates (i.e., flow rates), while keeping the influent substrate concentration constant ($S_0 = 3000$ mg/L). When the reactor attains steady-state at the selected dilution rate, X and S will be measured and the parameters θ_1, θ_2, and θ_3 will be estimated. Because several weeks may be needed to start a reactor and bring it to steady-state conditions, the experimenter naturally wants to get as much information as possible from each run. Here is how the iterative approach can be used to do this.

Assume that the experimenter has only two reactors and can test only two dilution rates simultaneously. Because two responses (X and S) are measured, the two experimental runs provide four data points (X_1 and S_1 at D_1; X_2 and S_2 at D_2), and this provides enough information to estimate the three parameters in the model. The first two runs provide a basis for another two runs, etc., until the model parameters have been estimated with sufficient precision.

Three iterations of the experimental cycle are shown in Table 42.1. An initial guess of parameter values is used to start the first iteration. Thereafter, estimates based on experimental data are used. The initial guesses of parameter values were $\theta_3 = 0.50$, $\theta_1 = 0.70$, and $\theta_2 = 200$. This led to selecting flow rate $D_1 = 0.66$ for one run and $D_2 = 0.35$ for the other.

The experimental design criterion for choosing efficient experimental settings of D is ignored for now because our purpose is merely to show the efficiency of iterative experimentation. We will simply say that it recommends doing two runs, one with the dilution rate set as near the critical value D_c as the experimenter dares to operate, and the other at about half this value. At any stage in the experimental cycle, the best current estimate of the critical flow rate is $D_c = \theta_1$. The experimenter must be cautious in using this advice because operating conditions become unstable as D_c is approached. If the actual critical dilution rate is exceeded, the experiment fails entirely and the reactor has to be restarted, at a considerable loss of time. On the other hand, staying too far on the safe side (keeping the dilution rate too low) will yield poor estimates of the parameters, especially of θ_1. In this initial stage of the experiment we should not be too bold.

TABLE 42.1
Three Iterations of the Experiment to Estimate Biokinetic Parameters

Iteration and Exp. Run	Best Current Estimates of Parameter Values			Controlled Dilution Rate	Observed Values		Parameter Values Estimated from New Data		
	θ_3	θ_1	θ_2	$D = V/Q$	S	X	$\hat{\theta}_3$	$\hat{\theta}_1$	$\hat{\theta}_2$
Iteration 1	(Initial guesses)								
Run 1	0.50	0.70	200	0.66	2800	100			
Run 2				0.35	150	1700	0.60	0.55	140
Iteration 2	(From iteration 1)								
Run 3	0.60	0.55	140	0.52	1200	70			
Run 4				0.27	80	1775	0.60	0.55	120
Iteration 3	(From iteration 2)								
Run 5	0.60	0.55	120	0.54	2998	2			
Run 6				0.27	50	1770	0.60	0.55	54

Source: Johnson, D. B. and P. M. Berthouex (1975). *Biotech. Bioengr.*, 18, 557–570.

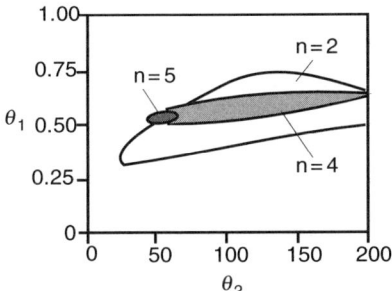

FIGURE 42.2 Approximate joint 95% confidence regions for θ_1 and θ_2 estimated after the first, second, and third experimental iterations. Each iteration consisted of experiments at two dilution rates, giving $n = 2$ after the first iteration, $n = 4$ after the second, and $n = 6$ after the third.

The parameter values estimated using data from the first pair of experiments were $\hat{\theta}_1 = 0.55$, $\hat{\theta}_2 = 140$, and $\hat{\theta}_3 = 0.60$. These values were used to design two new experiments at $D = 0.52$ and $D = 0.27$, and a second experimental iteration was done. The parameters were estimated using all data from the first and second iterations. The estimates $\hat{\theta}_1$ and $\hat{\theta}_3$ did not change from the first iteration, but $\hat{\theta}_2$ was reduced from 140 to 120. Because θ_1 and θ_3 seem to be estimated quite well (because they did not change from one iteration to the next), the third iteration essentially focuses on improving the estimate of θ_2.

In run 5 of the third iteration, we see $S = 2998$ and $X = 2$. This means that the dilution rate ($D = 0.54$) was too high and washout occurred. This experimental run therefore provides useful information, but the data must be handled in a special way when the parameters are estimated. (Notice that run 1 had a higher dilution rate but was able to maintain a low concentration of bacterial solids in the reactor and remove some substrate.)

At the end of three iterative steps — a total of only six experiments — the experiment was ended. Figure 42.2 shows how the approximate 95% joint confidence region for θ_1 and θ_2 decreased in size from the first to the second to the third set of experiments. The large unshaded region is the approximate joint 95% confidence region for the parameters after the first set of $n = 2$ experiments. Neither θ_1 nor θ_2 was estimated very precisely. At the end of the second iteration, there were $n = 4$ observations at four settings of the dilution rate. The resulting joint confidence region (the lightly shaded area) is horizontal, but elongated, showing that θ_1 was estimated with good precision, but θ_2 was not. The third iteration invested in data that would more precisely define the value of θ_2. Fitting the model to the $n = 5$ valid tests gives the estimates $\hat{\theta}_1 = 0.55$, $\hat{\theta}_2 = 54$, and $\hat{\theta}_3 = 0.60$. The final joint confidence region is small, as shown in Figure 42.2.

The parameters were estimated using multiresponse methods to fit S and X simultaneously. This contributes to smaller confidence regions; the method is explained in Chapter 46.

Comments

The iterative experimental approach is very efficient. It is especially useful when measurements are difficult or expensive. It is recommended in almost all model building situations, whether the model is linear or nonlinear, simple or complicated.

The example described in this case study was able to obtain precise estimates of the three parameters in the model with experimental runs at only six experimental conditions. Six runs are not many in this kind of experiment. The efficiency was the result of selecting experimental conditions (dilution rates) that produced a lot of information about the parameter values. Chapter 44 will show that making a large number of runs can yield poorly estimated parameters if the experiments are run at the wrong conditions.

Factorial and fractional factorial experimental designs (Chapters 27 to 29) are especially well suited to the iterative approach because they can be modified in many ways to suit the experimenter's need for additional information.

References

Box, G. E. P. (1991). "Quality Improvement: The New Industrial Revolution," Tech. Rep. No. 74, Madison, WI, Center for Quality and Productivity Improvement, University of Wisconsin–Madison.

Box, G. E. P. and W. G. Hunter (1965). "The Experimental Study of Physical Mechanisms," *Technometrics*, 7, 23.

Johnson, D. B. and P. M. Berthouex (1975). "Efficient Biokinetic Designs," *Biotech. Bioengr.*, 18, 557–570.

Exercises

42.1 **University Research I.** Ask a professor to describe a problem that was studied (and we hope solved) using the iterative approach to experimentation. This might be a multi-year project that involved several graduate students.

42.2 **University Research II.** Ask a Ph.D. student (a graduate or one in-progress) to explain their research problem. Use Figures 1.1 and 42.1 to structure the discussion. Explain how information gained in the initial steps guided the design of later investigations.

42.3 **Consulting Engineer.** Interview a consulting engineer who does industrial pollution control or pollution prevention projects and learn whether the iterative approach to investigation and design is part of the problem-solving method. Describe the project and the steps taken toward the final solution.

42.4 **Reaction Rates.** You are interested in destroying a toxic chemical by oxidation. You hypothesize that the destruction occurs in three steps.

Toxic chemical –> Semi-toxic intermediate –> Nontoxic chemical –> Nontoxic gas

You want to discover the kinetic mechanisms and the reaction rate coefficients. Explain how the iterative approach to experimentation could be useful in your investigations.

42.5 Adsorption. You are interested in removing a solvent from contaminated air by activated carbon adsorption and recovering the solvent by steam or chemical regeneration of the carbon. You need to learn which type of carbon is most effective for adsorption in terms of percent contaminant removal and adsorptive capacity, and which regeneration conditions give the best recovery. Explain how you would use an iterative approach to investigate the problem.

43

Seeking Optimum Conditions by Response Surface Methodology

KEY WORDS *composite design, factorial design, iterative design, optimization, quadratic effects, response surface methodology, regression, star point, steepest ascent, two-level design.*

The response of a system or process may depend on many variables such as temperature, organic carbon concentration, air flow rate, etc. An important problem is to discover settings for the critical variables that give the best system performance. Response surface analysis is an experimental approach to optimizing the performance of systems. It is the ultimate application of the iterative approach to experimentation.

The method was first demonstrated by Box and Wilson (1951) in a paper that Margolin (1985) describes as follows:

> The paper...is one of those rare, truly pioneering papers that completely shatters the existing paradigm of how to solve a particular problem.... The design strategy...to attain an optimum operating condition is brilliant in both its logic and its simplicity. Rather than exploring the entire continuous experimental region in one fell swoop, one explores a sequence of subregions. Two distinct phases of such a study are discernible. First, in each subregion a classical two-level fractional factorial design is employed...This first-phase process is iterative and is terminated when a region of near stationarity is reached. At this point a new phase is begun, one necessitating radically new designs for the successful culmination of the research effort.

Response Surface Methodology

The strategy is to explore a small part of the experimental space, analyze what has been learned, and then move to a promising new location where the learning cycle is repeated. Each exploration points to a new location where conditions are expected to be better. Eventually a set of optimal operating conditions can be determined. We visualize these as the peak of a hill (or bottom of a valley) that has been reached after stopping periodically to explore and locate the most locally promising path. At the start we imagine the shape of the hillside is relatively smooth and we worry mainly about its steepness. Figure 43.1 sketches the progress of an iterative search for the optimum conditions in a process that has two active independent variables. The early explorations use two-level factorial experimental designs, perhaps augmented with a center point. The main effects estimated from these designs define the path of steepest ascent (descent) toward the optimum. A two-level factorial design may fail near the optimum because it is located astride the optimum and the main effects appear to be zero. A quadratic model is needed to describe the optimal region. The experimental design to fit a quadratic model is a two-level factorial augmented with stars points, as in the optimization stage design shown in Figure 43.1.

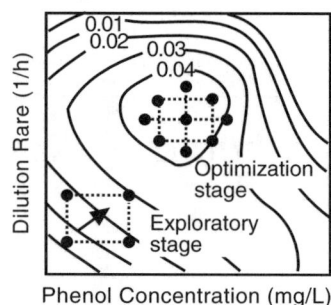

FIGURE 43.1 Two stages of a response surface optimization. The second stage is a two-level factorial augmented to define quadratic effects.

Case Study: Inhibition of Microbial Growth by Phenol

Wastewater from a coke plant contains phenol, which is known to be biodegradable at low concentrations and inhibitory at high concentrations. Hobson and Mills (1990) used a laboratory-scale treatment system to determine how influent phenol concentration and the flow rate affect the phenol oxidation rate and whether there is an operating condition at which the removal rate is a maximum. This case study is based on their data, which we used to create a response surface by drawing contours. The data given in the following sections were interpolated from this surface, and a small experimental error was added.

To some extent, a treatment process operated at a low dilution rate can tolerate high phenol concentrations better than a process operating at a high dilution rate. We need to define "high" and "low" for a particular wastewater and a particular biological treatment process and find the operating conditions that give the most rapid phenol oxidation rate (R). The experiment is arranged so the rate of biological oxidation of phenol depends on only the concentration of phenol in the reactor and the dilution rate. Dilution rate is defined as the reactor volume divided by the wastewater flow rate through the reactor. Other factors, such as temperature, are constant.

The iterative approach of experimentation, as embodied in response surface methodology, will be illustrated. The steps in each iteration are *design, data collection*, and *data analysis*. Here, only design and data analysis are discussed.

First Iteration

Design — Because each run takes several days to complete, the experiment was performed in small sequential stages. The first was a two-level, two-factor experiment — a 2^2 factorial design. The two experimental factors are dilution rate (D) and residual phenol concentration (C). The response is phenol oxidation rate (R). Each factor was investigated at two levels and the observed phenol removal rates are given in Table 43.1.

Analysis — The data can be analyzed by calculating the effects, as done in Chapter 27, or by linear regression (Chapter 30); we will use regression. There are four observations so the fitted model cannot have more than four parameters. Because we expect the surface to be relatively smooth, a reasonable model is $R = b_0 + b_1C + b_2D + b_{12}CD$. This describes a hyperplane. The terms b_1C and b_2D represent the main effects of concentration and dilution rate; $b_{12}CD$ is the interaction between the two factors.

The fitted model is $R = -0.022 - 0.018C + 0.2D + 0.3CD$. The response as a function of the two experimental factors is depicted by the contour map in Figure 43.2. The contours are values of R, in units of g/h. The approximation is good only in the neighborhood of the 2^2 experiment, which is indicated by the four dots at the corner of the rectangle. The direction toward higher removal rates is clear. The direction of steepest ascent, indicated by an arrow, is perpendicular to the contour line at the point of interest. Of course, the experimenter is not compelled to move along the line of steepest ascent.

TABLE 43.1

Experimental Design and Results for Iteration 1

C (g/L)	D (1/h)	R (g/h)
0.5	0.14	0.018
0.5	0.16	0.025
1.0	0.14	0.030
1.0	0.16	0.040

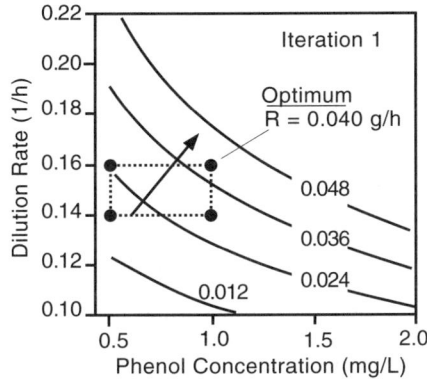

FIGURE 43.2 Response surface computed from the data collected in exploratory stage 1 of the optimizing experiment.

Second Iteration

Design — The first iteration indicates a promising direction but does not tell us how much each setting should be increased. Making a big step risks going over the peak. Making a timid step and progressing toward the peak will be slow. How bold — or how timid — should we be? This usually is not a difficult question because the experimenter has prior experience and special knowledge about the experimental conditions. We know, for example, that there is a practical upper limit on the dilution rate because at some level all the bacteria will wash out of the reactor. We also know from previously published results that phenol becomes inhibitory at some level. We may have a fairly good idea of the concentration at which this should be observed. In short, the experimenter knows something about limiting conditions at the start of the experiment (and will quickly learn more). To a large extent, we trust our judgment about how far to move.

The second factor is that iterative factorial experiments are so extremely efficient that the total number of experiments will be small regardless of how boldly we proceed. If we make what seems in hindsight to be a mistake either in direction or distance, this will be discovered and the same experiments that reveal it will put us onto a better path toward the optimum.

In this case of phenol degradation, published experience indicates that inhibitory effects will probably become evident with the range of 1 to 2 g/L. This suggests that an increase of 0.5 g/L in concentration should be a suitable next step, so we decide to try $C = 1.0$ and $C = 1.5$ as the low and high settings. Going roughly along the line of steepest ascent, this would give dilution rates of 0.16 and 0.18 as the low and high settings of D. This leads to the second-stage experiment, which is the 2^2 factorial design shown in Figure 43.2. Notice that we have not moved the experimental region very far. In fact, one setting ($C = 1.0$, $D = 0.16$) is the same in iterations 1 and 2. The observed rates (0.040 and 0.041) give us information about the experimental error.

Analysis — Table 43.2 gives the measured phenol removal rates at the four experimental settings. The average performance has improved and two of the response values are larger than the maximum observed in the first iteration. The fitted model is $R = 0.047 - 0.014C + 0.05D$. The estimated coefficient for the

TABLE 43.2

Experimental Design and Results for Iteration 2

C (g/L)	D (1/h)	R (g/h)
1.0	0.16	0.041
1.0	0.18	0.042
1.5	0.16	0.034
1.5	0.18	0.035

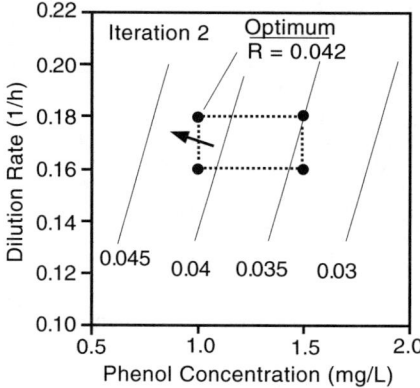

FIGURE 43.3 Approximation of the response surface estimated from the second-stage exploratory experiment.

CD term was zero. Figure 43.3, the response surface, indicates that C should be reduced and that further increase in D may be beneficial.

Before moving the experimental settings, consider the available information more carefully. The fitted model describes a plane and the plane is almost horizontal, as indicated by the small coefficients of both C and D. One way we can observe a nearly zero effect for both variables is if the four corners of the 2^2 experimental design straddle the peak of the response surface. Also, the direction of steepest ascent has changed from Figure 42.2 to 42.3. This suggests that we may be near the optimum. To check on this we need an experimental design that can detect and describe the increased curvature at the optimum. Fortunately, the design can be easily augmented to detect and model curvature.

Third Iteration: Exploring for Optimum Conditions

Design — We anticipate needing to fit a model that contains some quadratic terms, such as $R = b_0 + b_1 C + b_2 D + b_{12} CD + b_{11} C^2 + b_{22} D^2$. The basic experimental design is still a two-level factorial but it will be augmented by adding "star" points to make a *composite design* (Box, 1999). The easiest way to picture this design is to imagine a circle (or ellipse, depending on the scaling of our sketch) that passes through the four corners of the two-level design.

Rather than move the experimental region, we can use the four points from iteration 2 and four more will be added in a way that maintains the symmetry of the original design. The augmented design has eight points, each equidistant from the center of the design. Adding one more point at the center of the design will provide a better estimate of the curvature while maintaining the symmetric design. The nine experimental settings and the results are shown in Table 43.3 and Figure 43.4. The open circles are the two-level design from iteration 2; the solid circles indicate the center point and star points that were added to investigate curvature near the peak.

TABLE 43.3

Experimental Results for the Third Iteration

C (g/L)	D (1/h)	R (g/h)	Notes
1.0	0.16	0.041	Iteration 2 design
1.0	0.18	0.042	Iteration 2 design
1.5	0.16	0.034	Iteration 2 design
1.5	0.18	0.035	Iteration 2 design
0.9	0.17	0.038	Augmented "star" point
1.25	0.156	0.043	Augmented "star" point
1.25	0.17	0.047	Center point
1.25	0.184	0.041	Augmented "star" point
1.6	0.17	0.026	Augmented "star" point

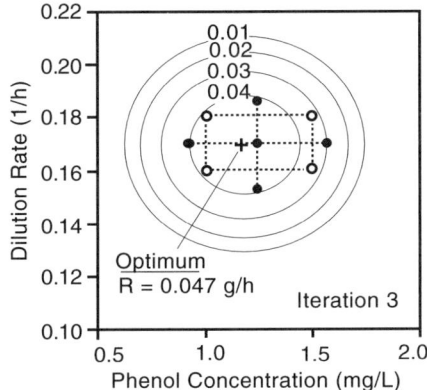

FIGURE 43.4 Contour plot of the quadratic response surface model fitted to an augmented two-level factorial experimental design. The open symbols are the two-level design from iteration 2; the solid symbols are the center and star points added to investigate curvature near the peak. The cross (+) locates the optimum computed from the model.

Analysis — These data were fitted to a quadratic model to get:

$$R = -0.76 + 0.28C + 7.54D - 0.12C^2 - 22.2D^2$$

The CD interaction term had a very small coefficient and was omitted. Contours computed from this model are plotted in Figure 43.4.

The maximum predicted phenol oxidation rate is 0.047 g/h, which is obtained at $C = 1.17$ g/L and $D = 0.17$ h^{-1}. These values are obtained by taking derivatives of the response surface model and simultaneously solving $\partial R/\partial C = 0$ and $\partial R/\partial D = 0$.

Iteration 4: Is It Needed?

Is a fourth iteration needed? One possibility is to declare that enough is known and to stop. We have learned that the dilution rate should be in the range of 0.16 to 0.18 h^{-1} and that the process seems to be inhibited if the phenol concentration is higher than about 1.1 or 1.2 mg/L. As a practical matter, more precise estimates may not be important. If they are, replication could be increased or the experimental region could be contracted around the predicted optimum conditions.

How Effectively was the Optimum Located?

Let us see how efficient the method was in this case. Figure 43.5a shows the contour plot from which the experimental data were obtained. This plot was constructed by interpolating the Hobson-Mills data with a simple contour plotting routine; no equations were fitted to the data to generate the surface. The location of their 14 runs is shown in Figure 43.5, which also shows the three-dimensional response surface from two points of view.

An experiment was run by interpolating a value of R from the Figure 43.5a contour map and adding to it an "experimental error." Although the first 2^2 design was not very close to the peak, the maximum was located with a total of only 13 experimental runs (4 in iteration 1, 4 in iteration 2, plus 5 in iteration 3). The predicted optimum is very close to the peak of the contour map from which the data were taken. Furthermore, the region of interest near the optimum is nicely approximated by the contours derived from the fitted model, as can be seen by comparing Figures 43.4 and Figure 43.5.

Hobson and Mills made 14 observations covering an area of roughly $C = 0.5$ to 1.5 mg/L and $D = 0.125$ to 0.205 h^{-1}. Their model predicted an optimum at about $D = 0.15$ h^{-1} and $C = 1.1$ g/L, whereas the largest removal rate they observed was at $D = 0.178$ h^{-1} and $C = 1.37$ g/L. Their model optimum differs from experimental observation because they tried to describe the entire experimental region using a quadratic model that could not describe the entire experimental region (i.e., all their data). A quadratic model gives a poor fit and a poor estimate of the optimum's location because it is not adequate to describe the irregular response surface. Observations that are far from the optimum can be useful in pointing us in a profitable direction, but they may provide little information about the location or value of the maximum. Such observations can be omitted when the region near the optimum is modeled.

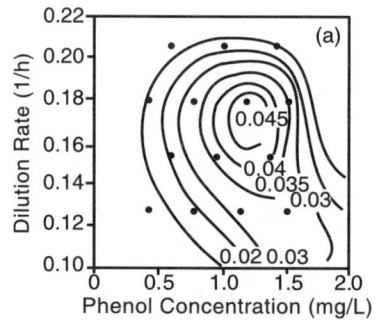

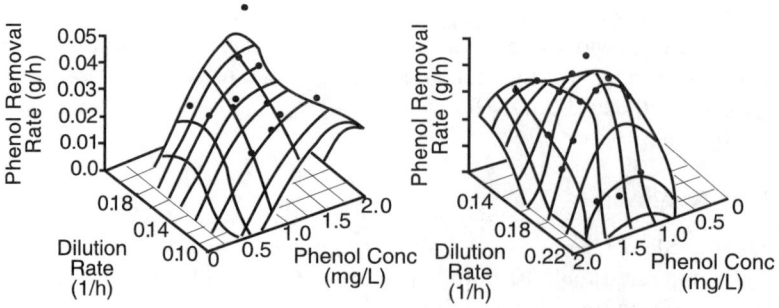

FIGURE 43.5 The location of the Hobson and Mills experimental runs are shown on the contour plot (top). Their data are shown from two perspectives in the three-dimensional plots.

Comments

Response surfaces are effective ways to empirically study the effect of explanatory variables on the response of a system and can help guide experimentation to obtain further information. The approach should have tremendous natural appeal to environmental engineers because their experiments (1) often take a long time to complete and (2) only a few experiments at a time can be conducted. Both characteristics make it attractive to do a few runs at a time and to intelligently use the early results to guide the design of additional experiments. This strategy is also powerful in process control. In most processes the optimal settings of control variables change over time and factorial designs can be used iteratively to follow shifts in the response surface. This is a wonderful application of the iterative approach to experimentation (Chapter 42).

The experimenter should keep in mind that response surface methods are not designed to faithfully describe large regions in the possible experimental space. The goal is to explore and describe the most promising regions as efficiently as possible. Indeed, large parts of the experimental space may be ignored.

In this example, the direction of steepest ascent was found graphically. If there are more than two variables, this is not convenient so the direction is found either by using derivatives of the regression equation or the main effects computed directly from the factorial experiment (Chapter 27). Engineers are familiar with these calculations and good explanations can be found in several of the books and papers referenced at the end of this chapter.

The composite design used to estimate the second-order effects in the third iteration of the example can only be used with quantitative variables, which are set at five levels ($\pm\alpha$, ± 1, and 0). Qualitative variables (present or absent, chemical A or chemical B) cannot be set at five levels, or even at three levels to add a center point to a two-level design. This creates a difficulty making an effective and balanced design to estimate second-order effects in situations where some variables are quantitative and some are qualitative. Draper and John (1988) propose some ways to deal with this.

The wonderful paper of Box and Wilson (1951) is recommended for study. Davies (1960) contains an excellent chapter on this topic; Box et al. (1978) and Box and Draper (1987) are excellent references. The approach has been applied to seeking optimum conditions in full-scale manufacturing plants under the name of Evolutionary Operation (Box, 1957; Box and Draper, 1969, 1989). Springer et al. (1984) applied these ideas to wastewater treatment plant operation.

References

Box, G. E. P. (1954). "The Exploration and Exploitation of Response Surfaces: Some General Considerations and Examples," *Biometrics,* 10(1), 16–60.

Box, G. E. P. (1957). "Evolutionary Operation: A Method for Increasing Industrial Productivity," *Applied Statistics,* 6(2), 3–23.

Box, G. E. P. (1982). "Choice of Response Surface Design and Alphabetic Optimality," *Utilitas Mathematica,* 21B, 11–55.

Box, G. E. P. (1999). "The Invention of the Composite Design," *Quality Engineering,* 12(1), 119–122.

Box, G. E. P. and N. R. Draper (1969). *Evolutionary Operation — A Statistical Method for Process Improvement,* New York, John Wiley.

Box, G. E. P. and N. R. Draper (1989). *Empirical Model Building and Response Surfaces,* New York, John Wiley.

Box, G. E. P. and J. S. Hunter (1957). "Multi-Factor Experimental Designs for Exploring Response Surfaces," *Annals Math. Stat.,* 28(1), 195–241.

Box, G. E. P., W. G. Hunter, and J. S. Hunter (1978). *Statistics for Experimenters: An Introduction to Design, Data Analysis, and Model Building,* New York, Wiley Interscience.

Box, G. E. P. and K. B. Wilson (1951). "On the Experimental Attainment of Optimum Conditions," *J. Royal Stat. Soc.,* Series B, 13(1), 1–45.

Davies, O. L. (1960). *Design and Analysis of Industrial Experiments,* New York, Hafner Co.

Draper, N. R. and J. A. John (1988). "Response-Surface Designs for Quantitative and Qualitative Variables," *Technometrics,* 20, 423–428.

Hobson, M. J. and N. F. Mills (1990). "Chemostat Studies of Mixed Culture Growing of Phenolics," *J. Water. Poll. Cont. Fed.,* 62, 684–691.

Margolin, B. H. (1985). "Experimental Design and Response Surface Methodology — Introduction," *The Collected Works of George E. P. Box,* Vol. 1, George Tiao, Ed., pp. 271–276, Belmont, CA, Wadsworth Books.

Springer, A. M., R. Schaefer, and M. Profitt (1984). "Optimization of a Wastewater Treatment Plant by the Employee Involvement Optimization System (EIOS)," *J. Water Poll. Control Fed.,* 56, 1080–1092.

Exercises

43.1 Sludge Conditioning. Sludge was conditioned with polymer (P) and fly ash (F) to maximize the yield (kg/m^2-h) of a dewatering filter. The first cycle of experimentation gave these data:

P (g/L)	40	40	60	60	50	50	50
F (% by wt.)	114	176	114	176	132	132	132
Yield	18.0	24	18.0	35	21	23	17

Source: Benitex, J. (1994). *Water Res.,* 28, 2067–2073.

(a) Fit these data by least squares and determine the path of steepest ascent. Plan a second cycle of experiments, assuming that second-order effects might be important.

(b) The second cycle of experimentation actually done gave these results:

P (g/L)	55	55	65	65	60	60	60	60	60	53	60	60	67
F (% by wt.)	140	160	140	160	150	150	150	150	150	150	165	135	150
Yield	29	100	105	108	120	171	118	120	118	77	99	102	97

The location of the experiments and the direction moved from the first cycle may be different than you proposed in part (a). This does not mean that your proposal is badly conceived, so don't worry about being wrong. Interpret the data graphically, fit an appropriate model, and locate the optimum dewatering conditions.

43.2 Catalysis. A catalyst for treatment of a toxic chemical is to be immobilized in a solid bead. It is important for the beads to be relatively uniform in size and to be physically durable. The desired levels are Durability > 30 and Uniformity < 0.2. A central composite-design in three factors was run to obtain the table below. The center point (0, 0, 0) is replicated six times. Fit an appropriate model and plot a contour map of the response surface. Overlay the two surfaces to locate the region of operating conditions that satisfy the durability and uniformity goals.

	Factor			Response	
Run	1	2	3	Durability	Uniformity
1	−1	−1	−1	8	0.77
2	+1	−1	−1	10	0.84
3	−1	+1	−1	29	0.16
4	+1	+1	−1	28	0.18
5	−1	−1	+1	23	0.23
6	+1	−1	+1	17	0.38
7	−1	+1	+1	45	0.1
8	+1	+1	+1	45	0.11
9	0	0	−1.682	14	0.32
10	0	0	1.682	29	0.15
11	0	−1.682	0	6	0.86
12	0	1.682	0	35	0.13
13	−1.682	0	0	23	0.18
14	1.682	0	0	7	0.77
15	0	0	0	22	0.21
16	0	0	0	21	0.23
17	0	0	0	22	0.21
18	0	0	0	24	0.19
19	0	0	0	21	0.23
20	0	0	0	25	0.19

43.3 Biosurfactant. Surfactin, a cyclic lipopeptide produced by *Bacillus subtilis*, is a biodegradable and nontoxic biosurfactant. Use the data below to find the operating condition that maximizes Surfactin production.

	Experimental Ranges and Levels				
Variables	−2	−1	0	+1	+2
Glucose (X_1), g/dm^3	0	20	40	60	80
NH$_4$NO$_3$ (X_2), g/dm^3	0	2	4	6	8
FeSO$_4$ (X_3), g/dm^3	0	6×10^{-4}	1.8×10^{-3}	3×10^{-3}	4.2×10^{-3}
MnSO$_4$ (X_4), g/dm^3	0	4×10^{-2}	1×10^{-2}	20×10^{-2}	28×10^{-2}

Run	X_1	X_2	X_3	X_4	y
1	−1	−1	−1	−1	23
2	+1	−1	−1	−1	15
3	−1	+1	−1	−1	16
4	+1	+1	−1	−1	18
5	−1	−1	+1	−1	25
6	+1	−1	+1	−1	16
7	−1	+1	+1	−1	17
8	+1	+1	+1	−1	20
9	−1	−1	−1	+1	26
10	+1	−1	−1	+1	16
11	−1	+1	−1	+1	18
12	+1	+1	−1	+1	21
13	−1	−1	+1	+1	36
14	+1	−1	+1	+1	24
15	−1	+1	+1	+1	33
16	+1	+1	+1	+1	24
17	−2	0	0	0	2
18	2	0	0	0	5
19	0	−2	0	0	14
20	0	2	0	0	20
21	0	0	−2	0	16
22	0	0	2	0	32
23	0	0	0	−2	15
24	0	0	0	2	34
25	0	0	0	0	35
26	0	0	0	0	36
27	0	0	0	0	35
28	0	0	0	0	35
29	0	0	0	0	34
30	0	0	0	0	36

43.4 Chrome Waste Solidification. Fine solid precipitates from lime neutralization of liquid effluents from surface finishing operations in stainless steel processing are treated by cement-based solidification. The solidification performance was explored in terms of water-to-solids ratio (W/S), cement content (C), and curing time (T). The responses were indirect tensile strength (ITS), leachate pH, and leachate chromium concentration. The desirable process will have high ITS, pH of 6 to 8, and low Cr. The table below gives the results of a central composite design that can be used to estimate quadratic effects. Evaluate the data. Recommend additional experiments if you think they are needed to solve the problem.

Run	W/S	C (%)	T (days)	ITS (MPa)	Leachate pH	Leachate Cr
1	1.23	14.1	37	0.09	5.6	1.8
2	1.23	25.9	37	0.04	7.5	4.6
3	1.23	25.9	61	0.29	7.4	3.8
4	1.13	25.9	61	0.35	7.5	4.8
5	1.23	14.1	61	0.07	5.5	2.5
6	1.13	25.9	37	0.01	7.9	6.2
7	1.13	14.1	61	0.12	5.8	1.8
8	1.13	14.1	37	0.08	5.7	2.9
9	1.10	20	49	0.31	6.7	2.1
10	1.26	20	49	0.22	6.4	1.8
11	1.18	10	49	0.06	5.5	3.8
12	1.18	30	49	0.28	8.2	6.4
13	1.18	20	28	0.18	6.9	2.2
14	1.18	20	70	0.23	6.5	1.9
15	1.18	20	49	0.26	6.7	2.1
16	1.18	20	49	0.30	6.8	2.3
17	1.18	20	49	0.29	6.7	2.0
18	1.18	20	49	0.24	6.6	2.0
19	1.18	20	49	0.30	6.5	2.3
20	1.18	20	49	0.28	6.6	2.1

44

Designing Experiments for Nonlinear Parameter Estimation

KEY WORDS *BOD, Box-Lucas design, derivative matrix, experimental design, nonlinear least squares, nonlinear model, parameter estimation, joint confidence region, variance-covariance matrix.*

The goal is to design experiments that will yield precise estimates of the parameters in a nonlinear model with a minimum of work and expense. Design means specifying what settings of the independent variable will be used and how many observations will be made. The design should recognize that each observation, although measured with equal accuracy and precision, will not contribute an equal amount of information about parameter values. In fact, the size and shape of the joint confidence region often depends more on where observations are located in the experimental space than on how many measurements are made.

Case Study: A First-Order Model

Several important environmental models have the general form $\eta = \theta_1(1 - \exp(-\theta_2 t))$. For example, oxygen transfer from air to water according to a first-order mass transfer has this model, in which case η is dissolved oxygen concentration, θ_1 is the first-order overall mass transfer rate, and θ_2 is the effective dissolved oxygen equilibrium concentration in the system. Experience has shown that θ_1 should be estimated experimentally because the equilibrium concentration achieved in real systems is not the handbook saturation concentration (Boyle and Berthouex, 1974). Experience also shows that estimating θ_1 by extrapolation gives poor estimates.

The BOD model is another familiar example, in which θ_1 is the ultimate BOD and θ_2 is the reaction rate coefficient. Figure 44.1 shows some BOD data obtained from analysis of a dairy wastewater specimen (Berthouex and Hunter, 1971). Figure 44.2 shows two joint confidence regions for θ_1 and θ_2 estimated by fitting the model to the entire data set ($n = 59$) and to a much smaller subset of the data ($n = 12$). An 80% reduction in the number of measurements has barely changed the size or shape of the joint confidence region. We wish to discover the efficient smaller design in advance of doing the experiment. This is possible if we know the form of the model to be fitted.

Method: A Criterion to Minimize the Joint Confidence Region

A model contains p parameters that are to be estimated by fitting the model to observations located at n settings of the independent variables (time, temperature, dose, etc.). The model is $\eta = f(\theta, x)$ where θ is a vector of parameters and x is a vector of independent variables. The parameters will be estimated by nonlinear least squares.

If we assume that the form of the model is correct, it is possible to determine settings of the independent variables that will yield precise estimates of the parameters with a small number of experiments. Our interest lies mainly in nonlinear models because finding an efficient design for a linear model is intuitive, as will be explained shortly.

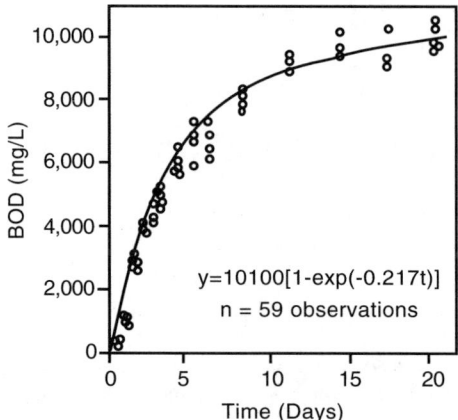

FIGURE 44.1 A BOD experiment with $n = 59$ observations covering the range of 1 to 20 days, with three to six replicates at each time of measurement. The curve is the fitted model with nonlinear least squares parameter estimates $\theta_1 = 10{,}100$ mg/L and $\theta_2 = 0.217$ day^{-1}.

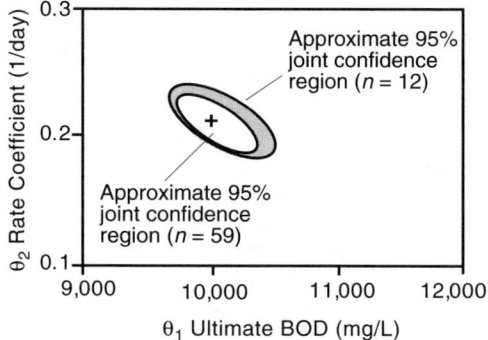

FIGURE 44.2 The unshaded ellipse is the approximate 95% joint confidence region for parameters estimated using all $n = 59$ observations. The cross locates the nonlinear least squares parameter estimates for $n = 59$. The shaded ellipse, which encloses the unshaded ellipse, is for parameters estimated using only $n = 12$ observations (6 on day 4 and 6 on day 20).

The minimum number of observations that will yield p parameter estimates is $n = p$. The fitted nonlinear model generally will not pass perfectly through these points, unlike a linear model with $n = p$ which will fit each observation exactly. The regression analysis will yield a residual sum of squares and a joint confidence region for the parameters. The goal is to have the joint confidence region small (Chapters 34 and 35); the joint confidence region for the parameters is small when their variances and covariances are small.

We will develop the regression model and the derivation of the variance of parameter estimates in matrix notation. Our explanation is necessarily brief; for more details, one can consult almost any modern reference on regression analysis (e.g., Draper and Smith, 1998; Rawlings, 1988; Bates and Watts, 1988). Also see Chapter 30.

In matrix notation, a linear model is:

$$y = X\beta + e$$

where y is an $n \times 1$ column vector of the observations, X is an $n \times p$ matrix of the independent variables (or combinations of them), β is a $p \times 1$ column vector of the parameters, and e is an $n \times 1$ column vector of the residual errors, which are assumed to have constant variance. n is the number of observations and p is the number of parameters in the model.

Designing Experiments for Nonlinear Parameter Estimation

The least squares parameter estimates and their variances and covariances are given by:

$$b = [X'X]^{-1}X'y$$

and

$$\text{Var}(b) = [X'X]^{-1}\sigma^2$$

The same equations apply for nonlinear models, except that the definition of the X matrix changes. A nonlinear model cannot be written as a matrix product of X and β, but we can circumvent this difficulty by using a linear approximation (Taylor series expansion) to the model. When this is done, the X matrix becomes a *derivative matrix* which is a function of the independent variables and the model parameters.

The variances and covariances of the parameters are given exactly by $[X'X]^{-1}\sigma^2$ when the model is linear. This expression is approximate when the model is nonlinear in the parameters. The minimum sized joint confidence region corresponds to the minimum of the quantity $[X'X]^{-1}\sigma^2$. Because the variance of random measurement error (σ^2) is a constant (although its value may be unknown), only the $[X'X]^{-1}$ matrix must be considered.

It is not necessary to compare entire variance-covariance matrices for different experimental designs. All we need to do is minimize the determinant of the $[X'X]^{-1}$ matrix or the equivalent of this, which is to maximize the determinant of $[X'X]$. This determinant design criterion, presented by Box and Lucas (1959), is written as:

$$\max \Delta = \max |X'X|$$

where the vertical bars indicate the determinant. Maximizing Δ minimizes the size of the approximate joint confidence region, which is inversely proportional to the square root of the determinant, that is, proportional to $\Delta^{-1/2}$.

$[X'X]^{-1}$ is the *variance-covariance matrix*. It is obtained from X, an n row by p column ($n \times p$) matrix, called the *derivative matrix*:

$$X = X_{ij} = \begin{bmatrix} X_{11} & X_{21} & \cdots & X_{p1} \\ X_{12} & X_{22} & \cdots & X_{p2} \\ \cdots & \cdots & \cdots & \cdots \\ X_{1n} & X_{2n} & \cdots & X_{pn} \end{bmatrix} \quad i = 1, 2, \ldots, p; \; j = 1, 2, \ldots, n$$

where p and n are the number of parameters and observations as defined earlier.

The elements of the X matrix are partial derivatives of the model with respect to the parameters:

$$X_{ij} = \frac{\partial f(\theta_i, x_j)}{\partial \theta_i} \quad i = 1, 2, \ldots, p; \; j = 1, 2, \ldots, n$$

For nonlinear models, however, the elements X_{ij} are functions of both the independent variables x_j and the unknown parameters θ_i. Thus, some preliminary work is required to compute the elements of the matrix in preparation for maximizing $|X'X|$.

For linear models, the elements X_{ij} do not involve the parameters of the model. They are functions only of the independent variables (x_j) or combinations of them. (This is the characteristic that defines a model as being linear in the parameters.) It is easily shown that the minimum variance design for a linear model spaces observations as far apart as possible. This result is intuitive in the case of fitting $\eta = \beta_0 + \beta_1 x$; the estimate of β_0 is enhanced by making an observation near the origin and the estimate of β_1 is enhanced by making the second observation at the largest feasible value of x. This simple example also points out the importance of the qualifier "if the model is assumed to be correct." Making measurements at two widely spaced settings of x is ideal for fitting a straight line, but it has terrible deficiencies if the correct model is quadratic. Obviously the design strategy is different when we know the form of the model compared to when we are seeking to discover the form of the model. In this chapter, the correct form of the model is assumed to be known.

Returning now to the design of experiments to estimate parameters in nonlinear models, we see a difficulty in going forward. To find the settings of x_j that maximize $|X'X|$, the values of the elements X_{ij} must be expressed in terms of numerical values for the parameters. The experimenter's problem is to provide these numerical values.

At first this seems an insurmountable problem because we are planning the experiment because the values of θ are unknown. Is it not necessary, however, to know in advance the answer that those experiments will give in order to design efficient experiments. The experimenter always has some prior knowledge (experienced judgment or previous similar experiments) from which to "guess" parameter values that are not too remote from the true values. These *a priori* estimates, being the best available information about the parameter values, are used to evaluate the elements of the derivative matrix and design the first experiments.

The experimental design based on maximizing $|X'X|$ is optimal, in the mathematical sense, with respect to the *a priori* parameter values, and based on the critical assumption that the model is correct. This does not mean the experiment will be perfect, or even that its results will satisfy the experimenter. If the initial parameter guess is not close to the true underlying value, the confidence region will be large and more experiments will be needed. If the model is incorrect, the experiments planned using this criterion will not reveal it. The so-called optimal design, then, should be considered as advice that should make experimentation more economical and rewarding. It is not a prescription for getting perfect results with the first set of experiments. Because of these caveats, an iterative approach to experimentation is productive.

If the parameter values provided are very near the true values, the experiment designed by this criterion will give precise parameter estimates. If they are distant from the true values, the estimated parameters will have a large joint confidence region. In either case, the first experiments provide new information about the parameter values that are used to design a second set of tests, and so on until the parameters have been estimated with the desired precision. Even if the initial design is poor, knowledge increases in steps, sometimes large ones, and the joint confidence region is reduced with each additional iteration. Checks on the structural adequacy of the model can be made at each iteration.

Case Study Solution

The model is $\eta = \theta_1(1 - \exp(-\theta_2 t))$. There are $p = 2$ parameters and we will plan an experiment with $n = 2$ observations placed at locations that are optimal with respect to the best possible initial guesses of the parameter values.

The partial derivatives of the model with respect to each parameter are:

$$X_{1j} = \frac{\partial[\theta_1(1 - e^{-\theta_2 t_j})]}{\partial \theta_1} = 1 - e^{-\theta_2 t_j} \quad j = 1, 2$$

$$X_{2j} = \frac{\partial[\theta_1(1 - e^{-\theta_2 t_j})]}{\partial \theta_2} = \theta_1 t_j e^{-\theta_2 t_j} \quad j = 1, 2$$

The derivative matrix X for $n = 2$ experiments is 2×2:

$$X = \begin{bmatrix} X_{11} & X_{21} \\ X_{12} & X_{22} \end{bmatrix} = \begin{bmatrix} 1 - e^{-\theta_2 t_1} & \theta_1 t_1 e^{-\theta_2 t_1} \\ 1 - e^{-\theta_2 t_2} & \theta_1 t_2 e^{-\theta_2 t_2} \end{bmatrix}$$

where t_1 and t_2 are the times at which observations will be made.

Premultiplying X by its transpose gives:

$$X'X = \begin{bmatrix} X_{11}^2 + X_{12}^2 & X_{11}X_{21} + X_{12}X_{22} \\ X_{11}X_{21} + X_{12}X_{22} & X_{21}^2 + X_{22}^2 \end{bmatrix}$$

The objective now is to maximize the determinant of the $X'X$ matrix:

$$\max \Delta = |X'X|$$

The vertical bars indicate a determinant. In a complicated problem, the matrix multiplication and the minimization of the determinant of the matrix would be done using numerical methods. The analytical solution for this example, and several other interesting models and provided by Box (1971), and can be derived algebraically. For the case where $n = p = 2$:

$$\Delta = (X_{11}X_{22} - X_{12}X_{21})^2$$

This expression is maximized when the absolute value of the quantity $(X_{11}X_{22} - X_{12}X_{21})$ is maximized. Therefore, the design criterion becomes:

$$\Delta^* = \max |X_{11}X_{22} - X_{12}X_{21}|$$

Here, the vertical bars designate the absolute value. The asterisk on delta merely indicates this redefinition of the Δ criterion.

Substituting the appropriate derivative elements gives:

$$\Delta^* = [(1 - e^{-\theta_2 t_1})(\theta_1 t_2 e^{-\theta_2 t_2}) - (1 - e^{-\theta_2 t_2})(\theta_1 t_1 e^{-\theta_2 t_1})]$$
$$= \theta_1[(1 - e^{-\theta_2 t_1})(t_2 e^{-\theta_2 t_2}) - (1 - e^{-\theta_2 t_2})(t_1 e^{-\theta_2 t_1})]$$

The factor θ_1 is a numerical constant that can be ignored. The quantity in brackets is independent of the value of θ_1.

Maximizing Δ can be done by taking derivatives and solving for the roots t_1 and t_2. The algebra is omitted. The solution is:

$$t_1 = 1/\theta_2 \quad \text{and} \quad t_2 = \infty$$

The value $t_2 = \infty$ is interpreted as advice to collect data with t_2 set at the largest feasible level. A measurement at this time will provide a direct estimate of θ_1 because η approaches the asymptote θ_1 as t becomes large (i.e., 20 days or more in a BOD test). This estimate of θ_1 will be essentially independent of θ_2. Notice that the value of θ_1 is irrelevant in setting the level of both t_2 and t_1 (which is a function only of θ_2).

The observation at $t_1 = 1/\theta_2$ is on the rising part of the curve. If we initially estimate $\theta_2 = 0.23$ day^{-1}, the optimal setting is $t_1 = 1/0.23 = 4.3$ days. As a practical matter, we might say that values of t_1 should be in the range of 4 to 5 days (because $\theta_2 = 0.2$ gives $t_1 = 5$ days and $\theta_2 = 0.25$ gives $t_1 = 4$ days). Notice that the optimal setting of t_1 depends only on the value of θ_2. Likewise, t_2 is set at a large value regardless of the value of θ_1 or θ_2.

Table 44.1 compares the three arbitrary experimental designs and the optimal design shown in Figure 44.3. The insets in Figure 44.3 suggest the shape and relative size of the confidence regions one expects from the designs. A smaller value of $\Delta^{-1/2}$ in Table 44.1 indicates a smaller joint confidence region. The absolute value of $\Delta^{-1/2}$ has no meaning; it depends upon the magnitude of the parameter values used,

TABLE 44.1
Relative Size of Joint Confidence Region for Several Experimental Designs

Design	Total Number of Observations	$\Delta^{-1/2}$ (10^{-4})
Designs shown in Figure 44.3		
Optimal design	2	27
Design A	6	28
Design B	10	7
Design C	20	3.5
Replicated optimal designs ($t_1 = 4.5$, $t_2 = 20$)		
2 replicates	4	14
3 replicates	6	9
4 replicates	8	7
5 replicates	10	5

Note: The optimal designs ($t_1 = 4.5$, $t_2 = 20$) are based on $\eta = 250(1 - \exp(-0.23t))$, $\theta_1 = 250$, and $\theta_2 = 0.23$.

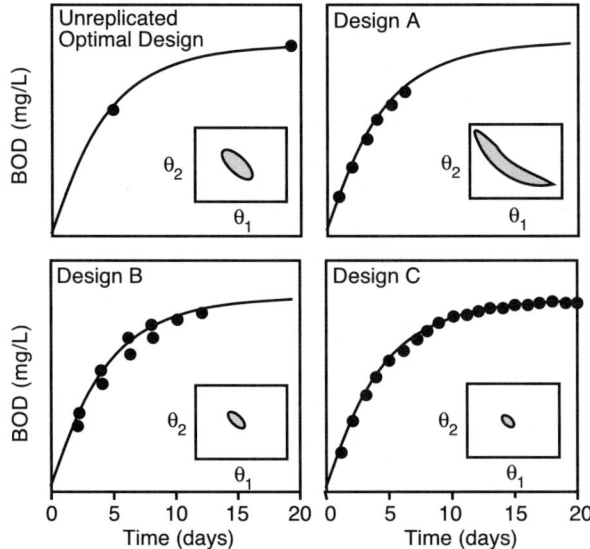

FIGURE 44.3 The optimal design and three arbitrary designs. Insets suggest the relative size and shape of the confidence regions.

the units, the time scale, etc., but the relative magnitude of $\Delta^{-1/2}$ for different designs indicates relative precision of the parameter estimates. The values of $\Delta^{-1/2}$ do not indicate the shape of the confidence region, but it happens that the best designs give well-conditioned regions because parameter correlation has been reduced.

The Optimal Design and Design A have confidence regions that are about the same size. What is not indicated by the $\Delta^{-1/2}$ values is that the region for optimal design will be elliptical while Design A will be elongated. Design A has been used very often in kinetic studies. It is inefficient. It does not give independent estimates of the parameters because all the observations are in the early part of the experiment and none are on the asymptote. It tends to estimate the product $\theta_1\theta_2$ rather than θ_1 and θ_2 independently. In fact, θ_1 and θ_2 are estimated so poorly as to be virtually useless; the joint confidence region is hyperbolic and elongated (banana-shaped) instead of the more elliptical shape we would like. This weakness in the design cannot be overcome merely by putting more observations into the same region. To improve, the observations must be made at times that yield more information.

Design B, with 10 observations, is similar to Design A, but the observations cover a wider range of time and four points are duplicated. This almost doubles the experimental work but it reduces the size of the confidence region by a factor of four. This reduction is due mainly to putting some observations on the asymptote. Adding five observations will do nothing if they are made in the wrong places. For example, duplicating four of the five points in Design A will not do much to improve the shape of the confidence region or to reduce its size.

Design C has 20 observations and yields a well-shaped confidence region that is half the size obtained by Design B. Also, it has the advantage of providing a check on the model over the range of the experimental space. If the first-order model is wrong, this design should reveal it. On the other hand, if the model is correct, the precise parameter estimates can be attained with less work by replicating the optimal design.

The simplest possible optimal design has only two observations, one at $t_1 = 1/\theta_2 = 5$ days and $t_2 = 20$ days.) Two well-placed observations are better than six badly located ones (Design A). The confidence region is smaller and the parameter estimates will be independent (the confidence region will tend to be elliptic rather than elongated). Replication of the optimal design quickly reduces the joint confidence region. The design with five replicates (a total of 10 observations) is about equal to Design C with 20 observations. This shows that Design C has about half its observations made at times that contribute little information about the parameter values. The extra experimental effort has gone to confirming the mathematical form of the model

Comments

This approach to designing experiments that are efficient for estimating parameters in nonlinear models depends on the experimenter assuming that the form of the model is correct. The goal is to estimate parameters in a known model, and not to discover the correct form of the model.

The most efficient experimental strategy is to start with simple designs, even as small as $n = p$ observations, and then to work iteratively. The first experiment provides parameter estimates that can be used to plan additional experiments that will refine the parameter estimates.

In many cases, the experimenter will not want to make measurements at only the p locations that are optimal based on the criterion of maximizing Δ. If setting up the experiment is costly but each measurement is inexpensive, it may preferable to use several observations at near optimal locations. If the experiment runs a long time (long-term BOD test) and it is difficult to store experimental material (wastewater), the initial design should be augmented, but still emphasizing the critical experimental regions. It may be desirable to add some observations at nonoptimal locations to check the adequacy of the model. Augmenting the optimal design is sensible. The design criterion, after all, provides advice — not orders.

References

Atkinson, A. C. and W. G. Hunter (1968). "Design of Experiments for Parameter Estimation," *Technometrics,* 10, 271.

Bates, D. M. and D. G. Watts (1988). *Nonlinear Regression Analysis & Its Applications,* New York, John Wiley.

Berthouex, P. M. and W. G. Hunter (1971). "Problems Associated with Planning BOD Experiments," *J. San. Engr. Div., ASCE,* 97, 333–344.

Berthouex, P. M. and W. G. Hunter (1971). "Statistical Experimental Design: BOD Tests," *J. San. Engr. Div., ASCE,* 97, 393–407.

Box, G. E. P. and W. G. Hunter (1965). "The Experimental Study of Physical Mechanisms," *Technometrics,* 7, 23.

Box, G. E. P. and H. L. Lucas (1959). "Design of Experiments in Nonlinear Situations," *Biometrika,* 45, 77–90.

Box, M. J. (1971). "Simplified Experimental Design," *Technometrics,* 13, 19–31.

Boyle, W. C. and P. M. Berthouex (1974). "Biological Wastewater Treatment Model Building — Fits and Misfits," *Biotech. Bioeng.,* 16, 1139–1159.

Draper, N. R. and H. Smith, (1998). *Applied Regression Analysis,* 3rd ed., New York, John Wiley.
Rawlings, J. O. (1988). *Applied Regression Analysis: A Research Tool,* Pacific Grove, CA, Wadsworth and Brooks/Cole.

Exercises

44.1 BOD. Recommend times at which BOD should be measured in order to estimate the ultimate BOD and rate coefficient of river water assuming θ_1 is in the range of 0.07 to 0.10 per day.

44.2 Groundwater Clean-Up. The groundwater and soil at a contaminated site must be treated for several years. It is expected that the contaminant concentrations will decrease exponentially and approach an asymptotic level $C_T = 4$ according to:

$$C = C_T + (C_0 - C_T)[1 - \exp(-kt)]$$

Several years of background data are available to estimate $C_0 = 88$. Recommend when samples should be taken in order to estimate whether it will be reached and, if so, when it will be reached. The value of k is expected to be in the range of 0.4 to 0.5 yr^{-1}.

44.3 Reactions in Series. The model for compound B, which is produced by a series of two first-order reactions, A→B→C, is:

$$\eta_B = \frac{\theta_1}{\theta_2 - \theta_1} C_{A_0}[\exp(-\theta_1 t) - \exp(-\theta_2 t)]$$

where θ_1 and θ_2 are the rate coefficients of the two reactions. (a) If you are going to make only two observations of B, which times would you observe to get the most precise estimates of θ_1 and θ_2? θ_1 is expected to be in the range 0.25 to 0.35 and θ_2 in the range 0.1 to 0.2. Assume $C_{A_0} = 10$ mg/L. (b) Explain how you would use the iterative approach to model building to estimate θ_1 and θ_2. (c) Recommend an iterative experiment to confirm the form of the model and to estimate the rate coefficients.

44.4 Compound C Production. Derive the model for compound C that is produced from A→B→C, again assuming first-order reactions. C is a function of θ_1 and θ_2. Assume the initial concentrations of A, B, and C are 2.0, 0.0, and 0.0, respectively. If only C is measured, plan experiments to estimate θ_1 and θ_2 for these conditions: (a) θ_1 and θ_2 are about the same, (b) θ_1 is larger than θ_2, (c) θ_2 is larger than θ_1, (d) θ_1 is much larger than θ_2, and (e) θ_2 is much larger than θ_1.

44.5 Temperature Effect. Assuming the model for adjusting rate coefficients for temperature ($k = \theta_1 \theta_2^{T-20}$) is correct, design an experiment to estimate θ_1 and θ_2.

44.6 Monod Model. You are designing an experiment to estimate the parameters θ_1 and θ_2 in the biological model $\mu = \frac{\theta_1 S}{S + \theta_2}$ when someone suggests that the model should be $\mu = \frac{\theta_1 S}{S + \theta_2 + \frac{S^2}{\theta_3}}$. Would this change your experimental plans? If so, explain how.

44.7 Oxygen Transfer. Manufacturers of aeration equipment test their products in large tanks of clean water. The test starts with a low initial dissolved oxygen concentration C_0 and runs until the concentration reaches 8 to 12 mg/L. Recommend a sampling plan to obtain precise parameter estimates of the overall oxygen transfer coefficient (K) and the equilibrium dissolved oxygen concentration (C_∞). The value of K is expected to be in the range of 1 to 2 h^{-1}. The model is:

$$C = C_\infty - (C_\infty - C_0)\exp(-Kt)$$

45

Why Linearization Can Bias Parameter Estimates

KEY WORDS bias, biological kinetics, linear model, linearization, Lineweaver-Burke, Michaelis-Menten, nonlinear least squares, nonlinear model, parameter estimation, precision, regression, transformations, Thomas slope method.

An experimenter, having invested considerable care, time, and money, wants to extract all the information the data contain. If the purpose is to estimate parameters in a nonlinear model, we should insist that the parameter estimation method gives estimates that are unbiased and precise. Generally, the best method of estimating the parameters will be nonlinear least squares, in which variables are used in their original form and units. Some experimenters transform the model so it can be fitted by linear regression. This can, and often does, give biased or imprecise estimates. The dangers of linearization will be shown by example.

Case Study: Bacterial Growth

Linearization may be helpful, as shown in Figure 45.1. The plotted data show the geometric growth of bacteria; x is time and y is bacterial population. The measurements are more variable at higher populations. Taking logarithms gives constant variance at all population levels, as shown in the right-hand panel of Figure 45.1. Fitting a straight line to the log-transformed data is appropriate and correct.

Case Study: A First-Order Kinetic Model

The model for BOD exertion as a function of time, assuming the usual first-order model, is $y_i = \theta_1[1 - \exp(-\theta_2 t_i)] + e_i$. The rate coefficient ($\theta_2$) and the ultimate BOD ($\theta_1$) are to be estimated by the method of least squares. The residual errors are assumed to be independent, normally distributed, and with constant variance. Constant variance means that the magnitude of the residual errors is the same over the range of observed values of y. It is this property that can be altered, either beneficially or harmfully, by linearizing a model.

One linearization of this model is the *Thomas slope method* (TSM). The TSM was a wonderful shortcut calculation before nonlinear least squares estimates could be done easily by computer. It should never be used today; it usually makes things worse rather than better.

Why is the TSM so bad? The method involves plotting $Y_i = y_i^{-1/3}$ on the ordinate against $X_i = y_i/t_i$ on the abscissa. The ordinate $Y_i = y_i^{-1/3}$ is badly distorted, first by the reciprocal and then by the cube root. The variance of the transformed variable Y is $\text{Var}(Y_i) = \frac{y^{-8/3}}{9} \text{Var}(y_i)$. Suppose that the measured values are $y_1 = 10$ at $t_1 = 4$ days and $y_2 = 20$ at $t_2 = 15$ days, and that the measurements have equal variance. Instead of having the desired condition of $\text{Var}(y_1) = \text{Var}(y_2)$, the transformation makes $\text{Var}(Y_i) = 6.4\text{Var}(Y_2)$. The transformation has obliterated the condition of constant variance.

Furthermore, in linear regression, the independent variable is supposed to be free of error, or at least have an error that is very small compared to the errors in the dependent variable. The transformed

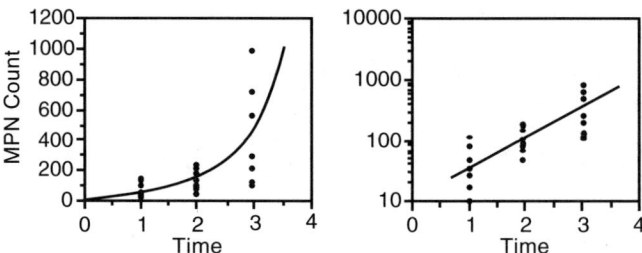

FIGURE 45.1 A log transformation of bacteria data linearizes the model and gives constant variance.

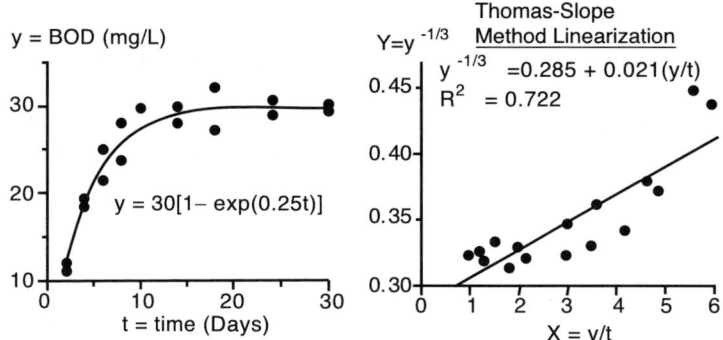

FIGURE 45.2 The Thomas slope method of linearizing the BOD model distorts the errors and gives biased parameter estimates.

abscissa ($X_i = y_i/t_i$) now contains the measurement error in y. Because it is scaled by t_i, each x_i contains a different amount of error. The results are $\text{Var}(X_i) = \text{Var}(y_i)/t_i^2$ and $\text{Var}(X_1) = 14\text{Var}(X_2)$.

Figure 45.2 shows an example. The original data, shown on the left, have constant variance. The fitted model is $\hat{y} = 30[1 - \exp(-0.25t)]$. The linearization, on the left, is not very linear but it gives a high coefficient of determination ($R^2 = 0.72$). The disturbing thing is that eliminating data at the longer times would make the plot of $y^{-1/3}$ vs. y/t more nearly linear. This would be a tragedy because the observations at large values of time contain almost all the information about the parameter θ_1, and failing to have observations in this region will make the estimates imprecise as well as biased.

Other linearization methods have been developed for transforming BOD data so they can be fitted to a straight line using linear regression. They should not be used because they all carry the danger of distortion illustrated with the Thomas method. This was shown by Marske and Polkowski (1972). The TSM estimates were often so badly biased that they did not fall within the approximate 95% joint confidence region for the nonlinear least squares estimates.

Case Study: Michaelis-Menten Model

The Michaelis-Menten model states, in biochemical terms, that an enzyme-catalyzed reaction has a rate $\eta = \theta_1 x/(\theta_2 + x)$ where x is the concentration of substrate (the independent variable). The maximum reaction rate (θ_1) is approached as x gets large. The saturation constant (θ_2) is the substrate concentration at which $\eta = \theta_1/2$. The observed values are:

$$y_i = \eta + e_i = \frac{\theta_1 x}{\theta_2 + x} + e_i$$

Three ways of estimating the two parameters in the model are:

1. Nonlinear least squares to fit the original form of the model:

$$\min S = \sum \left(y_i - \frac{\theta_1 x}{\theta_2 + x} \right)^2$$

2. Linearization using a Lineweaver-Burke plot (double reciprocal plot). The model is rearranged to give:

$$\frac{1}{y_i} = \frac{1}{\theta_1} + \frac{\theta_2}{\theta_1 x_i}$$

A straight line is fitted by ordinary linear regression to estimate the intercept $1/\theta_1$ and slope θ_2/θ_1.

3. Linearization using y against y/x gives:

$$y_i = \theta_1 - \theta_2 \frac{y_i}{x_i}$$

A straight line is fitted by ordinary linear regression to estimate the intercept θ_1 and slope $-\theta_2$.

Assuming there is constant variance in the original measurements of y, only the method of nonlinear least squares gives unbiased parameter estimates. The Lineweaver-Burke plot will give the most biased estimates. The y vs. y/x linearization has somewhat less bias.

The effectiveness of the three methods is demonstrated with simulated data. The simulated data in Table 45.1 were generated using the true parameter values $\theta_1 = 30.0$ and $\theta_2 = 15.0$. The observed y's are the true values plus random error (with mean $= 0$ and variance $= 1$). The nonlinear least squares parameter estimates are $\hat{\theta}_1 = 31.4$ and $\hat{\theta}_2 = 15.8$, which are close to the underlying true values. The two linearization methods give estimates (Table 45.2) that are distant from the true values; Figure 45.3 shows why.

TABLE 45.1

Simulated Results of a Biochemical Experiment to Fit the Michaelis-Menten Model

Substrate Conc. x	"True" Rate $\eta = 30x/(15 + x)$	Observed Rate $y = \eta + e_i$
2.5	4.28	5.6
5.0	7.50	7.3
10.0	12.00	12.5
20.0	17.14	16.1
40.0	21.82	23.2

TABLE 45.2

Michaelis-Menten Parameter Estimates

Estimation Method	θ_1	θ_2
"True" parameter values	30.0	15.0
Nonlinear least squares	31.4	15.8
Lineweaver-Burke	22.5	8.1
Plot y against y/x	25.6	10.0

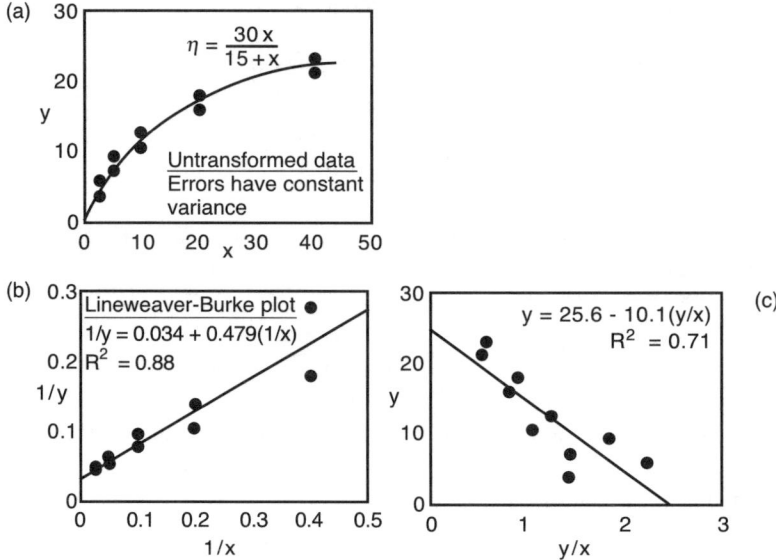

FIGURE 45.3 An illustration of how linearization of the Michaelis-Menten equation distorts the error structure.

Figure 45.3a was drawn using the values from Table 45.1 with five additional replicate observations that were arbitrarily chosen to make the spread the same between "duplicates" at each setting of x. Real data would not look like this, but this simplicity will help illustrate how linearization distorts the errors. Figure 45.3b shows the Lineweaver-Burke plot. Values at large x (small $1/x$) are squeezed together in the plot, making these values appear more precise than the others and literally fixing one end of the regression line. The values associated with small x (large $1/x$) seem to be greatly in error. The consequence of this is that the least squares fit of a straight line will be strongly influenced by the large values of x, whereas according to the true error structure they should not be given any more weight than the other values. The plot of y against y/x, Figure 45.3c, shows some distortion, but less than the Lineweaver-Burke plot.

One simulated experiment may not be convincing, but the 750 simulated experiments of Colquhoun (1971) are dramatic proof. An experiment was performed by adding normally distributed random experimental errors with mean = 0 and constant variance $\sigma^2 = 1$ to "true" values of y. Thus, unlike what happens in a real experiment, the distribution, mean, and standard deviation of the errors are known, and the true parameter values are known. Each set of data was analyzed to estimate the parameters by nonlinear least squares and by the two linearization methods.

The resulting 750 estimates of θ_1 were grouped to form histograms and are shown in Figure 45.4. The distributions for θ_2 (not reproduced here) were similarly biased because the estimates of θ_1 and θ_2 are highly correlated; that is, experiments that yield an estimate of θ_1 that is too high tend to give an estimate of θ_2 that is too high also, whichever method of estimation is used. This parameter correlation is a result of the model structure and the settings of the independent variables.

The average value of the nonlinear least squares (NLLS) estimates is 30.4, close to the true value of $\theta_1 = 30.0$. They have little bias and have the smallest variance. By comparison, the Lineweaver-Burke method gives terrible estimates, including some that were negative and some greater than 100. Near infinite estimates are obtained when the plot goes nearly through the origin, giving $1/\theta_1 = 0$. These estimates are so distorted that no realistic estimate of the bias is possible.

Plotting y against y/x gives estimates falling between the NLLS and the Lineweaver-Burke methods. Their standard deviation is only about 28% greater than that of the NLLS estimates. About 73% of the estimates were too low (below 30) and the average was 28.0. They have a negative bias. This bias is purely a consequence of the parameter estimation method.

Why Linearization Can Bias Parameter Estimates

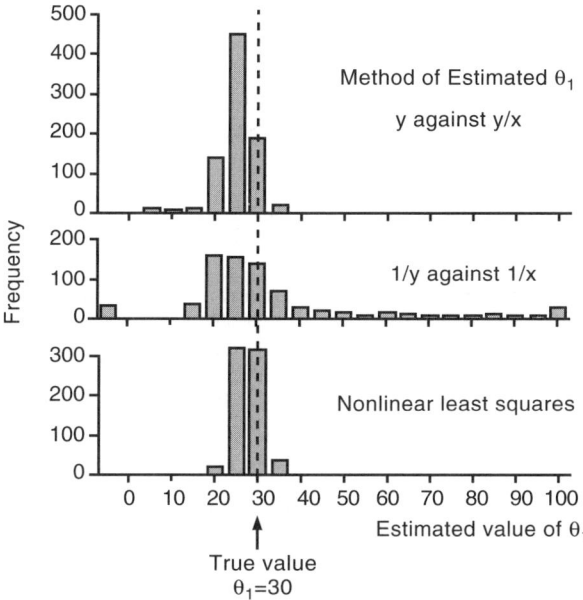

FIGURE 45.4 Distributions of 750 estimates of θ_1 (true value $\theta_1 = 30$) obtained using three methods in 750 simulated experiments. (Plotted using data from Colquhoun, 1971).

Colquhoun's simulations are a persuasive warning about linearization to facilitate parameter estimation, but they are not intended to prove that linearization is always bad. It can be helpful in the case of nonconstant errors. Dowd and Riggs (1965) found the y against y/x method better than NLLS for the case where the coefficient of variation (standard deviation/mean) was the same at each setting of x, but even in this case the Lineweaver-Burke method was still awful.

Comments

Transformation always distorts the error structure of the original measurements. Sometimes this is beneficial, but this can only happen when the original variances are not constant. If the original errors have constant variance, linearization will make things worse instead of better and the parameter estimates will be biased.

Do not linearize merely to facilitate using linear regression. Learn to use nonlinear least squares. It is more natural, more likely to be the appropriate method, and easy to do with readily available software.

References

Colquhoun, D. (1971). *Lectures in Biostatistics,* Oxford, Clarendon Press.

Dowd, J. E. and D. S. Riggs (1965). "A Comparison of Estimates of Michaelis-Menten Kinetic Constants from Various Linear Transformations," *J. Biol. Chem.,* 210, 863–872.

Marske, D. M. and L. B. Polkowski (1972). "Evaluation of Methods for Estimating Biochemical Oxygen Demand Parameters," *J. Water Pollution Control Fed.,* 44, 1987–1992.

Exercises

45.1 Log Transformations. Given below are three models that can be linearized by taking logarithms. Discuss the distortion of experimental error that this would introduce. Under what conditions would the linearization be justified or helpful?

$$y = \alpha x^{\beta} \qquad y = \alpha x^{-\beta} + \gamma \qquad y = \alpha \beta^{x}$$

45.2 Reciprocals. Each of the two sets of data on x and y are to be fitted to the model $\frac{1}{y} = \alpha + \beta x$. What is the effect of taking the reciprocal of y?

Data Set 1:	$x =$	2	4	6	8	10				
	$y =$	11.2	1.3	0.9	0.5	0.4				

Data Set 2:	$x =$	2	2	4	4	6	6	8	8	10	10
	$y =$	11.2	10.0	1.3	1.8	0.9	0.55	0.5	0.55	0.4	0.48

45.3 Thomas Slope Method. For the BOD model $y_i = \theta_1[1 - \exp(-\theta_2 t_i)] + e_i$, do a simulation experiment to evaluate the bias caused by the Thomas slope method linearization. Display your results as histograms of the estimated values of θ_1 and θ_2.

45.4 Diffusion Coefficient. The diffusion coefficient D in the model $N = c\sqrt{\frac{D}{\pi t}}$ is to be estimated using measurements on $N =$ the amount of chemical absorbed (mg/min), $c =$ concentration of chemical (mg/L), and $t =$ time (min). Explain how you would estimate D, giving the equations you would use. How would you calculate a confidence interval on D?

45.5 Monod Model. The growth of microorganisms in biological treatment can be described by the Monod model, $y = \theta_1 S/(\theta_2 + S) + e$. Data from two identical laboratory reactors are given below. Estimate the parameters using (a) nonlinear least squares and (b) linear regression after linearization. Plot the fitted models and the residuals for each method of estimation. Explain why the estimates differ and explain which are best.

Reactor 1

S (mg/L)	7	9	15	25	40	75	100	150
y (1/day)	0.25	0.37	0.48	0.65	0.80	0.97	1.06	1.04

Reactor 2

S (mg/L)	7	9	15	25	40	75	100	150
y (1/day)	0.34	0.29	0.42	0.73	0.85	1.05	0.95	1.11

46

Fitting Models to Multiresponse Data

KEY WORDS *biokinetics, Box-Draper criterion, covariance, determinant criterion, joint confidence regions, Monod model, multiresponse experiments, nonlinear least squares, parameter estimation.*

Frequently, data can be obtained simultaneously on two or more responses at given experimental settings. Consider, for example, two sequential first-order reactions by which species A is converted to B which in turn is converted to C, that is $A \to B \to C$. This reaction occurs in a batch reactor so the concentrations of A, B, or C can be measured at any time during the experiment. If only B is measured, the rate constants for each step of the reaction can be estimated from the single equation that describes the concentration of B as a function of time. The precision will be much better if all three species (A, B, and C) are measured and the three equations that describe A, B, and C are fitted simultaneously to all the data. A slightly less precise result will be obtained if two species (A and B, or B and C) are measured. To do this, it is necessary to have a criterion for simultaneously fitting multiple responses.

Case Study: Bacterial Growth Model

The data in Table 46.1 were collected on a continuous-flow completely mixed biological reactor with no recycle (Ramanathan and Gaudy, 1969). At steady-state operating conditions, the effluent substrate and biomass concentrations will be:

$$S = \frac{\theta_2 D}{\theta_1 - D} \qquad \text{material balance on substrate}$$

$$X = \theta_3 \left(S_0 - \frac{\theta_2 D}{\theta_1 - D} \right) = \theta_3 (S_0 - S) \qquad \text{material balance on biomass}$$

These equations define the material balance on a well-mixed continuous-flow reactor with constant liquid volume V and constant liquid feed rate Q. The reactor dilution rate $D = Q/V$.

The reactor contents and effluent have substrate concentration S and biomass concentration X. The rate of biomass production as substrate is destroyed is described by the Monod model $\mu = \theta_1 S/(\theta_2 + S)$ where θ_1 and θ_2 are parameters. Each gram of substrate destroyed produces θ_3 grams of biomass. The liquid detention time in the reactor is V/Q. The feed substrate concentration is $S_0 = 3000$ mg/L.

Experiments are performed at varied settings of D to obtain measurements of X and S in order to estimate θ_1, θ_2, and θ_3. One approach would be to fit the model for X to the biomass data and to independently fit the model for S to the substrate data. The disadvantage is that θ_1 and θ_2 appear in both equations and two estimates for each would be obtained. These estimates might differ substantially. The alternative is to fit both equations simultaneously to data on both X and S and obtain one estimate of each parameter. This makes better use of the data and will yield more precise parameter estimates.

TABLE 46.1

Data from an Experiment on Bacterial Growth in a Continuous-Flow, Steady-State Reactor

Dilution Rate (D) (1/h)	Substrate Conc. (S) (mg/L COD)	Biomass Conc. (X) (mg/L)
0.042	221	1589
0.056	87	2010
0.083	112	1993
0.167	120	1917
0.333	113	1731
0.500	224	1787
0.667	1569	676
1.000	2745	122

Note: The feed substrate concentration is $S_0 = 3000$ mg/L.

Source: Ramanathan, M. and A. F. Gaudy (1969). *Biotech. and Bioengr.*, 11, 207.

Method: A Multiresponse Least Squares Criterion

A logical criterion for simultaneously fitting three measured responses (y_A, y_B, and y_C) would be a simple extension of the least squares criterion to minimize the combined sums of squares for all three responses:

$$\min \sum (y_A - \hat{y}_A)^2 + (y_B - \hat{y}_B)^2 + (y_C - \hat{y}_C)^2$$

where y_A, y_B, and y_C are the measured responses and $\hat{y}_A$, $\hat{y}_B$, and $\hat{y}_C$ are values computed from the model. This criterion holds only if three fairly restrictive assumptions are satisfied, namely: (1) the errors of each response are normally distributed and all data points for a particular response are independent of one another, (2) the variances of all responses are equal, and (3) there is no correlation between data for each response for a particular experiment (Hunter, 1967).

Assumption 2 is violated when certain responses are measured more precisely than others. This condition is probably more common than all responses being measured with equal precision. If assumption 2 is violated, the appropriate criterion is to minimize the weighted sums of squares, where weights are inversely proportional to the variance of each response. If both assumptions 2 and 3 are violated, the analysis must account for variances and covariances of the responses. In this case, Box and Draper (1965) have shown that minimizing the determinant of the variance-covariance matrix of the residuals gives the maximum likelihood estimates of the model parameters. The Box-Draper *determinant criterion* is especially attractive because it is not restricted to linear models, and the estimates are invariant to scaling or linear transformations of the observations (Bates and Watts, 1985).

The criterion is written as a combined sum of squares function augmented by the covariances between responses. For the example reaction A → B → C, with three responses measured, the Box-Draper determinant criterion has the following form:

$$\min |\mathbf{V}| = \begin{vmatrix} \sum (y_A - \hat{y}_A)^2 & \sum (y_A - \hat{y}_A)(y_B - \hat{y}_B) & \sum (y_A - \hat{y}_A)(y_C - \hat{y}_C) \\ \sum (y_A - \hat{y}_A)(y_B - \hat{y}_B) & \sum (y_B - \hat{y}_B)^2 & \sum (y_B - \hat{y}_B)(y_C - \hat{y}_C) \\ \sum (y_A - \hat{y}_A)(y_C - \hat{y}_C) & \sum (y_B - \hat{y}_B)(y_C - \hat{y}_C) & \sum (y_C - \hat{y}_C)^2 \end{vmatrix}$$

where $\sum$ indicates summation over all observations. We assume that each response has been measured the same number of times. The vertical lines indicate the determinant of the matrix. The best parameter estimates, analogous to the least squares estimates, are those which minimize the determinant of this matrix.

Fitting Models to Multiresponse Data

The diagonal elements correspond to the residual (error) sum of squares for each of the three responses. The off-diagonal terms account for measurements on the different responses being correlated. If the residual errors of the three responses are independent of each other, the expected values of the off-diagonal terms will be zero and the parameter estimation criterion simplifies to:

$$\min \sum (y_A - \hat{y}_A)^2 + (y_B - \hat{y}_B)^2 + (y_C - \hat{y}_C)^2$$

For the special case of a single response, the determinant criterion simplifies to the method of least squares: minimize the sum of the squared residuals for the response of interest.

$$\min \sum (y_A - \hat{y}_A)^2 \quad \text{or} \quad \min \sum (y_B - \hat{y}_B)^2 \quad \text{or} \quad \min \sum (y_C - \hat{y}_C)^2$$

There are three concerns when using the determinant criterion (Box et al., 1973). All responses must be measured independently. If we know, for example from a mass balance, that the reaction was started with 1 mole of A and that the total amount of A, B, and C must equal 1 mole throughout the course of the experiment, it is tempting to reduce the experimental burden by measuring only A and B and then computing C by difference; that is, $y_C = 1 - y_A - y_B$. If we then proceed as though three concentrations had been measured independently, we invite trouble. The computed y_C is not new information. It is redundant because it is linearly dependent on the sum of $y_A + y_B$. The mathematical consequence is that the **V** matrix will be singular and severe computational problems will arise. The practical consequence is that the experimenter has failed to collect data that could demonstrate model adequacy by verifying the mass balance constraint.

The second concern is linear dependence among the expected values of the responses. This happens if one of the model equations is a linear combination of one or more of the other model equations. This situation is more subtle than the similar one described above with data, but its consequence is the same: namely, a singular **V** matrix.

A third concern is always present and it involves model lack-of-fit. The analysis assumes the model (the set of equations) is correct. Model inadequacy will inflate the residual errors, which will cause the off-diagonal elements in |**V**| to be large even when true correlation among the measurement errors is small.

To guard against these potential difficulties, the experimenter should:

1. Check for model adequacy by plotting the residuals of the fitted model against the predicted values and against the independent variables. Inadequacies revealed by the residual plots usually provide clues about how the model can be improved.
2. Pay particular attention to |**V**| for indications of singularity. If there is such a problem, remove the linearly dependent data or the linearly dependent equation from the matrix.

In general, the variance-covariance matrix **V** is square with dimensions determined by the number of responses being fitted. Because the number of responses is usually small (usually two to four), it is easy to create the proper pattern of diagonal and off-diagonal terms in the matrix by following the three-response example given above. Any nonlinear optimization program should be able to handle the minimization calculations, but they will not provide any diagnostic statistical information. The GREG program developed by Stewart et al. (1992) offers powerful methods for multiresponse parameter estimation.

The precision of the parameter estimates can be quantified by determining the approximate joint confidence region. The minimum value of the determinant criterion |**V**| takes the place of the minimum residual sum of squares value in the usual single-response parameter estimation problem (Chapter 33). The approximate $(1 - \alpha)100\%$ joint confidence region is bounded by:

$$|\mathbf{V}|_{1-\alpha} = |\mathbf{V}|_{\min} \exp\left(\frac{\chi^2_{p,1-\alpha}}{n}\right)$$

where p = number of parameters estimated, n = number of observations, and $\chi^2_{p,1-\alpha}$ is the chi-square value for p degrees of freedom and $(1 - \alpha)100\%$ probability level.

Case Study Solution

The data were fitted simultaneously to the two equations for substrate and biomass (Johnson and Berthouex, 1975). In this example $p = 3$ and $n = 8$. The determinant criterion is:

$$\min |V| = \begin{vmatrix} \sum(X_i - \theta_3(S_0 - S_i))^2 & \sum(X_i - \theta_3(S_0 - S_i))\left(S_i - \frac{\theta_2 D_i}{\theta_1 - D_i}\right) \\ \sum(X_i - \theta_3(S_0 - S_i))\left(S_i - \frac{\theta_2 D_i}{\theta_1 - D_i}\right) & \sum\left(S_i - \frac{\theta_2 D_i}{\theta_1 - D_i}\right)^2 \end{vmatrix}$$

The minimization must be constrained to prevent the term $(\theta_1 - D)$ in the denominator from becoming zero or negative during the numerical search.

The determinant of the matrix was minimized to find least squares parameter estimates $\hat{\theta}_1 = 0.71$, $\hat{\theta}_2 = 114$, and $\hat{\theta}_3 = 0.61$. The data, the fitted model, and the residuals are plotted in Figure 46.1. The residuals for substrate tend to decrease as the predicted value of S increases. This is because the model predicts that S goes to zero as dilution rate becomes small, but this is impossible because the measurements are COD, and some COD is not biodegradable. A model of the form $S = \theta_4 + [\theta_2 D/(\theta_1 - D)]$, where θ_4 is refractory substrate, might fit better.

The next step is to evaluate the precision of the estimated parameters. This involves understanding the shape of the three-dimensional function $|V|$, which is a function of θ_1, θ_2, and θ_3. Figure 46.2 shows the general shape of $|V|$ for this set of data. Figure 46.3 shows contours of $|V|$ at the cross-section where $\theta_3 = 0.61$. The approximate 95% joint confidence region is shaded. The shape of the contours shows that θ_1 is estimated more precisely than θ_2 and there is slight correlation between the parameters.

The estimated parameters from fitting the X response alone were $\hat{\theta}_1 = 0.69$, $\hat{\theta}_2 = 60$, and $\hat{\theta}_3 = 0.62$. Fitting S alone gave estimates $\hat{\theta}_1 = 0.71$, $\hat{\theta}_2 = 112$, and $\hat{\theta}_3 = 0.61$. The precision of the parameters estimated using the single responses is so poor that no segment of the 95% joint confidence regions falls within the borders of Figure 46.3. This is shown by Figure 46.4, in which the entire area of Figure 46.3 occupies a small space in the lower-left corner of the diagram. Also, the correlation between the parameters is high.

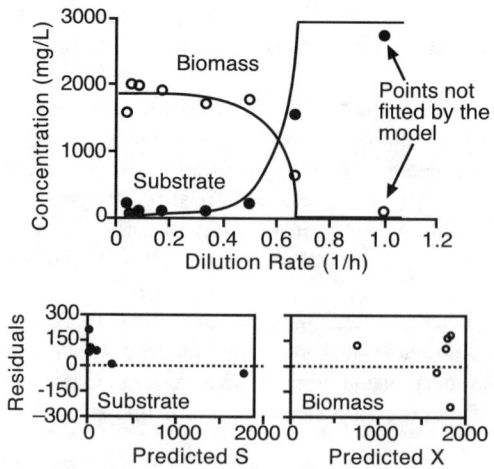

FIGURE 46.1 Fitted model and residual plots.

Fitting Models to Multiresponse Data

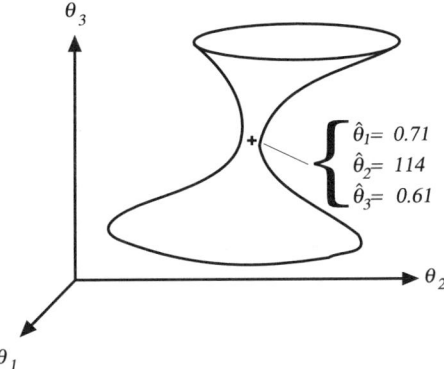

FIGURE 46.2 Three-dimensional representation of the variance-covariance matrix for the two-response, three-parameter bacterial growth model.

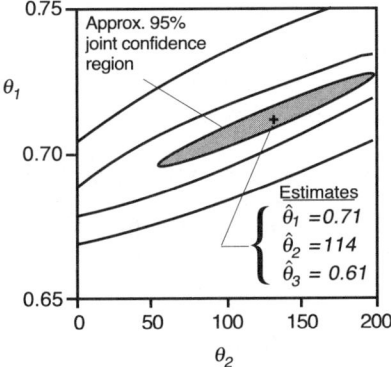

FIGURE 46.3 Contours of the determinant of the variance-covariance matrix at $\theta_3 = 0.61$. The shaded region is the approximate 95% joint confidence region.

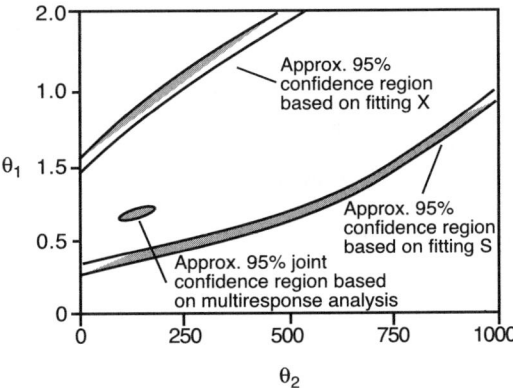

FIGURE 46.4 The 95% joint confidence regions at $\theta_3 = 0.61$ for the model fitted to both X and S simultaneously (dark shaded region), the model fitted to X only (unshaded region), and the model fitted to S only (region with shaded boundary).

The single-response parameter estimates are not imprecise because the individual models fit the data very poorly. Actually, the curves calculated using the individually estimated values fit nicely and fall near those obtained from the multiresponse estimation. The dramatic gain in the precision of the parameter estimates comes from constraining the parameters to fit both data sets simultaneously.

Comments

When multiple responses have been measured in an experiment, parameters will be estimated more precisely if all responses are fitted simultaneously. This is accomplished by minimizing the determinant of a matrix that contains the residual sums of squares of each response along the diagonal and off-diagonal terms that account for correlation between the residuals. If the measurement errors of each response are independent of other responses, the determinant criterion simplifies to minimizing the sum of the individual sums of squares. In the case of a single response, the criterion simplifies to the usual least squares calculation. A comprehensive article of multiresponse estimation, with several excellent discussions, is Bates and Watts (1985). The software of Stewart et al. (1992) allows the different responses to be weighted in accordance with their variances. It also provides detailed diagnostic information about the fitted models.

References

Bates, D. M. and D. G. Watts (1985). "Multiresponse Estimation with Special Application to Linear Systems of Differential Equations," *Technometrics,* 27, 329–339.

Bates, D. M. and D. G. Watts (1988). *Nonlinear Regression Analysis & Its Applications,* New York, John Wiley.

Box, G. E. P. and N. R. Draper (1965). "Bayesian Estimation of Common Parameters from Several Responses," *Biometrika,* 52, 355.

Box, G. E. P., W. G. Hunter, J. F. MacGregor, and J. Erjavac (1973). "Some Problems Associated with the Analysis of Multiresponse Data," *Technometrics,* 15, 33.

Hunter, W. G. (1967). "Estimation of Unknown Constants from Multiresponse Data," *Ind. & Engr. Chem. Fund.,* 6, 461.

Johnson, D. B. and P. M. Berthouex (1975). "Using Multiresponse Data to Estimate Biokinetic Parameters," *Biotech. and Bioengr.,* 17, 571–583.

Ramanathan, M. and A. F. Gaudy (1969). "Effect of High Substrate Concentration and Cell Feedback on Kinetic Behavior of Heterogeneous Populations in Completely Mixed Systems," *Biotech. and Bioengr.,* 11, 207.

Stewart, W. E., M. Caracotsios, and J. P. Sorensen (1992). *GREG Software Package Documentation,* Madison, WI, Department of Chemical Engineering, University of Wisconsin–Madison.

Exercises

46.1 Consecutive Reactions. Reactant A decomposes to form Product B which in turn decomposes to form Product C. That is, A → B → C. Both reactions are first-order with respect to the reactant. The models for the concentrations of A and B are:

$$\eta_A = C_{A0} \exp(-\theta_1 t)$$

$$\eta_B = \frac{\theta_1}{\theta_2 - \theta_1} C_{A0}[\exp(-\theta_1 t) - \exp(-\theta_2 t)] + C_{B0} \exp(-\theta_2 t)$$

where θ_1 is the reaction rate coefficient for the transformation of A to B and θ_2 is the reaction rate coefficient for the transformation of B to C. $C_{A0} = 5.0$ mg/L and $C_{B0} = 0.5$ mg/L are the initial concentrations of A and B. An experiment to measure C_A and C_B was performed with the results given in the table below.

Time (days)	0	2	4	6	8	10	12	14	16	18	20
C_A (mg/L)	5.5	3.12	1.88	0.97	0.57	0.16	1.05	0.55	0.45	0.05	0.18
C_B (mg/L)	0.24	2.17	1.89	2.37	1.21	1.99	1.35	0.16	0.74	0.76	0.53

(a) Fit the model for C_B to estimate θ_1 and θ_2. Plot the 95% joint confidence region for the parameters. (b) Fit the models for C_A and C_B simultaneously to estimate θ_1 and θ_2. Plot the 95% joint confidence region for the parameters. (c) Comment on the improvements gained by fitting the two models simultaneously.

46.2 **Bacterial Growth.** Formulate the Box-Draper determinant criterion for the case study using $S = \theta_4 + [\theta_2 D/(\theta_1 - D)]$ as the model for substrate degradation. Estimate the four parameters in the models.

46.3 **Biokinetics.** A solution with substrate concentration $S_0 = 100$ mg/L was fed to a batch reactor and the following data were obtained. The proposed model is $S_t = S_0 \exp(-\theta_1 t)$ and $X_t = \theta_2 (S_0 - S)$, where X is microorganism concentration. Estimate θ_1 and θ_2.

Time (min)	10	30	50	60	100	120	130	140
S (mg/L)	78.7	54.7	33.2	32.6	13	8.5	9.5	3.5
X (mg/L)	5	17.3	29.8	31.9	36.4	43.3	39.2	44.5

46.4 **Nitrification.** Nitrification occurs in long-term BOD tests, and must be corrected for. One method is to separately model the nitrification oxygen demand by measuring ammonia and nitrate concentrations with time. Assume that nitrification occurs as consecutive first-order reactions A → B → C, where A, B, and C represent ammonia-N, nitrite-N and nitrate-N concentrations, respectively. Use the following ammonia and nitrate data to estimate the two first-order rate coefficients (θ_1 and θ_2) and the initial ammonia and nitrate concentrations (C_{A0} and C_{C0}). Assume the initial nitrite concentration is zero.

Time (days)	Ammonia-N, C_A (mg/L)	Nitrate-N, C_C (mg/L)
0	—	0.05
1	1.4	—
3	1.3	—
8	—	—
10	0.8	0.55
15	0.3	—
18	0.05	1.8
23	—	1.8

The models for C_A and C_C are:

$$C_A = C_{A0} \exp(-\theta_1 t)$$

and

$$C_C = C_{A0}\left[1 - \frac{\theta_2 \exp(-\theta_1 t) - \theta_1 \exp(-\theta_2 t)}{\theta_2 - \theta_1}\right] + C_{B0}(1 - \exp(-\theta_2 t)) + C_{C0}.$$

47

A Problem in Model Discrimination

KEY WORDS biological kinetics, least squares, mechanistic models, model building, model discrimination, Monod kinetics, posterior probability, posterior odds ratio, prior probability, residual sum of squares, Tiessier model.

When several rival models are tentatively considered, it is common to find more than one model that seems to fit the data. For some applications there is no need to select one model as the best because, over the range of interest, either model will serve the intended purpose. In other situations, knowing which model is best may throw light on fundamental questions about reaction mechanisms, catalysis, inhibition, and other phenomena under investigation.

A fundamental concept in model discrimination is that rival models often diverge noticeably only at extreme conditions. It follows that extremes must be studied or discrimination on a statistical basis is impossible. This gives the researcher the choice of restricting the experiment to a limited range of conditions and accepting any plausible model that fits the data over this range, or testing at more stressful conditions to obtain evidence in favor of one among several rival models. To discriminate between models (and between mechanisms), the models must be put in jeopardy of failing.

Figure 47.1 shows two examples of this. The two first-order reactions in series, with a reversible second step, cannot be discriminated from the irreversible series A → B → C unless observations are made at long times. The Haldane inhibition model cannot be distinguished from the simpler Monod model unless studies are made at high concentrations. One might expect the problem of model discrimination to be fairly obvious with models such as these, but examples of weak experimental designs do exist even for these models.

Given models that seem adequate on statistical grounds, we might try to select the best model on the basis of (1) minimum sum of squares, (2) lack of fit tests (F tests), (3) fewest parameters (parsimony), (4) simplest functional form, and (5) estimated parameter values consistent with the mechanistic premise of the model. In practice, all of these criteria come into consideration.

Case Study: Two Rival Biological Models

We consider two models for bacterial growth in industrial biotechnology and biological wastewater treatment. Both assume that the specific growth rate of bacterial solid is limited by the concentration of a single substrate. The two models are by Monod (1942) and Tiessier (1936):

$$\text{Monod model:} \quad \mu_i = \frac{\theta_1 S_i}{\theta_2 + S_i} + e_i$$

$$\text{Tiessier model:} \quad \mu_i = \theta_3(1 - \exp(\theta_4 S_i)) + e_i$$

where μ_i is the growth rate observed at substrate concentration S_i. Each experimental setting of the reactor's dilution rate (flow rate divided by reactor volume) produces a pair of values of growth rate (μ_i) and substrate concentration S_i.

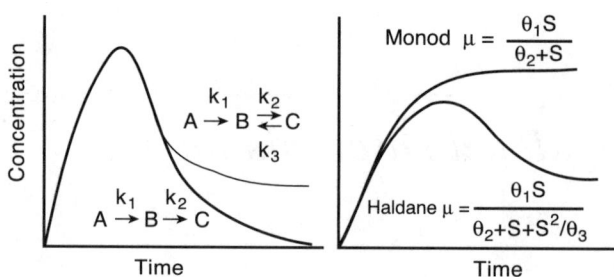

FIGURE 47.1 Examples of two models where discrimination is impossible unless observations are made at high levels of the independent variable.

TABLE 47.1

Bacterial Growth Rate Data of Schulze and Lipe (1964)

$\mu =$	0.059	0.091	0.124	0.177	0.241	0.302	0.358	0.425
$S =$	5.1	8.3	13.3	20.3	30.4	37.0	43.1	58.0
$\mu =$	0.485	0.546	0.61	0.662	0.725	0.792	0.852	
$S =$	74.5	96.5	112	161	195	266	386	

Note: μ = growth rate (1/h) and S = substrate concentration (mg/L).

The parameter θ_2 controls the shape of the hyperbolic Monod model, and θ_4 has this function in the exponential Tiessier model. The parameters θ_1 and θ_3 have units of 1/h. They are asymptotic (i.e., maximum) growth rates that are approached when the system is operated at high substrate concentration and high dilution rate. Unfortunately, direct observation of θ_1 and θ_3 is impossible because the system tends to become unstable and fail at high dilution rates because bacteria are "washed out" of the reactor faster than they can grow.

The Monod and Tiessier models were fitted to the $n = 15$ observations of Schulze and Lipe (1964), given in Table 47.1, who pushed the experimental conditions toward the asymptotic limits. We might then expect these data to "stress the models" and be advantageous for discriminating between them.

Nonlinear least squares was used to estimate the parameters with the following results:

Monod model: $\hat{\mu} = \dfrac{1.06 S}{90.4 + S}$ RSS = 0.00243 ($\nu = 13$)
   RMS = 0.000187

Tiessier model: $\hat{\mu} = 0.83(1 - \exp(-0.012 S))$ RSS = 0.00499 ($\nu = 13$)
   RMS = 0.000384

The hats (^) indicate the predicted values. RSS is the residual sum of squares of the fitted model, $\nu = 13$ is the degrees of freedom, and RMS is the residual mean square (RMS = RSS/ν). (The estimated parameter values have been rounded so the values shown will not give exactly the RSS shown.)

Having fitted two models, can we determine that one is better? Some useful diagnostic checks are to plot the data against the predictions, plot the residuals, plot joint confidence regions (Chapter 35), make a lack-of-fit test, and examine the physical meaning of the parameters.

The fitted models are plotted against the experimental data in Figure 47.2. Both appear to fit the data. Plots of the residuals (Figure 47.3) show a slightly better pattern for the Monod model, but neither model is inadequate. Figure 47.4 shows joint confidence regions that are small and well-conditioned for both models.

These statistical criteria do not disqualify either model, so we next consider the physical significance of the parameters. (In practice, an engineer would probably consider this before anything else.) If θ_1 represents

A Problem in Model Discrimination

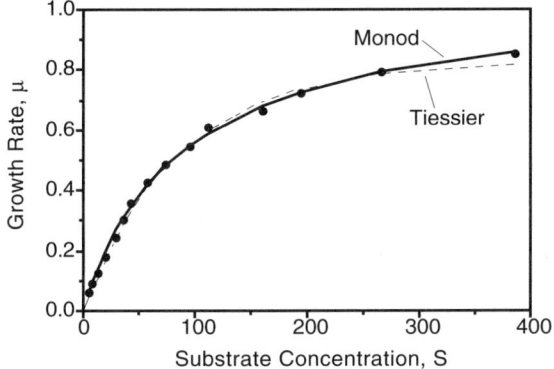

FIGURE 47.2 The Monod and Tiessier models fitted to the data of Schulze and Lipe (1964).

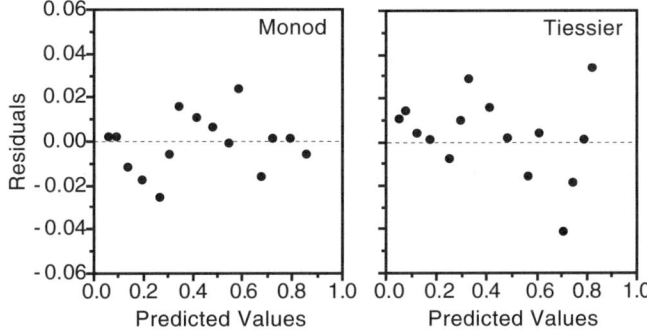

FIGURE 47.3 Residual plots for the Monod and Tiessier models.

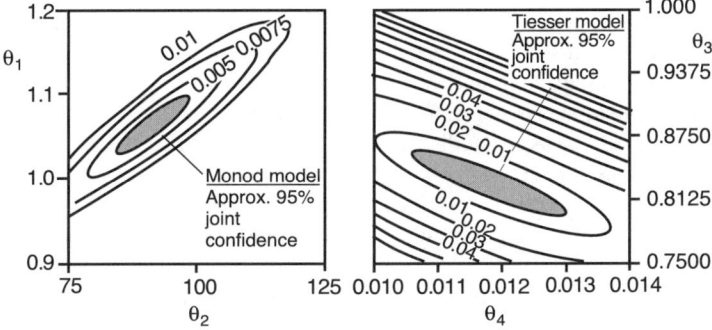

FIGURE 47.4 Joint confidence regions for the Monod and Tiessier models.

the maximum operable dilution rate, its estimated value should agree with the operating experience. Schulze and Lipe (1964) were not able to operate their reactor at dilution rates higher than about 0.85 h^{-1}. At higher rates, all the bacteria were washed out of the reactor and the system failed. The Tiessier model predicts the maximum growth rate of 0.83 h^{-1}, which is consistent with experience in the reaction system studied. The Monod model, on the other hand, estimates $\theta_1 = 1.06$, which is higher than the experimenters could attain. Based on this, one might conclude that the Monod model is inadequate because the estimated value of θ_1 is not consistent experience in operating the biological reactor.

A formal lack-of-fit test compares the ratio $F_{\text{model}} = \text{RMS}/\hat{\sigma}_e^2$ for the model with the tabulated value of F having the appropriate degrees of freedom. RMS = RSS/v is the residual mean square of the fitted model and $\hat{\sigma}_e^2$ is the estimated measurement error variance. If $F_{model} < F_{table}$, we conclude that the data do not disqualify the model.

Ideally, $\hat{\sigma}_e^2$ would be estimated using data from replicate experiments at some or all experimental settings. There were no replicates in this experiment so we have no independent estimate of the variance of measurement error. Lacking replication, the residual mean square estimates the measurement error variance if the model is correct.

To illustrate the lack-of fit-test, assume that the Monod model is correct and that $\hat{\sigma}_e^2 = 0.000187$ is a valid estimate of the measurement error variance. Assume this same measurement error variance applies to the Tiessier model. For the Tiessier model, $F_{model} = 0.000384/0.000187 = 2.05$. The numerator and denominator of the F ratio each have $v = n - p = 15 - 2 = 13$ degrees of freedom, so the critical F_{table} value is $F_{0.05,13,13} = 2.57$ for a test at the 95% confidence level. F_{model} is smaller than the F_{table}; that is, $2.05 < 2.57$. The interpretation of this result is the RMS of the Tiessier model is within the range that might be expected due to chance, given measurement errors with variance of $\hat{\sigma}_e^2 = 0.000187$.

There is no clear statistical evidence (lack-of-fit test, residual plots, etc.) to reject either the Tiessier or the Monod model. However, the Tiessier model might be preferred on mechanistic grounds because the estimated asymptotic level (θ_4) is in better agreement with experience.

Case Study: Four Rival Models

We now contrive a competition of four rival models. The four models below were fitted to a subset of the Schulze-Lipe data (the first 13 values). They are ranked according to their residual sums of squares (RSS).

Model 1: $\hat{\mu} = 0.766(1 - \exp(-0.0135S))$ RSS = 0.00175

Model 2: $\hat{\mu} = 0.5619 \tan^{-1}(0.01614S)$ RSS = 0.00184

Model 3: $\hat{\mu} = \dfrac{1.07S}{92.5 + S}$ RSS = 0.00237

Model 4: $\hat{\mu} = 0.7056 \tanh(0.0119S)$ RSS = 0.00420

Two have no mechanistic basis in biological kinetics but have shapes similar to the Monod model (model 3) and the Tiessier model (model 1). All four models fit the data very well for low values of S and diverge only at higher levels. If they had been seriously put forth as mechanistic models, it would be difficult to select the best among the four rival models.

Biological wastewater treatment systems are supposed to operate at low concentrations, and for these systems it may be satisfactory to use a model that fits the rising portion of the curve but does not fit the asymptote. Many models can fit the rising part of the curve, so it is understandable why some wastewater treatment experiments are successfully described by the Monod model, while other experiments successfully use other models. The typical biological treatment data set provides very little information for statistical model discrimination.

On the other hand, if the objective were to operate at high substrate levels (as in industrial biochemicals where the goal is to maximize cell growth), it could be important to predict performance where the curve bends and becomes asymptotic. Experiments in this region would be essential. With enough observations at the right conditions, modeling and model discrimination might be moot.

A Model Discrimination Function

A statistical method for model discrimination based on calculating the posterior probability in favor of each model being correct has been proposed (Box and Hill, 1967). This criterion should not be allowed to select a correct model, but it may be used to eliminate one or more models as being unworthy of additional investigation.

Posterior probability is the probability computed after the models have been fitted to the data. Before the models are fitted, we may have no particular reason to favor one model over another; that is, each model

TABLE 47.2
Posterior Odds for Four Rival Models ($n = 13$, $p = 2$)

Model	RSS$_i$ (× 100)	RSS Ratio (RSS$_{min}$/RSS$_i$)	R_i	Posterior Odds ($P_i = R_i/\Sigma R_i$)
1 (Tiessier)	0.175	1.00	1.00	0.50
2	0.184	0.96	0.78	0.39
3 (Monod)	0.237	0.74	0.19	0.10
4	0.420	0.42	0.01	0.01
		$\Sigma R_i = 3.12$		$\Sigma P_i = 1.00$

is equally likely to be correct. Therefore we say that the *prior probabilities* are equal for the four models. This means that, before we fit any of the models, we give them equal standing on scientific grounds and are willing to accept any one of them that fits the available data.

The criterion to be used assumes that the residual sums of squares of the rival models can be used to discriminate, with the better models having the lower RSS values. The posterior probability that each model is the best is calculated as a function of the RSS, adjusted for the number of parameters in the model.

Here we illustrate a special case where all models have the same number of parameters. There are k models. The posterior probability in favor of model i being correct is:

$$P_i = \frac{R_i}{\Sigma R_i}$$

where the summation is over the k models and:

$$R_i = \left(\frac{RSS_{min}}{RSS_i}\right)^{0.5(n-p)}$$

where RSS_i = residual sum of squares of model i, RSS_{min} = the smallest residual sum of squares in the set of k models, n = the number of observations (all models fitted to the same data), and p = number of estimated parameters in each model (p equal for all k models).

The residual sums of squares (RSS$_i$), RSS ratios, and posterior probabilities for each model are presented in Table 47.2. The Tiessier model has the smallest RSS; $RSS_{min} = 0.175$. An example computation for Model 2 gives $R_2 = (0.175/0.184)^{0.5(13-2)} = 0.78$. The computed posterior probabilities indicate a 50% probability for the Tiessier model being correct and a 10% probability for Monod model.

A warning is in order about using this discrimination criterion. A key assumption was that all models are equally likely to be correct before they are fitted (i.e., equal prior probability for each model). This is not the case in our example because the Tiessier and Monod models have some mechanistic justification, while the others were concocted for the sake of example. Therefore, the prior odds are not equal and the posterior odds given in Table 47.2 are just numbers that suggest how a model discrimination criterion might be used. Reducing a difficult problem to a few numbers should not replace careful thought, common sense, and making diagnostic plots of various kinds, especially plots of residuals. Furthermore, it is not a substitute for doing more experiments at wisely selected critical conditions.

Comments

Discriminating among rival models is difficult. Goodness of fit is a necessary, but not sufficient condition, for accepting a model as correct. One or more other models may also be adequate when judged by several common statistical criteria. If discrimination is to be attempted on statistical grounds, it is necessary to design experiments that will put the model in jeopardy by exposing it to conditions where its inadequacies will be revealed.

References

Box, G. E. P. and W. J. Hill (1967). "Discrimination among Mechanistic Models," *Technometrics,* 9(1), 57–71.
Boyle, W. C. and P. M. Berthouex (1974). "Biological Wastewater Treatment Model Building — Fits and Misfits," *Biotech. Bioeng.,* 16, 1139–1159.
Hill, W. J., W. G. Hunter, and D. Wichern (1968). "A Joint Design Criterion for the Dual Problem of Models Discrimination and Parameter Estimation," *Technometrics,* 10(1), 145–160.
Kittrel, J. R. and R. Mezaki (1967). "Discrimination among Rival Hougan-Watson Models through Intrinsic Parameters," *Am. Inst. Chem. Engr. J.,* 13(2), 389–470.
Monod, J (1942). *Researches sur la Croissance des Cultures Bacteriennes,* Paris, Herman & Cie.
Schulze, K. L. and R. S. Lipe (1964). "Relationship between Substrate Concentration, Growth Rate, and Respiration Rate of *Eschericia coli* in Continuous Culture," *Arch. fur Microbiol.,* 48, 1.
Tiessier, C. (1936). "Les Lois Quantitatives de la Croissance," *Ann. Physiol. Physicochim. Biol.,* 12, 527.

Exercises

47.1 Six Rival Models. Fit $\mu = \theta_1 \exp(-\theta_2/S)$ and $\mu = \theta_1 \left[1 - \frac{1}{\theta_2 + S}\right]$ to the data in Table 47.1 and redo the model discrimination analysis for six models.

47.2 Chlorine Residual. The following equations have been proposed to describe the decay of chlorine residual in wastewater (Haas, C. N. and S. B. Karra, *J. WPCF,* 56, 170–173). Explain how you would design an experiment that could discriminate between these models. The parameters are k, k_1, k_2, C_0, C^*, θ, and n.

$$C = C_0 - kt^n \qquad C = C_0 \exp(-kt) \qquad C = [kt(n-1) + (1/C_0)^{n-1}]^{-\left(\frac{1}{n-1}\right)}$$

$$C = C^* + (C_0 - C^*)\exp(-kt) \qquad C = C_0 \theta \exp(-k_1 t) + C_0(1-\theta)\exp(-k_2 t)$$

47.3 Trickling Filters. Given below are six equations that have been used to describe BOD removal by trickling filters and some performance data. Fit the six models in order to compare and classify them.

$$y = x\exp(-kQ^n) \qquad y = x\exp(-kD/Q^n) \qquad y = x\exp(-kD^m/Q^n)$$

$$y = \frac{x}{1 + kD^m/Q^n} \qquad y = x\left[1 - \frac{1}{1 + a\sqrt{Qx/D}}\right] \qquad y = a + (x-a)\exp(-kD^m/Q^n).$$

where y = effluent BOD, x = influent BOD, Q = hydraulic loading, and D = depth. a, k, m, and n are parameters to be estimated. In all models, k is a function of temperature (T): $k_T = k_{20} \, \theta^{T-20}$.

Inf. BOD (mg/L)	Hyd. Load. (mgd/ac)	Temp. (°C)	Effluent BOD (mg/L) Depth (ft)						
			2	3	4	5	6	7	8
100	10	17.2	45	37	33	33	32	29	26
119	15	20.8	43	35	31	29	27	26	24
108	20	18.1	58	53	49	46	44	41	33
61	20	17	27	22	19	18	17	17	16
93	25	15.7	39	32	30	29	28	25	23
85	35	17.9	56	46	41	36	35	32	31
101	10	2.3	61	52	44	38	33	30	29
60	15	2.3	33	28	25	23	22	21	19
106	20	3.1	70	60	52	46	40	36	33
130	20	2.9	79	68	58	50	46	42	40
97	25	3.1	53	44	38	34	33	31	28
93	35	2.3	60	52	47	44	40	36	34

47.4 Groundwater. Remedial pumping of a Superfund site has accomplished a steady decrease in the groundwater contamination level. It is desired to fit a model that will indicate (a) whether the contaminate concentration will decline to a target level (C_T) and (b) if so, when that will be accomplished. Below are four possible models. Compare them using the data in the table. Propose and test other models of your own.

$$C = C_T + (C_0 - C_T)\exp(-kt) \qquad C = C_T + (C_0 - C_T)t^{-k} \qquad C = C_T\left[\frac{a}{a + b\exp(-kt)}\right]$$

$$C = C_T + (C_0 - C_T)\theta\exp(-k_1 t) + (C_0 - C_T)(1 - \theta)\exp(-k_2 t)$$

Time (days)	0	30	100	200	300	350	400	500	600	700	800	900	1000
Conc. (μg/L)	2200	2200	1400	850	380	500	600	800	230	130	120	200	180

Time (days)	1100	1200	1300	1400	1500	1600	1700	1800	1900	2000
Conc. (μg/L)	230	200	190	280	200	250	290	200	140	180

Source: Spreizer, G. M. (1990). *ASTM STP*, 1053, 247–255.

47.5 Nitrification Inhibition. The rate of biological nitrification in streams and treatment systems is inhibited at low dissolved oxygen concentrations. Data from two independent studies are available below to test candidate models. Two have been proposed to fit the experimental data.

$$\text{Model 1:} \quad y = \theta_1[1 - \exp(-\theta_2 C)] + e \qquad \text{Model 2:} \quad y = \frac{\theta_3 C}{\theta_4 + C} + e$$

where y is the nitrification rate coefficient (day^{-1}) and C is the dissolved oxygen concentration (mg/L). Which model better describes the measured data? Do both studies favor the same model? Consider alternate models that may be suggested by the data.

Study A:

C (mg/L)	0.6	1.4	2	2.2	3.5	3.5	4.7	5.1	6.6	8.3	8.4
y (day^{-1})	0.6	1.0	1.3	1.1	1.2	1.3	1.5	1.2	1.5	1.4	1.6

Study B:

C (mg/L)	0.8	1.7	2.7	3.7	4.6	9.2
y (day^{-1})	0.4	0.8	1.0	1.1	1.4	1.5

48

Data Adjustment for Process Rationalization

KEY WORDS   *mass balance, measurement errors, data adjustment, Lagrange multipliers, least squares, process rationalization, weighted least squares.*

Measurement error often produces process monitoring data that are inconsistent with known constraints, such as the law of conservation of mass. If the measurements contain some redundant information, it is possible to test the measured world against the real world and make rational adjustments that bring the measured values into agreement with the constraints.

For example, a land surveyor may find that the three measured angles of a triangle do not add up to 180 degrees. Such an inconsistency can be discovered only if all three angles are measured. If only two angles of the triangle are measured, the third can be computed by difference, but there is no way to check the accuracy of the measurements. If all three angles are measured, a check on closure is available, and a rational adjustment can be made to bring the sum to exactly 180 degrees. If repeated measurements of the angles are made, the relative precision of the three angles can be determined and the adjustment becomes more rational and defensible.

The inputs and outputs of a treatment process or manufacturing system must obey the law of conservation of mass. The sulfur entering a power plant in coal must leave in the stack emissions or in the ash. Chromium entering a plating process leaves on the plated parts, in wastewater, or as a solid waste. Real inputs must equal real outputs (after accounting for internal storage), but measured inputs cannot be expected to equal measured outputs. If enough measurements are made, the material balance can be rationalized to satisfy conservation of mass.

Measurements are adjusted in the same spirit that a collection of observations is averaged to estimate a mean value. The measured values are used to extract the best possible estimates about the process being studied. The rationalization should be done so that the most precisely measured quantities receive the smallest adjustments. The method of least squares provides a common-sense adjustment procedure that is statistically valid and simple.

There are three types of errors (Deming, 1943; Ripps, 1961). Gross errors are caused by overt blunder, such as a misplaced decimal or instrument malfunction. Bias is a consistent displacement from the true value caused by improper instrument calibration or not following the correct analytical procedure. The purpose of quality assurance programs is to eliminate these kinds of errors. They should not exist in a well-run measurement program. The third type is unavoidable random error. The sign and magnitude of the errors vary randomly, but with definite statistical properties that can be described if repeated measurements are made. We will assume that the random errors have a mean value of zero and are normally distributed. We do not assume constant variance in all measured quantities.

Data that contain only random errors are said to be in statistical control. Data that are not in statistical control cannot be usefully adjusted. The first step, then, is to purge data containing gross errors. Second, remove known bias by making properly calibrated corrections. The data should now contain only random errors. These data can be rationally adjusted using the method of least squares.

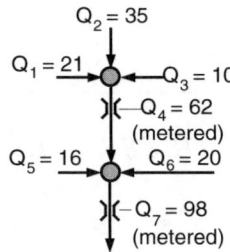

FIGURE 48.1 Seven measured flows are inconsistent because inflows do not equal outflows

Case Study: Errors in Flow Measurements

Seven flows have been measured, as shown in Figure 48.1. The measured values do not satisfy the physical constraint that inflow must equal outflow. The measured inflows to the upstream node (Q_1, Q_2, and Q_3) sum to 66, which does not agree with the value metered output of $Q_4 = 62$. The total of the five measured inputs to the downstream node sum to 102, which does not agree with the metered outflow of $Q_7 = 98$. We wish to use the available data to derive estimates of the flows that conform to the conservation of mass constraints that the real physical system must obey. The adjustment should be done so that the most precise measurements receive the smallest correction.

Data Adjustment

Some options for using or adjusting inconsistent data include:

1. Make no adjustment and use the measured values if the deficiencies fall within a predetermined tolerance.
2. Adjust one of the measured values. This might be proper if the accuracy of one measurement is much less than the others.
3. Distribute the discrepancy equally among the observed values.
4. Assign unequal adjustments to each observation according to some reasonable rule. One such rule is to make adjustments that are proportional to the precision of each observation (i.e., inversely proportional to the variance). This strategy is formulated as a least squares problem.

No one adjustment procedure is always the proper choice, but the method of least squares is the best method in the long run. In complex problems it may be the only satisfying method and it is often the simplest as well.

The objective of the method of least squares is to minimize the sum of weighted squares of the residuals:

$$S = \sum w_i (x_i - m_i)^2$$

where x_i and m_i are the calculated and measured values of quantity i. The summation is over all observations that are subject to random error. The weight w_i indicates the relative precision of measurement m_i. The special case where all weights are equal (constant variance) gives the familiar case of ordinary least squares (Chapter 33). If the weights are unequal, they might be inversely proportional to the variance of the measured value ($w_i = 1/\sigma_i^2$). If replicate measurements have been made, σ_i^2 is estimated by the sample variance (s_i^2). If there is no replication, knowledge of the measurement process will help in selecting reasonable weights. For example, if the error is expected to be proportional to the measured value, reasonable weights would be $w_i = 1/x_i$.

There is a natural tendency to overlook imprecise measurements in favor of those that are most precise. The weighted least squares model is in sympathy with this common-sense approach. A measurement with

TABLE 48.1

Measured Flow Rates and the Adjusted Values that Conform to the Conservation of Mass Constraints

Quantity	Measured Value	Adjusted Value	Adjustment
Q1	21	19.93	−1.07
Q2	35	33.93	+1.07
Q3	10	8.93	+1.07
Q4	62	62.80	+0.80
Q5	16	15.73	−0.27
Q6	20	19.73	−0.27
Q7	98	98.27	+0.27

Source: Schellpfeffer, J. W. and P. M. Berthouex (1972). "Rational Adjustment of Imbalances in Plant Survey Data," *Proc. 29th Ind. Waste Conf.*, Purdue University.

a large standard deviation has a small weight and contributes relatively little to the weighted sum of squares, thus leaving the precise measurements to dominate the overall adjustment procedure.

If the variances are unknown and cannot be estimated, subjective weights can be assigned arbitrarily to indicate the relative precision of the measurements as measured by the variance. Weights might be assigned using knowledge of meter characteristics, chemical analysis, or other technical information. It is the relative values of the weights and not the values that are important. The least reliable measurement can be assigned a weight $w = 1$, with proportionately larger weights being assigned to the more reliable measurements. If x_1 is twice as reliable as x_2, then specify $w_1 = 2$ and $w_2 = 1$.

Formulation of the solution will be illustrated for M constraining relations involving N variables, with $N \geq M$. If $N = M$, the system has no degrees of freedom and there is a unique solution to the set of constraint equations. If $N > M$, the system has $N - M = F$ degrees of freedom. Specifying values for F variables gives M equations in M unknowns and a unique solution can be found. If more than F variables are measured (for example, all N variables might be measured), there is some redundant information and inconsistencies in the measured values can be detected. This redundancy is a necessary condition for data adjustment.

The simplest problem is considered here. If we assumed that all n variables have been measured and all constraint equations are linear, the problem is to:

$$\text{minimize } S = \sum w_i(x_i - m_i)^2 \quad i = 1, 2, \ldots, N$$

subject to linear constraints of the form:

$$\sum a_{ij} x_i = b_j \quad j = 1, 2, \ldots, M$$

For this case of linear constraint equations, one method of solution is *Lagrange multipliers,* which are used to create a new objective function:

$$\text{minimize } G = \sum_{i=1}^{N} w_i(x_i - m_i)^2 + \sum_{j=1}^{M} \lambda_j (a_{ij} x_i - b_j)$$

There are now $N + M$ variables, the original N unknowns (the x_i's) plus the M Lagrange multipliers (λ_j's). The minimum of this function is located at the values where the partial derivatives with respect to each unknown variable vanish simultaneously. The derivative equations have two forms:

$$\frac{\partial G}{\partial x_i} = 2w_i(x_i - m_i) + \sum_{j=1}^{M} \lambda_j a_{ij} = 0 \quad i = 1, 2, \ldots, N$$

$$\frac{\partial G}{\partial \lambda_j} = \sum_{i=1}^{N} a_{ij} x_i - b_j = 0 \quad j = 1, 2, \ldots, M$$

The quantities m_i, w_i, and a_{ji} are known and the resulting set of linear equations are solved to compute x_i's and the λ_j's:

$$2w_i x_i + \sum_{j=1}^{M} a_{ij}\lambda_j = 2m_i w_i \qquad i = 1, 2, \ldots, N$$

$$\sum_{i=1}^{M} a_{ij} x_i = b_j \qquad j = 1, 2, \ldots, M$$

If the constraint equations are nonlinear, such as $\sum a_{ij} x_i^n = b_j$ or $\sum a_{ij} \exp(-x_i) = b_j$, the problem can be solved using methods in Madron (1992) and Schrage (1999).

Case Study: Solution

The system shown in Figure 48.1 will be solved for the case where all seven flows have been measured. Each value is measured with the same precision, so the weights are $w_i = 1$ for all values. The sum of squares equation is:

$$\text{minimize } G = (Q_1 - 21)^2 + (Q_2 - 35)^2 + (Q_3 - 10)^2 + (Q_4 - 62)^2$$
$$+ (Q_5 - 16)^2 + (Q_6 - 20)^2 + (Q_7 - 98)^2$$

subject to satisfying the two conservation of mass constraints:

$$Q_1 + Q_2 + Q_3 - Q_4 = 0 \qquad \text{and} \qquad Q_4 + Q_5 + Q_6 - Q_7 = 0$$

Incorporating the constraints gives a function of nine unknowns (seven Q's and two λ's):

$$\text{minimize } G = (Q_1 - 21)^2 + (Q_2 - 35)^2 + (Q_3 - 10)^2 + (Q_4 - 62)^2$$
$$+ (Q_5 - 16)^2 + (Q_6 - 20)^2 + (Q_7 - 98)^2$$
$$+ \lambda_1(Q_1 + Q_2 + Q_3 - Q_4) + \lambda_2(Q_4 + Q_5 + Q_6 - Q_7)$$

The nine partial derivative equations are all linear:

$2Q_1 + \lambda_1 = 21(2) = 42 \qquad 2Q_2 + \lambda_1 = 35(2) = 70 \qquad 2Q_3 + \lambda_1 = 10(2) = 20$
$2Q_4 - \lambda_1 + \lambda_2 = 62(2) = 124 \qquad 2Q_5 + \lambda_2 = 16(2) = 32 \qquad 2Q_6 + \lambda_2 = 20(2) = 40$
$2Q_7 - \lambda_2 = 98(2) = 196 \qquad Q_1 + Q_2 + Q_3 - Q_4 = 0 \qquad Q_4 + Q_5 + Q_6 - Q_7 = 0$

Notice that taking the derivatives has restored the material balance constraints that inflows equal outflows.

The nine linear equations are solved simultaneously for the flows that satisfy the material balance. The adjusted values are given in Table 48.1. The calculated values of λ_1 and λ_2 have no physical meaning and are not reported. The adjustments range from about 10% for Q_3 to a fraction of a percent for Q_7.

This example assumed that all flow measurements were equally precise. It might well happen that the inflows are less reliably estimated than the outflows. This could be reflected in the calculation by assigning weights. If the metered quantities Q_4 and Q_7 are known four times more precisely than the other quantities, the assigned weights would be $w_4 = 4$ and $w_7 = 4$, with all other weights being equal to 1.0. This would cause the flows Q_4 and Q_7 to be adjusted less than is shown in Table 48.1, whiles flows 1, 2, 3, 5, and 6 would be adjusted by larger amounts.

Comments

The method of least squares gives the smallest possible adjustments that make the measured survey data conform to the conservation of mass constraints. Also, it agrees with common sense in that unreliable measurements are adjusted more than precise measurements. The solution for linear constraining relations is particularly simple, even for a problem with many variables, because the set of equations can be written on inspection and are easily solved using standard matrix algebra. For nonlinear constraining equations, the least squares computations are manageable, but more difficult, because the partial derivatives are nonlinear. A nonlinear programming algorithm must be used to solve the problem.

Measurements must be taken at enough locations to test the measured world against the real world; that is, to check whether the conservation of mass has been satisfied. Redundancies in the set of data and the constraining relations are used to adjust the inconsistencies.

The example assumed that each measured value was equally reliable. This is not usually the case. Instead, the measurement errors might be proportional to the magnitude of the flow, or some measurements might have been repeated whereas others were done only once, or instruments (or personnel) with different precision might have been used at different locations. Such problem-specific information is included by using appropriate weights that could be estimated subjectively if replicate measurements are not available from which to estimate the variances.

In the case where $N = M$ (no degrees of freedom), errors in the measured values will propagate into the calculated values. This is discussed in Chapter 49.

References

Deming, W. E. (1943). *Statistical Adjustment of Data,* New York, John Wiley (Dover, Ed., 1964).
Madron, F (1992). *Process Plant Performance: Measurement and Data Processing for Optimization and Retrofits,* New York, Ellis Horwood.
Ripps, D. L. (1961). "Adjustment of Experimental Data," *Chem. Engr. Prog. Symp. Ser.,* 61(55), 8–13.
Schellpfeffer, J. W. and P. M. Berthouex (1972). "Rational Adjustment of Imbalances in Plant Survey Data," *Proc. 29th Ind. Waste Conf.,* Purdue University.
Schrage, Linus (1999). *Optimization Modeling with LINGO,* Chicago, IL, Lindo Systems, Inc.

Exercises

48.1 Flow Adjustment. Solve the flow adjustment problem assuming that flows 4 and 7 are measured four times more precisely than the other flows (i.e., $w_1 = 1$, $w_2 = 1$, $w_3 = 1$, $w_4 = 4$, $w_5 = 1$, $w_6 = 1$, and $w_7 = 4$)

48.2 Stream Flows. A certain reach of a stream receives two point-source discharges between the upstream and downstream monitoring stations. Average flows with their standard deviations are given below. Adjust the flows to achieve conservation of mass.

Flow	Mean (m^3/sec)	Std. Dev. (m^3/sec)
Upstream	1.3	0.10
Point source 1	0.6	0.10
Point source 2	2.1	0.15
Downstream	4.5	0.15

48.3 Mass Balance. The material balance equations for the process shown are $X_1 + X_4 - X_2 = 0$ and $X_2 - X_3 - X_4 - X_5 = 0$. The measured values are $X_1 = 10.3$, $X_2 = 15.2$, $X_3 = 8.8$, $X_4 = 4.4$, and $X_5 = 1.5$. Assuming the variance is the same for each X, adjust the data using the method of least squares.

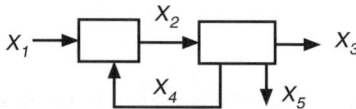

48.4 Reverse Osmosis. A technician has monitored the performance of an experimental two-stage reverse osmosis system and has given you the total dissolved solids data (kg/d TDS) shown below. Because a very imprecise method was used to measure the solids concentration, the conservation of mass balance is not satisfied. Adjust the measured quantities to conform to the requirement that mass in equals mass out.

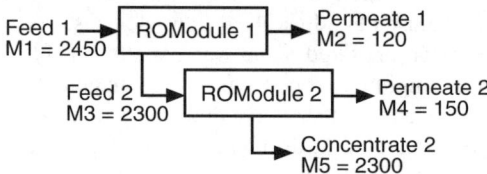

48.5 Stream Dilution. A stream with measured flow $Q_1 = 9$ and measured concentration $C_1 = 4$ joins another stream with measured flow $Q_2 = 18$ and measured concentration $C_2 = 7$. The confluent stream has measured flow $Q_3 = 32$ and measured concentration $C_3 = 6$. The true values must satisfy $Q_1 + Q_2 - Q_3 = 0$ and $Q_1C_1 + Q_2C_2 - Q_3C_3 = 0$. The variances are $\sigma^2 = 1$ for all flows and concentrations. Adjust the measured quantities so inflows equal outflows.

49

How Measurement Errors Are Transmitted into Calculated Values

KEY WORDS *error transmission, experimental planning, linear approximation, propagation of errors, propagation of uncertainty, reactor kinetics, sensitivity coefficients, titration errors, variance, uncertainty analysis.*

When observations that contain errors are used to calculate other values, the errors are transmitted, to a greater or lesser extent, into the calculated values. It is useful to know whether the errors in a particular measured variable will be magnified or suppressed. This could help in selecting measurement methods and planning the number of replications that will be required to control error propagation in the variables that most strongly transmit their errors. Here are three such situations.

1. A process material balance is described by m equations that involve n variables, say $m = 8$ and $n = 14$. If $n - m = 14 - 8 = 6$ variables are measured, the remaining eight variables can be calculated. Errors in the measured variables will be reflected in the calculated values.

2. An air pollution discharge standard might be based on the weight of NO_x emitted per dry standard cubic foot of gas corrected to a specified CO_2 content. To calculate this quantity, one needs measurements of NO_x concentration, velocity, temperature, pressure, moisture content, and CO_2 concentration. The precision and bias of each measurement will affect the accuracy of the calculated value. Which variables need to be measured most precisely?

3. An experiment to estimate the reaction rate coefficient k involves measuring reactor solids content X and effluent organic concentration S over a series of runs in which detention time $t = V/Q$ is controlled at various levels. The influent organic concentration S_0 can be prepared precisely at the desired value. The calculation of k from the known values is $k = (S_0 - S)Q/SXV$. Both X and S are difficult to measure precisely. Replicate samples must be analyzed to improve precision. Questions during the experimental planning stage include: "How many replicate samples of X and S are needed if k is to be estimated within plus or minus 3%? Must both X and S be measured precisely, or is one more critical than the other? Measuring S is expensive; can money be saved if errors in S are not strongly transmitted into the calculated value of k?"

Many experiments would be improved by answering questions like these during the planning rather than the analysis phase of the experiments.

Theory

The simplest case is for a linear function relating the calculated value y to the values of x_1, x_2, and x_3 obtained from measured data. The model has the form:

$$y = \theta_0 + \theta_1 x_1 + \theta_2 x_2 + \theta_3 x_3$$

where the θ's are known constants and the errors in x_1, x_2, and x_3 are assumed to be random. The expected value of y is:

$$E(y) = \theta_0 + \theta_1 E(x_1) + \theta_2 E(x_2) + \theta_3 E(x_3)$$

The variance of y is:

$$\begin{aligned}\text{Var}(y) = &\, \theta_1^2 \text{Var}(x_1) + \theta_2^2 \text{Var}(x_2) + \theta_3^2 \text{Var}(x_3) \\ &+ 2\theta_1\theta_2 \text{Cov}(x_1 x_2) + 2\theta_1\theta_3 \text{Cov}(x_1 x_3) + 2\theta_2\theta_3 \text{Cov}(x_2 x_3)\end{aligned}$$

The covariance terms ($\text{Cov}(x_i, x_j)$) can be positive, negative, or zero. If x_1, x_2, and x_3 are uncorrelated, the covariance terms are all zero and:

$$\text{Var}(y) = \theta_1^2 \text{Var}(x_1) + \theta_2^2 \text{Var}(x_2) + \theta_3^2 \text{Var}(x_3)$$

The terms on the right-hand side represent the separate contributions of each x variable to the overall variance of the calculated variable y. The derived $\text{Var}(y)$ estimates the true variance of y (σ_y^2) and can be used to compute a confidence interval for y.

This same approach can be applied to a linear approximation of nonlinear function. Figure 49.1 shows a curved surface $y = f(x_1, x_2)$ approximated by a Taylor series expansion of y where higher-order terms have been ignored:

$$Y = \theta_0 + \theta_1(x_1 - \bar{x}) + \theta_2(x_2 - \bar{x})$$

This expansion can be used to linearize any continuous function $y = f(x_1, x_2, \ldots, x_n)$. The approximation is centered at ($\bar{x}_1, \bar{x}_2$) and $\bar{y} = f(\bar{x}_1, \bar{x}_2) = \theta_0$. The linear coefficients are the derivatives of y with respect to x_1 and x_2, evaluated at $\bar{x}_1$ and $\bar{x}_2$:

$$\theta_1 = \left(\frac{\partial y}{\partial x_1}\right)_{\bar{x}_1} \quad \text{and} \quad \theta_2 = \left(\frac{\partial y}{\partial x_2}\right)_{\bar{x}_2}$$

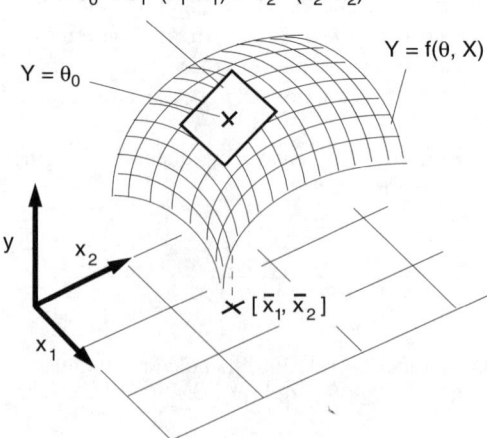

FIGURE 49.1 A nonlinear function $y = f(\theta, x)$ can be approximated by a linear function centered at the expected values $\bar{x}_1$ and $\bar{x}_2$.

Following the example of the linear function presented earlier, the variance of y is:

$$\text{Var}(Y) = \left(\frac{\partial y}{\partial x_1}\right)^2_{\bar{x}_1} \text{Var}(x_1) + \left(\frac{\partial y}{\partial x_2}\right)^2_{\bar{x}_2} \text{Var}(x_2) + 2\left(\frac{\partial^2 y}{\partial x_1 \partial x_2}\right)_{\bar{x}_1, \bar{x}_2} \text{Cov}(x_1 x_2)$$

This can be generalized to n measured variables:

$$\text{Var}(Y) = \sum_{i=1}^{n}\left(\frac{\partial y}{\partial x_i}\right)^2_{\bar{x}_i} \text{Var}(x_i) + 2\sum_{i=1}^{n}\sum_{j=i-1}^{n}\left(\frac{\partial^2 y}{\partial x_i \partial x_j}\right)_{\bar{x}_i \bar{x}_j} \text{Cov}(x_i x_j)$$

There will be $n(n-1)/2$ covariance terms. If $x_1, x_2, \ldots, x_n$ are independent of each other, all covariance terms will be zero. It is tempting to assume independence automatically, but making this simplification needs some justification. Because $\text{Cov}(x_i, x_j)$ can be positive or negative, correlated variables might increase or decrease the variance of Y.

If taking derivatives analytically is too unwieldy, the linear approximation can be derived from numerical estimates of the derivatives at $\bar{x}_1$, $\bar{x}_2$, etc. This approach may be desirable for even fairly simple functions. The region over which a linearization is desired is specified by $\bar{x}_1 + \Delta x_1$ and $\bar{x}_2 + \Delta x_2$. The function values $y_0 = f(\bar{x}_1, \bar{x}_2)$, $y_1 = f(\bar{x}_1 + \Delta x_1, \bar{x}_2)$, $y_2 = f(\bar{x}_1, \bar{x}_2 + \Delta x_2)$, etc. are used to compute the coefficients of the approximating linear model:

$$Y = \theta_0 + \theta_1(x_1 - \bar{x}_1) + \theta_2(x_2 - \bar{x}_2)$$

The coefficients are:

$$\theta_0 = y_0 = \bar{y} = f(\bar{x}_1, \bar{x}_2), \quad \theta_1 = \frac{y_1 - y_0}{\Delta x_1}, \quad \text{and} \quad \theta_2 = \frac{y_2 - y_0}{\Delta x_2}$$

The linearization will be valid over a small region about the centering point, where "small" can be reasonably defined as about $\bar{x} \pm 2s_x$.

Case Study: Reactor Kinetics

A lab-scale reactor is described by the model:

$$S = \frac{S_0}{1 + kX(V/Q)}$$

where S_0 is influent BOD, S is effluent BOD, X is mixed-liquor suspended solids, V is reactor volume, Q is volumetric flow rate, and k is a reaction rate coefficient. Assume that the volume is fixed at $V = 1.0$. The independent variables Q, X, S_0, and S will be measured for each experiment. Their expected measurement errors, given as standard deviations, are $\sigma_Q = 0.02$, $\sigma_X = 100$, $\sigma_S = 2$, $\sigma_{S_0} = 20$. How precisely will k be estimated?

The sensitivity of k to changes in the other variables can be calculated analytically if we solve the model for k:

$$k = \frac{(S_0 - S)Q}{SXV}$$

The linearized model is:

$$k = \theta_0 + \theta_{S_0}(S_0 - \bar{S}_0) + \theta_S(S - \bar{S}) + \theta_X(X - \bar{X}) + \theta_Q(Q - \bar{Q})$$

The θ's are *sensitivity coefficients* that quantify how much a unit change in variable i will change the value of k from the centered value $\theta_0 = 0.0045$. Because the θ's are evaluated at particular values of the variables, they have meaning only near this particular condition. The coefficients are evaluated at the centered values: $\bar{S}_0 = 200$, $\bar{X} = 2000$, $\bar{Q} = 1$, and $\bar{S} = 20$. For convenience, the overbars are omitted in the following equations.

The value of k at the given values of the variables is:

$$\theta_0 = k = \frac{(S_0 - S)Q}{SXV} = \frac{(200 - 20)(1)}{20(2000)(1)} = 0.0045$$

The other coefficients of the linearized model are:

$$\theta_{S_0} = \frac{\partial k}{\partial S_0} = \frac{Q}{SXV} = \frac{1}{20(2000)(1)} = 2.5 \times 10^{-5}$$

$$\theta_S = \frac{\partial k}{\partial S} = -\frac{Q}{SXV} = -\frac{1}{20(2000)(1)} = -2.5 \times 10^{-5}$$

$$\theta_X = \frac{\partial k}{\partial X} = -\frac{S_0 - S}{SX^2V} = -\frac{200 - 20}{20(2000)^2(1)} = -2.25 \times 10^{-6}$$

$$\theta_Q = \frac{\partial k}{\partial Q} = \frac{S_0 - S}{SXV} = \frac{200 - 20}{20(2000)(1)} = 0.0045$$

The variance of k, assuming all covariances are zero, is:

$$\sigma_k^2 = \theta_{S_0}^2 \sigma_{S_0}^2 + \theta_S^2 \sigma_S^2 + \theta_X^2 \sigma_X^2 + \theta_Q^2 \sigma_Q^2 = 3.11 \times 10^{-7}$$

which is computed from the separate contributions of the variables using the values and variances given earlier, and as shown in Table 49.1.

The estimated total variance is $\sigma_k^2 = 3.11 \times 10^{-7}$ and the standard deviation is $\sigma_k = 0.00056$. These numbers are small, but the expected value of $k = 0.0045$ is also small. A better perspective is obtained by finding an approximate 95% confidence interval for the estimated value of k, based on the linear approximation of the true model:

$$k = 0.0045 \pm 2(0.00056) = 0.0045 \pm 0.0011$$

TABLE 49.1

Contribution of the Variance of Each Variable to the Total Variance in k

Variable	Std. Deviation	Variance Component	% of σ_k^2
$S_0 = 200$	$\sigma_{S_0} = 20$	$\theta_{S_0}^2 \sigma_{S_0}^2 = 2.5 \times 10^{-7}$	80.3
$S = 20$	$\sigma_S = 2$	$\theta_S^2 \sigma_S^2 = 2.5 \times 10^{-9}$	0.8
$X = 2000$	$\sigma_X = 100$	$\theta_X^2 \sigma_X^2 = 0.51 \times 10^{-7}$	16.3
$Q = 1$	$\sigma_Q = 0.02$	$\theta_Q^2 \sigma_Q^2 = 0.08 \times 10^{-7}$	2.6

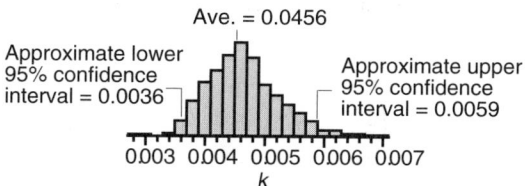

FIGURE 49.2 Histogram showing the distribution of 1000 values of k that were simulated using values given in the case study reactor problem.

and

$$0.0034 < k < 0.0057$$

The confidence interval for the linearized model is symmetric and normal, but this is not exactly right for the original model, in which the change in k is linear with respect to changes in Q and X but slightly nonlinear with respect to S and S_0. Even if there is some asymmetry, the analysis serves its purpose well. The purpose is to plan the experiment and not to predict the confidence interval of k.

We may be disappointed that the precision is not better, that is, something like $k = 0.0045 \pm 0.0004$. It is better to be disappointed before doing the experiment than after, so that plans may be modified at little loss of time or money.

The contributions of each variable to σ_k^2 suggest where substantial improvements might be made. At the current settings, S_0 contributes about 80% of the total variance in k. This factor should be targeted first for variance reduction to decrease the error in the estimated value of k. Perhaps the measurement method can be improved, but often a more practical approach is to replicate the measurements. Making duplicate measurements ($n = 2$) would reduce σ_i to $\sigma_i/\sqrt{n}$; using $n = 4$ would give $\sigma_i/\sqrt{4} = \sigma_i/2$. The corresponding values of $\theta^2\sigma^2$ would be reduced to one-half (for $n = 2$) and one-quarter (for $n = 4$) of the original values. For example, if the standard deviation in S_0 could be halved to 10 mg/L, the variance in k would be reduced by 60%.

Notice that the values of the sensitivity coefficients depend on the settings of the independent variables (S_0, S, X, and Q). If these are changed by moving the experiment to different settings, the values of θ_i will change. It takes some effort to see by how much and in which direction they change but, nevertheless, changing the experimental design is another way to reduce the variance of the estimated k.

The approximate linear analysis can be checked by simulation. Figure 49.2 shows the distribution of 1000 simulated k values calculated from randomly selected values of Q, X, S, and S_0 having the means and variances given in the previous section. The simulated average is $k = 0.00456$. The distribution is slightly skewed toward high values of k. The upper and lower confidence limits of [0.0036, 0.0059] were estimated by counting 2.5% from the top and from the bottom of the 1000 ranked values of k. The simulated values agree nicely with those calculated from the linear approximation.

Case Study: A Finite Difference Model

In the example above, the kinetic model was a simple algebraic equation. Frequently, the model of interest consists of one or more differential equations that must be solved numerically (Brown, 1987). Without worrying about the details of the system being modeled, or the specific form of the equations, assume that time-consuming and expensive experiments are needed to estimate a reaction rate coefficient (k) and that preliminary tests indicate that two of the independent variables cannot be measured with a high degree of precision. We seek to understand how errors in the measurement of these variables will be transmitted into the estimate of k. The first step is to compute some numerical solutions of the model in the region where the experimenter thinks the first run should be done.

For our two-variable example, the estimate of variance based on the Taylor series expansion shown earlier is:

$$\text{Var}(k) = \left(\frac{\Delta k}{\Delta X_1}\right)^2 \text{Var}(X_1) + \left(\frac{\Delta k}{\Delta X_2}\right)^2 \text{Var}(X_2)$$

We will estimate the sensitivity coefficients $\theta_1 = \Delta k/\Delta X_1$ and $\theta_2 = \Delta k/\Delta X_2$ by evaluating k at distances ΔX_1 and ΔX_2 from the center point.

Assume that the center of region of interest is located at $X_{1_0} = 200$, $X_{2_0} = 20$, and that $k_0 = 0.90$ at this point. Further assume that $\text{Var}(X_1) = 100$ and $\text{Var}(X_2) = 1$. A reasonable choice of ΔX_1 and ΔX_2 is from one to three standard deviations of the error in X_1 and X_2. We will use $\Delta X_1 = 2\sigma_{X_1} = 20$ and $\Delta X_2 = 2\sigma_{X_2} = 2$. Suppose that $k = 1.00$ at $[X_{1_0} + \Delta X_1 = 200 + 20, X_{2_0} = 20]$ and $k = 0.70$ at $[X_{1_0} = 200, X_{2_0} + \Delta X_2 = 20 + 2]$. The sensitivity coefficients are:

$$\frac{\Delta k}{\Delta X_1} = \frac{1.00 - 0.90}{20} = 0.0050$$

$$\frac{\Delta k}{\Delta X_2} = \frac{0.70 - 0.90}{2} = -0.10$$

These sensitivity coefficients can be used to estimate the expected variance of k:

$$\sigma_k^2 = (0.0050)^2 100 + (-0.10)^2 1 = 0.0025 + 0.010 = 0.0125$$

and

$$\sigma_k = 0.11$$

An approximate 95% confidence interval would be $k = 0.90 \pm 2(0.11) = 0.90 \pm 0.22$, or $0.68 < k < 1.12$.

Unfortunately, at these specified experimental settings, the precision of the estimate of k depends almost entirely upon X_2; 80% of the variance in k is contributed by X_2. This may be surprising because X_2 has the smallest variance, but it is such failures of our intuition that merit this kind of analysis. If the precision of k must be improved, the options are (1) try to center the experiment in another region where variation in X_2 will be suppressed, or (2) improve the precision with which X_2 is measured, or (3) make replicate measures of X_2 to average out the random variation.

Propagation of Uncertainty in Models

The examples in this chapter have been about the propagation of measurement error, but the same methods can be used to investigate the *propagation of uncertainty* in design parameters. Uncertainty is expressed as the variance of a distribution that defines the uncertainty of the design parameter. If only the range of parameter values is known, the designer should use a uniform distribution. If the designer can express a "most likely" value within the range of the uncertain parameter, a triangular distribution can be used. If the distribution is symmetric about the expected value, the normal distribution might be used. The variance of the distribution that defines the uncertainty in the design parameter is used in the propagation of error equations (Berthouex and Polkowski, 1970).

The simulation methods used in Chapter 51 can also be used to investigate the effect of uncertainty in design inputs on design outputs and decisions. They are especially useful when real variability in inputs exists and the variability in output needs to be investigated (Beck, 1987; Brown, 1987).

Comments

It is a serious disappointment to learn after an experiment that the variance of computed values is too large. Avoid disappointment by investigating this before running the experiment. Make an analysis of how measurement errors are transmitted into calculated values. This can be done when the model is a simple equation, or when the model is complicated and must be solved by numerical approximation.

References

Beck, M. B. (1987). "Water Quality Modeling: A Review of the Analysis of Uncertainty," *Water Resour. Res.*, 23(5), 1393–1441.

Berthouex, P. M. and L. B. Polkowski (1970). "Optimum Waste Treatment Plant Design under Uncertainty," *J. Water Poll. Control Fed.*, 42(9), 1589–1613.

Box, G. E. P., W. G. Hunter, and J. S. Hunter (1978). *Statistics for Experimenters: An Introduction to Design, Data Analysis, and Model Building*, New York, Wiley Interscience.

Brown, L. C. (1987). "Uncertainty Analysis in Water Quality Modeling Using QUAL2E," in *Systems Analysis in Water Quality Measurement*, (Advances in Water Pollution Control Series), M. B. Beck, Ed., Pergamon Press, pp. 309–319.

Mandel, J. (1964). *The Statistical Analysis of Experimental Data*, New York, Interscience Publishers.

Exercises

49.1 Exponential Model. The model for a treatment process is $y = 100 \exp(-kt)$. You wish to estimate k with sufficient precision that the value of y is known within ±5 units. The expected value of k is 0.2. How precisely does k need to be known for $t = 5$? For $t = 15$? (Hint: It may help if you draw y as a function of k for several values of t.)

49.2 Mixed Reactor. The model for a first-order kinetic reaction in a completely mixed reactor is $y = \frac{x}{1 + kV/Q}$. (a) Use a Taylor series linear approximation and evaluate the variance of y for $k = 0.5, V = 10, Q = 1$, and $x = 100$, assuming the standard deviation of each variable is 10% of its value (i.e., $\sigma_k = 0.1(0.05) = 0.005$). (b) Evaluate the variance of k for $V = 10, Q = 1, x = 100$, and $y = 20$, assuming the standard deviation of each variable is 10% of its value. Which variable contributes most to the variance of k?

49.3 Simulation of an Exponential Model. For the exponential model $y = 100 \exp(-kt)$, simulate the distribution of y for k with mean 0.2 and standard deviation 0.02 for $t = 5$ and for $t = 15$.

49.4 DO Model. The Streeter-Phelps equation used to model dissolved oxygen in streams is:

$$D = \frac{k_1 L_a}{k_2 - k_1}[\exp(-k_1 t) - \exp(k_2 t)] + D_a \exp(-k_2 t)$$

where D is the dissolved oxygen deficit (mg/L), L_a is the initial BOD concentration (mg/L), D_a is the initial dissolved oxygen deficit (mg/L), and k_1 and k_2 are the bio-oxidation and reaeration coefficients (1/day). For the following conditions, estimate the dissolved oxygen deficit and its standard deviation at travel times (t) of 1.5 and 3.0 days.

Parameter	Average	Std. Deviation
L_a (mg/L)	15	1.0
D_a (mg/L)	0.47	0.05
k_1 (day^{-1})	0.52	0.1
k_2 (day^{-1})	1.5	0.2

49.5 Chloroform Risk Assessment. When drinking water is chlorinated, chloroform (a trihalomethane) is inadvertently created in concentrations of approximately 30 to 70 µg/L. The model for estimating the maximum lifetime risk of cancer, for an adult, associated with the chloroform in the drinking water is:

$$\text{Risk} = \frac{PF \times C \times IR \times ED \times AF}{BW \times LT}$$

Use the given values to estimate the mean and standard deviation of the lifetime cancer risk. Under these conditions, which variable is the largest contributor to the variance of the cancer risk?

Variable	Mean	Std. Deviation
Chloroform concentration, C (µg/L)	50	15
Intake rate, IR (L/day)	2	0.4
Exposure duration, ED (yr)	70	12
Absorption factor, AF	0.8	0.05
Body weight, BW (kg)	70	20
Lifetime, LT (yr)	70	12
Potency factor, PF (kg-day/µg)	0.004	0.0015

50

Using Simulation to Study Statistical Problems

KEY WORDS *bootstrap, lognormal distribution, Monte Carlo simulation, percentile estimation, random normal variate, random uniform variate, resampling, simulation, synthetic sampling, t-test.*

Sometimes it is difficult to analytically determine the properties of a statistic. This might happen because an unfamiliar statistic has been created by a regulatory agency. One might demonstrate the properties or sensitivity of a statistical procedure by carrying through the proposed procedure on a large number of synthetic data sets that are similar to the real data. This is known as *Monte Carlo simulation*, or simply *simulation*.

A slightly different kind of simulation is *bootstrapping*. The bootstrap is an elegant idea. Because sampling distributions for statistics are based on repeated samples with replacement (*resamples*), we can use the computer to simulate repeated sampling. The statistic of interest is calculated for each resample to construct a simulated distribution that approximates the true sampling distribution of the statistic. The approximation improves as the number of simulated estimates increases.

Monte Carlo Simulation

Monte Carlo simulation is a way of experimenting with a computer to study complex situations. The method consists of sampling to create many data sets that are analyzed to learn how a statistical method performs.

Suppose that the model of a system is $y = f(x)$. It is easy to discover how variability in x translates into variability in y by putting different values of x into the model and calculating the corresponding values of y. The values for x can be defined as a probability density function. This process is repeated through many trials (1000 to 10,000) until the distribution of y values becomes clear.

It is easy to compute uniform and normal random variates directly. The values generated from good commercial software are actually *pseudorandom* because they are derived from a mathematical formula, but they have statistical properties that cannot be distinguished from those of true random numbers. We will assume such a random number generating program is available.

To obtain a random value $Y_U(\alpha, \beta)$ from a uniform distribution over the interval (α, β) from a random uniform variate R_U over the interval $(0, 1)$, this transformation is applied:

$$Y_U(\alpha, \beta) = \alpha + (\beta - \alpha) R_U(0, 1)$$

In a similar fashion, a normally distributed random value $Y_N(\eta, \sigma)$ that has mean η and standard deviation σ is derived from a standard normal random variate $R_N(0, 1)$ as follows:

$$Y_N = (\eta, \sigma) = \eta + \sigma R_N(0, 1)$$

Lognormally distributed random variates can be simulated from random normal variates using:

$$Y_{LN}(\alpha, \beta) = \exp(\eta + \sigma R_N(0, 1))$$

Here, the logarithm of Y_{LN} is normally distributed with mean η and standard deviation σ. The mean (α) and standard deviation (β) of the lognormal variable Y_{LN} are:

$$\alpha = \exp(\eta + 0.5\sigma^2)$$

and

$$\beta = \exp(\eta + 0.5\sigma^2)\sqrt{\exp(\sigma^2) - 1}$$

You may not need to make the manipulations described above. Most statistics software programs (e.g., MINITAB, Systat, Statview) will generate standard uniform, normal, t, F, chi-square, Beta, Gamma, Bernoulli, binomial, Poisson, logistic, Weibull, and other distributions. Microsoft EXCEL will generate random numbers from uniform, normal, Bernoulli, binomial, and Poisson distributions. Equations for generating random values for the exponential, Gamma, Chi-square, lognormal, Beta, Weibull, Poisson, and binomial distributions from the standard uniform and normal variates are given in Hahn and Shapiro (1967). Another useful source is Press et al. (1992).

Case Study: Properties of a Computed Statistic

A new regulation on chronic toxicity requires enforcement decisions to be made on the basis of 4-day averages. Suppose that preliminary sampling indicates that the daily observations x are lognormally distributed with a geometric mean of 7.4 mg/L, mean $\eta_x = 12.2$, and variance $\sigma_x = 16.0$. If $y = \ln(x)$, this corresponds to a normal distribution with $\eta_y = 2$ and $\sigma_y^2 = 1$. Averages of four observations from this system should be more nearly normal than the parent lognormal population, but we want to check on how closely normality is approached. We do this empirically by constructing a distribution of simulated averages. The steps are:

1. Generate four random, independent, normally distributed numbers having $\eta = 2$ and $\sigma = 1$.
2. Transform the normal variates into lognormal variates $x = \exp(y)$.
3. Average the four values to estimate the 4-day average ($\bar{x}_4$).
4. Repeat steps 1 and 2 one thousand times, or until the distribution of $\bar{x}_4$ is sufficiently clear.
5. Plot a histogram of the average values.

Figure 50.1(a) shows the frequency distribution of the 4000 observations actually drawn in order to compute the 1000 simulated 4-day averages represented by the frequency distribution of Figure 50.1(b). Although 1000 observations sounds likes a large number, the frequency distributions are still not smooth, but the essential information has emerged from the simulation. The distribution of 4-day averages is skewed, although not as strongly as the parent lognormal distribution. The median, average, and standard deviation of the 4000 lognormal values are 7.5, 12.3, and 16.1. The average of the 1000 4-day averages is 12.3; the standard deviation of the 4-day averages is 11.0; 90% of the 4-day averages are in the range of 5.0 to 26.5; and 50% are in the range of 7.2 to 15.4.

Case Study: Percentile Estimation

A state regulation requires the 99th percentile of measurements on a particular chemical to be less than 18 μg/L. Suppose that the true underlying distribution of the chemical concentration is lognormal as shown in the top panel of Figure 50.2. The true 99th percentile is 13.2 μg/L, which is well below the standard value of 18.0. If we make 100 random observations of the concentration, how often will the 99th percentile

Using Simulation to Study Statistical Problems

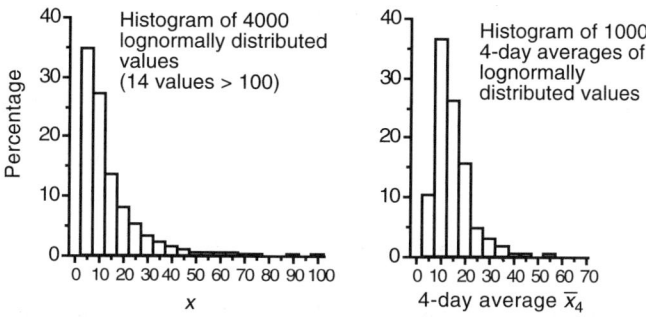

FIGURE 50.1 Left-hand panel: frequency distribution of 4000 daily observations that are random, independent, and have a lognormal distribution $x = \exp(y)$, where y is normally distributed with $\eta = 2$ and $\sigma = 1$. Right-hand panel: frequency distribution of 1000 4-day averages, each computed from four random values sampled from the lognormal distribution.

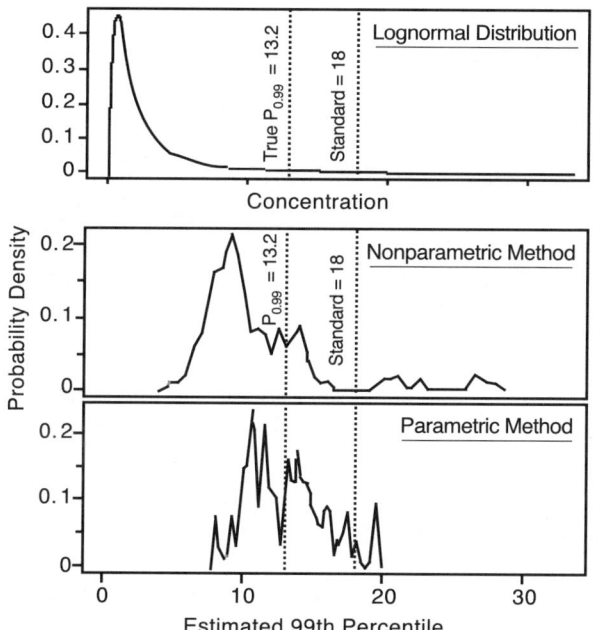

FIGURE 50.2 Distribution of 100 nonparametric estimates and 100 parametric estimates of the 99th percentile, each computed using a sample of $n = 100$ from the lognormal distribution shown in the top panel.

"violate" the 18-μg/L limit? Will the number of violations depend on whether the 99th percentile is estimated parametrically or nonparametrically? (These two estimation methods are explained in Chapter 8.)
These questions can be answered by simulation, as follows.

1. Generate a set of $n = 100$ observations from the "true" lognormal distribution.
2. Use these 100 observations to estimate the 99th percentile parametrically and nonparametrically.
3. Repeat steps 1 and 2 many times to generate an empirical distribution of 99th percentile values.

Figure 50.2 shows the empirical distribution of 100 estimates of the 99th percentiles made using a nonparametric method, each estimate being obtained from 100 values drawn at random from the log-normal distribution. The bottom panel of Figure 50.2 shows the distribution of 100 estimates made with the parametric method.

One hundred estimates gives a rough, but informative, empirical distribution. Simulating one thousand estimates would give a smoother distribution, but it would still show that the parametric estimates are less variable than the nonparametric estimates and they are distributed more symmetrically about the true 99th percentile value of $p_{0.99} = 13.2$. The parametric method is better because it uses the information that the data are from a lognormal distribution, whereas the nonparametric method assumes no prior knowledge of the distribution (Berthouex and Hau, 1991).

Although the true 99th percentile of 13.2 μg/L is well below the 18 μg/L limit, both estimation methods show at least 5% violations due merely to random errors in sampling the distribution, and this is with a large sample size of $n = 100$. For a smaller sample size, the percentage of trials giving a violation will increase. The nonparametric estimation gives more and larger violations.

Bootstrap Sampling

The bootstrap method is random resampling, with replacement, to create new sets of data (Metcalf, 1997; Draper and Smith, 1998). Suppose that we wish to determine confidence intervals for the parameters in a model by the bootstrap method. Fitting the model to a data set of size n will produce a set of n residuals. Assuming the model is an adequate description of the data, the residuals are *random errors*. We can imagine that in a repeat experiment the residual of the original eighth observation might happen to become the residual for the third new observation, the original third residual might become the new sixth residual, etc. This suggests how n residuals drawn at random from the original set can be assigned to the original observations to create a set of new data. Obviously this requires that the original data be a random sample so that residuals are independent of each other.

The resampling is done with replacement, which means that the original eighth residual can be used more than once in the bootstrap sample of new data.

The boostrap resampling is done many times, the statistics of interest are estimated from the set of new data, and the empirical reference distributions of the statistics are compiled. The number of resamples might depend on the number of observations in the pool that will be sampled. One recommendation is to resample $B = n \, [\ln(n)]^2$ times, but it is common to round this up to 100, 500, or 1000 (Peigorsch and Bailer, 1997).

The resampling is accomplished by randomly selecting the mth observation using a uniformly distributed random number between 1 and n:

$$m_i = \text{round}[nR_U(0,1) + 0.501]$$

where $R_U(0,1)$ is uniformly distributed between 0 and 1. The resampling continues with replacement until n observations are selected. This is the *bootstrap sample*.

The bootstrap method will be applied to estimating confidence intervals for the parameters of the model $y = \beta_0 + \beta_1 x$ that were obtained by fitting the data in Table 50.1. Of course, there is no need to bootstrap this problem because the confidence intervals are known exactly, but using a familiar example makes it easy to follow and check the calculations.

The fitted model is $\hat{y} = 49.13 + 0.358x$. The bootstrap procedure is to resample, with replacement, the 10 residuals given in Table 50.1. Table 50.2 shows five sets of 10 random numbers that were used to generate the resampled residuals and new y values listed in Table 50.3. The model was fitted to each set of new data to obtain the five pairs of parameter estimates shown Table 50.4, along with the parameters from the original fitting. If this process were repeated a large number of times (i.e., 100 or more), the distribution of the intercept and slope would become apparent and the confidence intervals could be inferred from these distribution. Even with this very small sample, Figure 50.3 shows that the elliptical joint confidence region is starting to emerge.

Using Simulation to Study Statistical Problems 437

TABLE 50.1

Data and Residuals Associated with the Model
$\hat{y} = 49.14 + 0.358x$

Observation	x	y	$\hat{y}$	Residual
1	23	63.7	57.37	6.33
2	25	63.5	58.09	5.41
3	40	53.8	63.46	−9.66
4	48	55.7	66.32	−10.62
5	64	85.5	72.04	13.46
6	94	84.6	82.78	1.82
7	118	84.9	91.37	−6.47
8	125	82.8	93.88	−11.08
9	168	123.2	109.27	13.93
10	195	115.8	118.93	−3.13

TABLE 50.2

Random Numbers from 1 to 10 that were Generated to Resample the Residuals in Table 50.1

Resample	Random Number (from 1 to 10)									
1	9	6	8	9	1	2	2	8	2	7
2	1	9	8	7	3	2	5	2	2	10
3	10	4	8	7	6	6	8	2	7	1
4	3	10	8	5	7	2	7	9	6	3
5	7	3	10	4	4	1	7	6	5	9

TABLE 50.3

New Residuals and Data Generated by Resampling, with Replacement, Using the Random Numbers in Table 50.2 and the Residuals in Table 50.1

Random No.	9	6	8	9	1	2	2	8	2	7
Residual	13.93	1.82	−11.08	13.93	6.33	5.41	5.41	−11.08	5.41	−6.47
New y	77.63	65.32	42.72	69.63	91.83	90.01	90.31	71.72	128.61	109.33
Random No.	1	9	8	7	3	2	5	2	2	10
Residual	6.33	13.93	−11.08	−6.47	−9.66	5.41	13.46	5.41	5.41	−3.13
New y	70.03	77.43	42.72	49.23	75.84	90.01	98.36	88.21	128.61	112.67
Random No.	10	4	8	7	6	6	8	2	7	1
Residual	−3.13	−10.62	−11.08	−6.47	1.82	1.82	−11.08	5.41	−6.47	6.33
New y	60.57	52.88	42.72	49.23	87.32	86.42	73.82	88.21	116.73	122.13
Random No.	3	10	8	5	7	2	7	9	6	3
Residual	−9.66	−3.13	−11.08	13.46	−6.47	5.41	−6.47	13.93	1.82	−9.66
New y	54.04	60.37	42.72	69.16	79.03	90.01	78.43	96.73	125.02	106.14
Random No.	7	3	10	4	4	1	7	6	5	9
Residual	−6.47	−9.66	−3.33	−10.62	−10.62	6.33	−6.47	1.82	13.46	13.93
New y	57.23	53.84	50.67	45.08	74.88	90.93	78.43	84.62	136.66	129.73

Comments

Another use of simulation is to test the consequences of violation the assumptions on which a statistical procedure rests. A good example is provided by Box et al. (1978) who used simulation to study how nonnormality and serial correlation affect the performance of the *t*-test. The effect of nonnormality was not very serious. In a case where 5% of tests should have been significant, 4.3% were significant for

TABLE 50.4

Parameter Estimates for the Original Data and for Five Sets of New Data Generated by Resampling the Residuals in Table 50.1

Data Set	b_0	b_1
Original	49.14	0.358
Resample 1	56.22	0.306
Resample 2	49.92	0.371
Resample 3	41.06	0.410
Resample 4	46.68	0.372
Resample 5	36.03	0.491

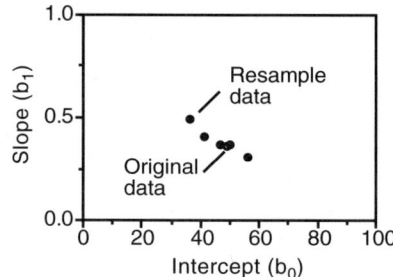

FIGURE 50.3 Emerging joint confidence region based the original data plus five new sets generated by resampling, with replacement.

normally distributed data, 6.0% for a rectangular parent distribution, and 5.9% for a skewed parent distribution. The effect of modest serial correlation in the data was much greater than these differences due to nonnormality. A positive autocorrelation of $r = 0.4$ inflated the percentage of tests found significant from the correct level of 5% to 10.5% for the normal distribution, 12.5% for a rectangular distribution, and 11.4% for a skewed distribution. They also showed that randomization would negate the autocorrelation and give percentages of significant results at the expected level of about 5%. Normality, which often causes concern, turns out to be relatively unimportant while serial correlation, which is too seldom considered, can be ruinous.

The bootstrap method is a special form of simulation that is based on resampling with replacement. It can be used to investigate the properties of any statistic that may have unusual properties or one for which a convenient analytical solution does not exist.

Simulation is familiar to most engineers as a design tool. Use it to explore and discover unknown properties of unfamiliar statistics and to check the performance of statistical methods that might be applied to data with nonideal properties. Sometimes we find that our worries are misplaced or unfounded.

References

Berthouex, P. M. and I. Hau (1991). "Difficulties in Using Water Quality Standards Based on Extreme Percentiles," *Res. J. Water Pollution Control Fed.*, 63(5), 873–879.

Box, G. E. P., W. G. Hunter, and J. S. Hunter (1978). *Statistics for Experimenters: An Introduction to Design, Data Analysis, and Model Building,* New York, Wiley Interscience.

Draper, N. R. and H. Smith, (1998). *Applied Regression Analysis,* 3rd ed., New York, John Wiley.

Hahn, G. J. and S. S. Shapiro (1967). *Statistical Methods for Engineers,* New York, John Wiley.

Metcalf, A. V. (1997). *Statistics in Civil Engineering,* London, Arnold.

Peigorsch, W. W. and A. J. Bailer (1997). *Statistics for Environmental Biology and Toxicology,* New York, Chapman & Hall.

Press, W. H., B. P. Flannery, S. A. Tenkolsky, and W. T. Vetterling (1992). *Numerical Recipes in FORTRAN: The Art of Scientific Computing,* 2nd ed., Cambridge, England, Cambridge University Press.

Exercises

50.1 Limit of Detection. The Method Limit of Detection is calculated using MDL = $3.143s$, where s is the standard deviation of measurements on seven identical aliquots. Use simulation to study how much the MDL can vary due to random variation in the replicate measurements if the true standard deviation is $\sigma = 0.4$.

50.2 Nonconstant Variance. Chapter 37 on weighted least squares discussed a calibration problem where there were three replicate observations at several concentration levels. By how much can the variance of triplicate observations vary before one would decide that there is nonconstant variance? Answer this by simulating 500 sets of random triplicate observations, calculating the variance of each set, and plotting the histogram of estimated variances.

50.3 Uniform Distribution. Data from a process is discovered to have a uniform distribution with mean 10 and range 2. Future samples from this process will be of size $n = 10$. By simulation, determine the reference distribution for the standard deviation, the standard error of the mean, and the 95% confidence interval of the mean for samples of size $n = 10$.

50.4 Regression. Extend the example in Table 50.3 and add five to ten more points to Figure 50.3.

50.5 Bootstrap Confidence Intervals. Fit the exponential model $y = \theta_1 \exp(-\theta_2 x)$ to the data below and use the bootstrap method to determine the approximate joint confidence region of the parameter estimates.

x	1	4	8	10	11
y	179	104	51	35	30

Optional: Add two observations ($x = 15$, $y = 14$ and $x = 18$, $y = 8$) to the data and repeat the bootstrap experiment to see how the shape of the confidence region is changed by having data at larger values of x.

50.6 Legal Statistics. Find an unfamiliar or unusual statistic in a state or U.S. environmental regulation and discover its properties by simulation.

50.7 99th Percentile Distribution. A quality measure for an industrial discharge (kg/day of TSS) has a lognormal distribution with mean 3000 and standard deviation 2000. Use simulation to construct a reference distribution of the 99th percentile value of the TSS load. From this distribution, estimate an upper 90% confidence limit for the 99th percentile.

51

Introduction to Time Series Modeling

KEY WORDS *ARIMA model, ARMA model, AR model, autocorrelation, autocorrelation function, autoregressive model, cross-correlation, integrated model, IMA model, intervention analysis, lag, linear trend, MA model, moving average model, nonstationary, parsimony, seasonality, stationary, time series, transfer function.*

A time series of a finite number of successive observations consists of the data $z_1, z_2, \ldots, z_{t-1}, z_t, z_{t+1}, \ldots, z_n$. Our discussion will be limited to discrete (sampled-data) systems where observations occur at equally spaced intervals. The z_t may be known precisely, as the price of IBM stock at the day's closing of the stock market, or they may be measured imperfectly, as the biochemical oxygen demand (BOD) of a treatment plant effluent. The BOD data will contain a component of measurement error; the IBM stock prices will not. In both cases there are forces, some unknown, that nudge the series this way and that. The effect of these forces on the system can be "remembered" to some extent by the process. This memory makes adjacent observations dependent on the recent past. Time series analysis provides tools for analyzing and describing this dependence. The goal usually is to obtain a model that can be used to forecast future values of the same series, or to obtain a transfer function to predict the value of an output from knowledge of a related input.

Time series data are common in environmental work. Data may be monitored frequently (pH every second) or at long intervals and at regular or irregular intervals. The records may be complete, or have missing data. They may be homogeneous over time, or measurement methods may have changed, or some intervention has shifted the system to a new level (new treatment plant or new flood control dam). The data may show a trend or cycle, or they may vary about a fixed mean value. All of these are possible complications in times series data. The common features are that time runs with the data and we do not expect neighboring observations to be independent. Otherwise, each time series is unique and its interpretation and modeling are not straightforward. It is a specialty. Most of us will want a specialist's help even for simple time series analysis. The authors have always teamed with an experienced statistician on these jobs.

Some Examples of Time Series Analysis

We have urged plotting the data before doing analysis because this usually provides some hints about the model that should be fitted to the data. It is a good idea to plot time series data as well, but the plots usually do not reveal the form of the model, except perhaps that there is a trend or seasonality. The details need to be dug out using the tools of time series analysis.

Figure 51.1 shows influent BOD data measured every 2 hours over a 10-day period in Madison, Wisconsin. Daytime to nighttime variation is clear, but this cycle is not a smooth harmonic (sine or cosine). One day looks pretty much like another although a long record of daily average data shows that Sundays are different than the other days of the week.

Figure 51.2 shows influent temperature at the Deer Island Treatment Plant in Boston. There is a smooth annual cycle. The positive correlation between successive observations is very strong, high temperatures being followed by more high temperatures. Figure 51.3 shows effluent pH at Deer Island; the pH drifts over a narrow range and fluctuates from day to day by several tenths of a pH unit. Figure 51.4 shows

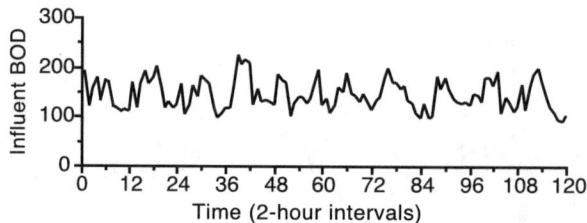

FIGURE 51.1 Influent BOD at the Nine Springs Wastewater Treatment Plant, Madison, WI.

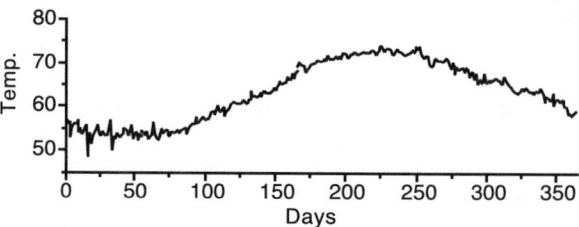

FIGURE 51.2 Influent temperature at Deer Island Wastewater Treatment Plant, Boston, for the year 2000.

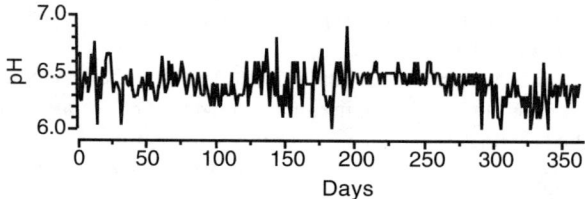

FIGURE 51.3 Effluent pH at the Deer Island Wastewater Treatment Plant for the year 2000.

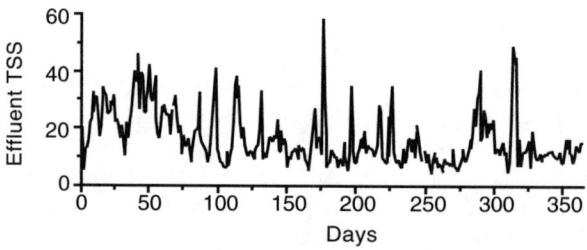

FIGURE 51.4 Effluent suspended solids at the Deer Island Wastewater Treatment Plant for the year 2000.

Deer Island effluent suspended solids. The long-term drift over the year is roughly the inverse of the temperature cycle. There is also "spiky" variation. The spikes occurred because of an intermittent physical condition in the final clarifiers (the problem has been corrected). An ordinary time series model would not be able to capture the spikes because they erupt at random.

Figure 51.5 shows a time series of phosphorus data for a Canadian river (Hipel et al., 1986). The data are monthly averages that run from January 1972 to December 1977. In February 1974, an important

Introduction to Time Series Modeling

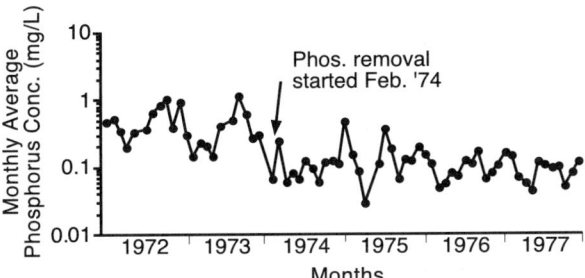

FIGURE 51.5 Phosphorus data for a Canadian river showing an intervention that reduced the P concentration after February 1974.

wastewater treatment plant initiated phosphorus removal and the nature of the time series changed abruptly. A time series analysis of this data needs to account for this intervention. Chapter 54 discusses *intervention analysis*.

Each of these time series has correlation of adjacent or nearby values within the time series. This is called *autocorrelation*. Correlation between two time series is *cross-correlation*. There is cross-correlation between the effluent suspended solids and temperature in Figures 51.2 and 51.4.

These few graphs show that a time series may have a trend, a cycle, an intervention shift, and a strong random component. Our eye can see the difference but not quantify it. We need some special tools to characterize and quantify the special properties of time series. One important tool is the *autocorrelation function* (ACF). Another is the *ARIMA* class of time series models.

The Autocorrelation Function

The *autocorrelation function* is the fundamental tool for diagnosing the structure of a time series. The correlation of two variables (x and y) is:

$$r(x,y) = \frac{\sum (x_i - \bar{x})(y_i - \bar{y})}{\sqrt{\sum (x_i - \bar{x})^2 \sum (y_i - \bar{y})^2}}$$

The denominator scales the correlation coefficient so $-1 \leq r(x, y) \leq 1$.

In a time series, adjacent and nearby observations are correlated, so we want a correlation of z_t and z_{t-k}, where k is the *lag* distance, which is measured as the number of sampling intervals between the observations. For lag = 2, we correlate z_1 and z_3, z_2 and z_4, etc. The general formula for the sample autocorrelation at lag k is:

$$r_k = \frac{\sum_{t=k+1}^{n} (z_t - \bar{z})(z_{t-k} - \bar{z})}{\sum_{t=1}^{n} (z_t - \bar{z})^2}$$

where n is the total number of observations in the time series. The sample autocorrelation (r_k) estimates the population autocorrelation (ρ_k). The numerator is calculated with a few less terms than n; the denominator is calculated with n terms. Again, the denominator scales the correlation coefficient so it falls in the range $-1 \leq r_x \leq 1$.

The autocorrelation function is the collection of r_k's for $k = 0, 1, 2, \ldots, m$, where m is not larger than about $n/4$. In practice, at least 50 observations are needed to estimate the autocorrelation function (ACF).

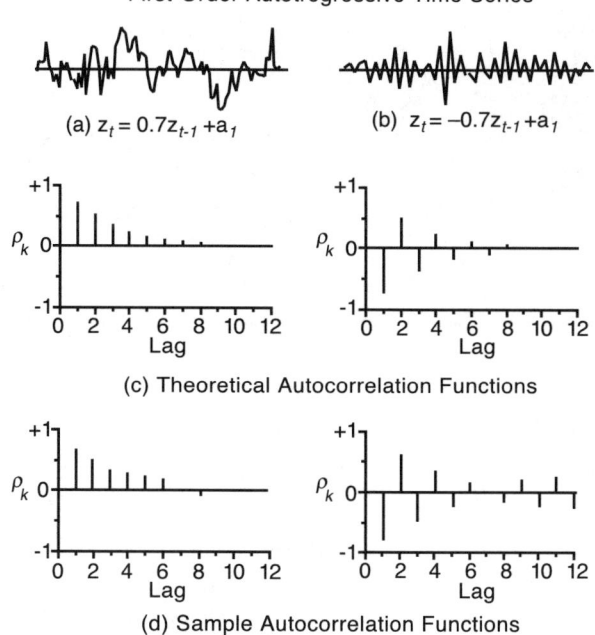

FIGURE 51.6 Two first-order autoregressive time series (a and b) with their theoretical (c) and sample (d) autocorrelation functions.

Figure 51.6 shows two first-order *autoregressive* time series, $z_t = 0.7z_{t-1} + a_t$ on the left and $z_t = -0.7z_{t-1} + a_t$ on the right. Autoregressive processes are regressions of z_t on recent past observations; in this case the regression is z_t on z_{t-1}. The theoretical and sample autocorrelation functions are shown. The alternating positive-negative-positive pattern on the right indicates a negative correlation and this is reflected in the negative coefficient (-0.7) in the model.

If we calculate the autocorrelation function for $k = 1$ to 25, we expect that many of the 25 r_k values will be small and insignificant (in both the statistical and practical sense). The large values of r_k give information about the structure of the time series process; the small ones will be ignored. We expect that beyond some value of k the correlation will "die out" and can be ignored.

We need a check on whether the r_k are small enough to be ignored. An approximation of the variance of r_k (for all k) is:

$$\text{Var}(r_k) \cong 1/n$$

where n is the length of the series. The corresponding standard error is $\sqrt{1/n}$ and the approximate 95% confidence interval for the r_k is $\pm 1.96/\sqrt{n}$. A series of $n \geq 50$ is needed to get reliable estimates, so the confidence interval will be ± 0.28 or less. Any r_k smaller than this can be disregarded.

The autocorrelation function produced by commercial software for time series analysis (e.g., Minitab) will give confidence intervals that are better than this approximation, but the equation above provides sufficient information for our introduction.

If the series is purely random noise, all the r_k will be small. This is what we expect of the residuals of any fitted model, and one check on the adequacy of the model is independent residuals (i.e., no correlation).

Figure 51.7 is the autocorrelation function for the first 200 observations in the Deer Island pH series of Figure 51.3. The dotted lines are the approximate 95% confidence limits. The ACF has almost all positive correlations and they do not die out within a reasonable number of lags. This indicates that the time series has a linear trend (upward because the correlations are positive). A later section (nonstationary processes) explains how this trend is handled.

Introduction to Time Series Modeling

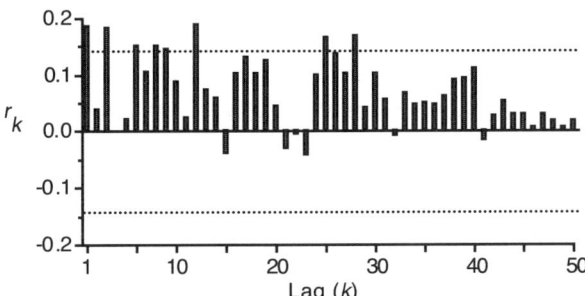

FIGURE 51.7 Autocorrelation function for the Deer Island pH series of Figure 53.3. The dotted lines are the approximate 95% confidence intervals.

The ARIMA Family of Time Series Models

We will examine the family of autoregressive integrated moving average (ARIMA) models. ARIMA describes a collection of models that can be very simple or complicated. The simple models only use the autoregressive (AR) part of the ARIMA structure, and some only use the moving average (MA) part. AR models are used to describe stationary time series — those that fluctuate about a fixed level. MA models are used to describe nonstationary processes — those that drift and do not have a fixed mean level, except as an approximation over a short time period. More complicated models integrate the AR and MA features. They also include features that deal with drift, trends, and seasonality. Thus, the ARIMA structure provides a powerful and flexible collection of models that can be adapted to all kinds of time series.

The ARIMA models are parsimonious; they can be written with a small number of parameters. The models are useful for forecasting as well as interpreting time series. Parameter estimation is done by an iterative minimization of the sum of squares, as in nonlinear regression.

Time Series Models for a Stationary Process

A time series that fluctuates about a fixed mean level is said to be *stationary*. The data describe, or comes from, a stationary process. The data properties are unaffected by a change of the time origin. The sample mean and sample variance of a stationary process are:

$$\bar{z} = \frac{1}{n}\sum_{t=1}^{n} z_t \quad \text{and} \quad \hat{\sigma}^2 = \frac{1}{n}\sum_{t=1}^{n}(z_t - \bar{z})$$

The correlation may be positive or negative, and it may change from positive to negative as the lag distance increases.

For a stationary process, the relation between z_t and z_{t+k} is the same for all times t across the series. That is, the nature of the relationship between observations separated by a constant time can be inferred by making a scatterplot using pairs of values (z_t, z_{t+k}) separated by a constant time interval or lag. Direction in time is unimportant in determining autocorrelation in a stationary process, so we could just as well plot (z_t, z_{t-k}).

Stationary processes are described by *autoregressive models*. For the case where p recent observations are relevant, the model is:

$$z_t = \phi_1 z_{t-1} + \phi_2 z_{t-2} + \cdots + \phi_p z_{t-p} + a_t$$

where $|\phi| < 1$ is necessary for stationarity. It represents z_t, the value of the time series at time t, as a weighted sum of p previous values plus a random component a_t. The mean of the actual time series has been subtracted out so that z_t has a zero mean value. This model is abbreviated as AR(p).

A first-order autoregressive process, abbreviated AR(1), is:

$$z_t = \phi z_{t-1} + a_t$$

A second-order autoregressive process, AR(2), has the form:

$$z_t = \phi_1 z_{t-1} + \phi_2 z_{t-2} + a_t$$

It can be shown that the autocorrelation function for an AR(1) model is related to the value of ϕ, as follows:

$$\rho_1 = \phi, \quad \rho_2 = \phi^2, \ldots, \rho_k = \phi^k$$

Knowing the autocorrelation function therefore provides an initial estimate of ϕ. This estimate is improved by fitting the model to the data. Because $|\phi| < 1$, the autocorrelation function is exponentially decreasing as the number of lags increases. For ϕ near ± 1, the exponential decay is quite slow; but for smaller ϕ, the decay is rapid. For example, if $\phi = 0.4$, $\rho_4 = 0.4^4 = 0.0256$ and the correlation has become negligible after four or five lags. If $0 < \phi < 1$, all correlations are positive; if $-1 < \phi < 0$, the lag 1 autocorrelation is negative ($\rho_1 = \phi$) and the signs of successive correlations alternate from positive to negative with their magnitudes decreasing exponentially.

Time Series Models for a Nonstationary Process

Nonstationary processes have no fixed mean. They "drift" and this drift is described by the nonstationary *moving average* models. A moving average model expresses the current value of the time series as a weighted sum of a finite number of previous random inputs.

For a moving average process[1] where only the first q terms are significant:

$$z_t = a_t - \theta_1 a_{t-1} - \theta_2 a_{t-2} - \cdots - \theta_q a_{t-q}$$

This is abbreviated as MA(q). Moving average models have been used for more than 60 years (e.g., Wold, 1938).

A first-order moving average model MA(1) is:

$$z_t = a_t - \theta a_{t-1}$$

The autocorrelation function for an MA(1) model is a single spike at lag 1, and zero otherwise. That is:

$$\rho_1 = \frac{-\theta}{1+\theta^2} \quad \text{for} \quad k = 1 \quad \text{and} \quad \rho_k = 0 \quad \text{for} \quad k > 2.$$

The largest value possible for ρ_1 is 0.5, which occurs when $\theta = -1$, and the smallest value is -0.5 when $\theta = 1$. Unfortunately, the value ρ_1 is the same for θ and $1/\theta$, so knowing ρ_1 does not tell us the precise value of θ.

---

[1] The name *moving average* is somewhat misleading because the weights $1, -\theta_1, -\theta_2, \ldots, \theta_\theta$ need not total unity nor need they be positive (Box et al., 1994).

Introduction to Time Series Modeling

The second-order model MA(2) is:

$$z_t = a_t - \theta_1 a_{t-1} - \theta_2 a_{t-2}$$

We have been explaining these models with a minimal amount of notation, but the "backshift operator" should be mentioned. The MA(1) model can also be represented as $z_t = (1 - \theta B)a_t$, where B is an operator that means "backshift"; thus, $Ba_t = a_{t-1}$ and $(1 - \theta B)a_t = a_t - \theta a_{t-1}$. Likewise, the AR(1) model could be written $(1 - \phi B)z_t = a_t$. The AR(2) model could be written as $(1 - \phi_1 B - \phi_2 B^2)z_t = a_t$.

The Principle of Parsimony

If there are two equivalent ways to express a model, we should choose the most parsimonious expression, that is, the form that uses the fewest parameters.

The first-order autoregressive process $z_t = \phi z_{t-1} + a_t$ is valid for all values of t. If we replace t with $t-1$, we get $z_{t-1} = \phi z_{t-2} + a_{t-1}$. Substituting this into the original expression gives:

$$z_t = \phi(\phi z_{t-2} + a_{t-1}) + a_t = a_t + \phi a_{t-1} + \phi^2 z_{t-2}$$

If we repeat this substitution $k - 1$ times, we get:

$$z_t = a_t + \phi a_{t-1} + \phi^2 a_{t-2} + \cdots + \phi^{k-1} a_{t-(k-1)} + \phi^k z_{t-k}$$

Because $|\phi| < 1$, the last term at some point will disappear and z_t can be expressed as an infinite series of present and past terms of a decaying white noise series:

$$z_t = a_t + \phi a_{t-1} + \phi^2 a_{t-2} + \phi^3 a_{t-3} + \cdots$$

This is an infinite moving average process. This important result allows us to express an infinite moving average process in the parsimonious form of an AR(1) model that has only one parameter.

In a similar manner, the finite moving average process described by an MA(1) model can be written as a infinite autoregressive series:

$$z_t = a_t - \theta_1 z_{t-1} - \theta_2 z_{t-2} - \theta_3 z_{t-3} - \cdots$$

Mixed Autoregressive–Moving Average Processes

We saw that the finite first-order autoregressive model could be expressed as an infinite series moving average model, and vice versa. We should express a first-order model with one parameter, either θ in the moving average form or ϕ in the autoregressive form, rather than have a large series of parameters. As the model becomes more complicated, the parsimonious parameterization may call for a combination of moving average and autoregressive terms. This is the so-called *mixed autoregressive–moving average process*. The abbreviated name for this process is ARMA(p,q), where p and q refer to the number of terms in each part of the model:

$$z_t = \phi_1 z_{t-1} + \cdots + \phi_p z_{t-p} + a_t - \theta_1 a_{t-1} - \cdots - \theta_q a_{t-q}$$

The ARMA(1,1) process is:

$$z_t = \phi_1 z_{t-1} + a_t - \theta_1 a_{t-1}$$

Integrated Autoregressive–Moving Average Processes

A *nonstationary process* has no fixed central level, except perhaps as an approximation over a short period of time. This drifting or trending behavior is common in business, economic, and environmental data, and also in controlled manufacturing systems. If one had to make an *a priori* prediction of the character of a times series, the best bet would be nonstationary. There is seldom a hard and fast reason to declare that a time series will continue a deterministic trend forever.

A nonstationary series can be converted into a stationary series by differencing. An easily understood case of nonstationarity is an upward trend. We can flatten the nonstationary series by taking the difference of successive values. This removes the trend and leaves a series of stationary, but still autocorrelated, values. Occasionally, a series has to be differenced twice (take the difference of the differences) to produce a stationary series. The differenced stationary series is then analyzed and modeled as a stationary series.

A time series is said to follow an *integrated autoregressive–moving average process*, abbreviated as ARIMA(p,d,q), if the dth difference is a stationary ARMA(p,q) process. In practice, d is usually 1 or 2.

We may also have an ARIMA($0,d,q$), or IMA(d,q), integrated moving average model. The IMA(1,1) model:

$$z_t = z_{t-1} + a_t - \theta a_{t-1}$$

satisfactorily represents many times series, especially in business and economics. The difference is reflected in the term $z_t - z_{t-1}$.

Another note on notation that is used in the standard texts on times series: a first difference is denoted by ∇. The IMA(1,1) model can be written as $\nabla z_t = a_t - \theta a_{t-1}$ where ∇ indicates the first difference ($\nabla z_t = z_t - z_{t-1}$). A even more compact notation is $\nabla z_t = (1 - \theta B)a_t$. A second difference is indicated as ∇^2.

The IMA(2,2) model is $z_t = 2z_{t-1} - z_{t-2} + a_t - \theta_1 a_{t-1} - \theta_2 a_{t-2}$. The term $2z_{t-1}$ comes from taking the second difference. The first difference of z_t is $\nabla z_t = z_t - z_{t-1}$, and the difference of this is:

$$\nabla^2 z_t = (z_t - z_{t-1}) - (z_{t-1} - z_{t-2}) = z_t - 2z_{t-1} + z_{t-1}$$

Seasonality

Seasonal models are important in environmental data analysis, especially when dealing with natural systems. They are less important when dealing with treatment process and manufacturing system data. What appears to be seasonality in these data can often be modeled as drift using one of the simple ARIMA models. Therefore, our discussion of seasonal models will be suggestive and the interested reader is directed to the literature for details.

Some seasonal patterns can be modeled by introducing a few lag terms. If there is a 7-day cycle (Sundays are like Sundays, and Mondays like Mondays), we could try a model with z_{t-7} or a_{t-7} terms. A 12-month cycle in monthly data would call for consideration of z_{t-12} or a_{t-12} terms.

In some cases, seasonal patterns can be modeled with cosine curves that incorporate the smooth change expected from one time period to the next. Consider the model:

$$y_t = \beta \cos(2\pi f t + \varphi)$$

where β is the amplitude, f is the frequency, and φ is the phase shift. As t varies, the curve oscillates between a maximum of β and a minimum of $-\beta$. The curve repeats itself every $1/f$ time units. For monthly data, $f = 1/12$. The phase shift serves to set the origin on the t-axis.

The form $y_t = \beta \cos(2\pi f t + \varphi)$ is inconvenient for model fitting because the parameters β and φ do not enter the expression linearly. A trigonometric identity that is more convenient is:

$$\beta \cos(2\pi f t + \varphi) = \beta_1 \cos(2\pi f t) + \beta_2 \sin(2\pi f t)$$

Introduction to Time Series Modeling

where β_1 and β_2 are related to β and φ. For monthly data with an annual cycle and average β_0, this model becomes:

$$y_t = \beta_0 + \beta_1 \cos(2\pi t/12) + \beta_2 \sin(2\pi t/12) + a_t$$

Cryer (1986) shows how this model is used.

To go beyond this simple description requires more than the available space, so we leave the interested reader to consult additional references. The classic text is Box, Jenkins and Reinsel (1994). Esterby (1993) discusses several ways of dealing with seasonal environmental data. Pandit and Wu (1983) discuss applications in engineering.

Fitting Time Series Models

Fitting time series models is much like fitting nonlinear mechanistic models by nonlinear least squares. The form of the model is specified, an initial estimate is made for each parameter value, a series of residuals is calculated, and the minimum sum of squares is located by an iterative search procedure. When convergence is obtained, the residuals are examined for normality, independence, and constant variance. If these conditions are satisfied, the final parameter values are unbiased least squares estimates.

Deer Island Effluent pH Series

The autocorrelation function (ACF) of the pH series does not die out and it has mostly positive correlations (Figure 51.7). This indicates a nonstationary series. Figure 51.8 shows that the ACF of the differenced series essentially dies out after one lag; this indicates that the model should include a first difference and a first-order dependence. The IMA(1,1) model is suggested.

The fitted IMA(1,1) model:

$$y_t = y_{t-1} + a_t - 0.9 a_{t-1}$$

has a mean residual sum of squares (MS) of 0.023. The fitted IMA(1,1) model is shown in Figure 51.9. The residuals (not shown) are random and independent. By comparison, an ARIMA(1,1,1) model has MS = 0.023, and an ARI(1,1) model has MS = 0.030.

We can use this example to see why we should often prefer time series models to deterministic trend models. Suppose the first 200 pH values shown in the left-hand panel of Figure 51.10 were fitted by a straight line to get pH = 6.49 + 0.00718t. The line looks good over the data. The t statistic for the slope is 4.2, R^2 = 0.09 is statistically significant, and the mean square error (MSE) is 0.344. From these statistics you might wrongly conclude that the linear model is adequate.

The flaw is that there is no reason to believe that a linear upward trend will continue indefinitely. The extended record (Figure 51.9) shows that pH drifts between pH 6 and 7, depending on rainfall, snow

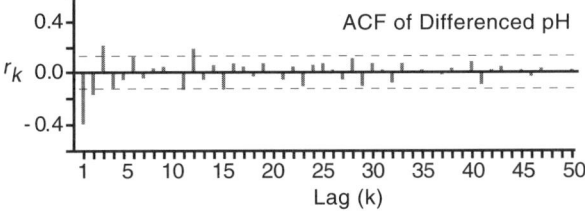

FIGURE 51.8 The autocorrelation function of the differenced pH series decays to negligible values after a few lags.

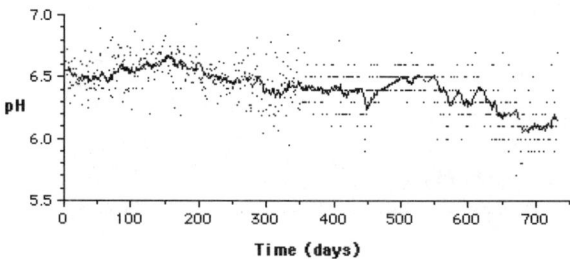

FIGURE 51.9 IMA(1,1) model $y_t = y_{t-1} + a_t - 0.9a_{t-1}$ fitted to the Deer Island pH series. Notice that the data recording format changed from two decimal places prior to day 365 and to one place thereafter.

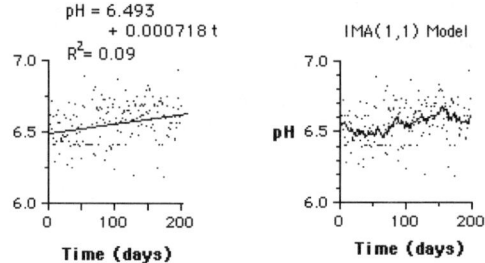

FIGURE 51.10 Comparison of deterministic and IMA(1,1) models fitted to the first 200 pH observations. The deterministic model appears to fit the data, but it in inadequate in several ways, mainly by not being able to follow the downward drift that occurs within a short time.

melt, and other factors and that around day 200 it starts to drift downward. A linear model fitted to the entire data series (Figure 51.9) has a negative slope. R^2 increases to 0.6 because there are so many data points. But the linear model is not sufficient. It is not adaptable, and the pH is not going to keep going up or down. It is going to drift.

The fitted IMA(1,1) model shown in the right-hand panel of Figure 51.10 will adapt and follow the drift. The mean square error of the IMA model is 0.018, compared with MSE = 0.344 for the linear model ($n = 200$).

Madison Flow Data

Figure 51.11 shows 20 years of monthly average wastewater flows to the Madison Nine Springs Wastewater Treatment Plant and the fitted time series model:

$$z_t = 0.65 z_{t-1} + 0.65 z_{t-12} - z_{t-13} + a_t - 0.96 a_{t-1}$$

Because April in one year tends to be like April in another year, and one September tends to be like another, there is a seasonality in the data. There is also an upward trend due to growth in the service area. The series is nonstationary and seasonal. Both characteristics can be handled by differencing, a one-lag difference to remove the trend and a 12-month lag difference for the seasonal pattern.

The residuals shown in Figure 51.11 are random. Also, they have constant variance except for a few large positive values near the end of the series. These correspond to unusual events that the time series model cannot capture. The sewer system is not intended to collect stormwater and under normal conditions it does not, except for small quantities of infiltration. However, stormwater inflow and infiltration increase noticeably if high groundwater levels are combined with a heavy rainfall. This is what caused the exceptionally high flows that produced the large residuals. If one wanted to try and model these extremes, a dummy variable could be used to indicate special conditions. It is true of most models, but especially of regression-type models that they tend to overestimate extreme low values and

Introduction to Time Series Modeling

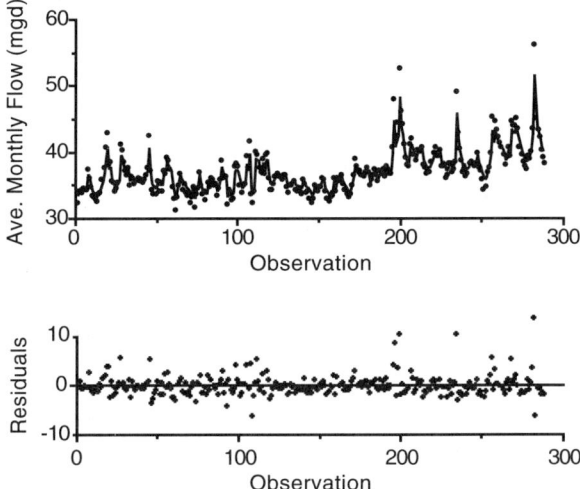

FIGURE 51.11 A seasonal nonstationary model fitted to a 20-yr record of average monthly wastewater flows for Madison, WI. The residual plot highlights a few high-flow events that the model cannot capture.

underestimate extreme high values. A mechanistic model may better predict extreme values, but only if the correct mechanisms have been included in the model.

Comments

Time series are one of the most common forms of environmental data. Time series analysis, unfortunately, can be frustrating and mysterious and therefore we tend to avoid it. Perhaps this chapter has removed some of the mystery.

This chapter is an introduction to a difficult subject. Only the simplest models and concepts have been mentioned. Many practical problems contain at least one feature that is "nonstandard." The spikes in the effluent suspended solids and Madison flow data are examples. Nonconstant variance or missing data are two others. Use common sense to recognize the special features of a problem. Then find expert help when you need it.

Some statistical methods can be learned quickly by practicing a few example calculations. Regression and t-tests are like this. Time series analysis is not. It takes a good deal of practice; experience is priceless. If you want to learn, start by generating time series and then progress to fitting time series that have been previously studied by specialists. Minitab is a good package to use.

When you need to do time series analysis, do not be shy about consulting a professional statistician who is experienced in doing time series modeling. You will save time and learn at the side of a master. That is a bargain not to be missed.

References

Box, G. E. P., G. M. Jenkins, and G. C. Reinsel (1994). *Time Series Analysis, Forecasting and Control*, 3rd ed., Englewood Cliffs, NJ, Prentice-Hall.

Cryer, J. D. (1986). *Time Series Analysis*, Boston, Duxbury Press.

Esterby, S. R. (1993). "Trend Analysis Methods for Environmental Data," *Environmetrics*, 4(4) 459–481.

Hipel, K. W., A. I. McLeod, and P. K. Fosu (1986). "Empirical Power Comparison of Some Tests for Trend," pp. 347–362, in *Statistical Aspects of Water Quality Monitoring*, A. H. El-Shaarawi and R. E. Kwiatowski Eds., Amsterdam, Elsevier.

Pandit, S. M. and S. M. Wu (1983). *Time Series and Systems Analysis with Applications*, New York, John Wiley.

Wold, H. O. A. (1938). *A Study in the Analysis of Stationary Time Series,* 2nd ed. (1954), Uppsala, Sweden, Almquist and Wiskell.

Exercises

51.1 BOD Trend Analysis. The table below gives 16 years of BOD loading (lb/day) data for a municipal wastewater treatment plant. Plot the data. Difference the data to remove the upward trend. Examine the differenced data for a seasonal cycle. Calculate the autocorrelation function of the differenced data. Fit some simple times series models to the data.

Month	1991	1992	1993	1994	1995	1996	1997	1998	1999	2000
Jan.	63435	61670	57963	61250	70430	64166	65286	70499	68341	74237
Feb.	60846	68151	58334	66376	69523	68157	68596	76728	74079	79884
Mar.	60688	62522	60535	68986	75072	64836	68787	71968	70185	75395
Apr.	58131	67914	66606	65781	64084	66916	69196	74188	76514	74362
May	62700	65339	62560	58731	64958	67086	68337	66269	71019	74906
June	51305	60336	64060	65650	68022	68241	60917	73624	70342	71035
July	54603	60078	61180	59509	70979	68366	61938	69896	69160	76591
Aug.	58347	57836	55476	59797	74727	73214	68481	65997	72799	78417
Sept.	57591	54395	57522	67213	67948	68794	80666	70790	69912	76859
Oct.	59151	56723	57563	60132	65956	71463	73281	72233	71734	78826
Nov.	57777	67361	61846	63791	67001	70107	67404	72166	73614	73718
Dec.	61707	57618	59829	66234	65105	63205	68807	71104	75573	73825

51.2 Phosphorus Loading. The table below gives 16 years of phosphorus loading (lb/day) data for a municipal wastewater treatment plant. Interpret the data using time series analysis.

Month	1993	1994	1995	1996	1997	1998	1999	2000
Jan.	2337	2089	2095	2094	2040	2108	2078	2101
Feb.	2187	2243	2059	2102	2049	2159	2171	2202
Mar.	2117	2368	2132	2191	2014	2216	2201	2234
Apr.	2218	2219	2069	2179	2024	2215	2157	2100
May	2055	2097	1935	2226	2094	2205	2089	2141
June	2088	2116	2023	2250	1946	2355	2151	2001
July	2138	2131	2041	2176	2020	2152	2038	2049
Aug.	2154	2082	2107	2202	1997	2101	1988	2104
Sept.	2226	2167	2063	2180	2202	2179	2067	2001
Oct.	2213	2111	2104	2236	2110	2117	2135	2108
Nov.	2154	2029	2037	2114	2036	2203	2059	2083
Dec.	2169	2097	2007	1906	1954	2097	2040	2066

52

Transfer Function Models

KEY WORDS   chemical reaction, CSTR, difference equations, discrete model, dynamic system, empirical model, impulse, material balance, mechanistic model, lag, step change, stochastic model, time constant, transfer function, time series.

The *transfer function models* in this chapter are simple dynamic models that relate time series output of a process to the time series input. Our examples have only one input series and one output series, but the basic concept can be extended to deal with multiple inputs.

In the practical sense, the only models we care about are those classified as *useful* and *adequate*. On another level we might try to classify models as mechanistic vs. empirical, or as deterministic vs. stochastic. The categories are not entirely exclusive. A simple power function $y = \theta_1 x^{\theta_2}$ looks like an empirical model (a mathematical French curve) until we see it in the form $e = mc^2$. We will show next how a mechanistic model could masquerade as an empirical stochastic model and explain why empirical stochastic models can be a good way to describe real processes.

Case Study: A Continuous Stirred Tank Reactor Model

The material balance on a continuous stirred tank reactor (CSTR) is a continuous *mechanistic model*:

$$V \frac{dy}{dt} = Qx - Qy$$

where x is the influent concentration and y is the effluent concentration. Notice that there is no reaction so the average input concentration and the average effluent concentration are equal over a long period of time. Defining a time constant $\tau = V/Q$, this becomes:

$$\tau \frac{dy}{dt} = x - y$$

We know that the response of a mixed reactor to an impulse (spike) in the input is exponential decay from a peak that occurs at the time of the impulse. Also, the response to a step change in the input is an exponential approach to an asymptotic value.

The left-hand panel of Figure 52.1 shows ideal input and output for the CSTR. "Ideal" means the x and y are perfectly observed. The ideal output y_t was calculated using $y_t = 0.2x_t + 0.8y_{t-1}$. The right-hand panel looks more like measured data from a real process. The input x_t was the idealized pattern shown in Figure 52.1 plus random noise from a normal distribution with mean zero and $\sigma = 2$ (i.e., N(0,2)). The output with noise was $y_t = 0.2x_t + 0.8y_{t-1} + a_t$, where a_t was N(0,4). Time series *transfer function* modeling deals with the kind of data shown in the right-hand panel.

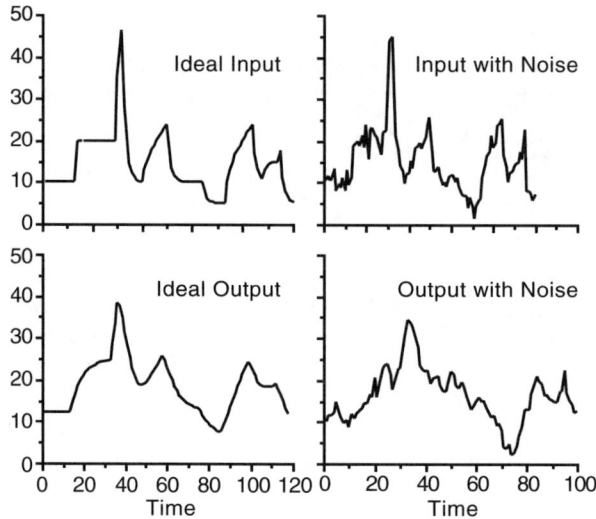

FIGURE 52.1 The left-hand panels show the ideal input and output of a CSTR. The right-hand panels show "data." The input is the ideal input plus a random variable with mean zero and standard deviation 2.0. The output was calculated using $y_t = 0.2x_t + 0.8y_{t-1} + a_t$, where a_t is N(0,4).

A Discrete Time Series Approximation

Suppose now that this process has been observed over time at a number of equally spaced intervals to obtain a times series of input and output data. Over one time step of size Δt, we can approximate $\frac{dy}{dt}$ by $\frac{\Delta y}{\Delta t} = y_t - y_{t-1}$. Substituting gives a *discrete mechanistic* model:

$$\tau(y_t - y_{t-1}) = x_t - y_t$$

and

$$y_t = \frac{x_t}{1+\tau} + \frac{\tau}{1+\tau}y_{t-1}$$

This is the *transfer function* model.

For $\tau = 1$, the model is $y_t = 0.5x_t + 0.5y_{t-1}$. Examine how this responds when an impulse of 2 units is put into the system at $t = 1$. At time $t = 1$, the model predicts $y_1 = 0.5(2) + 0.5(0) = 1$; at time $t = 2$, it predicts $y_2 = 0.5(0) + 0.5(1) = 0.5$, etc. as shown by the values below.

t	0	1	2	3	4	5	6	7
x_t	0	2	0	0	0	0	0	0
y_t	0	1	0.5	0.25	0.125	0.0675	...	...

The effluent concentration jumps to a peak and then decreases in an exponential decay. This is exactly the expected response of a CSTR process to a pulse input.

The table below shows how this model responds to an step change of one unit in the input.

t	0	1	2	3	4	5	6	7
x_t	0	1	1	1	1	1	1	1
y_t	0	0.5	0.75	0.875	0.94	0.97	0.98	0.99

Transfer Function Models

TABLE 52.1

Arrangement of the Input and Output Data to Fit the Model $y_t = \theta_1 x_t + \theta_2 y_{t-1} + a_t$ by Linear Regression

x(t)	y(t−1)	y(t)
6.7	10.1	12.0
7.3	12.0	11.6
7.3	11.6	12.9
11.4	12.9	16.7
8.5	16.7	12.7
10.0	12.7	11.7
12.1	11.7	9.4
8.3	9.4	10.8
10.9	10.8	8.5
12.1	8.5	13.4
7.0	13.4	11.2
7.5	11.2	12.6
9.0	12.6	10.1
8.4	10.1	8.2
6.4	8.2	8.1
...	...	...

Note: $y(t)$ is the dependent variable and $x(t)$ and $y(t-1)$ are the independent (predictor) variables.

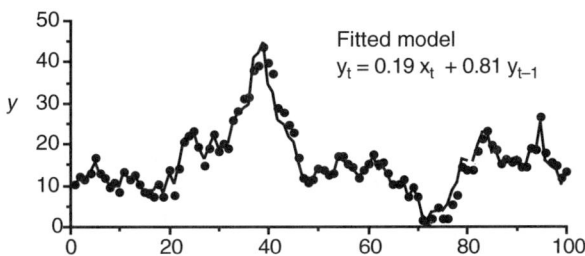

FIGURE 52.2 The fitted transfer function model $y_t = 0.19 x_t + 0.81 y_{t-1}$ and the effluent values, which were simulated using $y_t = 0.2 x_t + 0.8 y_{t-1} + a_t$. The input x_t was the idealized pattern shown in Figure 53.6 plus random noise from a normal distribution with mean zero and $s = 2$ (i.e., N(0,2)). The output was calculated from $y_t = 0.2 x_t + 0.8 y_{t-1} + a_t$, where a_t was N(0,4).

The step change in the input forces the effluent to rise exponentially and approach the new steady-state level of 1.0. Once again, this meets our expectation for the process.

Fitting the CSTR Model

This example has been constructed so the discrete form of the transfer function for the CSTR ($y_t = \theta_1 x_t + \theta_2 y_{t-1} + a_t$) can be fitted by linear regression. A new predictor variable (y_{t-1}) is created by lagging the y_t one step, as shown in Table 52.1. Then linear regression is done using x_t and y_{t-1} as predictors. The regression uses y_t as the dependent variable.

Linear regression can be used only when the error term is a_t. If the error structure were $a_t + \theta a_{t-1}$, or some other form of moving average structure, the a_t's would be obtained by a recursive calculation and the parameter estimation would be done by nonlinear regression. The linear regression approach can be used for autoregressive models and for transfer function models that do not need a moving average component.

Figure 52.2 shows the fitted transfer function model $y_t = 0.19x_t + 0.81y_{t-1}$ and the effluent concentration data. This model is very close to the model $y_t = 0.2x_t + 0.8y_{t-1} + a_t$ that was used to generate the output data series.

Comments

Many real processes are easier to describe with a discrete time series model than with differential equations. The advantage is that pure mechanistic models do not capture the memory, capacitance, and randomness of real processes. The disadvantage is that the parameters of a time series models are not easily related to mechanistic parameters such as reaction rate coefficient and mass transfer coefficients.

An expanded mechanistic model may still have a simple discrete time-series-like form. The mechanistic model for a CSTR with a first-order chemical reaction, with inputs and outputs that are functions of time, is:

$$V \frac{dy(t)}{dt} = Qx(t) - Qy(t) - kVy(t)$$

which can be written as:

$$y_t = \frac{1}{1+\tau+k\tau}x_t + \frac{\tau}{1+\tau+k\tau}y_{t-1} + a_t$$

$$y_t = \theta_1 x_t + \theta_2 y_{t-1} + a_t$$

For $k = 1$ and $\tau = 2$, this becomes $y_t = 0.111x_t + 0.4y_{t-1} + a_t$.

Imagine now a process that has some characteristics of the CSTR reactor but which is more complicated. Perhaps the mixing is not efficient, or there is some time delay in the system, or the input contains interfering impurities in random amounts, or that the temperature changes so the reaction rate coefficient (k) is not constant. This makes it more difficult to write an adequate differential equation. Still, it should be possible to derive a dynamic model in terms of the inputs and outputs. Instead of simply $y_t = \theta_1 x_t + \theta_2 y_{t-1} + a_t$, we might add more terms, perhaps a time delay (x_{t-d}) or more memory of past conditions (z_{t-2} or a_{t-2}), to get $y_t = \theta_1 x_t + \theta_2 x_{t-d} + \theta_3 y_{t-1} + \theta_4 y_{t-2} + a_t - \phi a_{t-1}$. With some gentle persuasion the time series data will reveal the necessary terms. A tentative model is fitted to the data. The residuals of the model are examined. If they are not random and independent, they contain additional information and the model is modified and refitted. This iterative process continues until an adequate model of the process emerges.

Mechanistic models of dynamic processes tend to get complicated, whereas time series type models tend have fewer parameters (Jeppsson, 1993, 1996; Olsson and Newell, 1999). Sometimes a time series model can be constructed using only the output series, with all input perturbations somehow being remembered by the process and reflected in the output. Wastewater treatment data can be like this (Berthouex et al., 1978, 1979). This sounds odd, and it is, but time series modeling holds surprises and our intuition is not error-free.

References

Berthouex, P. M., W. G. Hunter, L. Pallesen, and C.Y. Shih (1978). "Dynamic Behavior of Activated Sludge Plants," *Water Res.,* 12, 957–972.

Berthouex, P. M., W. G. Hunter, and L. Pallesen (1979). "Analysis of the Dynamic Behavior of an Activated Sludge Plant: Comparison of Results with Hourly and Bi-hourly Data," *Water Res.,* 13, 1281–1284.

Jeppsson, U. (1996). *Modelling Aspects of Wastewater Treatment Processes,* Lund, Sweden, Lund Institute of Technology.

Transfer Function Models

Jeppsson, U. (1993). *On the Verifiability of the Activated Sludge System Dynamics,* Lund, Sweden, Lund Institute of Technology.

Olsson, G. and B. Newell (1999). *Wastewater Treatment Systems; Modelling, Diagnosis and Control,* London, IWA Publishing.

Exercises

52.1 Time Constant. Using the model $y_t = \frac{x_t}{1+\tau} + \frac{\tau}{1+\tau} y_{t-1}$ with $\tau = 1$, calculate the response to (a) a pulse input of $x_1 = 4$ and (b) a step input of 2 units that occurs at $t = 1$. Repeat this for $\tau = 3$.

52.2 Input–Output. For the discrete transfer function model $y_t = 0.2 x_t + 0.8 y_{t-1}$, calculate the output y_t for the inputs given below.

Time	x(t)	Time	x(t)	Time	x(t)	Time	x(t)	Time	x(t)
1	10	11	10	21	20	31	10	41	10
2	10	12	10	22	50	32	15	42	10
3	10	13	20	23	50	33	15	43	10
4	15	14	20	24	50	34	10	44	10
5	10	15	20	25	50	35	10	45	20
6	10	16	20	26	50	36	20	46	30
7	10	17	20	27	50	37	20	47	25
8	10	18	20	28	10	38	20	48	10
9	10	19	20	29	10	39	25	49	10
10	10	20	20	30	10	40	20	50	10

52.3 Model Fitting. Fit a model of the form $y_t = \theta_1 x_t + \theta_2 y_{t-1} + a_t$ to the data below. Evaluate the fit by (a) plotting the predicted values over the data, (b) plotting the residuals against the predicted values, and (c) calculating the autocorrelation function of the residuals.

Time	x(t)	y(t)	Time	x(t)	y(t)	Time	x(t)	y(t)	Time	x(t)	y(t)
1	6.7	10.1	13	19	14.8	25	50.2	43.7	37	22.9	29.7
2	7.3	12	14	18.4	13.8	26	47.2	41.7	38	21.5	26.7
3	7.3	11.6	15	16.4	13.2	27	51.4	38.6	39	23.5	28.5
4	16.4	13.9	16	18.7	14.1	28	9.7	35.9	40	19.9	30
5	8.5	17.5	17	23.2	14.1	29	8.2	34.1	41	10.1	29.2
6	10	13.4	18	21.5	17.5	30	9	25.5	42	9.4	22.6
7	12.1	12.3	19	18	15.3	31	7.5	23.9	43	9.3	22.7
8	8.3	9.8	20	18.7	21.9	32	17.8	20.9	44	6.5	20.6
9	10.9	11.1	21	18.5	14.3	33	14.1	26.4	45	19.9	21.3
10	12.1	8.7	22	48.5	25.2	34	10.9	26.5	46	27.1	19.8
11	9.7	13.6	23	52.4	35.7	35	8.4	27.7	47	22.1	17.5
12	7.5	11.4	24	47.7	40	36	20.4	28.8	48	7.1	12.7

52.4 Activated Sludge Models. The dynamic behavior of an activated sludge process was studied by sampling the influent and effluent BOD every 2 h. The influent BOD data are similar to Figure 51.1. The flow had the same pattern, so the mass BOD loading varied by a factor of 9 to 10 with a similar diurnal pattern over each 24-h period. The treatment process was operating beyond the design loading and the return activated sludge flow was constant. Therefore, one might expect the effluent to vary considerably, perhaps with a diurnal pattern similar to the influent loading. It did not. Given below are a few models that were fitted to the log-transformed effluent BOD data (Berthouex et al., 1978). The residual mean square (RMS = residual sum of squares/degrees of freedom) can be used to compare the fit of the models. The first model is a horizontal line at the level of the process mean. Models A and B, which model the process entirely in terms of effluent data, have an RMS nearly as low as model D, which incorporates information about the influent BOD (x_t). Several three-parameter and four-parameter models with x_t also had RMS values of 0.03. How is it possible that the transfer function model (D) is only slightly better than the autoregressive models B and C?

Model	RMS
(A) $y_t = a_t$	0.101
(B) $y_t = y_{t-1} + a_t$	0.037
(C) $y_t = 0.819 y_{t-1} + a_t$	0.034
(D) $y_t = 0.193 y_{t-1} + 0.778 x_t + a_t$	0.030

53

Forecasting Time Series

KEY WORDS *autoregressive, AR(1) model, EWMA, exponential smoothing, forecast, forecast error, IMA(0,1,1) model, MA(1) model, moving average, lead time, random walk, variance.*

A primary objective of building a time series model is to forecast the series at one or more steps into the future. A statement of the precision of the forecasts should also be reported. Naturally, the expected forecast error increases as the forecast extends from one time step ahead to larger steps into the future. When making the forecasts we assume that the model is known exactly and that the parameter values are not changing.

Forecasting an AR(1) Process

Autoregressive (AR) processes are used to model stationary time series. Before fitting the data, the mean η is subtracted from each observation Z_t to give $z_t = Z_t - \eta$. The mean of the z_t is zero. The AR(1) process with a zero mean is:

$$z_t = \phi z_{t-1} + a_t$$

Consider the model $z_t = 0.7 z_{t-1} + a_t$. Substituting $t + 1$ for t gives:

$$z_{t+1} = 0.7 z_t + a_{t+1}$$

The forecast of z_{t+1} is $\hat{z}_t(1)$. The hat (^) indicates a predicted value and the lead time of the forecast is in parentheses. The forecast $\hat{z}_t(1)$ is the expected value of z_{t+1}. Because the expected value of a_t is zero, the forecast is:

$$\hat{z}_t(1) = 0.7 z_t$$

The two-step-ahead forecast is:

$$\hat{z}_t(2) = 0.7 \hat{z}_t(1) = 0.7(0.7 z_t) = 0.7^2 z_t$$

In general:

$$\hat{z}_t(\ell) = 0.7^\ell z_t \quad \text{for } \ell > 1.$$

This converges to the mean of zero as the lead time gets large. This result holds for all stationary ARMA models.

This is easier to understand if we write this with a nonzero mean:

$$Z_t - \eta = \phi(Z_{t-1} - \eta) + a_t$$

459

At large lead times, the forecast value is simply the mean of the process (η).

The one-step-ahead forecast is:

$$\hat{Z}_t(1) = \eta + \phi(Z_t - \eta)$$

and the forecast is obtained by adding to the process mean a proportion ϕ of the current deviation from the process mean. For lead time ℓ, replace t by $t + \ell$ and take the expected value of both sides to get:

$$Z_t(\ell) = \eta + \phi[\hat{Z}_t(\ell - 1) - \eta] \quad \text{for } \ell \geq 2$$

This shows how the forecast for any lead time ℓ can be built up from the forecasts for shorter lead times. Doing this gives the recursive solution:

$$Z_t(\ell) = \eta + \phi^\ell(\hat{Z}_t - \eta)$$

The current deviation from the mean is discounted by a factor ϕ^ℓ. This discounted deviation is added to the process mean to produce the lead ℓ forecast. Because $\phi < 1$, this factor decreases as the lead time increases.

Example 53.1

Forecast from the origin $t = 50$, five values ahead for the AR(1) process shown in Figure 53.1. The fitted model is $z_t = 0.72 z_t + a_t$, with $\hat{\sigma}_a^2 = 0.89$. The observations are shown as dots and the fitted model is the line. The mean ($\eta = 10$) was subtracted before fitting the series and was added back to plot the series.

The predictions and forecast errors shown in Table 53.1 were made as follows. The observed value at $t = 50$ is $Z_{50} = 10.6$. The one-step-ahead forecast is:

$$\hat{Z}_t(1) = 10.0 + 0.72(10.6 - 10.0) = 10.43$$

This has been rounded to 10.4 in Table 53.1.

For lead time $\ell = 2$:

$$\hat{Z}_t(2) = 10.0 + 0.72(10.43 - 10.0) = 10.3$$

or the equivalent:

$$\hat{Z}_t(2) = 10.0 + 0.72^2(10.6 - 10.0) = 10.3$$

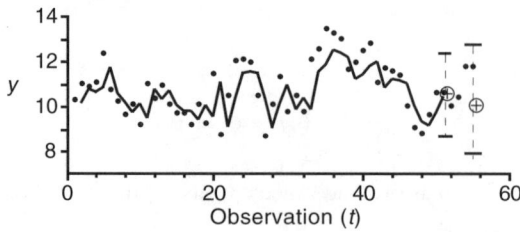

FIGURE 53.1 Forecasts from origin $t = 50$ for an AR(1) model. The forecasts converge exponentially toward the mean level. The approximate 95% confidence intervals for lead one and lead five forecasts are shown by dotted lines capped with bars. The confidence intervals become a constant bandwidth at long lead times.

TABLE 53.1

Forecasts and Forecast Errors for the AR(1) Process Shown in Figure 53.1

Time	50	51	52	53	54	55
Predicted		10.4	10.3	10.2	10.1	10.1
Observed	10.6	10.6	10.0	10.3	11.7	11.8
Forecast error		0.18	−0.3	−0.1	1.6	1.7

For lead time $\ell = 5$:

$$\hat{Z}_t(5) = 10.0 + 0.72^5(10.6 - 10.0) = 10.10$$

As ℓ increases, the forecasts will exponentially converge to:

$$\hat{Z}_t(\ell) \approx \eta = 10$$

A statement of the forecast precision is needed. The one-step-ahead forecast error is:

$$e_t(1) = z_{t+1} - \hat{z}_t(1) = \phi z_t + a_{t+1} - \phi z_t = a_{t+1}$$

The a_{t+1} is white noise. The expected value of the forecast error is zero so the forecasts are unbiased. Because a_t is white noise, the forecast error $e_t(1)$ is independent of the history of the process. (If this were not so, the dependence could be used to improve the forecast.) This result also implies that the one-step-ahead forecast error variance is given by:

$$\text{Var}[e_t(1)] = \sigma_a^2$$

The AR(1) model can be written as the infinite MA model:

$$z_t = a_t + \phi a_{t-1} + \phi^2 a_{t-2} + \cdots$$

which leads to the general forecast error:

$$e_t(\ell) = a_{t+\ell} + \phi a_{t+\ell-1} + \phi^2 a_{t+\ell-2} + \cdots + \phi^{\ell-1} a_{t+1}$$

This shows that the forecast error increases as the lead ℓ increases. The forecast error variance also increases:

$$\text{Var}[e_t(\ell)] = \sigma_a^2(1 + \phi + \phi^2 + \cdots + \phi^{\ell-1})^2$$

which simplifies to:

$$\text{Var}[e_t(\ell)] = \sigma_a^2 \frac{1 - \phi^{2\ell}}{1 - \phi^2}$$

For long lead times, the numerator becomes 1.0 and:

$$\text{Var}[e_t(\ell)] = \sigma_a^2 \frac{1}{1 - \phi^2}$$

Because the lag 1 autocorrelation coefficient $\rho = \phi$ for an AR(1) process, this is equivalent to $\sigma_z^2 = \sigma_a^2 \frac{1}{1-\rho^2}$, as used in Chapter 41.

It is also true for long lead times that:

$$\text{Var}[e_t(\ell)] \approx \text{Var}(z_t)$$

which says that the forecast error variance equals the variance of the z_t about the mean value. This result is valid for all stationary ARMA time series processes.

Example 53.2

Compute the variance and approximate 95% confidence intervals for the lead one and lead five forecast errors for the AR(1) process in Example 53.1. The variance of the lead one forecast is $\hat{\sigma}_a^2 = 0.89$. The approximate 95% confidence interval is $\pm 2\hat{\sigma}_a = \pm 2(0.94) = \pm 1.88$. The variance of the lead five forecast error is:

$$\text{Var}[e_t(5)] = \sigma_a^2 \frac{1-\phi^{2(5)}}{1-\phi^2} = 0.89\left(\frac{1-0.72^{10}}{1-0.72^2}\right) = 0.89\left(\frac{0.9626}{0.4816}\right) = 1.78.$$

The approximate 95% confidence interval is $\pm 2(1.33) = \pm 2.66$. These confidence intervals are shown in Figure 53.1.

Forecasting an AR(2) Process

The forecasts for an AR(2) model are a damped sine that decays exponentially to the mean. This is illustrated with the AR(2) model:

$$z_t = \phi_1 z_{t-1} - \phi_2 z_{t-2} + a_t$$

The series of $n = 100$ observations in Figure 53.2 (top) is described by the fitted model $\hat{z}_t = 0.797 z_{t-1} - 0.527 z_{t-2} + a_t$. The forecasted values from origin $t = 25$ are shown as open circles connected by a line to define the damped sine. The forecasts converge to the mean, which is zero here because the z_t are deviations from the long-term process mean.

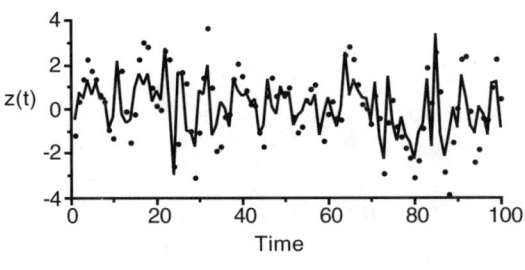

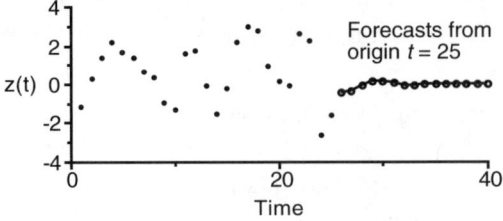

FIGURE 53.2 The top panel shows an AR(2) process fitted to a time series of size $n = 100$. The bottom panel shows the forecasts from origin $t = 25$ for this series. The forecasts converge to the mean value as a damped sine.

Forecasting the ARIMA(0,1,1) Process

The ARIMA (0,1,1) model is:

$$z_{t+1} = z_t + a_{t+1} - \theta a_t$$

The expected value of z_{t+1} is the lead one forecast:

$$\hat{z}_t(1) = z_t - \theta a_t$$

The lead ℓ forecast is:

$$\hat{z}_t(\ell) = \hat{z}_t(\ell - 1) \text{ for } \ell \geq 2$$

This can also be written as:

$$\hat{z}_t(\ell) = (1 - \theta)z_t + \theta \hat{z}_{t-1}(\ell)$$

This implies that the new forecast is a linear interpolation between the old forecast and the new observation.

The one-step-ahead forecast error is:

$$e_t(1) = z_{t+1} - \hat{z}_t(1) = a_{t+1}$$

and also

$$e_t(\ell) = z_{t+\ell} - \hat{z}_t(\ell) = \sum_{j=0}^{\ell-1} a_{t+\ell-1}$$

The variance for the forecast is:

$$\text{Var}[e_t(\ell)] = \sigma_a^2 [1 - (\ell - 1)(1 - \theta)]$$

For the lead one forecast, the variance is simply σ_a^2. Details are found in Box et al. (1994).

Exponential Smoothing

One of the most useful time series models is the exponentially weighted moving average (EWMA). It is the forecast for the ARIMA(0,1,1) model, and it often does quite well in forecasting other nonstationary time series.

The EWMA for z_t is:

$$\tilde{z}_t = (1 - \theta)z_t + \theta \tilde{z}_{t-1}$$

The tilde (~) indicates a moving average. In more detail:

$$\tilde{z}_t = (1 - \theta)(z_t + \theta z_{t-1} + \theta^2 z_{t-2} + \theta^3 z_{t-3} + \cdots)$$

For $\theta = 0.6$:

$$\tilde{z}_t = 0.4(z_t + 0.6z_{t-1} + 0.6^2 z_{t-2} + 0.6^3 z_{t-3} + \cdots)$$

The one-step-ahead forecast for the EWMA is the EWMA $\hat{z}_{t+1} = \tilde{z}_t$. This is also the forecast for several days ahead; the forecast from origin t for any distance ahead is a straight horizontal line.

To update the forecast as new observations become available, use the forecast updating formula:

$$\hat{z}_{t+1} = (1 - \theta)z_t + \theta\hat{z}_t$$

The choice of θ is a question of finding the right compromise between achieving fast reaction to change and averaging out the noise. If we choose θ to be too large (e.g., 0.8), then the averaging process reaches back a long way into the series so that the smoothing out the noise (local variation) will be considerable. But such a forecast will be slow to react to a permanent change in level. On the other hand, if we take a smaller θ (e.g., 0.4), the reaction to change will be quicker, but if the series is noisy, the forecast will be volatile, reacting to changes that may just be temporary fluctuations.

The best compromise can be found by trial-and-error calculation on a run of past data for the particular process ($n \geq 100$ if possible).

Example 53.3

An EWMA model with $\theta = 0.5$ was developed from a series with $n = 120$ observations. Use the model to forecast the next ten observations, starting from time $t = 121$, at which $z_{121} = 260$. The forecast updating model is $\hat{z}_{t+1} = 0.5z_t + 0.5\hat{z}_t$. To start, set $\hat{z}_{121}$ equal to the actual value observed at $t = 121$; that is, $\hat{z}_{121} = 260$. Then use the updating model to determine the one-step-ahead forecast for $t = 122$:

$$\hat{z}_{122} = 0.5(260) + 0.5(260) = 260$$

When the observation for $t = 122$ is available, use the updating model to forecast $t = 123$, etc.:

$$\hat{z}_{123} = 0.5(240) + 0.5(260) = 250$$

$$\hat{z}_{124} = 0.5(220) + 0.5(250) = 235$$

The forecast error is the difference between what we forecast one step ahead and what was actually observed:

$$e_t = z_t - \hat{z}_t$$

The observations, forecasts, and forecast errors for $t = 121$ through 130 are given in the Table 53.2.

TABLE 53.2

Ten Forecasts for the EWMA Model with $\theta = 0.5$

t	121	122	123	124	125	126	127	128	129	130
z_t	260	240	220	240	260	260	280	270	240	250
$\hat{z}_t$	260	260	250	235	237	249	254	267	269	254
e_t	(0)	−20	−30	5	23	11	26	3	−29	−4

Note: The updating model is $\hat{z}_{t+1} = 0.5z_t + 0.5\hat{z}_t$.

The precision of the forecast is the average of the sum of squares of the forecast errors:

$$SS_{\theta=0.5} = \frac{(-20)^2 + (-30)^2 + (5)^2 + \cdots + (-29)^2}{10} = 351.7$$

Fitting the EWMA model to obtain the best value of θ is also a matter of evaluating the sum of squares of the forecast errors. This series is too short to give a reliable estimate of θ but it illustrates the approach. For a longer series (i.e., $n \geq 50$), we could calculate the sum of squares of the errors SS_θ for $\theta = 0.0$, 0.1, 0.2,..., 0.9, 1.0 and choose the value that gives the smallest value of SS_θ. Or we could make a plot of SS_θ against θ, draw a smooth curve through the points, and read off the value of θ that gives the minimum.

Once the best value of θ has been found, the model must be evaluated to see whether the EWMA is an adequate model. Does it give good forecasts? Plot the forecast errors and see if they look like a random series. "If they don't and it looks as if each error might be forecast to some extent from its predecessors, then your forecasting method is not predicting good forecasts (for if you can forecast forecasting errors, the you can obviously obtain a better forecast than the one you've got)" (Box, 1991). Forecast errors from a good forecasting method must produce a random error series (a nonforecastable error series). If this does not happen, it may mean you have simply chosen a bad value for θ or you may need a more sophisticated method of forecasting, such as those described by Box et al. (1994). Often, however, we are lucky. "It is surprising how far you can get with this very simple EWMA technique...." (Box, 1991).

Comments

Forecasts of the value of a time series at one or more steps into the future depend on the form of the time series model (AR, MA, or ARIMA). The variance of the forecast errors depends on the form of the model and also on how far into the future the forecast is made. The forecast error variance is not the variance of the purely random (white noise) component of the time series.

The forecasts for stationary processes (i.e., AR processes) converge to the process mean as exponential or damped sine functions. The variance of the forecast error converges to the variance of the process about the mean value.

The forecasts for nonstationary processes (i.e., MA processes) do not converge to a mean value because there is no long-term mean for a nonstationary process). The forecast from origin t for the useful IMA(0,1,1) model (EMWA forecasts) is a horizontal line projected from the forecast origin. The forecast variance increases as the forecasts are extended farther into the future.

Box et al. (1994) gives details for several process models. It also gives general methods for deriving forecasting weights, similar to the exponential decaying weights of the EWMA. Minitab is a convenient statistical software package for fitting time series models and forecasting.

References

Box, G. E. P. (1991). "Understanding Exponential Smoothing: A Simple Way to Forecast Sales and Inventory," *Quality Engineering,* 3, 561–566.

Box, G. E. P. and L. Luceno (2000). "Six Sigma, Process Drift, Capability Indices, and Feedback Adjustment," *Quality Engineering,* 12(3), 297–302.

Box, G. E. P., G. M. Jenkins, and G. C. Reinsel (1994). *Time Series Analysis, Forecasting and Control,* 3rd ed., Englewood Cliffs, NJ, Prentice-Hall.

Exercises

53.1 AR(1) Forecasting. Assume the current value of Z_t is 165 from a process that has the AR(1) model $z_t = 0.4z_{t-1} + a_t$ and mean 162. (a) Make one-step-ahead forecasts for the 10 observations.

t	121	122	123	124	125	126	127	128	129	130
z_t	2.1	2.8	1.5	1.2	0.4	2.7	1.3	−2.1	0.4	0.9

(b) Calculate the 50 and 95% confidence intervals for the forecasts in part (a).

(c) Make forecasts from origin $t = 130$ for days 131 through 140.

53.2 AR(1) and MA(1) Processes. The latest observation of level in a treatment process is 8 units above the desired performance level. As process operator, would you act differently if you know the process behaves according to an AR(1) model than if it behaves according to an MA(1) model? Explain your answer.

53.3 EWMA. Assume that the current value $\hat{z}_t = 18.2$ is from an EWMA with $\theta = 0.25$. The next 10 observations are given below. Use the EWMA to make one-step-ahead forecasts.

18.2   19.5   20.1   16.8   18.8   17.3   22   19.8   18.3   16.6

53.4 AR(1) Process. Simulate an AR(1) time series with $\phi = 0.7$ and $\sigma_a^2 = 1.00$ for $n = 100$ observations. Fit the simulated data to estimate ϕ and σ_a^2 for the actual series. Then, using the estimated values, forecast the values for $t = 101$ to $t = 105$. Calculate the forecast errors and the approximate 95% confidence intervals of the forecasts.

54

Intervention Analysis

KEY WORDS *exponentially weighted average, intervention analysis, detergent ban, phosphorus, random walk, time series, white noise.*

Environmental regulations are intended to cause changes in environmental quality. Often it is not easy to detect the resulting change, let alone estimate its magnitude, because the system does not vary randomly about a fixed level before or after the intervention. Serial correlation must be accounted for in the intervention analysis. When the system exhibits drift (nonstationarity) or seasonality, that also must be taken into account. *Intervention analysis* estimates the effect of a known change in conditions affecting a time series of serially correlated data (Box and Tiao, 1965). We will present a simple model that might describe environmental time series that have a slow drift but no regular seasonality.

Case Study: Ban on Phosphate Detergents

Wisconsin passed a law in 1978 that resulted in two interventions that were intended to change the amount of phosphorus entering into the environment. The law required that after July 1, 1979, household laundry detergents could contain no more than 0.5% phosphorus (P) by weight. Before this virtual ban on phosphate detergents went into effect, detergents contained approximately 5% phosphorus. In 1982, the phosphate ban lapsed and in July 1982 detergents containing phosphates started to reappear, although product reformulation and marketing changes had reduced the average detergent phosphorus content from pre-ban levels. A few years later, a new ban was imposed.

From the mid-1970s until the mid-1980s, there was controversy about how much the ban actually reduced the phosphate loading to wastewater treatment plants. Of course, the detergent manufacturers knew how much less phosphate had been sold in detergents, but confirmation of their estimates from treatment plant data was desired. Making this estimate was difficult because the potential reduction was relatively small compared to the natural random fluctuations of the relevant environmental series.

The largest treatment plants in Wisconsin are the Jones Island and South Shore plants in Milwaukee. Both plants have expert management and reliable measurement processes. The combined mass load of influent phosphorus to these two plants is plotted in Figure 54.1. The Wisconsin phosphate detergent ban stretched from July 1979 to June 1982, and indeed the lowest phosphorus concentrations are recorded during this period. The record shows a downward trend starting long before the imposition of the ban, and it is not obvious from the plot how to distinguish the effect of the ban from this general trend. If an average value for the ban period is to be compared to a pre-ban average, the difference would depend heavily on how far the pre-ban average is extended into the past.

Somehow, allowance must be made for the seemingly random drift in influent P load that is unrelated to the ban. Intuition suggests that to minimize the influence of random drift or trends on the estimation, the averages of observations made to estimate the effect of the ban should not stretch too far away from time of the intervention. On the other hand, relying on too few observations around the intervention ban might impair the estimation because there was too little averaging of random fluctuations. It would be appealing if the observations closest to the expected shift were given more weight than observations more distant in time.

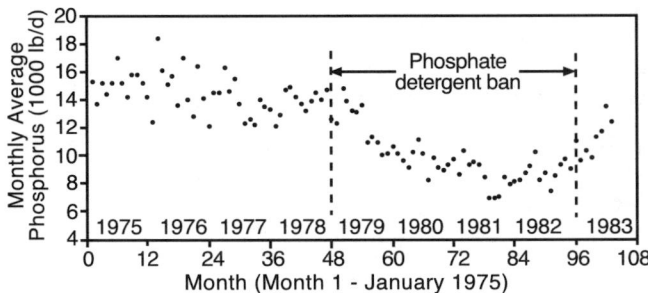

FIGURE 54.1 Monthly average mass flow of total phosphorus for the Milwaukee South Shore and Jones Island plants combined.

Before presenting a model that is useful for analysis of many environmental series, a brief discussion is given of two other models that have often been relied upon (explicitly or implicitly) to estimate the effects of change. These are the (1) *white noise model* and the (2) *random walk model*. The white noise model is the basis for estimating the change in level when the data series consists of random independent variation about a fixed mean level. The intervention in the fixed-mean-levels case would be estimated by taking the difference of the mean levels before and after the shift. The random walk model is used when the data series consists of pure random walk in which there is no fixed level. In the random walk case, only the observations immediately before and after the intervention would be used to estimate the shift. In the first case, all the data contribute to the estimation; in the second case, almost all the data are disregarded in making the estimation. In both cases, we assume that the intervention shifts the data series upward or downward and without changing the pattern of variation.

The White Noise Model

When the data vary about a fixed level and contain nonnegligible random measurement error, sampling error, etc., the statistical analysis often relies on the model:

$$y_t = \eta + e_t$$

where

y_t = observation at $t = 1, 2, \ldots, n$
η = true (and unobservable) mean value of y_i
e_t = random error, assumed to be independently distributed according to normal distribution with mean 0 and variance σ_e^2; that is, $e_i \approx N(0, \sigma_e^2)$.

This is called the *white noise model* because the random errors (e_t) are white noise.

Figure 54.2a illustrates n_1 observations made at one level, followed by n_2 observations at a new level, and represents a case where the white noise model would be appropriate. A deliberate intervention has caused the change in conditions at time T. The intervention model is:

$$y_t = \eta_1 + \delta I + e_t$$

where

y_t = value observed at time t
T = time the intervention takes place
δ = effect of the intervention, $\delta = \eta_2 - \eta_1$
I = an indicator function: $I = 0$ before the intervention; $I = 1$ after the intervention
η_1 = mean value of y_t for $t \leq T$

Intervention Analysis

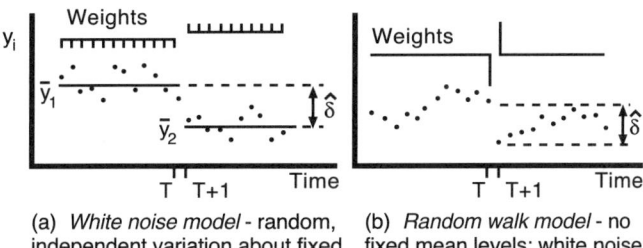

FIGURE 54.2 Illustration of the weights used to estimate a shift caused by an intervention according to two models that represent extreme assumptions regarding the pattern of random noise in the data.

η_2 = mean value of y_t for $t > T$
e_t = independently distributed error, $e_i \approx N(0, \sigma_e^2)$.

Based on this white noise model, the effect of the intervention is estimated as the difference of the averages before and after the intervention:

$$\hat{\delta} = \bar{y}_2 - \bar{y}_1 = \frac{1}{n_2}\sum_{t=T+1}^{T+n_2} y_t - \frac{1}{n_1}\sum_{t=1}^{T} y_t$$

The estimated variance of δ is:

$$\mathrm{Var}(\hat{\delta}) = \mathrm{Var}(\bar{y}_1) + \mathrm{Var}(\bar{y}_2) = \hat{\sigma}_e^2\left(\frac{1}{n_1} + \frac{1}{n_2}\right)$$

where $\hat{\sigma}_e^2$ is the pooled variance of the deviations from the two average levels. This computation is like the comparison of two averages using a t-test (Chapter 18). The ratio $\hat{\delta}/\sqrt{\mathrm{Var}(\hat{\delta})}$ can be compared to a t distribution with $v = n_1 + n_2 - 2$ degrees of freedom to test whether δ is significantly different from zero.

Estimating δ from the white noise model means giving equal weight to all observations before the intervention, regardless of their nearness to time T. Also, all observations after the intervention are given equal weight. This uniform weighting, shown in Figure 54.2a, is a natural consequence of assuming statistical independence among observations. The random walk model accounts for the more realistic situation when there is correlation between nearby values in the time series.

The Random Walk Model

Figure 54.2b illustrates observations made in time sequence, but where the true value of the response variable Y_t is observable at any time. This true value tends to drift randomly over time. Each value is independent of the values near it. Note the subtle but important difference from the usual statistical assumptions about observed values Y_t. In this case, Y_t is assumed to have no error (which is why we represent it by capital Y in contrast to lower case y which is a value observed with error). It is the difference $Y_t - Y_{t-1}$ (the drift) that is subject to random variation. Data series of this nature can often be described by a *random walk model*:

$$Y_t = Y_{t-1} + \varepsilon_t$$

where Y_t = true level of the response at time t and ε_t = random drift between time $t-1$ and t.

The random drift terms ε_t are independent $N(0, \sigma_\varepsilon^2)$.

The estimated effect of the intervention at time T is:

$$\hat{\delta} = Y_{T+1} - Y_T$$

Y_{T+1} and Y_T are known without error, but $\hat{\delta}$ is subject to uncertainty because it represents the intervention *and* some random drift ε_{T+1}. Hence:

$$\text{Var}(\hat{\delta}) = \text{Var}(\varepsilon_{T+1}) = \sigma_\varepsilon^2$$

and σ_ε^2 is estimated as:

$$\hat{\sigma}_\varepsilon^2 = \frac{\sum_{t=1}^{T}(Y_T - Y_{T-1})^2 + \sum_{t=T+1}^{T+n_2}(Y_T - T_{T-1})^2}{n_1 + n_2 - 2}$$

The ratio $\hat{\delta}/\hat{\sigma}_\varepsilon$ can be compared to a t distribution with $\nu = n_1 + n_2 - 2$ degrees of freedom to test whether δ is significantly different than zero.

The random walk estimate of δ gives full weight to the observations just before and just after the intervention. All other observations are given zero weight, as illustrated in Figure 54.2b.

The White Noise–Random Walk Model

For most environmental time series, both the white noise and the random walk models are unrealistic. It is usually unreasonable to assume that the mean level of the variable in question remains constant for all times except for a shift caused by the deliberate intervention. Hence, observations in the remote past carry little information about the level of the response prior to a particular intervention. Similarly, developments immediately after the intervention gradually become irrelevant to future observations. On the other hand, any single observation is usually affected by temporary random fluctuations, which implies that some sort of averaging would increase the precision of estimation.

A model that suitably reflect these conditions is a *combination* of the white noise and random walk models:

$$y_t = Y_t + e_t$$

$$Y_t = Y_{t-1} + \varepsilon_t$$

where
- y_t = observation at time t
- Y_t = underlying (but unobservable) "true" value for y_t
- e_t = white noise component of time series with $e_i \approx N(0, \sigma_e^2)$
- ε_t = random walk component of the time series with $\varepsilon_t = N(0, \sigma_\varepsilon^2)$

The two error series e_t and ε_t are independent of each other. If the true values of y_T and y_{T+1} were known, the effect of the intervention would be estimated as $\hat{\delta} = y_T - y_{T+1}$ with $\text{Var}(\hat{\delta}) = \sigma_e^2$. Because the true values are unknown, their estimates are used:

$$\hat{\delta} = \bar{y}_T - \bar{y}_{T+1} + \varepsilon_T$$

The overbar is used to indicate the estimated values because they are weighted averages, as will be shown shortly.

Intervention Analysis

Sometimes the effect of the intervention is not fully realized within the interval T and $T+1$. A practical consideration then is how to represent the transition period over which the intervention is realized. Here, we will simply say that the transition takes place during an interval of length G, being fully realized in the gap between T and $T+G$. In this case, the estimate of the shift becomes:

$$\hat{\delta} = \bar{y}_T - \bar{y}_{T+G} + \sum_{j=1}^{G} \varepsilon_{T+j}$$

The last term represents the expected magnitude of the random walk during the transition period over which the intervention is fully realized.

The averages $\bar{y}_T$ and $\bar{y}_{T+G}$ that represent the levels before and after the intervention are exponentially weighted averages. In the equations below, θ is a weighting factor that has a value between 0 and 1. As θ approaches 1, observations near the intervention are emphasized and observations farther away are forgotten (Chapters 4 and 53):

$$\bar{y}_T = \theta y_T + \theta(1-\theta)y_{T-1} + \theta(1-\theta)^2 y_{T-2} + \cdots$$
$$\bar{y}_{T+G} = \theta y_{T+G} + \theta(1-\theta)y_{T+G-1} + \theta(1-\theta)^2 y_{T+G-2} + \cdots$$

The variance of the estimated shift is:

$$\text{Var}(\hat{\delta}) = \text{Var}(\bar{y}_T) + \text{Var}(\bar{y}_{T+G}) + G\sigma_\varepsilon^2$$

If $G = 1$, the intervention is modeled as being fully realized in the interval between T and $T+1$. If a longer transition period is needed, one or more observations will be omitted. The value of G will be one plus the number of omitted observations. The number of random walk steps acting over the gap will be $G + 1$ and this is accounted for by the $\sum_{j=1}^{G} \varepsilon_{T+1}$ and $G\sigma_\varepsilon^2$ termos in the equations.

The exponential weighting factor θ is related to the variances in the model according to:

$$\frac{\sigma_e^2}{\sigma_\varepsilon^2} = \frac{\theta}{1-\theta}$$

When θ approaches 0 ($\sigma_e^2 \ll \sigma_\varepsilon^2$), the weights die out and the model approaches the white noise model of Figure 54.2a. If $\theta = 0$, each observation would be given equal weight and the time series would be considered to have no drift. On the other hand, θ approaching 1 ($\sigma_e^2 \gg \sigma_\varepsilon^2$) implies observations away from the intervention rapidly become irrelevant to estimating the intervention. In this case, the model approaches the random walk model described by Figure 54.2b.

If we accept this model, the intervention estimation problem becomes how to determine the weighting factor θ. Fortunately, an alternate formulation of the model makes this reasonably simple. The *white noise–random walk model* is equivalent to an ARIMA (0,1,1) model (Box et al., 1994):

$$y_t = y_{t-1} + a_t - \theta a_{t-1}$$

where $0 \leq \theta \leq 1$, and a_t = independent random noise distributed as $N(0,\sigma_a^2)$.

The white noise and the random walk error components are related to σ_a^2 as follows:

$$\sigma_e^2 = \theta \sigma_a^2$$
$$\sigma_\varepsilon^2 = (1-\theta)^2 \sigma_a^2$$

The equivalence of the two forms of the model and the derivations are given in Pallesen et al. (1985).

A recursive iteration is done separately for each section of data before and after the intervention to estimate θ. The method is to:

1. Choose a starting value for θ and use $a_t = y_t - y_{t-1} + \theta a_{t-1}$ to recursively calculate the residuals at $t = 2, 3,\ldots, T$. Assume $a_{t=0} = 0$ to start the calculations.
2. Calculate the residual sum of squares, $\text{RSS}(\theta) = \Sigma a_t^2$ for each section. Add these to get the total RSS for the entire series. If a gap has been used to account for a transition period, data in the gap are omitted from these calculations.
3. Search over a range of θ to minimize $\text{RSS}(\theta)$.
4. Use the minimum RSS to estimate $\hat{\sigma}_a^2 = \frac{\text{RSS}(\theta)}{n}$, where n is the total number of residuals used to compute $\text{RSS}(\theta)$.
5. Use the estimated variance $\hat{\sigma}_a^2$ to estimate σ_e^2 and σ_ε^2.

The variance of the estimated shift is:

$$\text{Var}(\hat{\delta}) = \text{Var}(\bar{y}_T) + \text{Var}(\bar{y}_{T+G}) + G\sigma_\varepsilon^2$$

The last term is the variance contributed by the random drift over the transition gap G. Because the minimum value of G is 1, this term cannot be omitted. Unless the series are short on either side of the intervention, we can assume that:

$$\text{Var}(\bar{y}_T) = \text{Var}(\bar{y}_{T+G}) = \sigma_e^2(1 - \theta)$$

and the variance of the estimated shift $\hat{\delta}$ is:

$$\text{Var}(\hat{\delta}) = \sigma_e^2(1 - \theta) + \sigma_e^2(1 - \theta) + G\sigma_\varepsilon^2 = 2(1 - \theta)\sigma_e^2 + G\sigma_\varepsilon^2$$

Case Study: Solution

The case study solution is presented in Pallesen et al. (1985). The data record is $n = 103$ monthly average phosphorus concentrations running from January 1975 to July 1983. The phosphorus ban imposed on July 1, 1979, may already have had some effect in June because shipment of phosphorus-free detergent to retailers in Wisconsin started in May. Hence, the analysis disregards the months of June, July, August, and September, which are taken as being transition months during which existing stocks of phosphate detergents were used up. The transition was handled by leaving a gap of four months; the data for June through September 1979 were disregarded in the calculations. The gap is $G = 5$. The ban was lifted in July 1982 and shipment of detergent containing phosphorus began during the month of July. Again, relying on information from the detergent manufacturers, a transition period of four months (July–October) was disregarded (again $G = 5$).

The calculations were done on the natural logarithms of the monthly average loads. The results are shown in Figure 54.3 and Table 54.1. Note that the variance of the November 1982 level is virtually identical to the others, although it is estimated from a very short series after the 1982 lifting of the ban. The estimated levels were transformed back to the original metric of lb/day. For example, the estimated level for May 1979 is 9.4835 on the logarithmic scale and the variance of this estimate is $\text{Var}(\text{May 1979}) = 0.002324$. The pre-ban level (May 1979) is $\exp(9.4835) = 13{,}141$ lb/day.

The variances of the effects are calculated using the values in Table 54.1:

$$\text{Var}(\hat{\delta}_{1979\ \text{ban}}) = 0.002324 + 0.002324 + 5(0.001312) = (0.106)^2$$

$$\text{Var}(\hat{\delta}_{1982\ \text{ban off}}) = 0.002324 + 0.002326 + 5(0.001312) = (0.106)^2$$

In both interventions, the standard deviation is $\sigma = 0.106$ in terms of the natural logarithm, which translates to a standard deviation of 10.6% of the geometric average of the two levels (pre- and post-)

Intervention Analysis

TABLE 54.1

Results of Intervention Analysis on Phosphorus Loads to Milwaukee Wastewater Treatment Plants

Event	n	Est. Levels (log scale)	Variances (log scale)	Est. Levels (lb/day)
1979 pre-ban, $\bar{y}_T$	53	9.4835	0.002324	13,141
1979 post-ban, $\bar{y}_{T+G}$	33	9.1889	0.002324	9,788
1982 pre-lift, $\bar{y}_T$	33	9.0427	0.002324	8,457
1982 post-lift, $\bar{y}_{T+G}$	9	9.1795	0.002326	9,696
Estimated effects		1979 ban		1982 ban lift
		$\hat{\delta} = -3353 \pm 1201$ lb/day		$\hat{\delta} = +1239 \pm 959$ lb/day
Variances (log scale)		$\sigma_e^2 = 0.006440$		$\sigma_\varepsilon^2 = 0.001312$
ARIMA model parameter		$\theta = 0.64$		

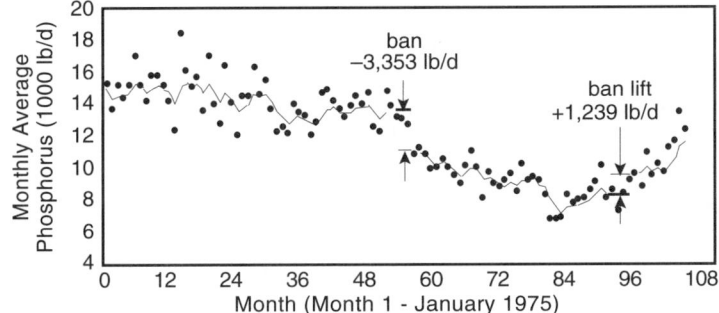

FIGURE 54.3 Estimated effects of the interventions and the model (solid line) fitted to the data.

involved on the original scale. For the 1979 ban, this is 10.6% of $(13{,}141 \times 9788)^{1/2} = 1200$ lb/day. For the 1982 ban lifting, 10.6% of $(8457 \times 9696)^{1/2} = 960$.

Based on the estimated shifts, phosphorus-based laundry detergents contributed about 25% of the influent P load prior to the ban and 13% after the ban was lifted. The post-ban shift is smaller because detergent products that reentered the market contained less phosphorus than those that were banned.

The weighting factor of the model was estimated to be $\theta = 0.64$. The weights used to estimate the pre-ban and post-ban averages and decay away from the intervention were as follows: 0.640, 0.230, 0.083, 0.03, etc., reducing each month by the factor $1 - \theta = 0.36$. The first 3 months on either side of the gap account for more than 95% of the estimated value of the weighted average. This is why the level after the ban was lifted could be estimated so precisely ($\sigma^2 = 0.002326$), although the November 1982 level was estimated using only nine observations.

Having noted that relatively few observations actually entered into the estimated effect of an intervention, it is equally important to make clear that a fairly long record is necessary for computing reliable estimates of θ, σ_e^2, and σ_ε^2.

Comments

Assuming independent random variation about fixed mean levels often provides a very poor model for assessing the magnitude of environmental interventions. Often, environmental data are not independent and often the time series does not vary about fixed mean levels. Lack of independence (serial correlation), drift, seasonality, and perhaps other patterns in the data will require using some sort of time series model or method. The time series model used in this chapter may fit a variety of environmental time series data. Box and Tiao (1965, 1975) provide a suite of intervention models that can be used when the ARIMA (0,1,1) model discussed in this chapter proves inadequate.

The technique presented here may be useful for objectives other than analysis of already collected data. Given a sufficiently long data record, it would be possible to find σ_e^2 and σ_ε^2 before a planned or proposed intervention is made. The values could be used to determine the variance of the effect of the expected intervention, thereby establishing beforehand how small an effect could be found statistically significant with reasonable probability. One could also decide ahead of time how many observations would be needed after the intervention in order to estimate the magnitude of its effect with essentially full precision.

References

Box, G. E. P., G. M. Jenkins, and G. C. Reinsel (1994). *Time Series Analysis, Forecasting and Control*, 3rd ed., Englewood Cliffs, NJ, Prentice-Hall.

Box, G. E. P. and G. C. Tiao (1965). "A Change in Level of a Non-Stationary Time Series," *Biometrika*, 52, 181–192.

Box, G. E. P. and G. C. Tiao (1975.). "Intervention Analysis with Applications to Economic and Environmental Problems," *J. Am. Stat. Assoc.*, 70, 70–79.

Pallesen, L., P. M. Berthouex, and K. Booman (1985). "Environmental Intervention Analysis: Wisconsin's Ban on Phosphate Detergents," *Water Res.*, 19, 353–362.

Exercises

54.1 White Noise Process. The data below represent 20 weekly averages before a change in polymer in a treatment process, and 20 weekly averages following the intervention. Time runs from left to right. Estimate the magnitude and 95% confidence interval of the intervention due to the polymer change.

Pre-intervention — Polymer A									
58	75	77	75	61	91	64	79	82	74
73	72	85	61	52	77	59	72	62	79

Post-intervention — Polymer B									
87	88	87	80	71	74	69	91	92	83
102	45	90	65	82	95	67	75	88	81

54.2 Random Walk Model. The data below represent 20 hourly readings before an intervention occurred in a random walk process, and 20 hourly readings following the intervention. Time runs left to right. Estimate the magnitude and 95% confidence interval of the intervention.

Pre-intervention									
10.7	10.9	11.2	10.5	9.7	9.0	8.9	9.6	9.0	8.4
7.6	8.9	9.5	8.6	5.9	5.7	5.2	6.8	6.9	8.0

Post-intervention									
10.5	9.9	9.6	7.6	6.9	7.4	7.8	6.7	7.4	8.8
8.0	6.6	6.9	5.9	7.0	8.5	9.2	8.8	10.1	10.2

54.3 Phosphorus Intervention. A new industry began wastewater discharge to a city sewer on January 1992 and the phosphorus loading increased suddenly. Estimate the lb/day of P added by the industry.

Month	1988	1989	1990	1991	1992	1993	1994	1995
Jan.	1537	1289	1295	1294	1375	1376	1377	1378
Feb.	1387	1443	1259	1302	1418	1486	1456	1479
Mar.	1317	1568	1332	1391	1427	1537	1549	1580
Apr.	1418	1419	1269	1379	1392	1594	1579	1612
May	1255	1297	1135	1426	1402	1593	1535	1478
June	1288	1316	1223	1450	1472	1583	1467	1519
July	1338	1331	1241	1376	1324	1733	1529	1379
Aug.	1354	1282	1307	1402	1398	1530	1416	1427
Sept.	1426	1367	1263	1380	1375	1479	1366	1482
Oct.	1413	1311	1304	1436	1580	1557	1445	1379
Nov.	1354	1229	1237	1314	1488	1495	1513	1486
Dec.	1369	1297	1207	1106	1414	1581	1437	1461

Appendix

Statistical Tables

TABLE A

Table of Normal Distribution Function

z	0.00	0.01	0.02	0.03	0.04	0.05	0.06	0.07	0.08	0.09
0.0	0.5000	0.5040	0.5080	0.5120	0.5160	0.5199	0.5239	0.5279	0.5319	0.5359
0.1	0.5398	0.5438	0.5478	0.5517	0.5557	0.5596	0.5636	0.5675	0.5714	0.5753
0.2	0.5793	0.5832	0.5871	0.5910	0.5948	0.5987	0.6026	0.6064	0.6103	0.6141
0.3	0.6179	0.6217	0.6255	0.6293	0.6331	0.6368	0.6406	0.6443	0.6480	0.6517
0.4	0.6554	0.6591	0.6628	0.6664	0.6700	0.6736	0.6772	0.6808	0.6844	0.6879
0.5	0.6915	0.6950	0.6985	0.7019	0.7054	0.7088	0.7123	0.7157	0.7190	0.7224
0.6	0.7257	0.7291	0.7324	0.7357	0.7389	0.7422	0.7454	0.7486	0.7517	0.7549
0.7	0.7580	0.7611	0.7642	0.7673	0.7704	0.7734	0.7764	0.7794	0.7823	0.7852
0.8	0.7881	0.7910	0.7939	0.7967	0.7995	0.8023	0.8051	0.8078	0.8106	0.8133
0.9	0.8159	0.8186	0.8212	0.8238	0.8264	0.8289	0.8315	0.8340	0.8365	0.8389
1.0	0.8413	0.8438	0.8461	0.8485	0.8508	0.8531	0.8554	0.8577	0.8599	0.8621
1.1	0.8643	0.8665	0.8686	0.8708	0.8729	0.8749	0.8770	0.8790	0.8810	0.8830
1.2	0.8849	0.8869	0.8888	0.8907	0.8925	0.8944	0.8962	0.8980	0.8997	0.9015
1.3	0.9032	0.9049	0.9066	0.9082	0.9099	0.9115	0.9131	0.9147	0.9162	0.9177
1.4	0.9192	0.9207	0.9222	0.9236	0.9251	0.9265	0.9279	0.9292	0.9306	0.9319
1.5	0.9332	0.9345	0.9357	0.9370	0.9382	0.9394	0.9406	0.9418	0.9429	0.9441
1.6	0.9452	0.9463	0.9474	0.9484	0.9495	0.9505	0.9515	0.9525	0.9535	0.9545
1.7	0.9554	0.9564	0.9573	0.9582	0.9591	0.9599	0.9608	0.9616	0.9625	0.9633
1.8	0.9641	0.9649	0.9656	0.9664	0.9671	0.9678	0.9686	0.9693	0.9699	0.9706
1.9	0.9713	0.9719	0.9726	0.9732	0.9738	0.9744	0.9750	0.9756	0.9761	0.9767
2.0	0.9772	0.9778	0.9783	0.9788	0.9793	0.9798	0.9803	0.9808	0.9812	0.9817
2.1	0.9821	0.9826	0.9830	0.9834	0.9838	0.9842	0.9846	0.9850	0.9854	0.9857
2.2	0.9861	0.9864	0.9868	0.9871	0.9875	0.9878	0.9881	0.9884	0.9887	0.9890
2.3	0.9893	0.9896	0.9898	0.9901	0.9904	0.9906	0.9909	0.9911	0.9913	0.9916
2.4	0.9918	0.9920	0.9922	0.9925	0.9927	0.9929	0.9931	0.9932	0.9934	0.9936
2.5	0.9938	0.9940	0.9941	0.9943	0.9945	0.9946	0.9948	0.9949	0.9951	0.9952
2.6	0.9953	0.9955	0.9956	0.9957	0.9959	0.9960	0.9961	0.9962	0.9963	0.9964
2.7	0.9965	0.9966	0.9967	0.9968	0.9969	0.9970	0.9971	0.9972	0.9973	0.9974
2.8	0.9974	0.9975	0.9976	0.9977	0.9977	0.9978	0.9979	0.9979	0.9980	0.9981
2.9	0.9981	0.9982	0.9982	0.9983	0.9984	0.9984	0.9985	0.9985	0.9986	0.9986
3.0	0.9987	0.9987	0.9987	0.9988	0.9988	0.9989	0.9989	0.9989	0.9990	0.9990
3.1	0.9990	0.9991	0.9991	0.9991	0.9992	0.9992	0.9992	0.9992	0.9993	0.9993
3.2	0.9993	0.9993	0.9994	0.9994	0.9994	0.9994	0.9994	0.9995	0.9995	0.9995
3.3	0.9995	0.9995	0.9995	0.9996	0.9996	0.9996	0.9996	0.9996	0.9996	0.9997
3.4	0.9997	0.9997	0.9997	0.9997	0.9997	0.9997	0.9997	0.9997	0.9997	0.9998
3.5	0.9998									
4.0	0.99997									
5.0	0.9999997									
6.0	0.999999999									

TABLE B

Probability Points of the t Distribution with v Degrees of Freedom

	Tail Area Probability						
v	0.4	0.25	0.1	0.05	0.025	0.01	0.005
1	0.325	1.000	3.078	6.314	12.706	31.821	63.657
2	0.289	0.816	1.886	2.920	4.303	6.965	9.925
3	0.277	0.765	1.638	2.353	3.182	4.541	5.841
4	0.271	0.741	1.533	2.132	2.776	3.747	4.604
5	0.267	0.727	1.476	2.015	2.571	3.365	4.032
6	0.265	0.718	1.440	1.943	2.447	3.143	3.707
7	0.263	0.711	1.415	1.895	2.365	2.998	3.499
8	0.262	0.706	1.397	1.860	2.306	2.896	3.355
9	0.261	0.703	1.383	1.833	2.262	2.821	3.250
10	0.260	0.700	1.372	1.812	2.228	2.764	3.169
11	0.260	0.697	1.363	1.796	2.201	3.718	3.106
12	0.259	0.695	1.356	1.782	2.179	2.681	3.055
13	0.259	0.694	1.350	1.771	2.160	2.650	3.012
14	0.258	0.692	1.345	1.761	2.145	2.624	2.977
15	0.258	0.691	1.341	1.753	2.131	2.602	2.947
16	0.258	0.690	1.337	1.746	2.120	2.583	2.921
17	0.257	0.689	1.333	1.740	2.110	2.567	2.898
18	0.257	0.688	1.330	1.734	2.101	2.552	2.878
19	0.257	0.688	1.328	1.729	2.093	2.539	2.861
20	0.257	0.687	1.325	1.725	2.086	2.528	2.845
21	0.257	0.686	1.323	1.721	2.080	2.518	2.831
22	0.256	0.686	1.321	1.717	2.074	2.508	2.819
23	0.256	0.685	1.319	1.714	2.069	2.500	2.807
24	0.256	0.685	1.318	1.711	2.064	2.492	2.797
25	0.256	0.684	1.316	1.708	2.060	2.485	2.787
26	0.256	0.684	1.315	1.706	2.056	2.479	2.779
27	0.256	0.684	1.314	1.703	2.052	2.473	2.771
28	0.256	0.683	1.313	1.701	2.048	2.467	2.763
29	0.256	0.683	1.311	1.699	2.045	2.462	2.756
30	0.256	0.683	1.310	1.697	2.042	2.457	2.750
40	0.255	0.681	1.303	1.684	2.021	2.423	2.704
60	0.254	0.679	1.296	1.671	2.000	2.390	2.660
120	0.254	0.677	1.289	1.658	1.980	2.358	2.617
∞	0.253	0.674	1.282	1.645	1.960	2.326	2.576

TABLE C

Percentage Points of the F Distribution: Upper 5% Points

v_2/v_1	1	2	3	4	5	6	7	8	9	10	12	15	20	24	30	40	60	120	∞
1	161.4	199.5	215.7	224.6	230.2	234.0	236.8	238.9	240.5	241.9	243.9	245.9	248.0	249.1	250.1	251.1	252.2	253.3	254.3
2	18.51	19.00	19.16	19.25	19.30	19.33	19.35	19.37	19.38	19.40	19.41	19.43	19.45	19.45	19.46	19.47	19.48	19.49	19.50
3	10.13	9.55	9.28	9.12	9.01	8.94	8.89	8.85	8.81	8.79	8.74	8.70	8.66	8.64	8.62	8.59	8.57	8.55	8.53
4	7.71	6.94	6.59	6.39	6.26	6.16	6.09	6.04	6.00	5.96	5.91	5.86	5.80	5.77	5.75	5.72	5.69	5.66	5.63
5	6.61	5.79	5.41	5.19	5.05	4.95	4.88	4.82	4.77	4.74	4.68	4.62	4.56	4.53	4.50	4.46	4.43	4.40	4.36
6	5.99	5.14	4.76	4.53	4.39	4.28	4.21	4.15	4.10	4.06	4.00	3.94	3.87	3.84	3.81	3.77	3.74	3.70	3.67
7	5.59	4.74	4.35	4.12	3.97	3.87	3.79	3.73	3.68	3.64	3.57	3.51	3.44	3.41	3.38	3.34	3.30	3.27	3.23
8	5.32	4.46	4.07	3.84	3.69	3.58	3.50	3.44	3.39	3.35	3.28	3.22	3.15	3.12	3.08	3.04	3.01	2.97	2.93
9	5.12	4.26	3.86	3.63	3.48	3.37	3.29	3.23	3.18	3.14	3.07	3.01	2.94	2.90	2.86	2.83	2.79	2.75	2.71
10	4.96	4.10	3.71	3.48	3.33	3.22	3.14	3.07	3.02	2.98	2.91	2.85	2.77	2.74	2.70	2.66	2.62	2.58	2.54
11	4.84	3.98	3.59	3.36	3.20	3.09	3.01	2.95	2.90	2.85	2.79	2.72	2.65	2.61	2.57	2.53	2.49	2.45	2.40
12	4.75	3.89	3.49	3.26	3.11	3.00	2.91	2.85	2.80	2.75	2.69	2.62	2.54	2.51	2.47	2.43	2.38	2.34	2.30
13	4.67	3.81	3.41	3.18	3.03	2.92	2.83	2.77	2.71	2.67	2.60	2.53	2.46	2.42	2.38	2.34	2.30	2.25	2.21
14	4.60	3.74	3.34	3.11	2.96	2.85	2.76	2.70	2.65	2.60	2.53	2.46	2.39	2.35	2.31	2.27	2.22	2.18	2.13
15	4.54	3.68	3.29	3.06	2.90	2.79	2.71	2.64	2.59	2.54	2.48	2.40	2.33	2.29	2.25	2.20	2.16	2.11	2.07
16	4.49	3.63	3.24	3.01	2.85	2.74	2.66	2.59	2.54	2.49	2.42	2.35	2.28	2.24	2.19	2.15	2.11	2.06	2.01
17	4.45	3.59	3.20	2.96	2.81	2.70	2.61	2.55	2.49	2.45	2.38	2.31	2.23	2.19	2.15	2.10	2.06	2.01	1.96
18	4.41	3.55	3.16	2.93	2.77	2.66	2.58	2.51	2.46	2.41	2.34	2.27	2.19	2.15	2.11	2.06	2.02	1.97	1.92
19	4.38	3.52	3.13	2.90	2.74	2.63	2.54	2.48	2.42	2.38	2.31	2.23	2.16	2.11	2.07	2.03	1.98	1.93	1.88
20	4.35	3.49	3.10	2.87	2.71	2.60	2.51	2.45	2.39	2.35	2.28	2.20	2.12	2.08	2.04	1.99	1.95	1.90	1.84
21	4.32	3.47	3.07	2.84	2.68	2.57	2.49	2.42	2.37	2.32	2.25	2.18	2.10	2.05	2.01	1.96	1.92	1.87	1.81
22	4.30	3.44	3.05	2.82	2.66	2.55	2.46	2.40	2.34	2.30	2.23	2.15	2.07	2.03	1.98	1.94	1.89	1.84	1.78
23	4.28	3.42	3.03	2.80	2.64	2.53	2.44	2.37	2.32	2.27	2.20	2.13	2.05	2.01	1.96	1.91	1.86	1.81	1.76
24	4.26	3.40	3.01	2.78	2.62	2.51	2.42	2.36	2.30	2.25	2.18	2.11	2.03	1.98	1.94	1.89	1.84	1.79	1.73
25	4.24	3.39	2.99	2.76	2.60	2.49	2.40	2.34	2.28	2.24	2.16	2.09	2.01	1.96	1.92	1.87	1.82	1.77	1.71
26	4.23	3.37	2.98	2.74	2.59	2.47	2.39	2.32	2.27	2.22	2.15	2.07	1.99	1.95	1.90	1.85	1.80	1.75	1.69
27	4.21	3.35	2.96	2.73	2.57	2.46	2.37	2.31	2.25	2.20	2.13	2.06	1.97	1.93	1.88	1.84	1.79	1.73	1.67
28	4.20	3.34	2.95	2.71	2.56	2.45	2.36	2.29	2.24	2.19	2.12	2.04	1.96	1.91	1.87	1.82	1.77	1.71	1.65
29	4.18	3.33	2.93	2.70	2.55	2.43	2.35	2.28	2.22	2.18	2.10	2.03	1.94	1.90	1.85	1.81	1.75	1.70	1.64
30	4.17	3.32	2.92	2.69	2.53	2.42	2.33	2.27	2.21	2.16	2.09	2.01	1.93	1.89	1.84	1.79	1.74	1.68	1.62
40	4.08	3.23	2.84	2.61	2.45	2.34	2.25	2.18	2.12	2.08	2.00	1.92	1.84	1.79	1.74	1.69	1.64	1.58	1.51
60	4.00	3.15	2.76	2.53	2.37	2.25	2.17	2.10	2.04	1.99	1.92	1.84	1.75	1.70	1.65	1.59	1.53	1.47	1.39
120	3.92	3.07	2.68	2.45	2.29	2.17	2.09	2.02	1.96	1.91	1.83	1.75	1.66	1.61	1.55	1.50	1.43	1.35	1.25
∞	3.84	3.00	2.60	2.37	2.21	2.10	2.01	1.94	1.88	1.83	1.75	1.67	1.57	1.52	1.46	1.39	1.32	1.22	1.00

Source: Merrington, M. and C. M. Thompson (1943). "Tables of Percentage Points of the Inverted Data (F) Distribution," *Biometrika*, 33, 73.

Index

A

Aberrant values, 4
Accuracy, 11, 77–85, see also Bias; Error; Precision
    bias, 79–80
    interlaboratory comparisons, 81–83
    multiple sources of variation, 81
    precision, 78–79
    relative error, 79
    ruggedness testing, 83–84
Action limits, 98, 104
ANOVA (analysis of variance), 215–220
    multiple factor, 233–237
        case study: dioxin and furan emissions, 233–234, 235–236
        method, 234–235
    one-way, 215–218
        case study: comparison of five laboratories, 215, 218–219
        sampling, 205–206
    six-variance empirical model, 340
Approximation, linear, 301
ARIMA models, 445, 448
ARIMA (0.1,1) process, 463
Arithmetic moving average (AMA), 114–115
AR(1) process, 459–462
AR(2) process, 462
Assignable causes, 104
Asymmetric distribution, 4
Autocorrelation, 4, 103, 370–371
    case study: serial dependence of BOD data, 289, 290–291
    correlation and autocorrelation coefficients, 289–290
    implications for sampling frequency, 290
Autocorrelation (serial correlation), 289–293, 365–372, see also Correlation
    case study: suspicious laboratory experiment, 365–366
    effect on regression, 366
    examples of errors, 368–369
    test statistic for, 369–370
    trend analysis and, 370–371
    variance estimate distortion mechanism, 367–368
Autocorrelation coefficient, 367–368
Autocorrelation function, 443–445
Autoregressive–moving average (ARMA) processes, 447, 448
Autoregressive (AR) processes, 459–466, see also AR entries
Average
    exponentially weighted (EMWA), 463–465
    moving, 56, 57, 107–110
        arithmetic (AMA), 114–115
        integrated (IMA), 117, 461–462
    sample, 10
    sampling distribution of, 17–18
    standard deviation of, 78
    weighted moving, 105

B

Between-run precision, 12
Bias, 11, 397–402
    assessment, 79–80
    case studies
        bacterial growth, 397
        Michaelis-Menton model, 398–401
        Thomas slope method, 397–398
    in process rationalization, 419
Binomial coefficients, 162
Binomial model, 161–164
Binomial population, sample size for estimating, 206–208
Binomial probability, 162
Bioassays, 161–167
Blocking, 150–151, 188
BOD model, 389–396
    bias in, 397–402, see also Bias
Bonferroni (family) error rate, 169
Bootstrap sampling/resampling, 436–437
Box-and-whisker plots, 29
Box-Belunken design, 194
Box-Cox power transformations, 67–69
Box-Lucas design, 389–396
Box plot, 29
Bumps, 114

C

Calculated values, precision of, 87–95, see also under Precision
Calibration, 319–325, see also Calibration curve and specific methods
    case study: HPLC calibration, 320, 322–323
    theory: straight-line calibration curve, 320–322
    weighted least squares method, 327–336
Calibration curve, 327, 331
    in prediction of concentrations, 322–323
Cancer potency factors, 1
Categorical variables, 355–364
    case study: stream acidification during storm, 355, 358–361
    regression method, 355–358
Cause-and-effect relationships, 4
Causes, assignable, 104
Censored data, 4, 129–140
    Cohen's maximum likelihood method, 134–137
    delta-lognormal method, 137
    graphical methods, 132–133
    median, 129–130
    regression in rankits (normal order scores, order statistics), 133–134
    trimmed mean, 130
    Winsorized mean, 131
Central limit effect, 18
Chart(s)
    Cusum (cumulative sum), 106–110, 113
    I, 105
    moving average, 107–110
    process control, 103–118, see also Control charts
        fundamentals, 103–112
        specialized, 113–118
    Range (R), 97–100, 106
    X-bar (Shewhart), 97–100, 106, 108–110
Chi-square distribution, scaled, 18
Clarity, in plotting data, 33–34
Coefficient(s)
    autocorrelation, 367–368
    binomial, 162
    determination (R^2), 339, 345–353
        description vs. prediction, 351
        high, 346–347
        interpretation of "explains," 345–346
        lack-of-fit, 350
        low, 347
        magnitude, 348–349
        overfitting, 349–350
        significant, 347–348
        standard error of the estimate and, 351
    sensitivity, 428–430
    variation (relative standard deviation, RSD), 79
Cohen's maximum likelihood method, 134–137
Collaborative trial (interlaboratory comparison), 81–83
Color, value in plotting data, 33
Composite design, 194, 382, 385

Confidence interval, 21–22, 64–66, 143, 145, 175, 305, 307
    for an interaction, 204–205
    around parameter, 351
    for dependent variable, 351
    for empirical model, 341
    for a mean (type 1 error), 198–201
    transformations and, 64–66
Confounding, 263–264
Constant variance, 333–334
    transformations for, 62–64
Control, statistical, 104–105
Control charts, 103–118
    fundamentals, 103–112, see also Chart(s)
        chart construction, 105–106
        comparison of types, 108–110
        decision errors, 105
        kinds of, 106–108
        standard concepts, 103–104
        variability and statistical control, 104–105
    specialized, Cuscore statistic, 113–117, see also Cuscore statistic
Correlation, 281–287
    auto-, see Autocorrelation
    case studies
        correlation of BOD and COD measurements, 282–283
        taste and odor, 284
    covariance and, 281–282
    parameter
        linear model, 303–309
        nonlinear model, 311–317
        problem of, 314
    serial (autocorrelation), see Autocorrelation
    Spearman rank-order, 283–284
Critical levels, 59
Critical sum of squares, 300, 305–306, 311
Cube plot, 240–241, 242
Cumulative frequency plots, see Probability plots
Cumulative probability plot, 132
Curve
    calibration, 320–323
    French, 337
Curve fitting (regression), see Regression
Cuscore statistic, 113–117
    change in rate of increase, 115
    linear model, 114
    parameter change in time series model, 116–117
    principle, 113–114
    rectangular bump detection, 114
    sine wave buried in noise, 115–116
    spike detection, 114
Cusum chart (cumulative sum), 106–110, 113

D

Data
    censored, 4, 129–140, see also Censored data
    flattening of, 32

happenstance, 4
observational vs. experimental, 4
plotting, 8–9, 25–40, see also Plots; Plotting data
Data density plot, 8
Data snooping (data dredging), 169
Decision errors, in control charts, 105
Defining relation, 263
Degrees of freedom, 10–11
Delta-lognormal method, 137
Dependent (response) variable, 295
Derivative matrix, 391
Design, experimental, 185–196, see also Experimental design and specific models
Design matrix, 241
Diagram
    dot, 47–49
    frequency (histogram), 8–9
    plot and dot, 8
Digidot plots, 25
Discrete mechanistic model, 454, see also Transfer function models
Discrete random variable, 9
Discrete time series approximation, 454–456
Distribution, 47–54
    asymmetric, 4
    case study: industrial wastewater survey, 47
    cumulative frequency, 49, see also Probability plots
    dot diagrams, 47–49
    F, see F distribution
    normal, 4, 8, 15–16
        approximating binomial, 163–164
    population frequency, 9
    probability, 9
    probability (cumulative frequency) plots, 49–52
    randomness and independence, 562
    reference, 55–60
        in comparing two mean values, 57
        constructing, 55–57
        critical levels, 59
        in monitoring, 57–58
    sampling, of average and variance, 17–18
    t, see t distribution
    uniform, 48–49
Distribution function, normal, 477
Dot diagrams, 47–49
Drift, process, 116–117
Dummy (indicator) variables, 355, 358
Dunnett's paired comparison method with control, 172–173
Durbin-Watson statistic, 369–370

E

Effect, main, 241
Empirical models, 295, 337–344
    case study: sedimentation, 337–338, 339–341
    method: linear regression, 338–339
    selecting "best" regression model, 339–340

Environment, statistical problems special to, 3–4
Environmental law, statistics and, 1–2
Environmental Protection Agency (EPA)
    guidance documents, 1–2
    method detection limit (MDL) approach, 119–127
Equation, normal, 297
Equivalence, of two means, 203–204
Error(s)
    autocorrelation, 368–369, see also Autocorrelation
    decision, 105
    defined, 7
    experimental, 7–8, 77, 300
    independent, 365, 368
    magnification of, 90–91
    in process rationalization, 419
    propagation of, 78, 87–95
    random, 93–94
    random measurement, 296–297
    relative, 79
    standard, 197–198, 341
        of mean (standard deviation of average), 78
    suppression of, 90–91
    systematic, 77, 93–94
    type 1, 198–201
    type 2, 201–203
Error bars, 29–31
Error rate, 169
Error transmission, 425–432
    case studies
        finite difference model, 429–430
        reactor kinetics, 427–429
    propagation of uncertainty, 430
    theory, 425–427
Estimate, least squares, 299
Estimation
    of percentiles, 71–75
    of quantiles
        nonparametric, 74
        parametric, 71–73
Experimental conditions, 2–3
Experimental design, 185–196, see also Sampling and specific models
    attributes of good, 188–189
    defining objectives, 185–186
    interactions and, 192–194
    iterative, 194
    nonlinear models, 389–396, see also Nonlinear least squares; Nonlinear models
    one-factor-at-a-time (OFAT) experiments, 189–192, 246
    principles, 186–188
        blocking, 188
        comparative design, 186–187
        randomization, 187–188
        replication, 187
    sample size, 197–213, see also Sampling
Experimental errors, 7–8, 77
Experimental error variance, 300
Exponentially weighted moving average (EMWA) chart, 107–110

Exponentially weighted moving average (EMWA) model, 463–465
Exponential smoothing (EMWA model), 463–465
External reference distribution, 55–60, see also Reference distribution

F

Factorial designs, 239–248
    case study: compaction of fly ash, 239–240, 244–246
    fractional, 240, 249–259
        case study: sampling high dissolved oxygen concentrations, 249–250, 252–256
        method, 251–252
        screening for important variables, 261–270
    full (saturated), 240–244
    method, 240–244
    regression analysis, 271–279
        case study: two methods for measuring nitrate, 271–272, 275–277
        method, 273–275
Family (Bonferroni) error rate, 169
F distribution
    probability points, 478
    upper 5% percentage points, 479
Finite difference model, 429–430
Flattening data, 32
Forecasting time series, 459–466
    ARIMA (0,1,1) process, 463
    AR(1) process, 459–462
    AR(2) process, 462
    exponential smoothing (EMWA model), 463–465
Fractional factorial designs, 240, 249–259
    case study: sampling high dissolved oxygen concentrations, 249–250, 252–256
    method, 251–252
    screening for important variables, 261–270
French curve, 337
Frequency diagram (histogram), 8–9, 29
F statistic, 300, 313, 360
Full (saturated) factorial designs, 240–244

G

Gaussian distribution, see Normal distribution
Geometric mean, 65
Graphs, see Plot(s); Plotting data

H

Haldane inhibition model, 411
Happenstance data, 4
Histogram, 29, see also Data density plot
    defined, 8–9

Hypothesis, null, 20
Hypothesis test, 20–21

I

I chart, 105
Independence, 14, 52
Independent errors, 365, 368
Independent t-test, 157–160
    case study: mercury in domestic wastewater, 157, 158–159
    for difference of two averages, 157–160
    theory, 157–158
Independent (input, predictor, regressor) variable, 295
Indicator (dummy) variables, 355, 358
Individual error rate, 169
Input (independent, predictor) variable, 295
Integrated–autoregressive moving average (IMA) model, 448, 462
Integrated moving average (IMA), 117, 462
Interaction
    confidence interval for, 204–205
    three-factor, 242
    two-factor, 240–244, 255–256
Interlaboratory comparison (collaborative trial), 81–83
Interquartile range, 29
Interval, see specific type
Interval testing approach, 203–204
Intervention analysis, 443, 467–475
    case study: ban on phosphate detergents, 467–468, 472–473
    random walk model, 469–470
    white noise model, 468–469
    white noise–random walk model, 470–472
Iterative designs, 194, 373–377
    case study: bacterial growth, 373–376

J

Joint confidence region, 405–407
    approximate $1 - \alpha$, 311
    concept, 303–304
    exact $(1 - \alpha)$ 100%, 311
    for linear model, 303–309
    minimization of, 389–392
    for Monod and Tiessier models, 413
    for nonlinear model, 311–317

L

Laboratory quality assuance, 97–101, see also Quality assurance
Lack-of-fit test, 414
Langlier saturation index (LSI), 92

Learning process, 2–3
Least squares, weighted, 327–336, see also Weighted least squares
Least squares criterion, multiresponse, 404–406
Least squares estimate, 299
Least squares method, 295–302
    linear and nonlinear models and, 295–297
    method, 297–299
    nonlinear, 389–396, 400
    precision of estimates, 299–301
    in process rationalization, 420–421, 423
Limit of detection, 1, 119–127, see also Method detection limit (MDL)
    as term, 120
Limits
    action, 98, 104
    warning, 98
Linear approximation, 301
Linearization
    bias produced by, 397–402, see also Bias
    transformations for, 61–70
Linear models, 295–297
    Cuscore statistic and, 114
    precision of parameter estimates, 303–309, see also Joint confidence region
Linear regression, in empirical model building, 337–344, see also Empirical models
Lineweaver-Burke plots, 399–400
Lurking variables, 4

M

Main effect, 241
Matrix (matrices)
    derivative, 391
    design, 241
    of independent variables, 273–275
    model, 242
    variance-covariance (V), 391, 405–406
Maximum likelihood method, Cohen's, 134–137
Mean(s)
    case study: laboratory study of dissolved oxygen (DO) measurement, 141, 143–145
    equivalence of two, 203–204
    geometric, 65
    population, 10
    standard error of (standard deviation of average), 17–18, 78
    standardized (t statistic), 18
    theory, 141–143
    trimmed, 130
    t-test for agreement with a standard, 141–146
    Winsorized, 131
Mean residual sum of squares, 306–307
Mechanistic model, 295
Median, of censored data, 129–130
Method detection limit (MDL), 119–127
    alternative (unofficial) model, 123–125
    calibration designs, 125

case study: lead measurements, 119
    discussion, 125–126
    EPA approach, 120–123
    general concepts, 119–120
Method of least squares, 295–202, see also Least squares method
Michaelis-Menton model, 398–401
Model(s), see also specific designs
    empirical, 295, 337–344, see also Empirical models
    mechanistic, 295
    Monod, see Monod model
    parsimonious, 337
Model discrimination, 411–417
    case studies
        four models, 414
        Monod vs. Tiessier model, 411–414
    method: posterior probability, 414–415
Model matrix, 242
Monod model, 313–314, 403
    for bacterial growth, 373–376
    vs. Tiessier model, 411–414
Monte Carlo simulation, 433–434
Moving average, 56–58, 446–447, see also Time series models
    arithmetic (AMA), 114–115
    exponentially weighted (EMWA), 107–110
    integrated (IMA), 117, 462, 465
Moving average chart, 107–110
Multiple factor ANOVA, 233–237, see also under ANOVA
Multiple paired comparisons, 169–174
    case study: lead measurements by five laboratories, 170
    data snooping and, 169
    Dunnett's method with control, 172–173
    Tukey's paired comparison method, 170–172
Multiplicative expressions, 89–90
Multiresponse experiments, 403–409
    case study: bacterial growth model, 403–404, 406–407
    method: multiresponse least squares criterion, 404–406
Multiresponse least squares criterion, 404–406

N

Noise
    background, 123
    as compared with signal, 120, see also Method detection limit (MDL)
    defined, 7
    sine wave buried in, Cuscore statistic for, 115–116
Non-additivity, 242
Noncontrast variance, 4
Nonlinear models, 295–297
    experimental design, 389–396
        case study: first-order models, 389, 392–395
        method: minimizing joint confidence region, 389–392

Nonparametric estimation, of quantiles, 74
Nonstationary processes, moving average model, 445–446
Normal distribution, 4, 8, 15–16
   approximating binomial, 163–164
Normal distribution function, 477
Normal equation, 297
Normality, 13–14
Normal order scores (rankits, order statistics), 50–52, 133–134
Normal parent distribution, 22
Null hypothesis, 20, 142–143, 145

O

Observational data, vs. experimental, 4
One-factor-at-a-time (OFAT) experiments, 189–192, 246
One-sided hypothesis test, 20–21
Overfitting, 337

P

Paired t-test, see also Multiple paired comparisons
   advantages, 150—151
   for average of differences, 147–155
   case study: benefits of paired design, 151–153
   case study: interlaboratory study of dissolved oxygen (DO), 148–150
   theory, 149
Palleson's method detection limit, 123–125
Parameter, defined, 10
Parameter correlation
   linear model, 303–309
   nonlinear model, 311–317
   problem of, 314
Parameter estimation (regression), see Regression
Parametric estimation, of quantiles, 71–73
Parent distribution, normal, 22
Parsimonious model, 337
Parsimony principle, 447
Percentile estimation, 434–436
Percentile plots, 28–29
Percentiles, see also Quantiles
   defined, 71
   estimation of, 71–75
pH measurement, 92–93
Plot(s), see also Plotting data
   box, 29
   box-and-whisker, 29
   cube, 240–241, 242
   cumulative probability, 132
   data density, 8
   digidot, 25
   frequency distribution, 9
   Lineweaver-Burke, 399–400
   percentile, 28–29
   probability (cumulative frequency), 9, 49–52, 53
   residual, 31–33
   scatterplots, 26–27, see also Scatterplots
   series, 132
   stem-and-leaf, 25
   time-sequence, 25–26
   time series, 27–28, 56
   Youden, 81–82
Plotting data, 8–9, 25–40
   censored data, 132–133
   clarity and style, 33–34
   necessity of, 34–35
   original data as plot, 25–26
   residuals, 31–33
   scatterplots, 26–27
   statistical value and precision, 29–31
   trend analysis, 27–29
Population
   binomial, sample size for estimating, 206–208
   defined, 7
Population frequency distribution, 9
Population mean, 10
Population standard deviation, 10–11
Population variance, 10
Posterior probability, 414–415
Precision, 12
   between-run, 12
   of calculated values, 87–95
      case study: calcium carbonate in water mains, 91–92
      error suppression and magnification, 90–91
      linear combination of variables, 87–89
      multiplicative expressions, 89–90
      random and systematic errors, 93–94
   of least squares estimates, 299–301
      linear model, 299–300
      nonlinear model, 301
   of measurements, 78–79, see also Accuracy; Bias; Error
   number of observations and, 313–314
   plotting of, 29–31
   Range (R) chart and, 98
   within-run, 12
Prediction interval, 177–178, 177–181, 305, 307
   for dependent variable, 351
   for standard deviation of normal distribution, 178–179
      case study: groundwater monitoring, 180
      case study: spare parts inventory, 179–180
      case study: water quality compliance, 180–181
   two-sided, 177
Predictor (independent, input, regressor) variable, 295
Probability
   binomial, 162
   posterior, 414–415
Probability density function, 9
Probability distribution, 9
Probability (cumulative frequency) plots, 9, 49–52, 53
Probability points
   of F distribution, 478
   of t distribution, 477

Process drift, 116–117
Process rationalization, 419–424
    case study: errors in flow measurement, 420, 421–422
    data adjustment, 420–421
Productivity improvement, 104
Propagation of errors, 78, 87–95, see also Precision, of calculated values
Propagation of uncertainty, 430, see also Error transmission
Proportions, difference of, 161–167
    binomial model, 161–164
    case study: toxicity of effluent, 161, 165–166

Q

Quality assurance, 97–101
    reacting to unfavorable conditions, 100
    X-bar and range charts
        construction, 97–99
        use, 99–100
Quality improvement, 104
Quantiles
    defined, 71
    nonparametric estimation, 74
    parametric estimation, 71–73
Quartiles, defined, 71

R

Random errors, 93–94
Randomization, 187–188
Random measurement errors, 296–297
Randomness, 13–14, 52
Random variable, 7
    discrete, 9
Random walk model, 469–470
Range, interquartile, 29
Range (R) charts, 97–100, 106
Rankits (normal order scores, order statistics), 50–52, 133–134
Rank order statistics, 71–75, see also Percentiles; Quartiles
Rank test, Youden's, 81
Rate of increase, Cuscore statistic for change in, 115
Reference distribution, 55–60
    in comparing two mean values, 57
    constructing, 55–57
    critical levels, 59
    in monitoring, 57–58
Regression, 295–297
    autocorrelation effect on, 366
    linear, in empirical model building, 337–344, see also Empirical models
    in rankits (normal order schored, order statistics), 133–134
    weighted, 327–336, see also Weighted least squares

Regression sum of squares, 341–342
Regressor (independent, input, predictor variable), 295
Relative error, 79
Repeatability, 12–13, 81
Replication, 12, 187
    sampling and, 197–198
Reproducibility, 12–13, 81
Residual plots, 31–33
Residuals, 296
Residual sum of squares, 296, 339, 340–341, 415, see also Least squares method
    mean, 306–307
Response surface methodology, 379–388
    case study: inhibition of micronial growth by phenol, 380–384
    efficiency, 384
    theory, 379–380
Response (dependent) variable, 295
Ruggedness testing, 83–84
Ryznar stability index (RSI), 92

S

Sample, see also Sampling
    bootstrap, 436
    defined, 7
    standard deviation of, 10–11
Sample average, 10
Sample size, 197–213, see also Experimental design; Sampling
Sample variance, 10, 164
Sampling, 197–213
    bootstrap, 436–437
    confidence interval for a mean (type I error), 198–201
    confidence interval for an interaction, 204–205
    for equivalence of two means, 203–204
    for estimating binomial population, 206–208
    for one-sided test, 202
    for one-way ANOVA, 205–206
    replication and experimental design, 197–198
    serial correlation and, 292
    simple random, 176
    stratified, 208–211
    three-stage, 211
    two-stage (subsampling), 211
    type II error, 201–203
    for variance components analysis, 229–231
Sampling distribution, of average and variance, 17–18
Saturated (full) factorial designs, 240–244
Scaled Chi-square distribution, 18
Scatterplots, 26–27
Scores, normal order (rankits, order statistics), 50–52, 133–134
Screening, of important variables, 261–270
    case study: using fly ash to make impenetrable barrier, 261–262, 264–266
    method, 261–264

Seasonality, time series models and, 448–449
Sensitivity coefficients, 428–430
Sequential design, 373–377
Serial correlation (autocorrelation), 4, 103, 289–293, 370–371, see also Autocorrelation
Series plot, 132
Shewhart (X-bar) chart, 97–100, 106, 108–110
Signal, as compared with noise, 120, see also Method detection limit (MDL)
Simple random sampling, 176
Simulation, 430, 433–439
    bootstrap sampling/resampling, 436–437
    case studies
        percentile estimation, 434–436
        properties of computed statistic, 434
    Monte Carlo, 433–434
Six Sigma programs, 103
Size, sample, 197–213, see also Experimental design; Sampling
Smoothing, exponential (EMWA model), 463–465
Spearman rank-order correlation, 283–284
Spike detection, Cuscore statistic for, 114
Squares, sum of, see Sum of squares
Standard deviation
    of average (standard error of mean), 78
    of normal distribution, 178–181
    population, 10–11
    relative (RSD, coefficient of variation), 79
    sample, 10–11, 17–18
Standard error, 197–198, 341
    of mean (standard deviation of average), 17–18, 78
Star points, 382
State of statistical control, 104–105
Stationary processes, time series models for, 445–446
Statistic, defined, 10
Statistical control, state of, 104–105
Statistical value, plotting of, 29–31
Statistics
    environmental law and, 1–2
    learning process and, 2–3
    problems special to environmental statistics, 3–4
    truth and, 2
Stem-and-leaf plots, 25
Stratified sampling, 208–211
Style, in plotting data, 33–34
Subsampling, 211
Sum of squares
    critical, 300, 305–306, 311
    mean residual, 306–307
    regression, 341–342
    residual, 296, 339, 340–341, 415
    total, 340
Systematic errors, 77, 93–94

T

t distribution, 16–17
    probability poionts, 477
Test, hypothesis and tolerance, 20–21

Thomas slope method, 397–398
Three-factor interaction, 242
Three-stage sampling, 211
Tiessier model, vs. Monod model, 411–414
Time-sequence plots, 25–26
Time series, forecasting and stationary, 459–466
Time series approximation, discrete, 454–456
Time series models, 441–452
    ARIMA models, 445
    autocorrelation function, 443–445
    autoregressive–moving average (ARMA) processes, 447
    examples, 441–443
    fitting, 449–451
    integrated–autoregressive moving average (IMA) model, 448
    for nonstationary process (moving average model), 446–447
    parameter change and Cuscore statistic, 116–117
    parsimony principle, 447
    seasonality, 448–449
    for stationary process, 445–446
Time series plots, 27–28, 56
Tolerance interval, 175–176, 179–181
    for standard deviation of normal distribution, 178–179
        case study:groundwater monitoring, 180
        case study: spare parts inventory, 179–180
        case study: water quality compliance, 180–181
    two-sided, 175–176
Tolerance test, 20–21
Total sum of squares, 340
Toxicity testing, 1, 161–167
Transfer function models, 453–458
    case study: continuous stirred tank reactor model, 453–454
    discrete time series approximation, 454–456
Transformations, 61–70
    Box-Cox power, 67–69
    confidence intervals and, 64–66
    for constant variance, 62–64
    for linearization, 61–62
Transmission, of errors, 425–432, see also Error transmission
t ratios, 341
Trend analysis, 27–29
    autocorrelation and, 370–371
Trimmed mean, 130
Truth as compared with inference, 2
t statistic, 16, 18
t-test
    comparing mean with a standard, 141–146
        case study: laboratory study of dissolved oxygen (DO) measurement, 141–143, 143–145
    independent, 157–160
        case study: mercury in domestic wastewater, 157, 158–159
        for difference of two averages, 157–160
        theory, 157–158
    paired, see also Multiple paired comparisons

Index

advantages, 150—151
for average of differences, 147–155
case study: benefits of paired design, 151–153
case study: interlaboratory study of dissolved oxygen (DO), 148–150
theory, 149
Tukey's paired comparison method, 170–172
Two-factor interaction, 240–244, 255–256
Two-sided hypothesis test, 20–21
Two-stage sampling (subsampling), 211
Type I error, 198–201
Type II error, 201–203

U

Uncertainty, propagation of, 430, see also Error transmission
Uniform distribution, 48–49

V

Values, aberrant, 4
Variability, 104
Variable(s)
   categorical, 355–364, see also Categorical variables
   dependent (response), 295
   dummy (indicator), 355, 358
   independent (input, predictor, regressor), 295
   linear combination of, 87–89
   lurking, 4
   random, 7, see Random variable
   screening of important, 261–270
      case study: using fly ash to make impenetrable barrier, 261–262, 264–266
      method, 261–264
Variance
   analysis of, see ANOVA
   of average, 17–18
   constant, 333–334
   experimental error, 300
   noncontrast, 4
   population, 10
   sample, 10, 164

   sampling distribution of, 17–18
   transformations for constant, 62–64
Variance components analysis, 223–232
   case study: foundry wastes, 224–226
   case study: sampling waste foundry sand, 226–229
   multiple sources of variance, 223–224
   testing and sampling, 229–231
Variance-covariance (**V**) matrix, 405–406
Variation, coefficient of (relative standard deviation, RSD), 79

W

Warning limits, 98
Weighted least squares, 327–336
   case study: ion chromatograph calibration, 327–330, 331–333
   constant variance, 333–334
   determination of appropriate weights, 330–331
   theory, 330
   without replicates, 333
Weighted moving average, see Moving average
Weighted regression, 327–336, see also Weighted least squares
White noise model, 468–469
White noise–random walk model, 470–472, see also ARIMA entries
Winsorized mean, 131
Wisconsin toxicity laws, 1
Within-run precision, 12

X

X-bar (Shewhart) charts, 97–100, 106, 108–110

Y

Yates continuity correction, 165–166
Youden pairs test, 81–83
Youden plot, 81–82
Youden's rank test, 81